De Gruyter Studies in Mathematics 37

René L. Schilling
Renming Song
Zoran Vondraček

Bernstein Functions

Theory and Applications

2nd Edition

De Gruyter

René L. Schilling
Institut für Stochastik
TU Dresden
01062 Dresden
Germany

Renming Song
Department of Mathematics
University of Illinois
Urbana, IL 61801
USA

Zoran Vondraček
Department of Mathematics
University of Zagreb
10000 Zagreb
Croatia

Mathematics Subject Classification 2010: 26-02, 30-02, 31Bxx, 31Cxx, 44Axx, 47Dxx, 60Exx, 60Gxx, 60Jxx.

ISBN 978-3-11-025229-3
e-ISBN 978-3-11-026933-8
ISSN 0179-0986

Library of Congress Cataloging-in-Publication Data

A CIP catalog record for this book has been applied for at the Library of Congress.

Bibliographic information published by the Deutsche Nationalbibliothek

The Deutsche Nationalbibliothek lists this publication in the Deutsche Nationalbibliografie; detailed bibliographic data are available in the Internet at http://dnb.dnb.de.

© 2012 Walter de Gruyter GmbH & Co. KG, Berlin/Boston

Typesetting: Da-TeX Gerd Blumenstein, Leipzig, www.da-tex.de
Printing and binding: Hubert & Co. GmbH & Co. KG, Göttingen
∞ Printed on acid-free paper

Printed in Germany

www.degruyter.com

MIX
Papier aus verantwor-
tungsvollen Quellen
FSC® C016439

Preface to the second edition

We were pleasantly surprised by the positive responses to the first edition of this monograph and the invitation of our publisher, de Gruyter, to prepare a second edition. We took this opportunity to revise the text and to rewrite some portions of the book. More importantly, we have added a substantial amount of new material.

Our aim was to retain the overall structure of the text with theoretical material in Chapters 1–11 and more specific applications in Chapters 12–15. The tables are now in Chapter 16. Let us highlight some of the most important additions: we included a new chapter on transformations of Bernstein functions, Chapter 10. This was motivated by the continuing interest in subclasses of infinitely divisible distributions. Our treatment focusses on the corresponding Laplace exponents. A new section on subordination and functional inequalities, Section 13.3, investigates the stability of Nash-type and related inequalities under subordination. Over the last decade this has been an active field of research. We have also updated the tables and added a new section on complete Bernstein functions which are given by an exponential representation, Section 16.11. These changes led to a shift in the numbering of the chapters.

Among the many smaller changes and additions let us mention multiply monotone functions and their integral representation in Chapter 1, which allow us to characterize multiply self-decomposable completely monotone functions in Chapter 5. Chapter 6 now contains a proof of the integral representation of Nevanlinna–Pick functions as well as a complete treatment of extended complete Bernstein functions. Because of this we decided to divide Chapter 6 into two sections.

We thank our readers for many comments and suggestions which helped us in the preparation of this new version. We are particularly grateful to Dave Applebaum, Christian Berg, Sonia Fourati, Wissem Jedidi, Mateusz Kwaśnicki, Michael Röckner, Ken-iti Sato and Thomas Simon. We also thank Christoph von Friedeburg and other de Gruyter staff for suggesting this new edition; and the copy-editors, typesetters and printers whose skills have impressed us. Having the continuing support and understanding of our families helped us to finish the book.

Dresden, Urbana and Zagreb
December 2011

René L. Schilling
Renming Song
Zoran Vondraček

Preface

Bernstein functions and the important subclass of complete Bernstein functions appear in various fields of mathematics – often with different definitions and under different names. Probabilists, for example, know Bernstein functions as Laplace exponents, and in harmonic analysis they are called negative definite functions. Complete Bernstein functions are used in complex analysis under the name Pick or Nevanlinna functions, while in matrix analysis and operator theory, the name operator monotone function is more common. When studying the positivity of solutions of Volterra integral equations, various types of kernels appear which are related to Bernstein functions. There exists a considerable amount of literature on each of these classes, but only a handful of texts observe the connections between them or use methods from several mathematical disciplines.

This book is about these connections. Although many readers may not be familiar with the name *Bernstein function*, and even fewer will have heard of *complete Bernstein functions*, we are certain that most have come across these families in their own research. Most likely only certain aspects of these classes of functions were important for the problems at hand and they could be solved on an *ad hoc* basis. This explains quite a few of the rediscoveries in the field, but also that many results and examples are scattered throughout the literature; the exceedingly rich structure connecting this material got lost in the process. Our motivation for writing this book was to point out many of these connections and to present the material in a unified way. We hope that our presentation is accessible to researchers and graduate students with different backgrounds. The results as such are mostly known, but our approach and some of the proofs are new: we emphasize the structural analogies between the function classes which we believe is a very good way to approach the topic. Since it is always important to know explicit examples, we took great care to collect many of them in the tables which form the last part of the book.

Completely monotone functions – these are the Laplace transforms of measures on the half-line $[0, \infty)$ – and Bernstein functions are intimately connected. The derivative of a Bernstein function is completely monotone; on the other hand, the primitive of a completely monotone function is a Bernstein function if it is positive. This observation leads to an integral representation for Bernstein functions: the Lévy–Khintchine formula on the half-line

$$f(\lambda) = a + b\lambda + \int_{(0,\infty)} (1 - e^{-\lambda t})\,\mu(dt), \quad \lambda > 0.$$

Although this is familiar territory to a probabilist, this way of deriving the Lévy–
Khintchine formula is not the usual one in probability theory. There are many more
connections between Bernstein and completely monotone functions. For example,
f is a Bernstein function if, and only if, for all completely monotone functions g
the composition $g \circ f$ is completely monotone. Since g is a Laplace transform, it is
enough to check this for the kernel of the Laplace transform, i.e. the basic completely
monotone functions $g(\lambda) = e^{-t\lambda}, t > 0$.

A similar connection exists between the Laplace transforms of completely monotone functions, that is, *double Laplace* or *Stieltjes transforms*, and *complete* Bernstein
functions. A function f is a complete Bernstein function if, and only if, for each
$t > 0$ the composition $(t + f(\lambda))^{-1}$ of the Stieltjes kernel $(t + \lambda)^{-1}$ with f is a
Stieltjes function. Note that $(t + \lambda)^{-1}$ is the Laplace transform of $e^{-t\lambda}$ and thus the
functions $(t + \lambda)^{-1}, t > 0$, are the basic Stieltjes functions. With some effort one can
check that complete Bernstein functions are exactly those Bernstein functions where
the measure μ in the Lévy–Khintchine formula has a completely monotone density
with respect to Lebesgue measure. From there it is possible to get a surprising geometric characterization of these functions: they are non-negative on $(0, \infty)$, have an
analytic extension to the cut complex plane $\mathbb{C} \setminus (-\infty, 0]$ and preserve upper and lower
half-planes. A familiar sight for a classical complex analyst: these are the Nevanlinna
functions. One could go on with such connections, delving into continued fractions,
continue into interpolation theory and from there to operator monotone functions ...

Let us become a bit more concrete and illustrate our approach with an example. The
fractional powers $\lambda \mapsto \lambda^\alpha, \lambda > 0, 0 < \alpha < 1$, are easily among the most prominent
(complete) Bernstein functions. Recall that

$$f_\alpha(\lambda) := \lambda^\alpha = \frac{\alpha}{\Gamma(1-\alpha)} \int_0^\infty (1 - e^{-\lambda t}) t^{-\alpha-1} \, dt. \tag{1}$$

Depending on your mathematical background, there are many different ways to derive
and to interpret (1), but we will follow probabilists' custom and call (1) the Lévy–
Khintchine representation of the Bernstein function f_α. At this point we do not want
to go into details, instead we insist that one should read this formula as an integral
representation of f_α with the kernel $(1 - e^{-\lambda t})$ and the measure $c_\alpha t^{-\alpha-1} \, dt$.

This brings us to negative powers, and there is another classical representation

$$\lambda^{-\beta} = \frac{1}{\Gamma(\beta)} \int_0^\infty e^{-\lambda t} t^{\beta-1} \, dt, \quad \beta > 0, \tag{2}$$

showing that $\lambda \mapsto \lambda^{-\beta}$ is a completely monotone function. It is no accident that the
reciprocal of the Bernstein function $\lambda^\alpha, 0 < \alpha < 1$, is completely monotone, nor
is it an accident that the representing measure $c_\alpha t^{-\alpha-1} \, dt$ of λ^α has a completely
monotone density. Inserting the representation (2) for $t^{-\alpha-1}$ into (1) and working out
the double integral and the constant, leads to the second important formula for the

fractional powers,

$$\lambda^{\alpha} = \frac{1}{\Gamma(\alpha)\Gamma(1-\alpha)} \int_0^{\infty} \frac{\lambda}{\lambda+t} t^{\alpha-1} dt. \tag{3}$$

We will call this representation of λ^{α} the Stieltjes representation. To explain why this is indeed an appropriate name, let us go back to (2) and observe that $t^{\alpha-1}$ is a Laplace transform. This shows that $\lambda^{-\alpha}$, $\alpha > 0$, is a double Laplace or Stieltjes transform. Another non-random coincidence is that

$$\frac{f_{\alpha}(\lambda)}{\lambda} = \frac{1}{\Gamma(\alpha)\Gamma(1-\alpha)} \int_0^{\infty} \frac{1}{\lambda+t} t^{\alpha-1} dt$$

is a Stieltjes transform and so is $\lambda^{-\alpha} = 1/f_{\alpha}(\lambda)$. This we can see if we replace $t^{\alpha-1}$ by its integral representation (2) and use Fubini's theorem:

$$\frac{1}{f_{\alpha}(\lambda)} = \lambda^{-\alpha} = \frac{1}{\Gamma(\alpha)\Gamma(1-\alpha)} \int_0^{\infty} \frac{1}{\lambda+t} t^{-\alpha} dt. \tag{4}$$

It is also easy to see that the fractional powers $\lambda \mapsto \lambda^{\alpha} = \exp(\alpha \log \lambda)$ extend analytically to the cut complex plane $\mathbb{C} \setminus (-\infty, 0]$. Moreover, z^{α} maps the upper half-plane into itself; actually it contracts all arguments by the factor α. Apart from some technical complications this allows to surround the singularities of f_{α} – which are all in $(-\infty, 0)$ – by an integration contour and to use Cauchy's theorem for the half-plane to bring us back to the representation (3).

Coming back to the fractional powers λ^{α}, $0 < \alpha < 1$, we derive yet another representation formula. First note that $\lambda^{\alpha} = \int_0^{\lambda} \alpha s^{-(1-\alpha)} ds$ and that the integrand $s^{-(1-\alpha)}$ is a Stieltjes function which can be expressed as in (4). Fubini's theorem and the elementary equality

$$\int_0^{\lambda} \frac{1}{t+s} ds = \log\left(1 + \frac{\lambda}{t}\right)$$

yield

$$\lambda^{\alpha} = \frac{\alpha}{\Gamma(\alpha)\Gamma(1-\alpha)} \int_0^{\infty} \log\left(1 + \frac{\lambda}{t}\right) t^{\alpha-1} dt. \tag{5}$$

This representation will be called the Thorin representation of λ^{α}. Not every complete Bernstein function has a Thorin representation. The critical step in deriving (5) was the fact that the derivative of λ^{α} is a Stieltjes function.

What has been explained for fractional powers can be extended in various directions. On the level of functions, the structure of (1) is characteristic for the class $\mathcal{BF}$ of Bernstein functions, (3) for the class $\mathcal{CBF}$ of complete Bernstein functions, and (5) for the Thorin–Bernstein functions $\mathcal{TBF}$. If we consider $\exp(-tf)$ with f from $\mathcal{BF}$,

$\mathcal{CBF}$ or $\mathcal{TBF}$, we are led to the corresponding families of completely monotone functions and measures. Apart from some minor conditions, these are the infinitely divisible distributions ID, the Bondesson class of measures BO and the generalized Gamma convolutions GGC. The diagrams in Remark 9.18 illustrate these connections. If we replace (formally) λ by $-A$, where A is a negative semi-definite matrix or a dissipative closed operator, then we get from (1) and (2) the classical formulae for fractional powers, while (3) turns into Balakrishnan's formula. Considering $\mathcal{BF}$ and $\mathcal{CBF}$ we obtain a fully-fledged functional calculus for generators and potential operators. Since complete Bernstein functions are operator monotone functions we can even recover the famous Heinz–Kato inequality.

Let us briefly describe the content and the structure of the book. It consists of three parts. The first part, Chapters 1–11, introduces the basic classes of functions: the positive definite functions comprising the completely monotone, Stieltjes and Hirsch functions, and the negative definite functions which consist of the Bernstein functions and their subfamilies – special, complete and Thorin–Bernstein functions. Two probabilistic intermezzi explore the connection between Bernstein functions and certain classes of probability measures. Roughly speaking, for every Bernstein function f the functions $\exp(-tf)$, $t > 0$, are completely monotone, which implies that $\exp(-tf)$ is the Laplace transform of an infinitely divisible sub-probability measure. This part of the book is essentially self-contained and should be accessible to non-specialists and graduate students.

In the second part of the book, Chapter 12 through Chapter 15, we turn to applications of Bernstein and complete Bernstein functions. The choice of topics reflects our own interests and is by no means complete. Notable omissions are applications in integral equations and continued fractions.

Among the topics are the spectral theorem for self-adjoint operators in a Hilbert space and a characterization of all functions which preserve the order (in quadratic form sense) of dissipative operators. Bochner's subordination plays a fundamental role in Chapter 13 where also a functional calculus for subordinate generators is developed. This calculus generalizes many formulae for fractional powers of closed operators. As another application of Bernstein and complete Bernstein functions we establish estimates for the eigenvalues of subordinate Markov processes. This is continued in Chapter 14 which contains a detailed study of excessive functions of killed and subordinate killed Brownian motion. Finally, Chapter 15 is devoted to two results in the theory of generalized diffusions, both related to complete Bernstein functions through Krein's theory of strings. Many of these results appear for the first time in a monograph.

The third part of the book is formed by extensive tables of complete Bernstein functions. The main criteria for inclusion in the tables were the availability of explicit representations and the appearance in mathematical literature.

In the appendix we collect, for the readers' convenience, some supplementary results.

We started working on this monograph in summer 2006, during a one-month work-shop organized by one of us at the University of Marburg. Over the years we were supported by our universities: Institut für Stochastik, Technische Universität Dresden, Department of Mathematics, University of Illinois, and Department of Mathematics, University of Zagreb. We thank our colleagues for a stimulating working environ-ment and for many helpful discussions. Considerable progress was made during the two week Research in Pairs programme at the Mathematisches Forschungsinstitut in Oberwolfach where we could enjoy the research atmosphere and the wonderful li-brary. Our sincere thanks go to the institute and its always helpful staff.

Panki Kim and Hrvoje Šikić read substantial parts of the manuscript. We are grate-ful for their comments which helped to improve the text. We thank the series editor Niels Jacob for his interest and constant encouragement. It is a pleasure to acknowl-edge the support of our publisher, Walter de Gruyter, and its editors Robert Plato and Simon Albroscheit.

Writing this book would have been impossible without the support of our families. So thank you, Herta, Jean and Sonja, for your patience and understanding.

Dresden, Urbana and Zagreb   *René L. Schilling*
October 2009   *Renming Song*
                                                                              Zoran Vondraček

Contents

Index of notation

This index is intended to aid cross-referencing, so notation that is specific to a single section is generally not listed. Some symbols are used locally, without ambiguity, in senses other than those given below; numbers following an entry are page numbers.

Unless otherwise stated, binary operations between functions such as $f \pm g$, $f \cdot g$, $f \wedge g$, $f \vee g$, comparisons $f \leq g$, $f < g$ or limiting relations $f_j \xrightarrow{j \to \infty} f$, $\lim_j f_j$, $\liminf_j f_j$, $\limsup_j f_j$, $\sup_j f_j$ or $\inf_j f_j$ are always understood pointwise.

Operations and operators

$a \vee b$	maximum of a and b
$a \wedge b$	minimum of a and b
$\mathfrak{D}, \mathfrak{R}$	domain and range
$\mathscr{L}$	Laplace transform, 1

Sets

$\mathbb{H}^{\uparrow}$	$\{z \in \mathbb{C} : \operatorname{Im} z > 0\}$
$\mathbb{H}^{\downarrow}$	$\{z \in \mathbb{C} : \operatorname{Im} z < 0\}$
$\overrightarrow{\mathbb{H}}$	$\{z \in \mathbb{C} : \operatorname{Re} z > 0\}$
$\mathbb{N}$	natural numbers: $1, 2, 3, \ldots$
positive	always in the sense > 0
negative	always in the sense < 0

Spaces of functions

$\mathcal{B}$	Borel measurable functions
C	continuous functions
$\mathbb{H}$	harmonic functions, 262
$\$$	excessive functions, 261
$\mathcal{BF}$	Bernstein functions, 21
$\mathcal{CBF}$	complete Bernstein fns, 69
$\mathcal{CM}$	completely monotone fns, 2
$\mathcal{H}$	Hirsch functions, 172
$\mathcal{PBF}$	61, 140, 155, 156
$\mathcal{P}$	potentials, 64
$\mathcal{S}$	Stieltjes functions, 16

$\mathcal{SBF}$	special Bernstein fns, 159
$\mathcal{TBF}$	Thorin–Bernstein fns, 109

Sub- and superscripts

$+$	*sets:* non-negative elements, *functions:* non-negative part
$*$	non-trivial elements ($\neq 0$)
$\perp$	orthogonal complement
b	bounded
c	compact support
f	subordinate w.r.t. the Bernstein function f
0	$f(0) = 0$, resp., $f(0+) = 0$

Spaces of distributions

BO	Bondesson class, 117
CE	convolutions of Exp, 125
Exp	exponential distributions, 125
GGC	generalized Gamma convolutions, 121
ID	infinitely divisible distr., 51
ME	mixtures of Exp, 118
SD	self-decomposable distr., 55
SD_∞	completely self-decomposable distributions, 59

Chapter 1

Laplace transforms and completely monotone functions

In this chapter we collect some preliminary material which we need later on in order to study Bernstein functions.

The *(one-sided) Laplace transform* of a function $m : [0, \infty) \to [0, \infty)$ or a measure μ on the half-line $[0, \infty)$ is defined by

$$\mathscr{L}(m; \lambda) := \int_0^\infty e^{-\lambda t} m(t)\, dt \quad \text{or} \quad \mathscr{L}(\mu; \lambda) := \int_{[0,\infty)} e^{-\lambda t}\, \mu(dt), \qquad (1.1)$$

respectively, whenever these integrals converge. Obviously, $\mathscr{L}m = \mathscr{L}\mu_m$ if $\mu_m(dt)$ denotes the measure $m(t)\, dt$.

The following real-analysis lemma is helpful in order to show that finite measures are uniquely determined in terms of their Laplace transforms.

Lemma 1.1. *We have for all* $t, x \geq 0$

$$\lim_{\lambda \to \infty} e^{-\lambda t} \sum_{k \leq \lambda x} \frac{(\lambda t)^k}{k!} = \mathbb{1}_{[0,x)}(t) + \frac{1}{2} \mathbb{1}_{\{x\}}(t). \qquad (1.2)$$

Proof. Let us rewrite (1.2) in probabilistic terms: if X is a Poisson random variable with parameter λt, (1.2) states that

$$\lim_{\lambda \to \infty} \mathbb{P}(X \leq \lambda x) = \mathbb{1}_{[0,x)}(t) + \frac{1}{2} \mathbb{1}_{\{x\}}(t).$$

Recall that the mean and variance of the Poisson random variable X are $\mathbb{E}X = \lambda t$ and $\operatorname{Var}X = \mathbb{E}((X - \lambda t)^2) = \lambda t$, respectively. For each $n \geq 1$ the Poisson random variable X has the same law as $S_n := X_1 + \cdots + X_n$ where X_j are independent and identically distributed Poisson random variables with parameter $\lambda t / n$. Without loss of generality we may assume that $\lambda = n$ is an integer. By the central limit theorem, the sequence $(S_\lambda - \lambda t)/\sqrt{\lambda t}$ converges weakly to a standard normal random variable G. Therefore,

$$\mathbb{P}(X \leq \lambda x)$$

$$= \mathbb{P}\left(\frac{S_\lambda - \lambda t}{\sqrt{\lambda t}} \leq \sqrt{\lambda}\, \frac{x - t}{\sqrt{t}} \right) \xrightarrow{\lambda \to \infty} \begin{cases} \mathbb{P}(G < \infty) = 1, & \text{if } x > t, \\ \mathbb{P}(G \leq 0) = \frac{1}{2}, & \text{if } x = t, \\ \mathbb{P}(G = -\infty) = 0, & \text{if } x < t, \end{cases}$$

and the claim follows. $\qquad\qquad \square$

Proposition 1.2. *A measure μ on $[0, \infty)$ is finite if, and only if, $\mathscr{L}(\mu; 0+) < \infty$. The measure μ is uniquely determined by its Laplace transform.*

Proof. The first part of the assertion follows from monotone convergence since we have $\mu[0, \infty) = \int_{[0,\infty)} 1 \, d\mu = \lim_{\lambda \to 0} \int_{[0,\infty)} e^{-\lambda t} \mu(dt)$.

For the uniqueness part we use first the differentiation lemma for parameter dependent integrals to get

$$(-1)^k \mathscr{L}^{(k)}(\mu; \lambda) = \int_{[0,\infty)} e^{-\lambda t} t^k \mu(dt).$$

Therefore,

$$\sum_{k \le \lambda x} (-1)^k \mathscr{L}^{(k)}(\mu; \lambda) \frac{\lambda^k}{k!} = \sum_{k \le \lambda x} \int_{[0,\infty)} \frac{(\lambda t)^k}{k!} e^{-\lambda t} \mu(dt)$$

$$= \int_{[0,\infty)} \sum_{k \le \lambda x} \frac{(\lambda t)^k}{k!} e^{-\lambda t} \mu(dt)$$

and we conclude with Lemma 1.1 and dominated convergence that

$$\lim_{\lambda \to \infty} \sum_{k \le \lambda x} (-1)^k \mathscr{L}^{(k)}(\mu; \lambda) \frac{\lambda^k}{k!} = \int_{[0,\infty)} \left(\mathbb{1}_{[0,x)}(t) + \frac{1}{2} \mathbb{1}_{\{x\}}(t) \right) \mu(dt) \tag{1.3}$$

$$= \mu[0, x) + \frac{1}{2} \mu\{x\}.$$

This shows that μ can be recovered from (all derivatives of) its Laplace transform. $\quad\square$

It is possible to characterize the range of Laplace transforms. For this we need the notion of complete monotonicity.

Definition 1.3. A function $f : (0, \infty) \to \mathbb{R}$ is a *completely monotone function* if f is of class C^∞ and

$$(-1)^n f^{(n)}(\lambda) \ge 0 \quad \text{for all } n \in \mathbb{N} \cup \{0\} \text{ and } \lambda > 0. \tag{1.4}$$

The family of all completely monotone functions will be denoted by $\mathcal{CM}$.

The conditions (1.4) are often referred to as *Bernstein–Hausdorff–Widder conditions*. The next theorem is known as *Bernstein's theorem*.

The version given below appeared for the first time in [45] and independently in [368]. Subsequent proofs were given in [116] and [102]. The theorem may be also considered as an example of the general integral representation of points in a convex cone by means of its extremal elements. See Theorem 4.8 and [85] for an elementary exposition. The following short and elegant proof is taken from [263].

Theorem 1.4 (Bernstein). *Let $f : (0, \infty) \to \mathbb{R}$ be a completely monotone function. Then it is the Laplace transform of a unique measure μ on $[0, \infty)$, i.e. for all $\lambda > 0$,*

$$f(\lambda) = \mathscr{L}(\mu; \lambda) = \int_{[0,\infty)} e^{-\lambda t} \, \mu(dt).$$

Conversely, whenever $\mathscr{L}(\mu; \lambda) < \infty$ for every $\lambda > 0$, $\lambda \mapsto \mathscr{L}(\mu; \lambda)$ is a completely monotone function.

Proof. Assume first that $f(0+) = 1$ and $f(+\infty) = 0$. Let $\lambda > 0$. For any $a > 0$ and any $n \in \mathbb{N}$, we see by Taylor's formula

$$f(\lambda) = \sum_{k=0}^{n-1} \frac{f^{(k)}(a)}{k!} (\lambda - a)^k + \int_a^\lambda \frac{f^{(n)}(s)}{(n-1)!} (\lambda - s)^{n-1} \, ds$$

$$= \sum_{k=0}^{n-1} \frac{(-1)^k f^{(k)}(a)}{k!} (a - \lambda)^k + \int_\lambda^a \frac{(-1)^n f^{(n)}(s)}{(n-1)!} (s - \lambda)^{n-1} \, ds. \quad (1.5)$$

If $a > \lambda$, then by the assumption all terms are non-negative. Let $a \to \infty$. Then

$$\lim_{a \to \infty} \int_\lambda^a \frac{(-1)^n f^{(n)}(s)}{(n-1)!} (s - \lambda)^{n-1} \, ds = \int_\lambda^\infty \frac{(-1)^n f^{(n)}(s)}{(n-1)!} (s - \lambda)^{n-1} \, ds$$

$$\leq f(\lambda).$$

This implies that the sum in (1.5) converges for every $n \in \mathbb{N}$ as $a \to \infty$. Thus, every term converges as $a \to \infty$ to a non-negative limit. For $n \geq 0$ let

$$\rho_n(\lambda) = \lim_{a \to \infty} \frac{(-1)^n f^{(n)}(a)}{n!} (a - \lambda)^n.$$

This limit does not depend on $\lambda > 0$. Indeed, for $\kappa > 0$,

$$\rho_n(\kappa) = \lim_{a \to \infty} \frac{(-1)^n f^{(n)}(a)}{n!} (a - \kappa)^n$$

$$= \lim_{a \to \infty} \frac{(-1)^n f^{(n)}(a)}{n!} (a - \lambda)^n \frac{(a - \kappa)^n}{(a - \lambda)^n} = \rho_n(\lambda).$$

Let $c_n = \sum_{k=0}^{n-1} \rho_k(\lambda)$. Then

$$f(\lambda) = c_n + \int_\lambda^\infty \frac{(-1)^n f^{(n)}(s)}{(n-1)!} (s - \lambda)^{n-1} \, ds.$$

Clearly, $f(\lambda) \geq c_n$ for all $\lambda > 0$. Let $\lambda \to \infty$. Since $f(+\infty) = 0$, it follows that $c_n = 0$ for every $n \in \mathbb{N}$. Thus we have obtained the following integral representation of the function f:

$$f(\lambda) = \int_\lambda^\infty \frac{(-1)^n f^{(n)}(s)}{(n-1)!} (s - \lambda)^{n-1} \, ds. \quad (1.6)$$

By the monotone convergence theorem

$$1 = \lim_{\lambda \to 0} f(\lambda) = \int_0^\infty \frac{(-1)^n f^{(n)}(s)}{(n-1)!} s^{n-1}\, ds. \tag{1.7}$$

Let

$$f_n(s) = \frac{(-1)^n}{n!} f^{(n)}\left(\frac{n}{s}\right) \left(\frac{n}{s}\right)^{n+1}. \tag{1.8}$$

Using (1.7) and changing variables according to s/t, it follows that for every $n \in \mathbb{N}$, f_n is a probability density function on $(0, \infty)$. Moreover, the representation (1.6) can be rewritten as

$$f(\lambda) = \int_0^\infty \left(1 - \frac{\lambda}{s}\right)_+^{n-1} \frac{(-1)^n f^{(n)}(s)}{(n-1)!} s^{n-1}\, ds$$

$$= \int_0^\infty \left(1 - \frac{\lambda t}{n}\right)_+^{n-1} f_n(t)\, dt. \tag{1.9}$$

By Helly's selection theorem, Corollary A.8, there exist a subsequence $(n_k)_{k\geq 1}$ and a probability measure μ on $(0, \infty)$ such that $f_{n_k}(t)\, dt$ converges weakly to $\mu(dt)$. Further, for every $\lambda > 0$,

$$\lim_{n \to \infty} \left(1 - \frac{\lambda t}{n}\right)_+^{n-1} = e^{-\lambda t}$$

uniformly in $t \in (0, \infty)$. By taking the limit in (1.9) along the subsequence $(n_k)_{k\geq 1}$, it follows that

$$f(\lambda) = \int_{(0,\infty)} e^{-\lambda t} \mu(dt).$$

Uniqueness of μ follows from Proposition 1.2.

Assume now that $f(0+) < \infty$ and $f(+\infty) = 0$. By looking at $f/f(0+)$ we see that the representing measure for f is uniquely given by $f(0+)\mu$.

Now let f be an arbitrary completely monotone function with $f(+\infty) = 0$. For every $a > 0$, define $f_a(\lambda) := f(\lambda + a)$, $\lambda > 0$. Then f_a is a completely monotone function with $f_a(0+) = f(a) < \infty$ and $f_a(+\infty) = 0$.

By what has been already proved, there exists a unique finite measure μ_a on $(0, \infty)$ such that $f_a(\lambda) = \int_{(0,\infty)} e^{-\lambda t} \mu_a(dt)$. It follows easily that for $b > 0$ we have $e^{at}\mu_a(dt) = e^{bt}\mu_b(dt)$. This shows that we can consistently define the measure μ on $(0, \infty)$ by $\mu(dt) = e^{at}\mu_a(dt)$, $a > 0$. In particular, the representing measure μ is uniquely determined by f. Now, for $\lambda > 0$,

$$f(\lambda) = f_{\lambda/2}(\lambda/2) = \int_{(0,\infty)} e^{(-\lambda/2)t} \mu_{\lambda/2}(dt)$$

$$= \int_{(0,\infty)} e^{-\lambda t} e^{(\lambda/2)t} \mu_{\lambda/2}(dt) = \int_{(0,\infty)} e^{-\lambda t} \mu(dt).$$

Finally, if $f(+\infty) = c > 0$, add $c\delta_0$ to μ.

For the converse we set $f(\lambda) := \mathscr{L}(\mu; \lambda)$. Fix $\lambda > 0$ and pick $\epsilon \in (0, \lambda)$. Since $t^n = \epsilon^{-n}(\epsilon t)^n \leq n!\epsilon^{-n}e^{\epsilon t}$ for all $t > 0$, we find

$$\int_{[0,\infty)} t^n e^{-\lambda t} \mu(dt) \leq \frac{n!}{\epsilon^n} \int_{[0,\infty)} e^{-(\lambda-\epsilon)t} \mu(dt) = \frac{n!}{\epsilon^n} \mathscr{L}(\mu; \lambda - \epsilon)$$

and this shows that we may use the differentiation lemma for parameter dependent integrals to get

$$(-1)^n f^{(n)}(\lambda) = (-1)^n \int_{[0,\infty)} \frac{d^n}{d\lambda^n} e^{-\lambda t} \mu(dt) = \int_{[0,\infty)} t^n e^{-\lambda t} \mu(dt) \geq 0. \quad \square$$

Remark 1.5. The last formula in the proof of Theorem 1.4 shows, in particular, that $f^{(n)}(\lambda) \neq 0$ for all $n \in \mathbb{N}$ and all $\lambda > 0$, unless $f \in \mathcal{CM}$ is identically constant.

Corollary 1.6. *The set $\mathcal{CM}$ of completely monotone functions is a convex cone, i.e.*

$$s f_1 + t f_2 \in \mathcal{CM} \quad \text{for all } s, t \geq 0 \text{ and } f_1, f_2 \in \mathcal{CM},$$

which is closed under multiplication, i.e.

$$\lambda \mapsto f_1(\lambda) f_2(\lambda) \text{ is in } \mathcal{CM} \text{ for all } f_1, f_2 \in \mathcal{CM},$$

and under pointwise convergence:

$$\mathcal{CM} = \overline{\{\mathscr{L}\mu \ : \ \mu \text{ is a finite measure on } [0, \infty)\}}$$

(the closure is taken with respect to pointwise convergence).

Proof. That $\mathcal{CM}$ is a convex cone follows immediately from the definition of a completely monotone function or, alternatively, from the representation formula in Theorem 1.4.

If μ_j denotes the representing measure of f_j, $j = 1, 2$, the convolution

$$\mu[0, u] := \mu_1 \star \mu_2[0, u] := \iint_{[0,\infty)\times[0,\infty)} \mathbb{1}_{[0,u]}(s + t) \mu_1(ds)\mu_2(dt)$$

is the representing measure of the product $f_1 f_2$. Indeed,

$$\int_{[0,\infty)} e^{-\lambda u} \mu(du) = \int_{[0,\infty)} \int_{[0,\infty)} e^{-\lambda(s+t)} \mu_1(ds)\mu_2(dt) = f_1(\lambda) f_2(\lambda).$$

Write $M := \{\mathscr{L}\mu \ : \ \mu \text{ is a finite measure on } [0, \infty)\}$. Theorem 1.4 shows that $M \subset \mathcal{CM} \subset \overline{M}$. We are done if we can show that $\mathcal{CM}$ is closed under pointwise convergence. For this choose a sequence $(f_n)_{n\in\mathbb{N}} \subset \mathcal{CM}$ such that the limit

$\lim_{n\to\infty} f_n(\lambda) = f(\lambda)$ exists for every $\lambda > 0$. If μ_n denotes the representing measure of f_n, we find for every $a > 0$

$$\mu_n[0,a] \leq e^{a\lambda} \int_{[0,a]} e^{-\lambda t} \mu_n(dt) \leq e^{a\lambda} f_n(\lambda) \xrightarrow{n\to\infty} e^{a\lambda} f(\lambda)$$

which means that the family of measures $(\mu_n)_{n\in\mathbb{N}}$ is bounded in the vague topology, hence vaguely sequentially compact, see Appendix A.1. Thus, there exist a subsequence $(\mu_{n_k})_{k\in\mathbb{N}}$ and a measure μ such that $\mu_{n_k} \to \mu$ vaguely. For every function $\chi \in C_c[0,\infty)$ with $0 \leq \chi \leq 1$, we find

$$\int_{[0,\infty)} \chi(t)e^{-\lambda t} \mu(dt) = \lim_{k\to\infty} \int_{[0,\infty)} \chi(t)e^{-\lambda t} \mu_{n_k}(dt) \leq \liminf_{k\to\infty} f_{n_k}(\lambda)$$
$$= f(\lambda).$$

Taking the supremum over all such χ, we can use monotone convergence to get

$$\int_{[0,\infty)} e^{-\lambda s} \mu(dt) \leq f(\lambda).$$

On the other hand, we find for each $a > 0$

$$f_{n_k}(\lambda) = \int_{[0,a)} e^{-\lambda t} \mu_{n_k}(dt) + \int_{[a,\infty)} e^{-\frac{1}{2}\lambda t} e^{-\frac{1}{2}\lambda t} \mu_{n_k}(dt)$$
$$\leq \int_{[0,a)} e^{-\lambda t} \mu_{n_k}(dt) + e^{-\frac{1}{2}\lambda a} f_{n_k}\left(\frac{1}{2}\lambda\right).$$

If we let $k \to \infty$ and then $a \to \infty$ along a sequence of continuity points of μ we get $f(\lambda) \leq \int_{[0,\infty)} e^{-\lambda t} \mu(dt)$ which shows that $f \in \mathcal{CM}$ and that the measure μ is actually independent of the particular subsequence. In particular, $\mu = \lim_{n\to\infty} \mu_n$ vaguely in the space of measures supported in $[0,\infty)$. $\qquad\square$

The seemingly innocuous closure assertion of Corollary 1.6 actually says that on the set $\mathcal{CM}$ the notions of *pointwise convergence*, *locally uniform convergence*, and even *convergence in the space* $C^\infty(0,\infty)$ coincide. This situation reminds remotely of the famous Montel's theorem from the theory of analytic functions, see e.g. Berenstein and Gay [31, Theorem 2.2.8].

Corollary 1.7. *Let* $(f_n)_{n\in\mathbb{N}}$ *be a sequence of completely monotone functions such that the limit* $\lim_{n\to\infty} f_n(\lambda) = f(\lambda)$ *exists for all* $\lambda \in (0,\infty)$. *Then* $f \in \mathcal{CM}$ *and* $\lim_{n\to\infty} f_n^{(k)}(\lambda) = f^{(k)}(\lambda)$ *for all* $k \in \mathbb{N} \cup \{0\}$ *locally uniformly in* $\lambda \in (0,\infty)$.

Proof. From Corollary 1.6 we know already that $f \in \mathcal{CM}$. Moreover, we have seen that the representing measures μ_n of f_n converge vaguely in $[0,\infty)$ to the representing measure μ of f. By the differentiation lemma for parameter dependent integrals we infer

$$f_n^{(k)}(\lambda) = (-1)^k \int_{[0,\infty)} t^k e^{-\lambda t} \mu_n(dt)$$

$$\xrightarrow{n \to \infty} (-1)^k \int_{[0,\infty)} t^k e^{-\lambda t} \mu(dt) = f^{(k)}(\lambda),$$

since $t \mapsto t^k e^{-\lambda t}$ is a function that vanishes at infinity, cf. (A.3) in Appendix A.1.

Finally, assume that $|\lambda - \kappa| \le \delta$ for some $\delta > 0$. Using the elementary estimate $|e^{-\lambda t} - e^{-\kappa t}| \le |\lambda - \kappa| t\, e^{-(\kappa \wedge \lambda)t}$, $\lambda, \kappa, t > 0$, we conclude that for $\kappa, \lambda \ge \epsilon$ and all $\epsilon > 0$

$$\left| f_n^{(k)}(\lambda) - f_n^{(k)}(\kappa) \right| \le \int_{(0,\infty)} \left| e^{-\lambda t} - e^{-\kappa t} \right| t^k \mu_n(dt)$$

$$\le \delta \int_{(0,\infty)} e^{-(\kappa \wedge \lambda)t}\, t^{k+1} \mu_n(dt)$$

$$= \delta \left| f_n^{(k+1)}(\kappa \wedge \lambda) \right|.$$

Using that $\lim_{n\to\infty} f_n^{(k+1)}(\kappa \wedge \lambda) = f^{(k+1)}(\kappa \wedge \lambda)$, we find for sufficiently large values of n

$$\left| f_n^{(k)}(\lambda) - f_n^{(k)}(\kappa) \right| \le 2\delta \sup_{\gamma \ge \epsilon} \left| f^{(k+1)}(\gamma) \right|.$$

This proves that the functions $f_n^{(k)}$ are uniformly equicontinuous on $[\epsilon, \infty)$. Therefore, the convergence $\lim_{n\to\infty} f_n^{(k)}(\lambda) = f^{(k)}(\lambda)$ is locally uniform on $[\epsilon, \infty)$ for every $\epsilon > 0$. Since $\epsilon > 0$ was arbitrary, we are done. □

Every family of completely monotone functions which is uniformly bounded at the origin is (weakly) sequentially compact.

Corollary 1.8. *Let $(f_n)_{n\in\mathbb{N}}$ be a sequence of completely monotone functions such that $\sup_{n\in\mathbb{N}} f_n(0+) < \infty$. Then there exist a subsequence $(f_{n_k})_{k\in\mathbb{N}}$ and a completely monotone function $f : (0,\infty) \to [0,\infty)$ such that $\lim_{k\to\infty} f_{n_k}(\lambda) = f(\lambda)$ for all $\lambda \in (0,\infty)$.*

Proof. Set $c := \sup_{n\in\mathbb{N}} f_n(0+)$ and denote for each $n \in \mathbb{N}$ the representing measure of f_n by μ_n. Since $f_n(0+) \le c$, we have $\mu_n[0,\infty) \le c$, implying that the sequence of measures $(\mu_n)_{n\in\mathbb{N}}$ is vaguely bounded, hence vaguely compact by Theorem A.5. Therefore, there exist a subsequence $(n_k)_{k\in\mathbb{N}}$ and a finite measure μ on $[0,\infty)$ such that $\mu_{n_k} \xrightarrow{k\to\infty} \mu$ vaguely. If we define $f := \mathscr{L}\mu$, then $\lim_{k\to\infty} f_{n_k}(\lambda) = f(\lambda)$ for every $\lambda > 0$ and, by Corollary 1.7, $f \in \mathcal{CM}$. □

Remark 1.9. The representation formula for completely monotone functions given in Theorem 1.4 has an interesting interpretation in connection with the Kreĭn–Milman theorem and the Choquet representation theorem. The set

$$\{f \in \mathcal{CM} \,:\, f(0+) = 1\}$$

is a basis of the convex cone $\mathcal{CM}_b$, and its extremal points are given by

$$e_t(\lambda) = e^{-\lambda t}, \quad 0 \le t < \infty, \quad \text{and} \quad e_\infty(\lambda) = \mathbb{1}_{\{0\}}(\lambda),$$

see Phelps [291, Lemma 2.2], Lax [239, p. 139] or the proof of Theorem 4.8. These extremal points are formally defined for $\lambda \in [0, \infty)$ with the understanding that $e_\infty|_{(0,\infty)} \equiv 0$. Therefore, the representation formula from Theorem 1.4 becomes a Choquet representation of the elements of $\mathcal{CM}_b$,

$$\int_{[0,\infty)} e^{-\lambda t} \, \mu(dt) = \int_{[0,\infty]} e^{-\lambda t} \, \mu(dt), \quad \lambda \in (0, \infty).$$

In particular, the functions

$$e_t|_{(0,\infty)}$$

are prime examples of completely monotone functions. Theorem 1.4 and Corollary 1.6 tell us that every $f \in \mathcal{CM}$ can be written as an 'integral mixture' of the extremal $\mathcal{CM}$-functions $\{e_t|_{(0,\infty)} \,:\, 0 \le t < \infty\}$.

In the remaining part of this chapter we study two generalizations of complete monotonicity: multiply monotone functions and completely monotone functions on the whole real line $\mathbb{R}$.

Definition 1.10. A function $f : (0, \infty) \to \mathbb{R}$ is *1-monotone* (also *monotone of order* 1) if $f(\lambda) \ge 0$ for all $\lambda > 0$ and f is non-increasing and right-continuous. A function $f : (0, \infty) \to \mathbb{R}$ is *n-monotone* (also *monotone of order n*), $n \ge 2$, if it is $n - 2$ times differentiable, $(-1)^j f^{(j)}(\lambda) \ge 0$ for all $j = 0, 1, \ldots, n - 2$ and $\lambda > 0$, and $(-1)^{n-2} f^{(n-2)}$ is non-increasing and convex.

It is not difficult to see from Definition 1.10 that an n times differentiable function $f : (0, \infty) \to \mathbb{R}$ is n-monotone if, and only if, $(-1)^j f^{(j)}(\lambda)$ is non-negative for all $j = 0, 1, \ldots, n$. In particular, a function $f : (0, \infty) \to \mathbb{R}$ is completely monotone if, and only if, it is n-monotone for every $n \ge 1$. In view of Theorem 1.4 it should not be surprising that n-monotone functions admit an integral representation.

Theorem 1.11. *Suppose that* $f : (0, \infty) \to \mathbb{R}$ *and* $n \in \mathbb{N}$. *Then the following statements are equivalent.*

(i) *f is n-monotone.*

(ii) *There exist a unique constant $c \geq 0$ and a unique measure v on $(0, \infty)$ such that*

$$f(\lambda) = c + \frac{1}{(n-1)!} \int_{(\lambda, \infty)} (t - \lambda)^{n-1} v(dt), \quad \lambda > 0. \tag{1.10}$$

(iii) *There exists a unique measure γ_n on $[0, \infty)$ such that*

$$f(\lambda) = \int_{[0,\infty)} (1 - \lambda t)_+^{n-1} \gamma_n(dt), \quad \lambda > 0. \tag{1.11}$$

Proof. (i)$\Rightarrow$(ii) Set $c := f(+\infty) := \lim_{\lambda \to \infty} f(\lambda)$. For $n = 1$, f is a right-continuous, non-increasing and non-negative function such that the limit $f(+\infty)$ exists. Every such function can be written as $f(\lambda) = c + v(\lambda, \infty)$ for a unique measure v on $(0, \infty)$ and a unique constant $c \geq 0$.

For $n = 2$, f is non-negative, non-increasing, convex and such that the limit $f(+\infty)$ exists. A convex function has at each point in the interior of its domain a right-hand derivative f'_+ which is non-decreasing and right-continuous. Since f is non-increasing it follows that $f'_+ \leq 0$. Therefore, $g(\lambda) := -f'_+(\lambda)$ is right-continuous, non-increasing and $f(\lambda) = f(+\infty) + \int_\lambda^\infty g(t)\, dt$. This proves that $f(\lambda) = c + \int_\lambda^\infty v(s, \infty)\, ds$ for the unique measure v on $(0, \infty)$ given by $v(t, \infty) := g(t)$ and the unique constant $c := f(+\infty) \geq 0$.

We proceed by induction. Assume that $n \geq 2$ and that the claim is proved for n, i.e. an n-monotone function admits the representation (1.10). Since the limit $f(+\infty)$ exists, it holds that $\lim_{\lambda \to \infty} f'(\lambda) = 0$. Hence $-f'$ is n-monotone. By the induction hypothesis, there is a unique measure v on $(0, \infty)$ such that

$$-f'(s) = \int_{(s,\infty)} \frac{1}{(n-1)!} (t - s)^{n-1} v(dt), \quad s > 0.$$

Using Fubini's theorem we obtain for some integration constant $c \in \mathbb{R}$

$$\begin{aligned}
f(\lambda) &= -\int_\lambda^\infty f'(s)\, ds + c \\
&= c + \int_\lambda^\infty \left(\int_{(s,\infty)} \frac{1}{(n-1)!} (t - s)^{n-1} v(dt) \right) ds \\
&= c + \int_{(\lambda,\infty)} \left(\int_\lambda^t \frac{1}{(n-1)!} (t - s)^{n-1}\, ds \right) v(dt) \\
&= c + \int_{(\lambda,\infty)} \frac{1}{n!} (t - \lambda)^n\, v(dt)
\end{aligned}$$

proving the claim. Since $c = f(+\infty)$, we see that c is non-negative and unique.

(ii)$\Rightarrow$(iii) Let $\nu^\star$ be the image measure of ν with respect to the mapping $t \mapsto t^{-1}$, and define the measure γ_n on $(0, \infty)$ by $\gamma_n(dt) = \frac{1}{(n-1)!} t^{1-n} \nu^\star(dt)$. Then

$$\frac{1}{(n-1)!} \int_{(0,\infty)} \mathbb{1}_{(\lambda,\infty)}(t) (t-\lambda)^{n-1} \nu(dt)$$

$$= \frac{1}{(n-1)!} \int_{(0,\infty)} \mathbb{1}_{(\lambda,\infty)}(t^{-1}) (t^{-1}-\lambda)^{n-1} \nu^\star(dt)$$

$$= \int_{(0,\infty)} \mathbb{1}_{(0,\lambda^{-1})}(t) (1-\lambda t)^{n-1} \gamma_n(dt)$$

$$= \int_{(0,\infty)} (1-\lambda t)_+^{n-1} \gamma_n(dt).$$

If we set $\gamma_n\{0\} = c$, the claim follows.

(iii)$\Rightarrow$(i) Assume that f is given by (1.11). If $n = 1$, $f(\lambda) = \gamma_n[0, 1/\lambda)$ which is a non-negative, non-increasing and right-continuous function. In case $n \geq 2$, the function $\lambda \mapsto (1-\lambda t)_+^{n-1}$ is $n-2$ times differentiable and for $k = 0, 1, \ldots, n-2$,

$$(-1)^k \frac{d^k}{d\lambda^k} (1-\lambda t)_+^{n-1} = (n-1)(n-2)\cdots(n-k)t^k (1-\lambda t)_+^{n-k-1}$$

is non-negative, non-increasing and convex. Moreover, $t \mapsto t^k(1-\lambda t)_+^{n-k-1}$, $t \geq 0$, is integrable with respect to γ_n which allows us to differentiate in (1.11) under the integral sign. Hence, f is $n-2$ times differentiable and

$$(-1)^k f^{(k)}(\lambda) = (n-1)(n-2)\cdots(n-k) \int_{[0,\infty)} t^k (1-\lambda t)_+^{n-k-1} \gamma_n(dt).$$

Since non-negativity, monotonicity and convexity are preserved under integration, we conclude that f is n-monotone. $\square$

Remark 1.12. (i) In the proof of Theorem 1.4 we have derived the representation (1.6) which requires only that f is n times differentiable and vanishes at infinity. If f is also n-monotone, we can compare this with (1.10), and conclude that the representing measure ν is given by $\nu(dt) = (-1)^n f^{(n)}(t)\, dt$; thus,

$$\gamma_n(dt) = \frac{(-1)^n}{(n-1)!} t^{-n-1} f^{(n)}(t^{-1})\, dt.$$

Let $\widetilde{\gamma}_n$ be the image measure of γ_n under the mapping $t \mapsto nt$. If $f(+\infty) = 0$, (1.11) becomes

$$f(\lambda) = \int_{(0,\infty)} \left(1 - \frac{\lambda t}{n}\right)_+^{n-1} \widetilde{\gamma}_n(dt).$$

Comparing this with (1.3) and (1.9), we see – if f is n times differentiable – that

$$\widetilde{\gamma}_n(dt) = \frac{(-1)^n}{n!} f^{(n)}\left(\frac{n}{t}\right)\left(\frac{n}{t}\right)^{n+1}.$$

(ii) The representation (1.10) allows an extension of the concept of monotonicity to non-integer order. A function $f : (0,\infty) \to \mathbb{R}$ is p-*monotone*, $p > 0$, (also *monotone of order p*) if there exist a constant $c \geq 0$ and a measure ν on $(0,\infty)$ such that

$$f(\lambda) = c + \frac{1}{\Gamma(p)} \int_{(\lambda,\infty)} (t - \lambda)^{p-1} \nu(dt), \quad \lambda > 0.$$

Equivalently, f is p-monotone, if (1.11) holds with n replaced by p.

The conditions in the definition of n-monotonicity have redundancies; this was shown in [372].

Proposition 1.13. *Let* $f : (0,\infty) \to \mathbb{R}$. *Then* f *is n-monotone, $n \geq 2$, if, and only if,*

(i) $f(+\infty) = \lim_{\lambda \to \infty} f(\lambda)$ *exists and is non-negative;*

(ii) $(-1)^{n-2} f^{(n-2)}$ *is non-negative, non-increasing and convex.*

Proof. Assume that (i) and (ii) hold. Then $f^{(n-2)}$ is convex, hence continuous, and by Taylor's formula, for every $a > 0$

$$f(\lambda) = \sum_{k=0}^{n-3} \frac{f^{(k)}(a)}{k!} (\lambda - a)^k + (-1)^{n-2} \int_a^\lambda \frac{(-1)^n f^{(n-2)}(s)}{(n-3)!} (\lambda - s)^{n-3} \, ds.$$

If $n - 2$ is even

$$f(\lambda) \geq \sum_{k=0}^{n-3} \frac{f^{(k)}(a)}{k!} (\lambda - a)^k,$$

while for $n - 2$ odd

$$f(\lambda) \leq \sum_{k=0}^{n-3} \frac{f^{(k)}(a)}{k!} (\lambda - a)^k.$$

Dividing by $(\lambda - a)^{n-3}$, letting $\lambda \to \infty$, and by using that $f(\lambda)$ stays bounded, we arrive at $f^{(n-3)}(a) \leq 0$ if $n - 2$ is even, and $f^{(n-3)}(a) \geq 0$ if $n - 2$ is odd. Thus, $(-1)^{n-3} f^{(n-3)}(a) \geq 0$. By induction it follows that $(-1)^k f^{(k)}(a) \geq 0$ for all $k = 1, 2, \ldots, n - 2$. Since $f(+\infty) \geq 0$, we see $f(a) \geq 0$, which means that f is n-monotone.   □

It was pointed out in [326] that the conditions (1.4) have redundancies. This follows immediately from the preceding proposition.

Corollary 1.14. *Let* $f : (0, \infty) \to \mathbb{R}$ *be a* C^∞ *function such that* $f \geq 0$, $f' \leq 0$ *and* $(-1)^n f^{(n)} \geq 0$ *for infinitely many* $n \in \mathbb{N}$. *Then* f *is a completely monotone function.*

Corollary 1.15. *The set of* n-*monotone functions is a convex cone which is closed under multiplication.*

Proof. It is clear from the definition that the set of n-monotone functions is a convex cone. We prove that it is closed under multiplication. The case $n = 1$ is immediate from definition. Let $n \geq 2$ and assume that the claim is proved for $n-1$. Let $g_j = -f_j'$ where f_j, $j = 1, 2$, is n-monotone. Then g_j, $j = 1, 2$, is $(n - 1)$-monotone. Noting that $f_1 f_2$ is the primitive of $f_1' f_2 + f_1 f_2' = -g_1 f_2 - f_1 g_2$, we see

$$f_1(\lambda) f_2(\lambda) - f_1(+\infty) f_2(+\infty) = \int_\lambda^\infty \Big(g_1(t) f_2(t) + f_1(t) g_2(t) \Big) dt.$$

By the induction hypothesis, $g_1 f_2$ and $f_1 g_2$ are $(n - 1)$-monotone, and the claim follows.   □

Note that Corollary 1.15 provides an alternative proof that $\mathcal{CM}$ is closed under multiplication.

The concepts of n-monotonicity and complete monotonicity can be extended to functions defined on the whole real line. Let $f : \mathbb{R} \to \mathbb{R}$. Then f is 1-*monotone* if it is non-negative, non-increasing and right-continuous. If $n \geq 2$ is an integer then f is n-*monotone* if it is $n - 2$ times differentiable, $(-1)^j f^{(j)}(\lambda) \geq 0$ for all $j = 0, 1, \ldots, n - 2$ and all $\lambda \in \mathbb{R}$, and $(-1)^{n-1} f^{(n-2)}$ is non-negative, non-increasing and convex. The analogue of Theorem 1.11 – with the same proof – is still valid but with $\lambda \in \mathbb{R}$ and ν and γ_n being measures on $\mathbb{R}$. The function $f : \mathbb{R} \to \mathbb{R}$ is *completely monotone on* $\mathbb{R}$ if it is n-monotone on $\mathbb{R}$ for all $n \in \mathbb{N}$; this is the same as to say that $f \in C^\infty(\mathbb{R})$ and $(-1)^n f^{(n)}(\lambda) \geq 0$ for all $n \geq 0$ and $\lambda \in \mathbb{R}$. Any such function f admits the representation

$$f(\lambda) = \int_{(0,\infty)} e^{-\lambda t} \mu(dt), \quad \lambda \in \mathbb{R}, \tag{1.12}$$

where μ is a measure on $(0, \infty)$. This follows easily from Theorem 1.4 and the fact for each $\lambda_0 \leq 0$, the function $\lambda \mapsto f(\lambda + \lambda_0)$ is completely monotone (on $(0, \infty)$, i.e. in the sense of Definition 1.3).

The relationship between higher-order monotonicity on $\mathbb{R}$ and on $(0, \infty)$ is explained in the following proposition.

Proposition 1.16. *Suppose that* $f : \mathbb{R} \to \mathbb{R}$ *is n-monotone, $n \in \mathbb{N}$, on $\mathbb{R}$. Then* $\lambda \mapsto f(\log \lambda)$, $\lambda > 0$, *is n-monotone. In particular, if* $f : \mathbb{R} \to \mathbb{R}$ *is completely monotone on $\mathbb{R}$, then* $\lambda \mapsto f(\log \lambda)$, $\lambda > 0$ *is completely monotone.*

Proof. For $n = 1$ the claim follows from the definition and the fact that the logarithm maps $(0, \infty)$ bijectively onto $\mathbb{R}$. Let $n \geq 2$ and assume that the claim is valid for $n - 1$. Assume that f is n-monotone on $\mathbb{R}$ and set $g(\lambda) = -f'(\lambda)$. Then g is $(n-1)$-monotone on $\mathbb{R}$ and $f(\lambda) = \int_\lambda^\infty g(s)\, ds$. Hence,

$$f(\log \lambda) = \int_{\log \lambda}^\infty g(s)\, ds = \int_\lambda^\infty g(\log t)\, t^{-1}\, dt.$$

By the induction hypothesis, $t \mapsto g(\log t)$ is an $(n-1)$-monotone function, hence by Corollary 1.15, $t \mapsto g(\log t)\, t^{-1}$ is also $(n-1)$-monotone. Thus, $\lambda \mapsto f(\log \lambda)$ is n-monotone. □

Remark 1.17. The converse of Proposition 1.16 is not valid. Indeed, let $n \geq 2$ and define $f : \mathbb{R} \to (0, \infty)$ by $f(\lambda) = (1 - e^\lambda)^{n-1} \mathbb{1}_{(-\infty,0)}(\lambda)$. Then

$$f(\log \lambda) = (1 - \lambda)^{n-1} \mathbb{1}_{(0,1)}(\lambda) = \int_{[0,\infty)} (1 - \lambda t)_+^{n-1}\, \delta_1(dt)$$

is n-monotone. On the other hand, f is not n-monotone on $\mathbb{R}$ since

$$f''(\lambda) = (n-1)(n-2)(1 - e^\lambda)^{n-3} e^{2\lambda} - (n-1)(1 - e^\lambda)^{n-2} e^\lambda < 0$$

for $\lambda < -\log(n-1)$. Similarly, let $f(\lambda) = \exp(-ce^\lambda)$ for some $c > 0$. Then $f(\log \lambda) = e^{-c\lambda}$ is completely monotone, while f is not completely monotone on $\mathbb{R}$. These examples come from [317].

Comments 1.18. Standard references for Laplace transforms include D. V. Widder's monographs [370, 371] and Doetsch's treatise [96]. For a modern point of view we refer to Berg and Forst [39] and Berg, Christensen and Ressel [38]. The most comprehensive tables of Laplace transforms are the Bateman manuscript project [107] and the tables by Prudnikov, Brychkov and Marichev [296].

The *concept* of complete monotonicity seems to go back to S. Bernstein [43] who studied functions on an interval $I \subset \mathbb{R}$ having positive derivatives of all orders. If $I = (-\infty, 0]$ this is, up to a change of sign in the variable, complete monotonicity. In later papers, Bernstein refers to functions enjoying this property as *absolument monotone*, see the appendix *première note*, [44, pt. IV, p. 190], and in [45] he states and proves Theorem 1.4 for functions on the negative half-axis.

Following Schur (probably [328]), Hausdorff [150, p. 80] calls a sequence $(\mu_n)_{n \in \mathbb{N}}$ *total monoton* – the literal translation *totally monotone* is only rarely used, e.g. in Hardy [145] – if all iterated differences $(-1)^k \Delta^k \mu_n$, $k \geq 1$, $n \geq 1$, are non-negative.

As usual, $\Delta \mu_n := \mu_{n+1} - \mu_n$. The modern terminology is *completely monotone*; for sequences the German equivalent *vollmonoton* was introduced by Jacobsthal [186], in connection with functions the notion of complete monotonicity appears for the first time in [368].

Hausdorff focusses in [150, 151] on the *moment problem*: the μ_n are of the form $\int_{(0,1]} t^n \, \mu(dt)$ for some measure μ on $(0, 1]$ if, and only if, the sequence $(\mu_n)_{n \in \mathbb{N} \cup \{0\}}$ is *total monoton*; moreover, he introduces the moment function $\mu_\lambda := \int_{(0,1]} t^\lambda \, \mu(dt)$ which he also calls *total monoton*. A simple change of variables $t \rightsquigarrow e^{-u}$, $t \in (0, 1]$, $u \in [0, \infty)$, shows that $\mu_\lambda = \int_{[0,\infty)} e^{-\lambda u} \, \tilde{\mu}(du)$ for some suitable image measure $\tilde{\mu}$ of μ. This means that every moment sequence gives rise to a unique completely monotone function. The converse is much easier since

$$\int_{[0,\infty)} e^{-\lambda u} \, \tilde{\mu}(du) = \sum_{n=0}^{\infty} \frac{(-1)^n \lambda^n}{n!} \int_{[0,\infty)} u^n \, \tilde{\mu}(du).$$

Many historical comments can be found in the second part, pp. 29–44, of Ky Fan's memoir [110]. For an up-to-date survey we recommend the scholarly commentary by Chatterji [78] written for Hausdorff's collected works [152]. More general *higher* or *multiple monotonicity properties* for a sequence were already considered in [186]. A sequence is *monotone of order m* if $(-1)^k \Delta^k \mu_n \geq 0$ for all $k = 0, 1, \ldots, m$ and $n \in \mathbb{N}$. This was extended by Knopp [211] to fractional orders and a series of papers by Fejér, Hausdorff, Kaluza and Knopp studies in depth the behaviour of power series and trigonometric series whose coefficients are completely monotone or monotone of some finite order. The paper [115] gives a detailed account on these developments. Royall [310] and, independently, Hartman [146] call a function f monotone of order m if it satisfies (1.4) only for $n \in \{0, 1, \ldots, m\}$, $m \in \mathbb{N} \cup \{0\}$. The present, more general, Definition 1.10 is due to Williamson [372]. Note that Löwner's notion of higher-order monotonicity introduced in [248] (*monoton nter Stufe*) and [249] (*monotonic of order n*) is different; it will play an important role in Chapter 12.2, cf. Definition 12.9. Further applications of $\mathcal{CM}$ and related functions to ordinary differential equations can be found, e.g. in Lorch *et al.* [247] and Mahajan and Ross [258], see also the study by van Haeringen [359] and the references given there. The connection between integral equations and $\mathcal{CM}$ are extensively covered in the monographs by Gripenberg *et al.* [136] and Prüss [297].

A by-product of the proof of Proposition 1.2 is an example of a so-called real inversion formula for Laplace transforms. Formula (1.3) is due to Dubourdieu [102] and Feller [116], see also Pollard [295] and Widder [370, p. 295] and [371, Chapter 6]. Our presentation follows Feller [118, VII.6].

The proof of Bernstein's theorem, Theorem 1.4, also contains a real inversion formula for the Laplace transform: (1.8) coincides with the operator $L_{k,y}(f(\lambda))$ of Widder [371, p. 140] and, up to a constant, also [370, p. 288]. Since the proof of Theorem 1.4 relies on a compactness argument using subsequences, the weak limit

$f_{n_k}(t)\,dt \rightarrow \mu(dt)$ might depend on the actual subsequence $(n_k)_{k\in\mathbb{N}}$. If we combine this argument with Proposition 1.2, we get at once that all subsequences lead to the same μ and that, therefore, the weak limit of the full sequence $f_n(t)\,dt \rightarrow \mu(dt)$ exists.

The representation (1.6) was also obtained in [103] in the following way: as $f(+\infty) = 0$ we can write $f(\lambda) = \int_\lambda^\infty (-1) f'(t_1)\,dt_1$. Since $-f'$ is again completely monotone and satisfies $-f'(+\infty) = 0$, the same argument proves that $f(\lambda) = \int_\lambda^\infty \int_{t_1}^\infty f''(t_2)\,dt_2\,dt_1$. By induction, for every $n \in \mathbb{N}$,

$$f(\lambda) = \int_\lambda^\infty \int_{t_1}^\infty \cdots \int_{t_{n-1}}^\infty (-1)^n f^{(n)}(t_n)\,dt_n \cdots dt_2\,dt_1.$$

The representation (1.6) follows by using Fubini's theorem and reversing the order of integration. The rest of the proof is now similar to our presentation.

It is possible to avoid the compactness argument in Theorem 1.4 and to give an 'intuitionistic' proof, see van Herk [361, Theorem 33] who did this for the class $\mathcal{S}$ (denoted by $\{F\}$ in [361]) of Stieltjes functions which is contained in $\mathcal{CM}$; his arguments work also for $\mathcal{CM}$.

A proof of Theorem 1.4 using Choquet's theorem or the Kreĭn–Milman theorem can be found in Kendall [202], Meyer [265] or Choquet [85, 84]. A modern textbook version is contained in Lax [239, Chapter 14.3, p. 138], Phelps [291, Chapter 2], Becker [28] and also in Theorem 4.8 below. The somewhat surprising compactness result Corollary 1.8 seems to be new.

Definition 1.10 and the results on multiply monotone functions are from [372], our proofs are adaptations of [317, Chapter 2]. The connection between complete monotonicity and complete monotonicity on $\mathbb{R}$ is also from [317]. Multiply monotone functions of non-integer order appear already in [372]; Sato [317] studies the interplay between fractional derivatives, fractional integrals and fractional-order monotone functions. Gneiting's short note [131] contains an example showing that one cannot weaken the Bernstein–Hausdorff–Widder conditions beyond what is stated in Corollary 1.14.

There is a deep geometric connection between completely monotone functions and the problem when a metric space can be embedded into a Hilbert space $\mathfrak{H}$. The basic result is due to Schoenberg [325]: a function f on $[0,\infty)$ with $f(0) = f(0+)$ is completely monotone if, and only if, $\xi \mapsto f(|\xi|^2)$, $\xi \in \mathbb{R}^d$, is positive definite for all dimensions $d \geq 0$, cf. Theorem 13.14.

Chapter 2
Stieltjes functions

Stieltjes functions are a subclass of completely monotone functions. They will play a central role in our study of complete Bernstein functions. In Theorem 7.3 we will see that f is a Stieltjes function if, and only if, $1/f$ is a complete Bernstein function. This allows us to study Stieltjes functions via the set of all complete Bernstein functions which is the focus of this tract. Therefore we restrict ourselves to the definition and a few fundamental properties of Stieltjes functions.

Definition 2.1. A (non-negative) *Stieltjes function* is a function $f : (0, \infty) \to [0, \infty)$ which can be written in the form

$$f(\lambda) = \frac{a}{\lambda} + b + \int_{(0,\infty)} \frac{1}{\lambda + t} \sigma(dt) \tag{2.1}$$

where $a, b \geq 0$ are non-negative constants and σ is a measure on $(0, \infty)$ such that $\int_{(0,\infty)} (1 + t)^{-1} \sigma(dt) < \infty$. We denote the family of all Stieltjes functions by $\mathcal{S}$.

The integral appearing in (2.1) is also called the *Stieltjes transform* of the measure σ. Using the elementary relation $(\lambda + t)^{-1} = \int_0^\infty e^{-tu} e^{-\lambda u} \, du$ and Fubini's theorem one sees that it is also a double Laplace transform. In view of the uniqueness of the Laplace transform, see Proposition 1.2, a, b and σ appearing in the representation (2.1) are uniquely determined by f. Since some authors consider measures σ on the compactification $[0, \infty]$, there is no marked difference between Stieltjes transforms and Stieltjes functions in the sense of Definition 2.1.

It is sometimes useful to rewrite (2.1) in the following form

$$f(\lambda) = \int_{[0,\infty]} \frac{1 + t}{\lambda + t} \bar{\sigma}(dt) \tag{2.2}$$

where $\bar{\sigma} := a\delta_0 + (1 + t)^{-1} \sigma(dt) + b\delta_\infty$ is a finite measure on the compact interval $[0, \infty]$.

Since for $z = \lambda + i\kappa \in \mathbb{C} \setminus (-\infty, 0]$ and $t \geq 0$

$$\left| \frac{1}{z + t} \right| = \frac{1}{\sqrt{(\lambda + t)^2 + \kappa^2}} \asymp \frac{1}{t + 1},$$

i.e. there exist two positive constants $c_1 < c_2$ (depending on λ and κ) such that

$$\frac{c_1}{t + 1} \leq \frac{1}{\sqrt{(\lambda + t)^2 + \kappa^2}} \leq \frac{c_2}{t + 1},$$

we can use (2.2) to extend $f \in \mathcal{S}$ uniquely to an analytic function on $\mathbb{C} \setminus (-\infty, 0]$. Note that

$$\operatorname{Im} z \cdot \operatorname{Im} \frac{1}{z+t} = \operatorname{Im} z \cdot \frac{-\operatorname{Im} z}{|z+t|^2} = -\frac{(\operatorname{Im} z)^2}{|z+t|^2}$$

which means that the mapping $z \mapsto f(z)$ swaps the upper and lower complex half-planes. We will see in Corollary 7.4 below that this property is also sufficient to characterize $f \in \mathcal{S}$.

Theorem 2.2. (i) *Every $f \in \mathcal{S}$ is of the form*

$$f(\lambda) = \mathscr{L}(a \cdot dt; \lambda) + \mathscr{L}\big(b \cdot \delta_0(dt); \lambda\big) + \mathscr{L}\big(\mathscr{L}(\sigma; t)\, dt; \lambda\big)$$

for the measure σ appearing in (2.1). In particular, $\mathcal{S} \subset \mathcal{CM}$ and $\mathcal{S}$ consists of all completely monotone functions having a representation measure with completely monotone density on $(0, \infty)$.

(ii) *The set $\mathcal{S}$ is a convex cone: if $f_1, f_2 \in \mathcal{S}$, then $s f_1 + t f_2 \in \mathcal{S}$ for all $s, t \geq 0$.*

(iii) *The set $\mathcal{S}$ is closed under pointwise limits: if $(f_n)_{n \in \mathbb{N}} \subset \mathcal{S}$ and if the limit $\lim_{n \to \infty} f_n(\lambda) = f(\lambda)$ exists for all $\lambda > 0$, then $f \in \mathcal{S}$.*

Proof. Since $(\lambda + t)^{-1} = \int_0^\infty e^{-(\lambda+t)u}\, du$, assertion (i) follows from (2.1) and Fubini's theorem; (ii) is obvious. For (iii) we argue as in the proof of Corollary 1.6: assume that the function f_n is given by (2.2) where we denote the representing measure by $\bar{\sigma}_n = a_n \delta_0 + (1+t)^{-1} \sigma_n(dt) + b_n \delta_\infty$. Since

$$\bar{\sigma}_n[0, \infty] = f_n(1) \xrightarrow{n \to \infty} f(1) < \infty,$$

the family $(\bar{\sigma}_n)_{n \in \mathbb{N}}$ is uniformly bounded. By the Banach–Alaoglu theorem, Corollary A.6, we conclude that $(\bar{\sigma}_n)_{n \in \mathbb{N}}$ has a weak* convergent subsequence $(\bar{\sigma}_{n_k})_{k \in \mathbb{N}}$ such that $\bar{\sigma} := \text{vague-}\lim_{k \to \infty} \bar{\sigma}_{n_k}$ is a bounded measure on the compact space $[0, \infty]$. Since $t \mapsto (1+t)/(\lambda+t)$ is in $C[0, \infty]$, we get

$$f(\lambda) = \lim_{k \to \infty} f_{n_k}(\lambda) = \lim_{k \to \infty} \int_{[0,\infty]} \frac{1+t}{\lambda+t} \bar{\sigma}_{n_k}(dt) = \int_{[0,\infty]} \frac{1+t}{\lambda+t} \bar{\sigma}(dt),$$

i.e. $f \in \mathcal{S}$. Since the limit $\lim_{n \to \infty} f_n(\lambda) = f(\lambda)$ exists – independently of any subsequence – and since the representing measure is uniquely determined by the function f, $\bar{\sigma}$ does not depend on any subsequence. In particular, $\sigma = \lim_{n \to \infty} \sigma_n$ vaguely in the space of measures supported in $(0, \infty)$. $\qquad\square$

Remark 2.3. Let $f, f_n \in \mathcal{S}, n \in \mathbb{N}$, where we write a, b, σ and a_n, b_n, σ_n for the constants and measures appearing in (2.1) and $\bar{\sigma}_n, \bar{\sigma}$ for the corresponding representation measures from (2.2). If $\lim_{n \to \infty} f_n(\lambda) = f(\lambda)$, the proof of Theorem 2.2 shows that

$$\text{vague-}\lim_{n \to \infty} \sigma_n = \sigma \quad \text{and} \quad \text{vague-}\lim_{n \to \infty} \bar{\sigma}_n = \bar{\sigma}.$$

Combining this with the portmanteau theorem, Theorem A.7, it is possible to show that

$$a = \lim_{\epsilon \to 0} \liminf_{n \to \infty} \left(a_n + \int_{(0,\epsilon)} \frac{\sigma_n(dt)}{1+t} \right)$$

and

$$b = \lim_{R \to \infty} \liminf_{n \to \infty} \left(b_n + \int_{(R,\infty)} \frac{\sigma_n(dt)}{1+t} \right);$$

in the above formulae we can replace $\liminf_n$ by $\limsup_n$. We do not give the proof here, but we refer to a similar situation for Bernstein functions which is worked out in Corollary 3.9.

In general, it is not true that $\lim_{n \to \infty} a_n = a$ or $\lim_{n \to \infty} b_n = b$. This is easily seen from the following examples: $f_n(\lambda) = \lambda^{-1+1/n}$ and $f_n(\lambda) = \lambda^{-1/n}$.

Remark 2.4. Just as in the case of completely monotone functions, see Remark 1.9, we can understand the representation formula (2.2) as a particular case of the Choquet or Kreĭn–Milman representation. The set

$$\{f \in \mathcal{S} : f(1) = 1\}$$

is a basis of the convex cone $\mathcal{S}$, and its extremal points are given by

$$e_0(\lambda) = \frac{1}{\lambda}, \quad e_t(\lambda) = \frac{1+t}{\lambda+t}, \quad 0 < t < \infty, \quad \text{and} \quad e_\infty(\lambda) = 1.$$

To see that the functions e_t, $0 \le t \le \infty$, are indeed extremal, note that the equality

$$e_t(\lambda) = \epsilon f(\lambda) + (1-\epsilon)g(\lambda), \quad f, g \in \mathcal{S},$$

and the uniqueness of the representing measures in (2.2) imply that

$$\delta_t = \epsilon \bar{\sigma}_f + (1-\epsilon)\bar{\sigma}_g$$

$(\bar{\sigma}_f, \bar{\sigma}_g$ are the representing measures) which is only possible if $\bar{\sigma}_f = \bar{\sigma}_g = \delta_t$. Conversely, since every $f \in \mathcal{S}$ is given by (2.2), the family $\{e_t\}_{t \in [0,\infty]}$ contains all extremal points. In particular,

$$1, \quad \frac{1}{\lambda}, \quad \frac{1}{\lambda+t}, \quad \frac{1+t}{\lambda+t}, \quad t > 0,$$

are examples of Stieltjes functions and so are their integral mixtures, e.g.

$$\lambda^{\alpha-1} \ (0 < \alpha < 1), \quad \frac{1}{\sqrt{\lambda}} \arctan \frac{1}{\sqrt{\lambda}}, \quad \frac{1}{\lambda} \log(1 + \lambda)$$

which we obtain if we choose $\frac{1}{\pi} \sin(\alpha\pi) t^{\alpha-1} dt$, $\mathbb{1}_{(0,1)}(t)\frac{dt}{2\sqrt{t}}$ or $\frac{1}{t}\mathbb{1}_{(1,\infty)}(t) dt$ as the representing measures $\sigma(dt)$ in (2.1).

The closure assertion of Theorem 2.2 says, in particular, that on the set S the notions of *pointwise convergence*, *locally uniform convergence*, and even *convergence in the space C^∞* coincide, cf. also Corollary 1.7.

Comments 2.5. The Stieltjes transform appears for the first time in the famous papers [340, 341] where T. J. Stieltjes investigates continued fractions in order to solve what we nowadays call the *Stieltjes Moment Problem*. For an appreciation of Stieltjes' achievements see the contributions of W. van Assche and W. A. J. Luxemburg in [342, Vol. 1].

The name Stieltjes transform for the integral (2.1) was coined by Doetsch [95] and, independently, Widder [369]. Earlier works, e.g. Perron [287], call $\int(\lambda + t)^{-1} \sigma(dt)$ *Stieltjes integral* (German: *Stieltjes'sches Integral*) but terminology has changed since then. Sometimes the name *Hilbert–Hankel transform* is also used in the literature, cf. Lax [239, p. 185]. A systematic account of the properties of the Stieltjes transform is given in [370, Chapter VIII].

Stieltjes functions are discussed by van Herk [361] as class $\{F\}$ in the wider context of moment and complex interpolation problems, see also the Comments 6.22, 7.24. Van Herk uses the integral representation

$$\int_{[0,1]} \frac{1}{1 - s + sz} \chi(ds)$$

which can be transformed by the change of variables $t = s^{-1} - 1 \in [0, \infty]$ for $s \in [0, 1]$ into the form (2.2); $\bar{\sigma}(dt)$ and $\chi(ds)$ are image measures under this transformation. Van Herk only observes that his class $\{F\}$ contains S, but comparing [361, Theorem 7.50] with our result (7.2) in Chapter 7 below shows that $\{F\} = S$.

In his paper [162] F. Hirsch introduces Stieltjes transforms into potential theory and identifies S as a convex cone operating on the abstract potentials, i.e. the densely defined inverses of the infinitesimal generators of C_0-semigroups. Hirsch establishes several properties of the cone S, which are later extended by Berg [32, 33]. A presentation of this material from a potential theoretic point of view is contained in the monograph [39, Chapter 14] by Berg and Forst, the connections to the moment problem are surveyed in [35].

Theorem 2.2 appears in Hirsch [162, Proposition 1], but with a different proof.

In contrast to completely monotone functions, S is not closed under multiplication. This follows easily since the (necessarily unique) representing measure of the function

$\lambda \mapsto (\lambda + a)^{-1}(\lambda + b)^{-1}, 0 < a < b$, is $(b-t)^{-1}\delta_a(dt) + (a-t)^{-1}\delta_b(dt)$ which is a signed measure. We will see in Proposition 7.13 that $\mathbb{S}$ is still logarithmically convex. It is known, see Hirschman and Widder [167, VII 7.4], that the product $\int_{(0,\infty)}(\lambda+t)^{-1}\sigma_1(dt) \int_{(0,\infty)}(\lambda+t)^{-1}\sigma_2(dt)$ is of the form $\int_{(0,\infty)}(\lambda+t)^{-2}\sigma(dt)$ for some measure σ. The latter integral is often called a *generalized Stieltjes transform*. The following related result is due to Srivastava and Tuan [335]: if $f \in L^p(0,\infty)$ and $g \in L^q(0,\infty)$ with $1 < p, q < \infty$ and $r^{-1} = p^{-1} + q^{-1} < 1$, then there is some $h \in L^r(0,\infty)$ such that $\mathscr{L}^2(f;\lambda)\mathscr{L}^2(g;\lambda) = \mathscr{L}^2(h;\lambda)$ holds. Since

$$h(t) = f(t) \cdot \text{p.v.} \int_0^\infty \frac{g(u)}{u-t}\,du + g(t) \cdot \text{p.v.} \int_0^\infty \frac{f(u)}{u-t}\,du,$$

h will, in general, change its sign even if f, g are non-negative.

Chapter 3

Bernstein functions

We are now ready to introduce the class of *Bernstein functions* which are closely related to completely monotone functions. The notion of Bernstein functions goes back to the potential theory school of A. Beurling and J. Deny and was subsequently adopted by C. Berg and G. Forst [39], see also [35]. S. Bochner [63] calls them *completely monotone mappings* (as opposed to completely monotone functions) and probabilists still prefer the term *Laplace exponents*, see e.g. Bertoin [47, 49]; the reason will become clear from Theorem 5.2.

Definition 3.1. A function $f : (0, \infty) \to \mathbb{R}$ is a *Bernstein function* if f is of class C^∞, $f(\lambda) \geq 0$ for all $\lambda > 0$ and

$$(-1)^{n-1} f^{(n)}(\lambda) \geq 0 \quad \text{for all } n \in \mathbb{N} \text{ and } \lambda > 0. \tag{3.1}$$

The set of all Bernstein functions will be denoted by $\mathcal{BF}$.

It is easy to see from the definition that, for example, the fractional powers $\lambda \mapsto \lambda^\alpha$, are Bernstein functions if, and only if, $0 \leq \alpha \leq 1$.

The key to the next theorem is the observation that a non-negative C^∞-function $f : (0, \infty) \to \mathbb{R}$ is a Bernstein function if, and only if, f' is a completely monotone function.

Theorem 3.2. *A function $f : (0, \infty) \to \mathbb{R}$ is a Bernstein function if, and only if, it admits the representation*

$$f(\lambda) = a + b\lambda + \int_{(0,\infty)} (1 - e^{-\lambda t}) \, \mu(dt), \tag{3.2}$$

where $a, b \geq 0$ and μ is a measure on $(0, \infty)$ satisfying $\int_{(0,\infty)} (1 \wedge t) \, \mu(dt) < \infty$.
In particular, the triplet (a, b, μ) determines f uniquely and vice versa.

Proof. Assume that f is a Bernstein function. Then f' is completely monotone. According to Theorem 1.4, there exists a measure ν on $[0, \infty)$ such that for all $\lambda > 0$

$$f'(\lambda) = \int_{[0,\infty)} e^{-\lambda t} \nu(dt).$$

Let $b := \nu\{0\}$. Then

$$f(\lambda) - f(0+) = \int_0^\lambda f'(y)\,dy = b\lambda + \int_0^\lambda \int_{(0,\infty)} e^{-yt}\,\nu(dt)\,dy$$

$$= b\lambda + \int_{(0,\infty)} \frac{1 - e^{-\lambda t}}{t}\,\nu(dt).$$

Write $a := f(0+)$ and define $\mu(dt) := t^{-1}\nu|_{(0,\infty)}(dt)$. Then the calculation from above shows that (3.2) is true. That $a, b \geq 0$ is obvious, and from the elementary (convexity) estimate

$$(1 - e^{-1})(1 \wedge t) \leq 1 - e^{-t}, \quad t \geq 0,$$

we infer

$$\int_{(0,\infty)} (1 \wedge t)\,\mu(dt) \leq \frac{e}{e-1} \int_{(0,\infty)} (1 - e^{-t})\,\mu(dt) = \frac{e}{e-1}\,f(1) < \infty.$$

Conversely, suppose that f is given by (3.2) with (a, b, μ) as in the statement of the theorem. Since $te^{-\lambda t} \leq t \wedge (e\lambda)^{-1}$, we can apply the differentiation lemma for parameter-dependent integrals for all λ from $[\epsilon, \epsilon^{-1}]$ and all $\epsilon > 0$. Differentiating (3.2) under the integral sign yields

$$f'(\lambda) = b + \int_{(0,\infty)} e^{-\lambda t}\,t\,\mu(dt) = \int_{[0,\infty)} e^{-\lambda t}\,\nu(dt),$$

where $\nu(dt) := t\mu(dt) + b\delta_0(dt)$. This formula shows that f' is a completely monotone function. Therefore, f is a Bernstein function.

Because $f(0+) = a$ and because of the uniqueness assertion of Theorem 1.4 it is clear that (a, b, μ) and $f \in \mathcal{BF}$ are in one-to-one correspondence.   □

Remark 3.3. (i) The representing measure μ and the characteristic triplet (a, b, μ) from (3.2) are often called the *Lévy measure* and the *Lévy triplet* of the Bernstein function f. The formula (3.2) is called the *Lévy–Khintchine representation* of f.

(ii) A useful variant of the representation formula (3.2) can be obtained by an application of Fubini's theorem. Since

$$\int_{(0,\infty)} (1 - e^{-\lambda t})\,\mu(dt) = \int_{(0,\infty)} \int_{(0,t)} \lambda e^{-\lambda s}\,ds\,\mu(dt)$$

$$= \int_0^\infty \int_{(s,\infty)} \lambda e^{-\lambda s}\,\mu(dt)\,ds$$

$$= \int_0^\infty \lambda e^{-\lambda s}\,\mu(s,\infty)\,ds$$

we get that any Bernstein function can be written in the form

$$f(\lambda) = a + b\lambda + \lambda \int_{(0,\infty)} e^{-\lambda s} M(s)\, ds \qquad (3.3)$$

where $M(s) = M_\mu(s) = \mu(s, \infty)$ is a non-increasing, right-continuous function. Integration by parts and the fact that $\int_0^\infty s e^{-\lambda s}\, ds = \lambda^{-2}$ and $\int_0^\infty e^{-\lambda s}\, ds = \lambda^{-1}$ yield

$$f(\lambda) = \lambda^2 \int_{(0,\infty)} e^{-\lambda s} k(s)\, ds = \lambda^2 \mathscr{L}(k; \lambda) \qquad (3.4)$$

with $k(s) = as + b + \int_0^s M(t)\, dt$, compare with Theorem 6.2 (iii). Note that k is positive, non-decreasing and concave.

(iii) The integrability condition $\int_{(0,\infty)} (1 \wedge t)\, \mu(dt) < \infty$ ensures that the integral in (3.2) converges for some, hence all, $\lambda > 0$. This is immediately seen from the convexity inequalities

$$\frac{t}{1+t} \leq 1 - e^{-t} \leq 1 \wedge t \leq 2\,\frac{t}{1+t}, \quad t > 0$$

and the fact that for $\lambda \geq 1$ [respectively for $0 < \lambda < 1$] and all $t > 0$

$$1 \wedge t \leq 1 \wedge (t\lambda) \leq \lambda(1 \wedge t) \qquad \left[\text{respectively } \lambda(1 \wedge t) \leq 1 \wedge (t\lambda) \leq 1 \wedge t\right].$$

In particular, if μ is a measure on $(0, \infty)$ such that $\int_{(0,\infty)} (1 - e^{-\lambda t})\, \mu(dt) < \infty$ for some $\lambda > 0$, then μ is a Lévy measure.

(iv) A useful consequence of the above estimate and the representation formula (3.2) are the following formulae to calculate the coefficients a and b:

$$a = f(0+) \quad \text{and} \quad b = \lim_{\lambda \to \infty} \frac{f(\lambda)}{\lambda}.$$

The first formula is obvious while the second follows from (3.2) and the dominated convergence theorem: $1 - e^{-\lambda t} \leq 1 \wedge (\lambda t)$ and $\lim_{\lambda \to \infty} (1 - e^{-\lambda t})/\lambda = 0$.

(v) Formula (3.3) shows, in particular, that

$$\int_0^1 \mu(t, \infty)\, dt = \int_0^1 M(t)\, dt < \infty. \qquad (3.5)$$

Since a non-increasing function which is integrable near zero is $o(1/t)$ as $t \to 0$, we conclude from (3.5) that

$$\lim_{t \to 0+} t\mu(t, \infty) = \lim_{t \to 0+} tM(t) = 0. \qquad (3.6)$$

(vi) Since the derivative of a Bernstein function f is completely monotone, Remark 1.5 shows that $f^{(n)}(\lambda) \neq 0$ for all $n \in \mathbb{N}$ and $\lambda > 0$, unless μ vanishes.

Lemma 3.4. *Let f be a Bernstein function given by (3.2) where $a = b = 0$. Set $I_\mu(\lambda) := \int_0^\lambda \mu(s, \infty)\, ds$. Then for $\lambda > 0$,*

$$\frac{e-1}{e}\, \lambda\, I_\mu\Big(\frac{1}{\lambda}\Big) \le f(\lambda) \le \lambda\, I_\mu\Big(\frac{1}{\lambda}\Big).$$

Proof. By Fubini's theorem we find

$$
\begin{aligned}
\lambda\, I_\mu\Big(\frac{1}{\lambda}\Big) &= \lambda \int_0^{1/\lambda} \mu(s, \infty)\, ds \\
&= \int_0^1 \mu\Big(\frac{t}{\lambda}, \infty\Big)\, dt \\
&= \int_0^1 \int_{t/\lambda}^\infty \mu(dr)\, dt \\
&= \int_0^\infty \big(1 \wedge (\lambda r)\big)\, \mu(dr).
\end{aligned}
$$

Using the elementary inequalities $\frac{e-1}{e}(1 \wedge r) \le 1 - e^{-r} \le 1 \wedge r$, $r \ge 0$, we get

$$\frac{e-1}{e}\, \lambda\, I_\mu\Big(\frac{1}{\lambda}\Big) = \int_0^\infty \frac{e-1}{e}\big(1 \wedge (\lambda r)\big)\, \mu(dr) \le \int_0^\infty (1 - e^{-\lambda r})\, \mu(dr) = f(\lambda).$$

The upper bound follows similarly.   □

The derivative of a Bernstein function is completely monotone. The converse is only true, if the primitive of a completely monotone function is positive. This fails, for example, for the completely monotone function λ^{-2} whose primitive, $-\lambda^{-1}$, is not a Bernstein function. The next proposition characterizes the image of $\mathcal{BF}$ under differentiation.

Proposition 3.5. *Let $g(\lambda) = b + \int_{(0,\infty)} e^{-\lambda t}\, v(dt)$ be a completely monotone function. It has a primitive $f \in \mathcal{BF}$ if, and only if, the representing measure v satisfies $\int_{(0,\infty)}(1 + t)^{-1}\, v(dt) < \infty$.*

Proof. Assume that f is a Bernstein function given in the form (3.2). Then

$$f'(\lambda) = b + \int_{(0,\infty)} e^{-\lambda t}\, t\, \mu(dt)$$

is completely monotone and the measure $v(dt) := t\mu(dt)$ satisfies

$$\int_{(0,\infty)} \frac{1}{1+t}\, v(dt) = \int_{(0,\infty)} \frac{t}{1+t}\, \mu(dt) \le \int_{(0,\infty)} (1 \wedge t)\, \mu(dt) < \infty.$$

Retracing the above steps reveals that $\int_{(0,\infty)}(1+t)^{-1}\,\nu(dt) < \infty$ is also sufficient to guarantee that $g(\lambda) := b + \int_{(0,\infty)} e^{-\lambda t}\,\nu(dt)$ has a primitive which is a Bernstein function. □

Theorem 3.2 allows us to extend Bernstein functions analytically onto the right complex half-plane $\overrightarrow{\mathbb{H}} := \{z \in \mathbb{C} : \operatorname{Re} z > 0\}$ and continuously onto the closure $\overrightarrow{\overline{\mathbb{H}}}$.

Proposition 3.6. *Every $f \in \mathcal{BF}$ has an extension $f : \overrightarrow{\overline{\mathbb{H}}} \to \overrightarrow{\overline{\mathbb{H}}}$ which is continuous for $\operatorname{Re} z \geq 0$, holomorphic for $\operatorname{Re} z > 0$ and satisfies $\overline{f(z)} = f(\bar{z})$ for all $z \in \overrightarrow{\overline{\mathbb{H}}}$. This extension has the following representation*

$$f(z) = a + bz + \frac{2}{\pi}\int_0^\infty \frac{z}{z^2+t^2}\operatorname{Re} f(it)\,dt, \qquad \operatorname{Re} z \geq 0. \tag{3.7}$$

In particular, f preserves angular sectors,

$$f(S_{[-\theta,\theta]}) \subset S_{[-\theta,\theta]} \quad\text{with}\quad S_{[-\theta,\theta]} = \{z \in \mathbb{C} : |\arg z| \leq \theta\}, \ \theta \in [0,\pi). \tag{3.8}$$

Proof. The function $\lambda \mapsto 1 - e^{-\lambda t}$ appearing in (3.2) has a unique holomorphic extension. If $z = \lambda + i\kappa$ is such that $\lambda = \operatorname{Re} z > 0$ we get

$$|1 - e^{-zt}| = \left|\int_0^{zt} e^{-\zeta}\,d\zeta\right| \leq t|z| \quad\text{and}\quad |1 - e^{-zt}| \leq 1 + |e^{-zt}| \leq 2.$$

Therefore the integral appearing in (3.2) converges uniformly even if we replace the real variable λ by $z \in \overrightarrow{\mathbb{H}}$. This means that $f(z)$ is well defined and holomorphic on $\overrightarrow{\mathbb{H}}$ such that $\overline{f(z)} = f(\bar{z})$. Moreover,

$$\operatorname{Re} f(z) = a + b\operatorname{Re} z + \int_{(0,\infty)}\operatorname{Re}(1-e^{-zt})\,\mu(dt)$$
$$= a + b\lambda + \int_{(0,\infty)}\left(1 - e^{-\lambda t}\cos(\kappa t)\right)\mu(dt)$$

which is positive since $\lambda = \operatorname{Re} z \geq 0$ and $1 - e^{-\lambda t}\cos(\kappa t) \geq 1 - e^{-\lambda t} \geq 0$.

Continuity up to the boundary follows from the estimate

$$|f(z) - f(w)| \leq b|z - w| + \int_{(0,\infty)}|e^{-wt} - e^{-zt}|\,\mu(dt)$$
$$\leq b|z-w| + \int_{(0,\infty)} 2 \wedge (t|w-z|)\,\mu(dt)$$

for all $z, w \in \overrightarrow{\overline{\mathbb{H}}}$ and the dominated convergence theorem. For $z = 0$ essentially the same calculation shows for all $w \in \overrightarrow{\overline{\mathbb{H}}}$

$$\left|\frac{f(w)}{w} - b - \frac{a}{w}\right| \leq \frac{1}{|w|}\int_{(0,\infty)} 2 \wedge (t|w|)\,\mu(dt) \xrightarrow{|w|\to\infty} 0.$$

Without loss of generality we may assume that $b = 0$, otherwise we could consider $f(z) - bz$ instead of f. Fix $z \in \overrightarrow{\mathbb{H}}$ and denote by $\gamma_{h,R}$ the closed integration contour consisting of a straight line from $h + iR$ to $h - iR$ which is closed by a semicircular arc of radius R centered at h. The point z is surrounded by $\gamma_{h,R}$ if $0 < h < \operatorname{Re} z$ and $R > |z|$. This is shown in the picture. Since f is holomorphic in $\overrightarrow{\mathbb{H}}$ we can use Cauchy's theorem to get

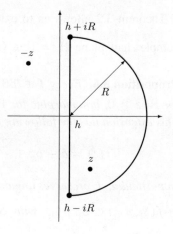

$$f(z) = \frac{1}{2\pi i} \int_{\gamma_{h,R}} \frac{f(\zeta)}{\zeta - z} \, d\zeta$$

$$= -\frac{1}{2\pi} \int_{-R}^{R} \frac{f(h - it)}{h - it - z} \, dt$$

$$+ \frac{1}{2\pi} \int_{-\pi/2}^{\pi/2} \frac{f(h + Re^{i\omega})}{h + Re^{i\omega} - z} Re^{i\omega} \, d\omega.$$

Since the integrands are continuous, we can let $h \to 0$ and find

$$f(z) = \frac{1}{2\pi} \int_{-R}^{R} \frac{f(-it)}{it + z} \, dt + \frac{1}{2\pi} \int_{-\pi/2}^{\pi/2} \frac{f(Re^{i\omega})}{Re^{i\omega} - z} Re^{i\omega} \, d\omega.$$

The same argument yields for the point $-z$ which is not surrounded by the contour $\gamma_{h,R}$

$$0 = \frac{1}{2\pi} \int_{-R}^{R} \frac{f(-it)}{it - z} \, dt + \frac{1}{2\pi} \int_{-\pi/2}^{\pi/2} \frac{f(Re^{i\omega})}{Re^{i\omega} + z} Re^{i\omega} \, d\omega.$$

If we subtract the last two equalities we arrive at

$$f(z) = \frac{1}{\pi} \int_{-R}^{R} \frac{z}{z^2 + t^2} f(-it) \, dt + \frac{1}{\pi} \int_{-\pi/2}^{\pi/2} \frac{Re^{i\omega}}{Re^{i\omega} - z} \frac{f(Re^{i\omega})}{Re^{i\omega} + z} z \, d\omega.$$

For large values of R and fixed $z \in \overrightarrow{\mathbb{H}}$, the integrand of the second integral is bounded. Since $\lim_{|\zeta| \to \infty, \zeta \in \overrightarrow{\mathbb{H}}} f(\zeta)/\zeta = 0$, we conclude that

$$f(z) = \frac{1}{\pi} \int_{-\infty}^{\infty} \frac{z}{z^2 + t^2} f(-it) \, dt$$

$$= \frac{1}{2\pi} \int_{-\infty}^{\infty} \frac{z}{z^2 + t^2} \left(f(-it) + f(it) \right) dt.$$

Finally, $f(-it) + f(it) = \overline{f(it)} + f(it) = 2\operatorname{Re} f(it)$ and the formula (3.7) follows.

From (3.7) we deduce immediately (3.8): if $z = re^{i\theta}, w = Re^{i\phi} \in \mathbb{C}$, the argument of $w + z$ lies between ϕ and θ. A similar statement is true for integrals:

$$\arg \int f(t, z)\, \mu(dt) \in \big[\inf_t \arg f(t, z),\ \sup_t \arg f(t, z)\big].$$

Note that for all $t > 0$

$$\frac{z}{z^2 + t^2} = \frac{re^{i\theta}}{r^2 e^{i2\theta} + t} = \frac{re^{i\theta}}{r' e^{i2\theta'}} \quad \text{where } 0 < |\theta'| < |\theta|.$$

Thus, $|\arg(\frac{z}{z^2+t^2})| \leq |\arg z|$, and the same is true for $f(z)$ because of the representation (3.7). $\square$

The following structural characterization comes from Bochner [63, pp. 83–84] where Bernstein functions are called completely monotone mappings.

Theorem 3.7. *Let f be a positive function on $(0, \infty)$. Then the following assertions are equivalent.*

(i) $f \in \mathcal{BF}$.

(ii) $g \circ f \in \mathcal{CM}$ *for every* $g \in \mathcal{CM}$.

(iii) $e^{-uf} \in \mathcal{CM}$ *for every* $u > 0$.

Proof. The proof relies on the following formula for the n-th derivative of the composition $h = g \circ f$ due to Faà di Bruno [109], see also [134, formula 0.430]:

$$h^{(n)}(\lambda) = \sum_{(m,i_1,\ldots,i_\ell)} \frac{n!}{i_1! \cdots i_\ell!}\, g^{(m)}\big(f(\lambda)\big) \prod_{j=1}^{\ell} \left(\frac{f^{(j)}(\lambda)}{j!}\right)^{i_j} \tag{3.9}$$

where $\sum_{(m,i_1,\ldots,i_\ell)}$ stands for summation over all $\ell \in \mathbb{N}$ and all $i_1, \ldots, i_\ell \in \mathbb{N} \cup \{0\}$ such that $\sum_{j=1}^{\ell} j \cdot i_j = n$ and $\sum_{j=1}^{\ell} i_j = m$.

(i)$\Rightarrow$(ii)Assume that $f \in \mathcal{BF}$ and $g \in \mathcal{CM}$. Then $h(\lambda) = g(f(\lambda)) \geq 0$. Multiply formula (3.9) by $(-1)^n$ and observe that $n = m + \sum_{j=1}^{\ell}(j-1)\cdot i_j$. The assumptions $f \in \mathcal{BF}$ and $g \in \mathcal{CM}$ guarantee that each term in the formula multiplied by $(-1)^n$ is non-negative. This proves that $h = g \circ f \in \mathcal{CM}$.

(ii)$\Rightarrow$(iii) This follows from the fact that $g(\lambda) := g_u(\lambda) := e^{-\lambda u}$, $u > 0$, is completely monotone.

(iii)$\Rightarrow$(i) The series $e^{-uf(\lambda)} = \sum_{j=0}^{\infty} \frac{(-1)^j u^j}{j!} [f(\lambda)]^j$ and all of its formal derivatives (w.r.t. λ) converge uniformly, so we can calculate $\frac{d^n}{d\lambda^n} e^{-uf(\lambda)}$ by termwise differentiation. Since e^{-uf} is completely monotone, we get

$$0 \leq (-1)^n \frac{d^n}{d\lambda^n} e^{-uf(\lambda)} = \sum_{j=1}^{\infty} \frac{u^j}{j!} (-1)^{n+j} \frac{d^n}{d\lambda^n} [f(\lambda)]^j.$$

Dividing by $u > 0$ and letting $u \to 0$ we see

$$0 \le (-1)^{n+1} \frac{d^n}{d\lambda^n} f(\lambda). \qquad \qquad \square$$

Theorems 3.2 and 3.7 have a few important consequences.

Corollary 3.8.   (i) *The set $\mathcal{BF}$ is a convex cone: $sf_1 + tf_2 \in \mathcal{BF}$ for all $f_1, f_2 \in BF$ and $s, t \ge 0$.*

(ii) *The set $\mathcal{BF}$ is closed under pointwise limits: if $(f_n)_{n\in\mathbb{N}} \subset \mathcal{BF}$ and if the limit $\lim_{n\to\infty} f_n(\lambda) = f(\lambda)$ exists for every $\lambda > 0$, then $f \in \mathcal{BF}$.*

(iii) *The set $\mathcal{BF}$ is closed under composition: if $f_1, f_2 \in \mathcal{BF}$, then $f_1 \circ f_2 \in \mathcal{BF}$. In particular, $\lambda \mapsto f_1(c\lambda)$ is in $\mathcal{BF}$ for any $c > 0$.*

(iv) *For all $f \in \mathcal{BF}$ the function $\lambda \mapsto f(\lambda)/\lambda$ is in $\mathcal{CM}$.*

(v) *$f \in \mathcal{BF}$ is bounded if, and only if, in (3.2) $b = 0$ and $\mu(0, \infty) < \infty$.*

(vi) *Let $f_1, f_2 \in \mathcal{BF}$ and $\alpha, \beta \in (0, 1)$ such that $\alpha + \beta \le 1$. Then $\lambda \mapsto f_1(\lambda^\alpha) f_2(\lambda^\beta)$ is again a Bernstein function.*

(vii) *Let $f \in \mathcal{BF}$. For every $\kappa > 0$ the function $f_\kappa(\lambda) := f(\lambda) + f(\kappa) - f(\lambda + \kappa)$, $\lambda > 0$, is a Bernstein function.*

In particular, $f \in \mathcal{BF}$ is subadditive, i.e. $f(\lambda + \kappa) \le f(\lambda) + f(\kappa)$ for all $\kappa, \lambda > 0$.

(viii) *For any $f \in \mathcal{BF}$, there exists $c > 0$ such that $f(\lambda) \le c\lambda$ for all $\lambda > 1$.*

Proof. (i) This follows immediately from Definition 3.1 or, alternatively, from the representation formula (3.2).

(ii) For every $u > 0$ we know that e^{-uf_n} is a completely monotone function and that $e^{-uf(\lambda)} = \lim_{n\to\infty} e^{-uf_n(\lambda)}$. Since $\mathcal{CM}$ is closed under pointwise limits, cf. Corollary 1.6, e^{-uf} is completely monotone and $f \in \mathcal{BF}$.

(iii) Let $f_1, f_2 \in \mathcal{BF}$. For any $g \in \mathcal{CM}$ we use the implication (i)$\Rightarrow$(ii) of Theorem 3.7 to get $g \circ f_1 \in \mathcal{CM}$, and then $g \circ (f_1 \circ f_2) = (g \circ f_1) \circ f_2 \in \mathcal{CM}$. The converse direction (ii)$\Rightarrow$(i) of Theorem 3.7 shows that $f_1 \circ f_2 \in \mathcal{BF}$.

(iv) Note that $(1 - e^{-\lambda t})/\lambda = \int_0^t e^{-\lambda s}\, ds$ is completely monotone. Therefore,

$$\frac{f(\lambda)}{\lambda} = \frac{a}{\lambda} + b + \int_{(0,\infty)} \frac{1 - e^{-\lambda t}}{\lambda}\, \mu(dt)$$

is the limit of linear combinations of completely monotone functions which is, by Corollary 1.6, completely monotone.

(v) That $b = 0$ and $\mu(0, \infty) < \infty$ imply the boundedness of f is clear from the representation (3.2). Conversely, if f is bounded, $b = 0$ follows from Remark 3.3 (iv), and $\mu(0, \infty) < \infty$ follows from (3.2) and Fatou's lemma.

(vi) We know that the fractional powers $\lambda \mapsto \lambda^\alpha$, $0 \le \alpha \le 1$, are Bernstein functions. Since $h(\lambda) := f_1(\lambda^\alpha) f_2(\lambda^\beta)$ is positive, it is enough to show that the derivative h' is completely monotone. We have

$$h'(\lambda) = \alpha f_1'(\lambda^\alpha) \lambda^{\alpha-1} f_2(\lambda^\beta) + \beta f_2'(\lambda^\beta) \lambda^{\beta-1} f_1(\lambda^\alpha)$$
$$= \lambda^{\alpha+\beta-1} \left(\alpha f_1'(\lambda^\alpha) \frac{f_2(\lambda^\beta)}{\lambda^\beta} + \beta f_2'(\lambda^\beta) \frac{f_1(\lambda^\alpha)}{\lambda^\alpha} \right).$$

Note that $f_1'(\kappa)$, $f_2'(\kappa)$ and, by part (iv), $\kappa^{-1} f_1(\kappa)$, $\kappa^{-1} f_2(\kappa)$ are completely monotone. By Theorem 3.7 (ii) the functions $f_1'(\lambda^\alpha)$, $f_2'(\lambda^\beta)$, $f_1(\lambda^\alpha)/\lambda^\alpha$ and $f_2(\lambda^\beta)/\lambda^\beta$ are again completely monotone. Since $\alpha + \beta \le 1$, $\lambda \mapsto \lambda^{\alpha+\beta-1}$ is completely monotone. As sums and products of completely monotone functions are in $\mathcal{CM}$, see Corollary 1.6, h' is completely monotone.

(vii) Using the representation (3.2) of the Bernstein function f we get

$$f(\lambda) = a + b\lambda + \int_{(0,\infty)} (1 - e^{-\lambda t}) \, \mu(dt),$$

and so

$$f_\kappa(\lambda) = a + \int_{(0,\infty)} (1 - e^{-\lambda t})(1 - e^{-\kappa t}) \, \mu(dt).$$

It is easy to see that $(1 - e^{-\kappa t}) \mu(dt)$ is, for each $\kappa > 0$, a Lévy measure. The subadditivity of f follows now from $f_\kappa(\lambda) \ge 0$.

(viii) Since f is concave, we have $f(\lambda) \le \lambda f(1)$ for all $\lambda \ge 1$. $\square$

Just as for completely monotone functions, the closure assertion of Corollary 3.8 says that on the set $\mathcal{BF}$ the notions of *pointwise convergence, locally uniform convergence*, and even *convergence in the space C^∞* coincide.

Corollary 3.9. *Let $(f_n)_{n\in\mathbb{N}}$ be a sequence of Bernstein functions such that the limit $\lim_{n\to\infty} f_n(\lambda) = f(\lambda)$ exists for all $\lambda \in (0, \infty)$. Then f is a Bernstein function and for all $k \in \mathbb{N} \cup \{0\}$ the convergence $\lim_{n\to\infty} f_n^{(k)}(\lambda) = f^{(k)}(\lambda)$ is locally uniform in $\lambda \in (0, \infty)$.*

If (a_n, b_n, μ_n) and (a, b, μ) are the Lévy triplets for f_n and f, respectively, see (3.2), we have

$$\lim_{n\to\infty} \mu_n = \mu \quad \text{vaguely in } (0, \infty),$$

and

$$a = \lim_{R \to \infty} \liminf_{n \to \infty} \left(a_n + \mu_n[R, \infty) \right),$$

$$b = \lim_{\epsilon \to 0} \liminf_{n \to \infty} \left(b_n + \int_{(0,\epsilon)} t \, \mu_n(dt) \right).$$

In both formulae we may replace $\liminf_n$ *by* $\limsup_n$.

Proof. From Corollary 3.8 (ii) we know that f is a Bernstein function. Obviously, $\lim_{n \to \infty} e^{-f_n} = e^{-f}$; by Theorem 3.7, the functions e^{-f_n}, e^{-f} are completely monotone and we can use Corollary 1.7 to conclude that

$$e^{-f_n} \xrightarrow{\ n \to \infty\ } e^{-f} \quad \text{and} \quad (-f_n') e^{-f_n} \xrightarrow{\ n \to \infty\ } (-f') e^{-f}$$

converge locally uniformly on $(0, \infty)$. In particular, $\lim_{n \to \infty} f_n'(\lambda) = f'(\lambda)$ for each $\lambda \in (0, \infty)$. Again by Corollary 1.7 and the complete monotonicity of f_n', f' we see that for $k \geq 1$ the derivatives $f_n^{(k)}$ converge locally uniformly to $f^{(k)}$. By the mean value theorem,

$$|f_n(\lambda) - f(\lambda)| = \left| \log e^{-f(\lambda)} - \log e^{-f_n(\lambda)} \right| \leq C \left| e^{-f(\lambda)} - e^{-f_n(\lambda)} \right|$$

with $C \leq e^{f(\lambda) + f_n(\lambda)}$. The locally uniform convergence of e^{-f_n} ensures that C is bounded for $n \in \mathbb{N}$ and λ from compact sets in $(0, \infty)$; this proves locally uniform convergence of f_n to f on $(0, \infty)$.

Differentiating the representation formula (3.2) we get

$$f_n'(\lambda) = b_n + \int_{(0,\infty)} t \, e^{-\lambda t} \mu_n(dt) = \int_{[0,\infty)} e^{-\lambda t} \left(b_n \delta_0(dt) + t \, \mu_n(dt) \right),$$

implying that $b_n \delta_0(dt) + t \, \mu_n(dt)$ converge vaguely to $b \delta_0(dt) + t \, \mu(dt)$. This proves at once that $\mu_n \to \mu$ vaguely on $(0, \infty)$ as $n \to \infty$.

Since $b_n \delta_0(dt) + t \, \mu_n(dt)$ converge vaguely to $b \delta_0(dt) + t \, \mu(dt)$, we can use the portmanteau theorem, Theorem A.7, to conclude that

$$\lim_{n \to \infty} \left(b_n + \int_{(0,\epsilon)} t \, \mu_n(dt) \right) = b + \int_{(0,\epsilon)} t \, \mu(dt)$$

at all continuity points $\epsilon > 0$ of μ. If $\epsilon_j > 0$ is a sequence of continuity points of μ such that $\epsilon_j \to 0$, we get

$$b = \lim_{j \to \infty} \lim_{n \to \infty} \left(b_n + \int_{(0,\epsilon_j)} t \, \mu_n(dt) \right).$$

For a sequence of arbitrary $\epsilon_j \to 0$ we find continuity points $\delta_j, \eta_j, j \in \mathbb{N}$, of μ such that $0 < \delta_j \leq \epsilon_j \leq \eta_j$ and $\delta_j, \eta_j \to 0$. Thus,

$$b_n + \int_{(0,\delta_j)} t\,\mu_n(dt) \leq b_n + \int_{(0,\epsilon_j)} t\,\mu_n(dt) \leq b_n + \int_{(0,\eta_j)} t\,\mu_n(dt),$$

and we conclude that

$$\lim_{j\to\infty}\lim_{n\to\infty}\left(b_n + \int_{(0,\delta_j)} t\,\mu_n(dt)\right) \leq \lim_{j\to\infty}\liminf_{n\to\infty}\left(b_n + \int_{(0,\epsilon_j)} t\,\mu_n(dt)\right)$$

$$\leq \lim_{j\to\infty}\limsup_{n\to\infty}\left(b_n + \int_{(0,\epsilon_j)} t\,\mu_n(dt)\right)$$

$$\leq \lim_{j\to\infty}\lim_{n\to\infty}\left(b_n + \int_{(0,\eta_j)} t\,\mu_n(dt)\right).$$

Since both sides of the inequality coincide, the claim follows.

Using $a = f(0+)$ we find for each $R > 1$

$$a = \lim_{\epsilon\to 0} f(\epsilon)$$

$$= \lim_{\epsilon\to 0}\lim_{n\to\infty} f_n(\epsilon)$$

$$= \lim_{\epsilon\to 0}\lim_{n\to\infty}\left(a_n + b_n\epsilon + \int_{(0,\infty)}(1 - e^{-\epsilon t})\,\mu_n(dt)\right)$$

$$= \lim_{\epsilon\to 0}\lim_{n\to\infty}\left(a_n + \int_{[R,\infty)}(1 - e^{-\epsilon t})\,\mu_n(dt)\right.$$

$$\left. + b_n\epsilon + \int_{(0,R)}(1 - e^{-\epsilon t})\,\mu_n(dt)\right).$$

From the convexity estimate

$$(1 - e^{-1})(1 \wedge t) \leq (1 - e^{-t}), \quad t \geq 0,$$

we obtain

$$(1 - e^{-\epsilon t})\mathbb{1}_{(0,R)}(t) \leq \epsilon t\,\mathbb{1}_{(0,R)}(t) \leq \epsilon(t \wedge R) \leq \epsilon R(1 \wedge t) \leq \epsilon R \frac{(1 - e^{-t})}{(1 - e^{-1})}$$

so that

$$b_n\epsilon + \int_{(0,R)}(1 - e^{-\epsilon t})\,\mu_n(dt) \leq b_n\epsilon + \frac{\epsilon R}{(1 - e^{-1})}\int_{(0,R)}(1 - e^{-t})\,\mu_n(dt)$$

$$\leq \frac{\epsilon R}{(1 - e^{-1})}\,f_n(1).$$

Since $\lim_{n\to\infty} f_n(1) = f(1)$, we get

$$a = \lim_{\epsilon\to0} \lim_{n\to\infty} \left(a_n + \int_{[R,\infty)} (1 - e^{-\epsilon t}) \mu_n(dt) \right).$$

For any continuity point $R > 1$ of μ, we have by vague convergence

$$\lim_{n\to\infty} \int_{[R,\infty)} e^{-\epsilon t} \mu_n(dt) = \int_{[R,\infty)} e^{-\epsilon t} \mu(dt) \xrightarrow{\epsilon\to0} \mu[R,\infty) \xrightarrow{R\to\infty} 0.$$

Letting $R \to \infty$ through a sequence of continuity points R_j, $j \in \mathbb{N}$, of μ we get

$$a = \lim_{j\to\infty} \lim_{n\to\infty} \left(a_n + \mu_n[R_j, \infty) \right).$$

That we do not need to restrict ourselves to continuity points R_j follows with a similar argument as for the coefficient b.   □

Example 3.10. The proof of Corollary 3.9 shows that the vague limit μ does not capture the accumulation of mass at $\lambda = 0$ and $\lambda = \infty$ of the μ_n as $n \to \infty$. These effects can cause the appearance of $a > 0$ and $b > 0$ in the Lévy triplet of the limiting function f, even if $a_n = b_n = 0$ for all functions f_n. Here are two extreme cases:

$$f_n(\lambda) = n(1 - e^{-\lambda/n}) \xrightarrow{n\to\infty} \lambda = f(\lambda),$$

i.e. $(a_n, b_n, \mu_n) = (0, 0, n\delta_{1/n})$ and $(a, b, \mu) = (0, 1, 0)$, and

$$f_n(\lambda) = 1 - e^{-n\lambda} \xrightarrow{n\to\infty} 1 = f(\lambda)$$

where $(a_n, b_n, \mu_n) = (0, 0, \delta_n)$ and $(a, b, \mu) = (1, 0, 0)$.

Remark 3.11. Let $(f_n)_{n\in\mathbb{N}}$ be a sequence of Bernstein functions and set $g_n := e^{-f_n}$. By Theorem 3.7, $(g_n)_{n\in\mathbb{N}}$ is a sequence of completely monotone functions satisfying $g_n(0+) \le 1$. It follows from Corollary 1.8 that there exist a subsequence $(g_{n_k})_{k\in\mathbb{N}}$ and some $g \in \mathcal{CM}$ such that $\lim_{k\to\infty} g_{n_k}(\lambda) = g(\lambda)$ for all $\lambda \in (0, \infty)$. Note that g may be identically equal to zero. Let $f := -\log g$ (where we use the convention that $\log 0 = -\infty$). Then $f : (0, \infty) \to (0, \infty]$ and if $f(\lambda) < \infty$ for some $\lambda > 0$, we see that $f \in \mathcal{BF}$.

There is a one-to-one correspondence between bounded Bernstein functions and bounded completely monotone functions.

Proposition 3.12. *If $g \in \mathcal{CM}$ is bounded, then $g(0+) - g \in \mathcal{BF}$. Conversely, if $f \in \mathcal{BF}$ is bounded, there exist some constant $c > 0$ and some bounded $g \in \mathcal{CM}$, $\lim_{\lambda\to\infty} g(\lambda) = 0$, such that $f = c - g$; then $c = f(0+) + g(0+)$.*

Proof. Assume that $g = \mathscr{L}\mu \in \mathcal{CM}$ is a bounded function. This means that $\mu(0,\infty) = g(0+) = \sup_{\lambda>0} g(\lambda) < \infty$. Hence, $f(\lambda) := g(0+) - g(\lambda) = \int_{(0,\infty)} (1 - e^{-\lambda t})\,\mu(dt)$ is a bounded Bernstein function.

Conversely, if $f \in \mathcal{BF}$ is bounded, $f(\lambda) = a + \int_{(0,\infty)}(1 - e^{-\lambda t})\mu(dt)$ for some bounded measure μ, cf. Corollary 3.8. Thus $f(\lambda) = c - g(\lambda)$ where $g(\lambda) = \mathscr{L}(\mu;\lambda)$ is in $\mathcal{CM}$, $\lim_{\lambda\to\infty} g(\lambda) = 0$ and $c = a + \mu(0,\infty) = f(0+) + g(0+) > 0$. $\qquad\square$

Remark 3.13. Just as in the case of completely monotone functions, see Remark 1.9, we can understand the representation formula (3.2) as a particular case of a Choquet or Kreĭn–Milman representation. The set

$$\left\{ f \in \mathcal{BF} : \int_{(0,\infty)} f(\lambda)e^{-\lambda}\,d\lambda = 1 \right\}$$

is a basis of the convex cone $\mathcal{BF}$, and its extremal points are given by

$$e_0(\lambda) = \lambda, \quad e_t(\lambda) = \frac{1+t}{t}(1 - e^{-\lambda t}), \quad 0 < t < \infty, \quad \text{and} \quad e_\infty(\lambda) = 1,$$

see Harzallah [149] or [320, Satz 2.9]. These functions are, of course, examples and building blocks for all Bernstein functions: every $f \in \mathcal{BF}$ can be written as an 'integral mixture' of the above extremal Bernstein functions.

For $\mu(dt) = \alpha/\Gamma(1-\alpha)\,t^{-1-\alpha}\,dt, \alpha \in (0,1)$, $\mu(dt) = e^{-t}$, or $\mu(dt) = t^{-1}e^{-t}$ we see that the functions

$$\lambda^\alpha \ (0 < \alpha < 1), \quad \text{or} \quad \frac{\lambda}{1+\lambda} \quad \text{or} \quad \log(1+\lambda)$$

are Bernstein functions.

Comments 3.14. The name Bernstein function is not universally accepted in the literature. It originated in the potential theory school of A. Beurling and J. Deny, but the name as such does not appear in Beurling's or Deny's papers. The earliest mentioning of Bernstein functions as well as a nice presentation of their properties is Faraut [111]. Bochner [63] calls Bernstein functions *completely monotone mappings* but this notion was only adopted in the 1959 paper by Woll [375]. Other names include *inner transformations* of $\mathcal{CM}$ (Schoenberg [325]), *Laplace* or *subordinator exponents* (Bertoin [47, 49]), *log-Laplace transforms* or *positive functions with completely monotone derivative* (Feller [118]).

Bochner pointed out the importance of Bernstein functions already in [62] and [63]; he defines them through the equivalent property (ii) of Theorem 3.7. The equivalence of (i) and (ii) in Theorem 3.7 is already present in Schoenberg's paper [325, Theorem 8]; Schoenberg defines Bernstein functions as primitives of completely monotone functions and calls them *inner transformations* (of completely monotone functions); his notation is T for the class $\mathcal{BF}$. Many properties as well as higher-dimensional analogues are given in [63, Chapter 4]. In particular, Theorems 3.2 and 3.7 can already be found there. An up-to-date account is given in the books by Berg and Forst [39, Chapter 9] (based on [111]) and Berg, Christensen and Ressel [38]. One can understand (3.2) as a Lévy–Khintchine formula for the semigroup $([0, \infty), +)$. This justifies the name Lévy measure and Lévy triplet for μ and (a, b, μ), respectively; this point of view is taken in [38]. The representation (3.4) is taken from Prüss [297, Chapter I.4.1] who calls the functions $k(s)$ (of the type needed in (3.4), i.e. positive, non-decreasing and concave) *creep functions*.

The estimate in Lemma 3.4 is from Bertoin [47, Proposition III.1.1, p. 74]. Bertoin's proof mimics the arguments of the de-Haan–Stadtmüller Tauberian theorem [58, Theorem 2.10.2, p. 118]. Our short proof is taken from [324] where techniques from Ôkura [284, Theorem 3.2] are used.

The representation formula (3.7) and the contraction property (3.8) were communicated to us by Sonia Fourati [119]. Our proof is based on techniques from the theory of harmonic functions in a half-plane, cf. Levin [241, Chapter V.§2, pp. 230–236] or Koosis [213, Chapter III.H.1, pp. 58–65]. Levin proves the following necessary and sufficient condition: Assume that g is harmonic[1] in $\overrightarrow{\mathbb{H}}$, continuous in $\overrightarrow{\mathbb{H}}$ and $\limsup_{|z| \to \infty} |g(z)|/|z| < \infty$. Then g has the following representation

$$g(z) = \frac{1}{\pi} \int_{-\infty}^{\infty} \frac{\operatorname{Re} z}{|z - it|^2} g(it)\, dt + k \operatorname{Re} z$$

with $k = \limsup_{\eta \to 0} g(\eta)/\eta$ if, and only if,

$$\int_{-\infty}^{\infty} \frac{|g(it)|}{1 + t^2}\, dt < \infty.$$

Using the harmonic conjugate, cf. [213], this leads to (3.7). The recent preprint [120] contains an alternative proof of Proposition 3.6 based on negative definite functions.

The assertions in Corollary 3.8 are mostly standard. We learned the nice argument in (vii) from Wissem Jedidi [119] who also pointed out the following converse of (vii): if for a smooth and increasing function $f : (0, \infty) \to [0, \infty)$ we know that $f_\kappa \in \mathcal{BF}$ for all $\kappa > 0$, then $f \in \mathcal{BF}$.

The characterization of Bernstein functions using Choquet's representation is due to Harzallah [147, 148, 149], a slightly different version is given in [320], reprinted in [183, Vol. 1, Theorem 3.9.20].

---

[1] The English edition states *holomorphic*, but this is a misprint, see the German translation.

Chapter 4

Positive and negative definite functions

Positive and negative definite functions appear naturally in connection with Fourier analysis and potential theory. In this chapter we will see that (bounded) completely monotone functions correspond to continuous positive definite functions on $(0, \infty)$ and that Bernstein functions correspond to continuous negative definite functions on the half-line. On $\mathbb{R}^d$ the notion of positive definiteness is familiar to most readers; for our purposes it is useful to adopt a more abstract point of view which includes both settings, $\mathbb{R}^d$ and the half-line $[0, \infty)$.

An *Abelian semigroup with involution* $(S, +, *)$ is a nonempty set equipped with a *commutative and associative addition* $+$, a *zero element* 0 and a mapping $* : S \to S$ called *involution*, satisfying

(a) $(s + t)^* = s^* + t^*$ for all $s, t \in S$;

(b) $(s^*)^* = s$ for all $s \in S$.

We are mainly interested in the semigroups $([0, \infty), +)$ and $(\mathbb{R}^d, +)$ where '$+$' is the usual addition; the respective involutions are the identity mapping $s \mapsto s$ in $[0, \infty)$ and the reflection at the origin $\xi \mapsto -\xi$ in $\mathbb{R}^d$. We will only develop those parts of the theory that allow us to characterize $\mathcal{CM}$ and $\mathcal{BF}$ as positive and negative definite functions on the half-line. Good expositions of the general case are the monographs by Berg, Christensen, Ressel [38] (for semigroups) and Berg, Forst [39] (for Abelian groups).

Definition 4.1. A function $f : S \to \mathbb{C}$ is *positive definite* if

$$\sum_{j,k=1}^{n} f(s_j + s_k^*) c_j \bar{c}_k \geq 0 \qquad (4.1)$$

holds for all $n \in \mathbb{N}$, all $s_1, \ldots, s_n \in S$ and all $c_1, \ldots, c_n \in \mathbb{C}$.

We need some simple properties of positive definite functions.

Lemma 4.2. *Let* $f : S \to \mathbb{C}$ *be positive definite. Then* $f(s + s^*) \geq 0$ *for all* $s \in S$ *and* f *is hermitian, i.e.* $f(s^*) = \overline{f(s)}$. *In particular,* $f(0) \geq 0$.

Lemma 4.2 shows also that positive definite functions on $[0, \infty)$ are non-negative: $f \geq 0$.

Proof of Lemma 4.2. That $f(s + s^*) \geq 0$ follows immediately from (4.1) if we take $n = 1$ and $s_1 = s$; $f(0) \geq 0$ is now obvious.

Again by (4.1) we see that the matrix $(f(s_j + s_k^*))$ is (non-negative definite) hermitian. Thus we find with $n = 2$ and $s_1 = 0$, $s_2 = s$ that

$$\begin{pmatrix} f(0) & f(s) \\ f(s^*) & f(s + s^*) \end{pmatrix} = \begin{pmatrix} f(0) & f(s^*) \\ f(s) & f(s + s^*) \end{pmatrix}^{\mathsf{T}} = \overline{\begin{pmatrix} f(0) & f(s^*) \\ f(s) & f(s + s^*) \end{pmatrix}}.$$

Comparing the entries of the matrices we conclude that $f(s^*) = \overline{f(s)}$.   □

Definition 4.3. A function $f : S \to \mathbb{C}$ is *negative definite* if it is hermitian, i.e. $f(s^*) = \overline{f(s)}$, and if

$$\sum_{j,k=1}^{n} \left(f(s_j) + \overline{f(s_k)} - f(s_j + s_k^*) \right) c_j \bar{c}_k \geq 0 \tag{4.2}$$

holds for all $n \in \mathbb{N}$, all $s_1, \ldots, s_n \in S$ and all $c_1, \ldots, c_n \in \mathbb{C}$.

Note that 'f is negative definite' does *not* mean that '$-f$ is positive definite'. The connection between those two concepts is given in the following proposition.

Proposition 4.4 (Schoenberg). *For a function $f : S \to \mathbb{C}$ the following assertions are equivalent.*

(i) *f is negative definite.*

(ii) *$f(0) \geq 0$, $f(s^*) = \overline{f(s)}$ and $-f$ is conditionally positive definite, i.e. for all $n \in \mathbb{N}$, all $s_1, \ldots, s_n \in S$ and all $c_1, \ldots, c_n \in \mathbb{C}$ satisfying $\sum_{j=1}^{n} c_j = 0$ one has*

$$\sum_{j,k=1}^{n} f(s_j + s_k^*) c_j \bar{c}_k \leq 0. \tag{4.3}$$

(iii) *$f(0) \geq 0$ and $s \mapsto e^{-tf(s)}$ is positive definite for all $t > 0$.*

Proof. (i)⇒(iii) The Schur (or Hadamard) product of two $n \times n$ matrices $A = (a_{jk})$ and $B = (b_{jk})$ is the $n \times n$ matrix C with entries $a_{jk} b_{jk}$.

If A, B are non-negative definite hermitian matrices, then C is non-negative definite hermitian. Indeed, writing $B = PP^*$ where $P = (p_{jk}) \in \mathbb{C}^{n \times n}$ and P^* is the adjoint matrix of P, we have

$$b_{jk} = \sum_{\ell=1}^{n} p_{j\ell} \, \bar{p}_{k\ell}.$$

For every choice of $c_1, \ldots, c_n \in \mathbb{C}$ we get

$$\sum_{j,k=1}^{n} a_{jk} b_{jk} c_j \bar{c}_k = \sum_{\ell=1}^{n} \left[\sum_{j,k=1}^{n} a_{jk} (p_{j\ell} c_j) \overline{(p_{k\ell} c_k)} \right] \geq 0.$$

In particular, if (a_{jk}) is a non-negative definite hermitian matrix, so is $(\exp(a_{jk}))$.

If we apply this to the non-negative definite hermitian matrices

$$\left(f(s_j) + \overline{f(s_k)} - f(s_j + s_k^*) \right)_{j,k=1,\ldots,n}, \quad n \in \mathbb{N}, \; s_1, \ldots, s_n \in \mathbb{C},$$

we conclude that

$$\left(\exp \left(f(s_j) + \overline{f(s_k)} - f(s_j + s_k^*) \right) \right)_{j,k=1,\ldots,n}$$

is a non-negative definite hermitian matrix. This implies that for all $c_1, \ldots, c_n \in \mathbb{C}$

$$\sum_{j,k=1}^{n} e^{-f(s_j + s_k^*)} c_j \bar{c}_k = \sum_{j,k=1}^{n} e^{f(s_j) + \overline{f(s_k)} - f(s_j + s_k^*)} e^{-f(s_j)} e^{-\overline{f(s_k)}} c_j \bar{c}_k$$

$$= \sum_{j,k=1}^{n} e^{f(s_j) + \overline{f(s_k)} - f(s_j + s_k^*)} \gamma_j \bar{\gamma}_k \geq 0,$$

where $\gamma_j := c_j e^{-f(s_j)} \in \mathbb{C}$. This proves that e^{-f} is positive definite. Replacing f by tf, $t > 0$, the same argument shows that e^{-tf} is positive definite.

That $f(0) \geq 0$ follows directly from the definition if we take $n = 1$, $s_1 = 0$ and $c_1 = 1$.

(iii)$\Rightarrow$(ii) Let $n \in \mathbb{N}$, $s_1, \ldots, s_n \in S$ and $c_1, \ldots, c_n \in \mathbb{C}$ with $\sum_{j=1}^{n} c_j = 0$. Because of (iii) we see that for all $t > 0$

$$\sum_{j,k=1}^{n} \frac{1}{t} \left(1 - e^{-tf(s_j + s_k^*)} \right) c_j \bar{c}_k = - \sum_{j,k=1}^{n} \frac{1}{t} e^{-tf(s_j + s_k^*)} c_j \bar{c}_k \leq 0.$$

Letting $t \to 0$ we obtain

$$\sum_{j,k=1}^{n} f(s_j + s_k^*) c_j \bar{c}_k \leq 0.$$

By Lemma 4.2, the functions e^{-tf} are hermitian and so is $f = \lim_{t \to 0} (1 - e^{-tf})/t$.

(ii)$\Rightarrow$(i) We use (4.3) with $n + 1$ instead of n and for $0, s_1, \ldots, s_n \in S$ and $c, c_1, \ldots, c_n \in \mathbb{C}$ where $c = - \sum_{k=1}^{n} c_k$. Then

$$f(0)|c|^2 + \sum_{j=1}^{n} f(s_j) c_j \bar{c} + \sum_{j=1}^{n} f(s_j^*) c \bar{c}_j + \sum_{j,k=1}^{n} f(s_j + s_k^*) c_j \bar{c}_k \leq 0.$$

Using the fact that $f(s_j^*) = \overline{f(s_j)}$ and inserting the definition of c, we can rewrite this expression and find

$$\sum_{j,k=1}^{n} \left(f(s_j) + \overline{f(s_k)} - f(s_j + s_k^*) \right) c_j \bar{c}_k \geq f(0)|c|^2 \geq 0. \qquad \square$$

We will now characterize the bounded continuous positive definite and continuous negative definite functions on the semigroup $([0, \infty), +)$ with involution $\lambda^* = \lambda$; it will turn out that these functions coincide with the families $\mathcal{CM}_b$ and $\mathcal{BF}$, respectively. Since we are working on the closed half-line $[0, \infty)$, it is useful to extend $f \in \mathcal{CM}_b$ or $f \in \mathcal{BF}$ continuously to $[0, \infty)$. Because of the monotonicity of f this can be achieved by

$$f(0) := f(0+) = \lim_{\lambda \to 0} f(\lambda).$$

Note that this extension is unique and that, for a completely monotone function f, $f(0+) < \infty$ if, and only if, f is bounded. Throughout this chapter we will tacitly use this extension whenever necessary.

Lemma 4.5. *Let $f \in \mathcal{CM}_b$. Then f is positive definite in the sense of Definition 4.1.*

Proof. From Theorem 1.4 we know that $f \in \mathcal{CM}_b$ is the Laplace transform of a finite measure μ on $[0, \infty)$. Therefore, we find for all $n \in \mathbb{N}$, all $\lambda_1, \ldots, \lambda_n \in [0, \infty)$ and all $c_1, \ldots, c_n \in \mathbb{C}$

$$\sum_{j,k=1}^{n} f(\lambda_j + \lambda_k) c_j \bar{c}_k = \sum_{j,k=1}^{n} \left(\int_{[0,\infty)} e^{-\lambda_j t} e^{-\lambda_k t} \mu(dt) \right) c_j \bar{c}_k$$

$$= \int_{[0,\infty)} \left(\sum_{j,k=1}^{n} e^{-\lambda_j t} c_j \overline{e^{-\lambda_k t} c_k} \right) \mu(dt)$$

$$= \int_{[0,\infty)} \left| \sum_{j=1}^{n} e^{-\lambda_j t} c_j \right|^2 \mu(dt) \geq 0. \qquad \square$$

Corollary 4.6. *Let $f \in \mathcal{BF}$. Then f is negative definite in the sense of Definition 4.3.*

Proof. Let $f \in \mathcal{BF}$. From Theorem 3.7 we know that $e^{-tf} \in \mathcal{CM}$ for all $t > 0$. Since e^{-tf} is bounded, we can use Lemma 4.5 to infer that for each $t > 0$ (the unique extension of) the function e^{-tf} is positive definite, and we conclude with Proposition 4.4 that (the unique extension of) f is negative definite. $\qquad \square$

We write $\Delta_a f(\lambda) := f(\lambda + a) - f(\lambda)$ for the difference of step $a > 0$. The iterated differences of step sizes $a_j > 0$, $j = 1, \ldots, n$, are defined by

$$\Delta_{a_n} \cdots \Delta_{a_1} f := \Delta_{a_n}(\Delta_{a_{n-1}} \cdots \Delta_{a_1} f).$$

Theorem 4.7. *Every bounded continuous positive definite function f on $([0,\infty),+)$ satisfies*

$$(-1)^n \Delta_{a_n} \cdots \Delta_{a_1} f \geq 0 \quad \text{for all } n \in \mathbb{N}, \ a_1,\ldots,a_n > 0. \tag{4.4}$$

Proof. Assume that f is positive definite. Then $f \geq 0$. Taking $n = 2$ in (4.1) we see that the matrix

$$\begin{pmatrix} f(2\lambda) & f(\lambda + a) \\ f(\lambda + a) & f(2a) \end{pmatrix}, \quad \lambda, a \geq 0,$$

is positive definite. Therefore its determinant is non-negative which implies

$$f(\lambda + a) \leq \sqrt{f(2\lambda)\,f(2a)}, \quad \lambda, a \geq 0. \tag{4.5}$$

Applying (4.5) repeatedly, we arrive at

$$f(\lambda) \leq f^{1/2}(0)\, f^{1/2}(2\lambda) \leq f^{3/4}(0)\, f^{1/4}(4\lambda) \leq \cdots \leq f^{1-2^{-n}}(0)\, f^{2^{-n}}(2^n\lambda).$$

Since f is bounded, we can take the limit $n \to \infty$ to get

$$f(\lambda) \leq f(0), \quad \lambda \geq 0.$$

For $N \in \mathbb{N}$, $c_1,\ldots,c_N \in \mathbb{C}$ and $\lambda_1,\ldots,\lambda_N > 0$ we introduce an auxiliary function

$$F(a) := \sum_{\ell,m=1}^{N} f(a + \lambda_\ell + \lambda_m)\, c_\ell \bar{c}_m.$$

For all choices of $n \in \mathbb{N}$, $b_1,\ldots,b_n \in \mathbb{R}$ and $a_1,\ldots,a_n \geq 0$ we find

$$\sum_{j,k=1}^{n} F(a_j + a_k)\, b_j \bar{b}_k = \sum_{j,k=1}^{n} \sum_{\ell,m=1}^{N} f(a_j + a_k + \lambda_\ell + \lambda_m)\, b_j \bar{b}_k c_\ell \bar{c}_m$$

$$= \sum_{(j,\ell),(k,m)} f\big((a_j + \lambda_\ell) + (a_k + \lambda_m)\big)\, (b_j c_\ell)\overline{(b_k c_m)}$$

$$\geq 0.$$

This proves that F is positive definite. Since f is bounded and continuous, so is F, and we conclude, as above, that $F(a) \leq F(0)$. This amounts to saying that

$$\sum_{\ell,m=1}^{N} \big(f(\lambda_\ell + \lambda_m) - f(a + \lambda_\ell + \lambda_m)\big) c_j \bar{c}_m \geq 0,$$

i.e. the function $-\Delta_a f$ is continuous and positive definite.

Repeating the above argument shows that $(-1)^n \Delta_{a_n} \ldots \Delta_{a_1} f$ is continuous and positive definite for all $n \in \mathbb{N}$ and $a_1,\ldots,a_n > 0$. $\qquad \square$

Theorem 4.8 (Bernstein). *For a measurable function* $f : (0, \infty) \to \mathbb{R}$ *the following conditions are equivalent.*

(i) $f \geq 0$ *and* $(-1)^n \Delta_{a_n} \ldots \Delta_{a_1} f \geq 0$ *for all* $n \in \mathbb{N}$ *and* $a_1, \ldots, a_n > 0$.

(ii) *There exists a unique non-negative measure* μ *on* $[0, \infty)$ *such that*

$$f(\lambda) = \int_{[0,\infty)} e^{-\lambda t} \, \mu(dt).$$

(iii) f *is infinitely often differentiable and* $(-1)^n f^{(n)} \geq 0$ *for all* $n \in \mathbb{N}$.

Proof. (i)$\Rightarrow$(ii) Denote by $\mathcal{B}(0, \infty)$ the real-valued Borel measurable functions on $(0, \infty)$. Then

$$\mathcal{C} := \{ f \in \mathcal{B}(0, \infty) \; : \; f \geq 0 \text{ and } (-1)^n \Delta_{a_n} \ldots \Delta_{a_1} f \geq 0$$
$$\text{for all } n \in \mathbb{N} \text{ and } a_1, \ldots, a_n > 0 \}$$

is a convex cone. If we equip $\mathcal{B}(0, \infty)$ with the topology of pointwise convergence, its topological dual separates points. Since every $f \in \mathcal{C}$ is non-increasing and non-negative, the limit $f(0+)$ exists, and f is bounded if, and only if, $f(0+) < \infty$.

For $f \in \mathcal{C}$ the second differences are non-negative; in particular $\Delta_a \Delta_a f \geq 0$ which we may rewrite as

$$f(\lambda) + f(\lambda + 2a) \geq 2f(\lambda + a), \quad \lambda, a > 0.$$

This means that f is mid-point convex. Since $f|_{[c,d]}$ is finite on every compact interval $[c, d] \subset (0, \infty)$, we conclude that $f|_{(c,d)}$ is continuous for all $d > c > 0$, i.e. f is continuous, see for example Donoghue [99, pp. 12–13].

Define $\mathcal{K} := \{ f \in \mathcal{C} \; : \; f(0+) = 1 \}$; this is a closed subset of $\mathcal{C}$. Since it is contained in the set $[0, 1]^{(0,\infty)} \cap \mathcal{B}(0, \infty)$ which is, by Tychonov's theorem, compact under pointwise convergence, it is also compact. Moreover, $\mathcal{K}$ is a basis of the cone $\mathcal{C}_b$ since for every $f \in \mathcal{C}_b$ the normalized function $f/f(0+) \in \mathcal{K}$.

Now the Kreĭn–Milman theorem applies and shows that $\mathcal{K}$ is the closed convex hull of its extreme points. For $f \in \mathcal{C}_b$ we write

$$f(\lambda) = f(\lambda + a) + \big(f(\lambda) - f(\lambda + a) \big) = f(\lambda + a) + (-1) \Delta_a f(\lambda).$$

It follows directly from the definition of the set $\mathcal{C}$ that $f(\cdot + a)$ and $-\Delta_a f$ are again in $\mathcal{C}_b$. Thus, if $f \in \mathcal{K}$ is extremal, both functions must be multiples of f. In particular,

$$f(\lambda + a) = c(a) f(\lambda), \quad \lambda \geq 0, \; a > 0.$$

Letting $\lambda \to 0$ we see that $f(a) = c(a)$.

Since $f(0+) = 1$ and since f is continuous, all solutions of the functional equation $f(\lambda + a) = f(a)f(\lambda)$ are of the form $f(\lambda) = e_t(\lambda) = e^{-t\lambda}$ where $t \in [0, \infty)$. This means that the extreme points of $\mathcal{K}$ are contained in the set $\{e_t : t \geq 0\}$. Since $\mathcal{K}$ is the closed convex hull of its extreme points, we conclude that there exists a measure μ supported in $[0, \infty)$ such that $\mu[0, \infty) = 1$ and

$$f(\lambda) = \int_{[0,\infty)} e^{-\lambda t}\, \mu(dt), \quad f \in \mathcal{K}.$$

This proves (ii) for $f \in \mathcal{C}_b$. By the uniqueness of the Laplace transform, see Proposition 1.2, the representing measure μ is unique.

If $f \in \mathcal{C}$ is not bounded, we argue as in the proof of Theorem 1.4. In this case, $f_a(\lambda) := f(\lambda + a)$ is in $\mathcal{C}_b$, and we find

$$f(\lambda + a) = \int_{[0,\infty)} e^{-\lambda t}\, \mu_a(dt) \quad \text{for all } a > 0.$$

By the uniqueness of the representation, it is easy to see that the measure $e^{at}\mu_a(dt)$ does not depend on $a > 0$. Writing $\mu(dt)$ for $e^{at}\mu_a(dt)$, we get the representation claimed in (ii) for all $f \in \mathcal{C}_b$.

(ii)$\Rightarrow$(iii) This is a simple application of the dominated convergence theorem, see for example the second part of the proof of Theorem 1.4.

(iii)$\Rightarrow$(i) By the mean value theorem we have for all $a > 0$

$$(-1)\Delta_a f(\lambda) = f(\lambda) - f(\lambda + a) = -af'(\lambda + \theta a), \quad \theta \in (0, 1),$$

hence $(-1)\Delta_a f \geq 0$. Iterating this argument yields

$$(-1)^n \Delta_{a_n} \ldots \Delta_{a_1} f(\lambda) = (-1)^n a_1 \cdots a_n\, f^{(n)}(\lambda + \theta_1 a_1 + \cdots + \theta_n a_n)$$

for suitable values of $\theta_1, \ldots, \theta_n \in (0, 1)$. This proves (i). $\qquad\square$

Corollary 4.9. *The family $\mathcal{CM}_b$ and the family of bounded continuous positive definite functions on $([0, \infty), +)$ coincide.*

Proof. From Lemma 4.5 we know already that (the extension of) every $f \in \mathcal{CM}_b$ is positive definite. Conversely, if f is continuous and positive definite, Theorem 4.8 shows that $f|_{(0,\infty)}$ is completely monotone. As f is continuous, we see that $f(0+) = f(0) < \infty$; this implies the boundedness of f, because f is non-increasing and non-negative on $(0, \infty)$. $\qquad\square$

Corollary 4.10. *The family $\mathcal{BF}$ and the family of continuous negative definite functions on $([0, \infty), +)$ coincide.*

Proof. Lemma 4.6 shows that (the extension of) every $f \in \mathcal{BF}$ is negative definite. Now let f be continuous and negative definite. Thus, for every $t \geq 0$ the (extension of the) function e^{-tf} is bounded, continuous and positive definite; by Corollary 4.9 it is in $\mathcal{CM}_b$, and by Theorem 3.7 we get $f \in \mathcal{BF}$. □

Theorem 4.8 has the following counterpart for multiply monotone functions.

Theorem 4.11. *For a measurable function $f : (0, \infty) \to \mathbb{R}$ and $n \geq 2$ the following conditions are equivalent.*

(i) $f \geq 0$ *and* $(-1)^j \Delta_{a_j} \ldots \Delta_{a_1} f \geq 0$ *for all* $j = 1, 2, \ldots n$ *and* $a_1, \ldots, a_n > 0$.

(ii) *There exists a unique measure γ_n on $[0, \infty)$ such that*

$$f(\lambda) = \int_{[0,\infty)} (1 - \lambda t)_+^{n-1} \, \gamma_n(dt), \quad \lambda > 0.$$

(iii) f *is n-monotone.*

In order to prove this theorem we will use the following lemma. For $a > 0$ and $m \in \mathbb{N}$, let Δ_a^m denote the m-times iterated operator Δ_a.

Lemma 4.12. *Let $f : (0, \infty) \to \mathbb{R}$ be a non-negative convex function satisfying $(-1)^{m+1} \Delta_a^{m+1} f(\lambda) \geq 0$ for all $\lambda > 0$, $a > 0$ and some $m \in \mathbb{N}$. Then the right-hand side derivative f'_+ exists and satisfies $(-1)^m \Delta_a^m f'_+(\lambda) \leq 0$ for all $\lambda > 0$ and $a > 0$.*

Proof. A convex function f has at each point in the interior of its domain a right-hand side derivative f'_+ which is non-decreasing. For each $l \geq 1$ we find from a telescoping argument

$$\Delta_a f(\lambda) = \sum_{j=0}^{l-1} \left[f\left(\lambda + (j+1)\frac{a}{l}\right) - f\left(\lambda + j\frac{a}{l}\right) \right]$$

$$= \sum_{j=0}^{l-1} \Delta_{a/l} f\left(\lambda + j\frac{a}{l}\right).$$

By expanding it follows that

$$\Delta_a^m f(\lambda) = \sum_{j_1=0}^{l-1} \cdots \sum_{j_m=0}^{l-1} \Delta_{a/l}^m f\left(\lambda + (j_1 + \cdots + j_m)\frac{a}{l}\right),$$

which implies

$$(-1)^m \Delta_{a/l} \Delta_a^m f(\lambda) = \sum_{j_1=0}^{l-1} \cdots \sum_{j_m=0}^{l-1} (-1)^m \Delta_{a/l}^{m+1} f\left(\lambda + (j_1 + \cdots + j_m)\frac{a}{l}\right)$$

$$\leq 0.$$

since, by assumption, $(-1)^m \Delta_{a/l}^{m+1} f \leq 0$. Therefore,

$$(-1)^m \Delta_a^m f(\lambda) \geq (-1)^m \Delta_a^m f \left(\lambda + \frac{a}{l} \right)$$

for all $l \in \mathbb{N}$. By iterating this inequality we obtain for all $k, l \in \mathbb{N}$ that

$$(-1)^m \Delta_a^m f(\lambda) \geq (-1)^m \Delta_a^m f \left(\lambda + \frac{a}{l} \right) \geq \ldots \geq (-1)^m \Delta_a^m f \left(\lambda + k \frac{a}{l} \right).$$

Since f is continuous, we conclude that $\lambda \mapsto (-1)^m \Delta_a^m f(\lambda)$ is non-increasing. In particular, for all $\lambda > 0$ and $\xi > 0$,

$$\frac{1}{\xi} \left((-1)^m \Delta_a^m f(\lambda) - (-1)^m \Delta_a^m f(\lambda + \xi) \right) \geq 0.$$

If we let $\xi \to 0$, it follows that $(-1)^m \Delta_a^m f'_+(\lambda) \leq 0$.   □

Proof of Theorem 4.11. (i)$\Rightarrow$(iii) If $n = 2$, then we have seen in the first part of the proof of Theorem 4.8 that f is non-increasing and convex. Assume that $n \geq 3$. Again, f is convex and therefore has left and right-hand side derivatives which satisfy

$$f'_+(\xi) \leq f'_-(\lambda) \leq f'_+(\lambda), \quad 0 < \xi < \lambda < \infty. \tag{4.6}$$

By Lemma 4.12, $-f'_+$ satisfies $(-1)^j \Delta_a^j (-f'_+)(\lambda) \geq 0$ for all $\lambda > 0$, $a > 0$ and $j = 0, 1, \ldots, n - 1$. Since $n - 1 \geq 2$, $-f'_+$ is continuous. Together with (4.6) we conclude that f' exists, $-f' \geq 0$ and $(-1)^j \Delta_a^j (-f')(\lambda) \geq 0$ for all $\lambda > 0$, $a > 0$ and $j = 0, 1, \ldots, n - 1$. Repeating this argument and using the assumption, we see that for $j = 0, 1, \ldots, n - 2$, $f^{(j)}$ exists, $(-1)^j f^{(j)}(\lambda) \geq 0$ and $(-1)^{n-2} f^{(n-2)}$ is non-increasing and convex.

(iii)$\Rightarrow$(i) If f is non-increasing and convex, then $f(\lambda) = c + \int_\lambda^\infty g(t)\, dt$ for some constant $c \geq 0$ and some non-negative, right-continuous and non-increasing function $g : (0, \infty) \to \mathbb{R}$, cf. proof of Theorem 1.11. This representation immediately implies that $-\Delta_{a_1} f \geq 0$ and $\Delta_{a_2} \Delta_{a_1} f \geq 0$ for all $a_1, a_2 > 0$ proving the claim when $n = 2$.

As in the proof of Theorem 4.8, we use the mean value theorem to see for all $a > 0$

$$(-1)\Delta_a f(\lambda) = f(\lambda) - f(\lambda + a) = -a f'(\lambda + \theta a), \quad \theta \in (0, 1),$$

hence $(-1)\Delta_a f \geq 0$. Iterating this argument yields for all $j = 1, 2, \ldots, n - 2$,

$$(-1)^j \Delta_{a_j} \cdots \Delta_{a_1} f(\lambda) = (-1)^j a_1 \cdots a_j\, f^{(j)}(\lambda + \theta_1 a_1 + \cdots + \theta_j a_j)$$

for suitable $\theta_1, \ldots, \theta_j \in (0, 1)$. In particular, $(-1)^{n-2} \Delta_{a_{n-2}} \cdots \Delta_{a_1} f(\lambda)$ is non-increasing and convex. Now we use the argument in the preceding paragraph to get the desired assertion.

(i)$\Rightarrow$(ii) For any $m \geq 1$, we define

$$\mathcal{C}_m := \Big\{ f \in \mathcal{B}(0, \infty) \,:\, f \geq 0, \, (-1)^j \Delta_a^j f \geq 0$$

$$\text{for all } j = 1, \ldots, m \text{ and } a > 0 \Big\},$$

and $\mathcal{K}_m := \{ f \in \mathcal{C}_m : f(0+) = 1 \}$. Then $\mathcal{K}_m$ is a closed convex subset of $\mathcal{C}_m$ with respect to pointwise convergence.

Let f be an extreme point in $\mathcal{K}_n$ and set $\zeta := \inf\{\lambda > 0 \,:\, f(\lambda) = 0\}$ with the usual convention that $\inf \emptyset = \infty$. Fix some $\alpha \in (0, \zeta)$. By Taylor's formula with integral remainder term,

$$f(\lambda) = f(\alpha) + \sum_{j=1}^{n-2} \frac{(-1)^j}{j!} f^{(j)}(\alpha)(\alpha - \lambda)^j$$

$$+ \int_\lambda^\alpha \frac{(-1)^{n-1}}{(n-2)!} f_+^{(n-1)}(t)(t - \lambda)^{n-2} \, dt,$$

where $f_+^{(n-1)}$ is the right-hand side derivative of $f^{(n-2)}$. Note that $f_+^{(n-1)}$ exists because $(-1)^{(n-2)} f^{(n-2)}$ is non-increasing and convex. Since $(-1)^{n-1} f_+^{(n-1)}$ is non-increasing, we see for all $\lambda \in (0, \alpha)$ that

$$\int_\lambda^\alpha \frac{(-1)^{n-1}}{(n-2)!} f_+^{(n-1)}(t)(t - \lambda)^{n-2} \, dt \geq \int_\lambda^\alpha \frac{(-1)^{n-1}}{(n-2)!} f_+^{(n-1)}(\alpha)(t - \lambda)^{n-2} \, dt$$

$$= \frac{(-1)^{n-1}}{(n-1)!} f_+^{(n-1)}(\alpha)(\alpha - \lambda)^{n-1}.$$

Define a new function $g : (0, \infty) \to \mathbb{R}$ by $g|_{[\alpha,\infty)} = f|_{[\alpha,\infty)}$ and, for all $\lambda \in (0, \alpha)$,

$$g(\lambda) = f(\alpha) + \sum_{j=1}^{n-2} \frac{(-1)^j}{j!} f^{(j)}(\alpha)(\alpha - \lambda)^j + \frac{(-1)^{n-1}}{(n-1)!} f_+^{(n-1)}(\alpha)(\alpha - \lambda)^{n-1}.$$

Then $g \leq f$ and, by construction, both g and $f - g$ are n-monotone, hence in $\mathcal{C}_n$. Set $q = g(0+)$ and observe that

$$f = g + (f - g) = q \frac{g}{g(0+)} + (1 - q) \frac{f - g}{1 - g(0+)}.$$

Since f is extremal, this means that $g = qf$ for some $q \in [0, 1]$. But $g = f$ on $[\alpha, \infty)$ implying $q = 1$, and thus $g = f$ on $(0, \alpha)$. Therefore, $f|_{(0,\alpha)}$ is a polynomial of degree at most $n - 1$. Since $\alpha < \zeta$ was arbitrary, we conclude that $f|_{(0,\zeta)}$ is a polynomial of degree at most $n - 1$.

If $\zeta < \infty$, then $f(0+) = 1$ and $f^{(j)}(\zeta) = 0$ for $j = 0, 1, \ldots, n - 2$, imply that $f(\lambda) = (1 - \lambda/\zeta)^{n-1}$ for $0 < \lambda < \zeta$. By setting $t = 1/\zeta > 0$ we see that $f(\lambda) = e_{n,t}(\lambda) := (1 - \lambda t)^{n-1}_+$. If $\zeta = \infty$, then clearly $f \equiv 1 =: e_{n,0}$. This means that the extreme points of $\mathcal{K}_n$ are contained in the set $\{e_{n,t} : t \geq 0\}$. As in the proof of Theorem 4.8 we can use the Kreĭn–Milman theorem to conclude that for any $f \in \mathcal{K}_n$ there exists a probability measure γ_n supported in $[0, \infty)$ such that

$$f(\lambda) = \int_{[0,\infty)} (1 - \lambda t)^{n-1}_+ \, \gamma_n(dt).$$

For a general $f \in \mathcal{C}_n$, we can use the argument in the last part of the proof of Theorem 4.8 or Theorem 1.4 to arrive at the above integral representation.

The uniqueness of the measure γ_n follows once we know that every function in the family $\{e_{n,t} : t \geq 0\} = \{(1 - t\lambda)^{n-1}_+ : t \geq 0\}$ is an extremal point of $\mathcal{K}_n$. We are going to show that the $e_{n,t}$ are extremal even for the larger set $\mathcal{C}_n$. Assume that $e_{n-1,t}$ is an extremal point of $\mathcal{C}_{n-1}$. Then $e_{n,t}$ is extremal for $\mathcal{C}_n$. Otherwise, there are $g, h \in \mathcal{C}_n$ not proportional to each other and $q \in (0, 1)$ such that

$$e_{n,t} = qg + (1 - q)h, \quad \text{hence,} \quad t(n - 1)e_{n-1,t} = q(-g') + (1 - q)(-h').$$

Note that $-g', -h' \in \mathcal{C}_{n-1}$ are not proportional to each other, otherwise g, h would be proportional. This implies that neither $t(n - 1)e_{n-1,t}$ nor its multiple $e_{n-1,t}$ are extremal in $\mathcal{C}_{n-1}$, contradicting our assumption.

Thus, it is enough to prove that $e_{1,t} = (1 - \lambda t)^0_+ = \mathbb{1}_{(0,1/t)}(\lambda), t \geq 0$, is extremal for $\mathcal{C}_1$. Assume that $\mathbb{1}_{(0,1/t)}$ is not extremal. Then there are $g, h \in \mathcal{C}_1, g \neq h$ and $q \in (0, 1)$ such that $\mathbb{1}_{(0,1/t)} = qg + (1 - q)h$. Since g and h are nonnegative and non-increasing, we conclude that g and h are constant in $(0, 1/t)$. Thus $\mathbb{1}_{(0,1/t)}$ is indeed extremal.

(ii)$\Rightarrow$(iii) This is proved in Theorem 1.11. $\square$

Remark 4.13. Let $f : (0, \infty) \to [0, \infty)$, and for each $a > 0$, $f_a : (0, \infty) \to [0, \infty)$. Suppose that $\Delta_a f(\lambda) = -f_a(\lambda)$ for all $a > 0$. Then f is n-monotone, $n \geq 2$, if, and only if, f_a is $(n - 1)$-monotone for all $a > 0$. This fact is a direct consequence of Theorem 4.11.

Analogues of Theorem 4.11 and Remark 4.13 are also valid for multiply monotone functions on $\mathbb{R}$.

Let us briefly review the notion of positive and negative definiteness on the group $(\mathbb{R}^d, +)$. Our standard reference is the monograph [39] by Berg and Forst.

On $\mathbb{R}^d$ the involution is given by $\xi^* = -\xi, \xi \in \mathbb{R}^d$. This means that f is positive definite (in the sense of Bochner) if

$$\sum_{j,k=1}^{n} f(\xi_j - \xi_k) \, c_j \bar{c}_k \geq 0, \tag{4.7}$$

and negative definite (in the sense of Schoenberg) if

$$\sum_{j,k=1}^{n} \left(f(\xi_j) + \overline{f(\xi_k)} - f(\xi_j - \xi_k) \right) c_j \bar{c}_k \geq 0 \tag{4.8}$$

hold for all $n \in \mathbb{N}$, $\xi_1, \ldots, \xi_n \in \mathbb{R}^d$ and $c_1, \ldots, c_n \in \mathbb{C}$.

The $\mathbb{R}^d$-analogue of Lemma 4.5 and Corollary 4.9 is known as Bochner's theorem.

Theorem 4.14 (Bochner). *A function $\phi : \mathbb{R}^d \to \mathbb{C}$ is continuous and positive definite if, and only if, it is the Fourier transform of a finite measure μ on $\mathbb{R}^d$,*

$$\phi(\xi) = \widehat{\mu}(\xi) = \int_{\mathbb{R}^d} e^{ix\xi} \, \mu(dx), \quad \xi \in \mathbb{R}^d.$$

The measure μ is uniquely determined by ϕ, and vice versa.

Corollaries 4.6 and 4.10 yield on $\mathbb{R}^d$ the Lévy–Khintchine representation.

Theorem 4.15 (Lévy; Khintchine). *A function $\psi : \mathbb{R}^d \to \mathbb{C}$ is continuous and negative definite if, and only if, there exist a number $\alpha \geq 0$, a vector $\beta \in \mathbb{R}^d$, a symmetric and positive semi-definite matrix $Q \in \mathbb{R}^{d \times d}$ and a measure ν on $\mathbb{R}^d \setminus \{0\}$ satisfying $\int_{y \neq 0} (1 \wedge |y|^2) \, \nu(dy) < \infty$ such that*

$$\psi(\xi) = \alpha + i\beta \cdot \xi + \frac{1}{2} \xi \cdot Q\xi + \int_{y \neq 0} \left(1 - e^{i\xi \cdot y} + \frac{i\xi \cdot y}{1 + |y|^2} \right) \nu(dy). \tag{4.9}$$

The quadruple (α, β, Q, ν) is uniquely determined by ψ, and vice versa.

Comments 4.16. Standard references for positive and negative definite functions are the monographs [38] by Berg, Christensen and Ressel and [39] by Berg and Forst. For a more probabilistic presentation we refer to Dellacherie and Meyer [93, Chapter X].

The natural notion of positive and negative definiteness of a (finite-dimensional) matrix was extended to functions by Mercer [264]. A function (or kernel) of *positive type* is a continuous and symmetric function $\kappa : [a, b] \times [a, b] \to \mathbb{R}$ satisfying

$$\int_a^b \int_a^b \kappa(s, t) \, \theta(s) \theta(t) \, ds \, dt \geq 0 \quad \text{for all } \theta \in C[a, b].$$

This condition is originally due to Hilbert [159] who refers to it as *Definitheit* (definiteness). It is shown in [264] that this is equivalent to saying that for all $n \in \mathbb{N}$ and $s_1, \ldots, s_n \in \mathbb{R}$ the matrices $(\kappa(s_j, s_k))$ are non-negative definite hermitian, which is in accordance with Definition 4.1. Mercer says that κ is of *negative type* if $-\kappa$ is of positive type. Note that this does *not* match our definition of negative definiteness which appears first in Schoenberg [325] and, independently, in Beurling [52];

the name *negative definite* was first used by Beurling [53] in connection with bounded negative definite functions – these are all of the form $c - f$ where f is positive definite and c is a constant $c \geq f(0)$, see e.g. [39, 7.11–7.13] – while Definition 4.3 seems to appear first in Beurling and Deny [54].

The notion of a *positive definite* function was introduced by Mathias [261] and further developed by Bochner [61, Chapter IV.20] culminating in his characterization of all positive definite functions, Theorem 4.14.

Note that, in general, a negative definite function is not the negative of a positive definite function – and vice versa. In order to avoid any confusion, many authors nowadays use 'positive definite (in the sense of Bochner)' and 'negative definite (in the sense of Schoenberg)'.

The abstract framework of positive (and negative) definiteness on semigroups is due to Ressel [302] and Berg, Christensen and Ressel [37]. Ressel observes that positive definiteness (in a semigroup sense) establishes the connection between two seemingly different topics: Fourier transforms and Bochner's theorem on the one hand and, on the other, Laplace transforms, completely monotone functions and Bernstein's theorem. Note that in the case of Abelian groups there are always two choices for involutions: $s^* := -s$, which is natural for groups, and $s^* = s$ which is natural for semigroups. Each leads to a different notion of positive and negative definiteness.

The proof of Proposition 4.4 is adapted from [39, Chapter II.7] and [38, Chapters 4.3, 4.4]; for Theorems 4.7 and 4.8 we used Dellacherie–Meyer [93, X.73 and X.75] as well as Lax [239, Chapter 14.3]. The best references for continuous negative definite functions on $\mathbb{R}^d$ are [39, Chapter II] and [93, Chapter X]. Further proofs of the Bochner–Bernstein theorem based on extreme point methods are discussed in Lukes *et al.* [252, Chapter 14.7] and Simon [330, Theorem 9.10].

Lemma 4.12 and the proof of the implication (i)⇒(iii) in Theorem 4.11 are taken from Wendland [366, Theorem 7.4]. The extreme point argument used in the proof of Theorem 4.11 is a simplified version of Pestana and Mendonça [288]. A more classical proof is contained, as a by-product, in Wendland's proof of the integral representation of completely monotone functions, cf. [366, Sections 7.1 and 7.2]. Note that Theorem 4.11 gives an independent proof of part of Theorem 1.11. Since $\lim_{n \to \infty} |e^{-nx} - (1 - x)_+^{n-1}| = 0$ uniformly for all $x \in [0, \infty)$, Theorem 4.11 can be easily made into a further proof of Theorem 1.4 and 4.8, cf. [288, Theorem 4].

Chapter 5

A probabilistic intermezzo

Completely monotone functions and Bernstein functions are intimately connected with vaguely continuous convolution semigroups of sub-probability measures on the half-line $[0, \infty)$. The probabilistic counterpart of such convolution semigroups are subordinators – processes with stationary independent increments with state space $[0, \infty)$ and right-continuous paths with left limits, i.e. Lévy processes with values in $[0, \infty)$. In this chapter we are going to describe this connection which will lead us in a natural way to the concept of infinite divisibility and the central limit problem.

Recall that a sequence $(\nu_n)_{n \in \mathbb{N}}$ of sub-probability measures on $[0, \infty)$ converges vaguely to a measure ν if

$$\lim_{n \to \infty} \int_{[0,\infty)} f(t) \, \nu_n(dt) = \int_{[0,\infty)} f(t) \, \nu(dt)$$

holds for every compactly supported continuous function $f : [0, \infty) \to \mathbb{R}$, see Appendix A.1. The convolution of the sub-probability measures μ and ν on $[0, \infty)$ is defined to be the sub-probability measure $\mu \star \nu$ on $[0, \infty)$ such that for every bounded continuous function $f : [0, \infty) \to \mathbb{R}$,

$$\int_{[0,\infty)} f(t)(\mu \star \nu)(dt) = \int_{[0,\infty)} \int_{[0,\infty)} f(t + s) \, \mu(dt)\nu(ds).$$

Definition 5.1. A vaguely continuous *convolution semigroup* of sub-probability measures on $[0, \infty)$ is a family of measures $(\mu_t)_{t \geq 0}$ satisfying the following properties:

(i) $\mu_t[0, \infty) \leq 1$ for all $t \geq 0$;

(ii) $\mu_{t+s} = \mu_t \star \mu_s$ for all $t, s \geq 0$;

(iii) vague-$\lim_{t \to 0} \mu_t = \delta_0$.

It follows from (ii) and (iii) that vague-$\lim_{t \to 0} \mu_{t+s} = \mu_s$ for all $s \geq 0$. Unless otherwise stated we will always assume that a convolution semigroup is vaguely continuous.

For all $\chi \in C_c[0, \infty)$ satisfying $0 \leq \chi \leq \mathbb{1}_{[0,\infty)}$ and $\chi(0) = 1$ we find

$$1 \geq \mu_t[0, \infty) \geq \int_{[0,\infty)} \chi(s) \, \mu_t(ds) \xrightarrow{t \to 0} \int_{[0,\infty)} \chi(s) \, \delta_0(ds) = 1.$$

This proves $\lim_{t \to 0} \mu_t[0, \infty) = 1 = \delta_0[0, \infty)$; from Theorem A.4 we conclude that in this case weak and vague continuity coincide.

Theorem 5.2. *Let $(\mu_t)_{t \geq 0}$ be a convolution semigroup of sub-probability measures on $[0, \infty)$. Then there exists a unique $f \in \mathcal{BF}$ such that the Laplace transform of μ_t is given by*

$$\mathscr{L}\mu_t = e^{-tf} \quad \text{for all } t \geq 0. \tag{5.1}$$

Conversely, given $f \in \mathcal{BF}$, there exists a unique convolution semigroup of sub-probability measures $(\mu_t)_{t \geq 0}$ on $[0, \infty)$ such that (5.1) holds true.

Proof. Suppose that $(\mu_t)_{t \geq 0}$ is a convolution semigroup of sub-probability measures on $[0, \infty)$. Fix $t \geq 0$. Since $\mathscr{L}\mu_t > 0$, we can define a function $f_t : (0, \infty) \to \mathbb{R}$ by $f_t(\lambda) = -\log \mathscr{L}(\mu_t; \lambda)$. In other words,

$$\mathscr{L}(\mu_t; \lambda) = e^{-f_t(\lambda)}.$$

By property (ii) of convolution semigroups, it holds that $f_{t+s}(\lambda) = f_t(\lambda) + f_s(\lambda)$ for all $t, s \geq 0$, i.e. $t \mapsto f_t(\lambda)$ satisfies Cauchy's functional equation. Vague continuity ensures that this map is continuous, hence there is a unique solution $f_t(\lambda) = tf(\lambda)$ where $f(\lambda) = f_1(\lambda)$. Therefore,

$$\mathscr{L}\mu_t = e^{-tf} \quad \text{for all } t \geq 0;$$

in particular, $e^{-tf} \in \mathcal{CM}$ for all $t \geq 0$. By Theorem 3.7, $f \in \mathcal{BF}$.

Conversely, suppose that $f \in \mathcal{BF}$. Again by Theorem 3.7 we have that for all $t \geq 0$, $e^{-tf} \in \mathcal{CM}$. Therefore, for every $t \geq 0$ there exists a measure μ_t on $[0, \infty)$ such that $\mathscr{L}\mu_t = e^{-tf}$. We check that the family $(\mu_t)_{t \geq 0}$ is a convolution semigroup of sub-probability measures. Firstly, $\mu_t[0, \infty) = \mathscr{L}(\mu_t; 0+) = e^{-tf(0+)} \leq 1$. Secondly, $\mathscr{L}(\mu_t \star \mu_s) = \mathscr{L}\mu_t \mathscr{L}\mu_s = e^{-tf} e^{-sf} = e^{-(t+s)f} = \mathscr{L}\mu_{t+s}$. By the uniqueness of the Laplace transform we get that $\mu_{t+s} = \mu_t \star \mu_s$.

Finally, $\lim_{t \to 0} \mathscr{L}\mu_t(\lambda) = \lim_{t \to 0} e^{-tf(\lambda)} = 1 = \mathscr{L}\delta_0(\lambda)$ for all $\lambda > 0$ and, by Lemma A.9, vague-$\lim_{t \to 0} \mu_t = \delta_0$. $\qquad\square$

Remark 5.3. Because of formula (5.1), probabilists often use the name *Laplace exponent* instead of Bernstein function.

Definition 5.4. A stochastic process $S = (S_t)_{t \geq 0}$ defined on a probability space $(\Omega, \mathscr{F}, \mathbb{P})$ with state space $[0, \infty)$ is called a *subordinator* if it has independent and stationary increments, $S_0 = 0$ a.s., and for almost every $\omega \in \Omega$, $t \mapsto S_t(\omega)$ is a right-continuous function with left limits. The measures $(\mu_t)_{t \geq 0}$ defined by

$$\mu_t(B) = \mathbb{P}(S_t \in B), \quad B \subset [0, \infty) \text{ Borel,}$$

are called the *transition probabilities* of the subordinator S.

Almost all paths $t \mapsto S_t(\omega)$ of a subordinator S are non-decreasing functions.

Let e_a be an exponentially distributed random variable with parameter $a \geq 0$, i.e. $\mathbb{P}(e_a > t) = e^{-at}$, $t > 0$. We allow that $a = 0$ in which case $e_a = +\infty$. Assume that e_a is independent of the subordinator S. We define a process $\widehat{S} = (\widehat{S}_t)_{t \geq 0}$ by

$$\widehat{S}_t := \begin{cases} S_t, & t < e_a, \\ +\infty, & t \geq e_a. \end{cases} \tag{5.2}$$

The process $\widehat{S}$ is the subordinator S killed at an independent exponential time. Any process with state space $[0, \infty]$ having the same distribution as $\widehat{S}$ will be called a *killed subordinator*.

The connection between (killed) subordinators and convolution semigroups of sub-probability measures on $[0, \infty)$ is as follows. For $t \geq 0$ let

$$\mu_t(B) := \mathbb{P}(\widehat{S}_t \in B), \quad B \subset [0, \infty) \text{ Borel},$$

be the transition probabilities of $(\widehat{S}_t)_{t \geq 0}$, and note that

$$\mathbb{E}[e^{-\lambda \widehat{S}_t}] = \int_{[0,\infty)} e^{-\lambda y} \, \mu_t(dy) = \mathscr{L}(\mu_t; \lambda).$$

Proposition 5.5. *The family $(\mu_t)_{t \geq 0}$ is a convolution semigroup of sub-probability measures.*

Proof. We check properties (i)–(iii) in Definition 5.1. It is clear that $\mu_t[0, \infty) \leq 1$. Let $g : [0, \infty) \to \mathbb{R}$ be a bounded continuous function. Then by the right-continuity of the paths, $\lim_{t \to 0} g(S_t) = g(S_0) = g(0)$ a.s. By the dominated convergence theorem it follows that $\lim_{t \to 0} \mathbb{E}[g(S_t)] = g(0)$, i.e.

$$\lim_{t \to 0} \int_{[0,\infty)} g \, d\mu_t = g(0) = \int_{[0,\infty)} g \, d\delta_0,$$

thus proving property (iii). In order to show property (ii), assume first that $a = 0$, that is there is no killing. Then for $s, t \geq 0$,

$$\begin{aligned} \mathscr{L}(\mu_{s+t}; \lambda) = \mathbb{E}[e^{-\lambda S_{s+t}}] &= \mathbb{E}[e^{-\lambda(S_{s+t} - S_s)} e^{-\lambda(S_s - S_0)}] \\ &= \mathbb{E}[e^{-\lambda S_t}] \mathbb{E}[e^{-\lambda S_s}] \\ &= \mathscr{L}(\mu_t; \lambda) \mathscr{L}(\mu_s; \lambda), \end{aligned} \tag{5.3}$$

which is equivalent to $\mu_{s+t} = \mu_s \star \mu_t$. If $a > 0$, we find by independence

$$\mathbb{E}[e^{-\lambda \widehat{S}_t}] = \mathbb{E}[e^{-\lambda S_t} \mathbb{1}_{\{t < e_a\}}] = \mathbb{P}(t < e_a) \, \mathbb{E}[e^{-\lambda S_t}] = e^{-at} \, \mathbb{E}[e^{-\lambda S_t}],$$

which, together with (5.3), concludes the proof.   □

The converse of Proposition 5.5 is also true: given a convolution semigroup $(\mu_t)_{t\geq0}$ of sub-probability measures on $[0, \infty)$, there exists a killed subordinator $(S_t)_{t\geq0}$ on a probability space $(\Omega, \mathcal{F}, \mathbb{P})$ such that $\mu_t = \mathbb{P}(S_t \in \cdot)$ for all $t \geq 0$. A proof of this fact can be found in [47].

Let us rewrite (5.1) as

$$\mathbb{E}[e^{-\lambda \widehat{S}_t}] = \mathscr{L}(\mu_t; \lambda) = e^{-tf(\lambda)}.$$

Recall that $f \in \mathcal{BF}$ has the representation

$$f(\lambda) = a + b\lambda + \int_{(0,\infty)} (1 - e^{-\lambda t}) \mu(dt).$$

It is clear from the proof of Theorem 5.2 that $a = f(0+) > 0$ if, and only if, $\mu_t[0, \infty) < 1$ for all $t > 0$; in this case, the associated stochastic process has a.s. finite lifetime. This is why a is usually called the killing term of the Laplace exponent f.

Let $(\mu_t)_{t\geq0}$ be a convolution semigroup of sub-probability measures on $[0, \infty)$ and let f be the corresponding Bernstein function. Then the completely monotone function $g(\lambda) := \mathscr{L}(\mu_1; \lambda) = e^{-f(\lambda)}$ has the property that for every $t > 0$, g^t is again completely monotone. This follows easily from

$$g^t(\lambda) = e^{-tf(\lambda)} = \mathscr{L}(\mu_t; \lambda).$$

From the probabilistic point of view, this means that S_t is an *infinitely divisible random variable*. To be more precise $S_t = \sum_{j=1}^{n}(S_{jt/n} - S_{(j-1)t/n})$ for every $n \in \mathbb{N}$, and the random variables $(S_{jt/n} - S_{(j-1)t/n})_{1\leq j\leq n}$ are independent and identically distributed. Hence $\mu_t = \mu_{t/n}^{\star n}$ is the n-fold convolution. This motivates the following definition.

Definition 5.6. A completely monotone function g is said to be *infinitely divisible* if for every $t > 0$ the function g^t is again completely monotone.

If g is infinitely divisible and $g(0+) \leq 1$, the sub-probability measure π on $[0, \infty)$ satisfying $\mathscr{L}\pi = g$ is said to be an *infinitely divisible distribution*; we write $\pi \in \mathrm{ID}$.

Remark 5.7. In Definition 5.6 it is enough to require that $g^t \in \mathcal{CM}$ for *small* t. In fact, $g \in \mathcal{CM}$ is infinitely divisible if, and only if, $g^{1/n} \in \mathcal{CM}$ for all $n \geq n_0$.

The necessity is obvious. Assume that $g^{1/n} \in \mathcal{CM}$ for all $n \geq n_0$. For any $t > 0$ there is a sequence $r_m = k_m/\ell_m \in \mathbb{Q}^+$ such that $\ell_m \geq n_0$ and $\lim_{m\to\infty} r_m = t$. Observe that $g^t(\lambda) = \lim_{m\to\infty} g^{k_m/\ell_m}(\lambda)$ for all $\lambda > 0$. Since $\mathcal{CM}$ is closed under pointwise limits, cf. Corollary 1.6, it is enough to show that $g^{k_m/\ell_m} \in \mathcal{CM}$ for each $m \geq 1$. This, however, follows since by assumption $g^{1/\ell_m} \in \mathcal{CM}$ and since $\mathcal{CM}$ is closed under pointwise multiplication: $g^{k_m/\ell_m} = g^{1/\ell_m} \cdots g^{1/\ell_m}$, cf. Corollary 1.6.

The discussion preceding Definition 5.6 shows that $g := e^{-f}$ is completely monotone and infinitely divisible if $f \in \mathcal{BF}$. Moreover, $g(0+) \leq 1$. This is already one
direction of the next result.

Lemma 5.8. *Suppose that* $g : (0, \infty) \to (0, \infty)$. *Then the following statements are
equivalent.*

(i) $g \in \mathcal{CM}$, g *is infinitely divisible and* $g(0+) \leq 1$.

(ii) $g = e^{-f}$ *where* $f \in \mathcal{BF}$.

Proof. Suppose that (i) holds. Since $g^t \in \mathcal{CM}$ and clearly $g^t(0+) \leq 1$, there exists a sub-probability measure μ_t on $[0, \infty)$ such that $g^t(\lambda) = \mathcal{L}(\mu_t; \lambda)$. Since
$\lim_{t \to 0} g^t(\lambda) = 1$, it follows that vague-$\lim_{t \to 0} \mu_t = \delta_0$. For $s, t \geq 0$ it holds that
$g^t g^s = g^{t+s}$, and consequently

$$\mathcal{L}(\mu_t; \lambda)\, \mathcal{L}(\mu_s; \lambda) = \mathcal{L}(\mu_{t+s}; \lambda).$$

By the uniqueness of the Laplace transform this means that $(\mu_t)_{t \geq 0}$ is a convolution
semigroup of sub-probability measures on $[0, \infty)$. By Theorem 5.2, there exists a
unique $f \in \mathcal{BF}$ such that $\mathcal{L}(\mu_t; \lambda) = e^{-tf(\lambda)}$; in particular, $g = e^{-f}$. ☐

Lemma 5.8 admits an extension to completely monotone functions g which need
not satisfy $g(0+) \leq 1$. For the purpose of stating this extension, let us say that a C^∞
function $f : (0, \infty) \to \mathbb{R}$ is an *extended Bernstein function* if

$$(-1)^{n-1} f^{(n)}(\lambda) \geq 0 \quad \text{for all } n \in \mathbb{N} \text{ and } \lambda > 0.$$

Remark 5.9. Note that a non-negative extended Bernstein function is actually a Bernstein function, and a non-positive extended Bernstein function is the negative of a
completely monotone function. In fact, f is an extended Bernstein function if, and
only if,

$$f = g - h \text{ for some } g \in \mathcal{BF}, \ h \in \mathcal{CM}, \text{ i.e. } f \in \mathcal{BF} - \mathcal{CM}.$$

Indeed, if $f = g - h$ where $g \in \mathcal{BF}$ and $h \in \mathcal{CM}$, it is obvious that for $n \in \mathbb{N}$

$$(-1)^{n-1} f^{(n)} = (-1)^{n-1}\big[g^{(n)} - h^{(n)}\big] = (-1)^{n-1} g^{(n)} + (-1)^n h^{(n)} \geq 0,$$

i.e. $g - h$ is an extended Bernstein function. Conversely, assume that f is an extended
Bernstein function. If $f(c) \geq 0$ for some $c > 0$, we set $g(\lambda) := f(\lambda + c)$ and
$h(\lambda) := f(\lambda + c) - f(\lambda)$. By construction, $f = g - h$ and $g, h \geq 0$; for all $n \in \mathbb{N}$

$$(-1)^{n-1} g^{(n)}(\lambda) = (-1)^{n-1} f^{(n)}(\lambda + c) \geq 0, \quad \text{and}$$

$$(-1)^n h^{(n)}(\lambda) = (-1)^n f^{(n)}(\lambda + c) - (-1)^n f^{(n)}(\lambda) \geq 0$$

where we used that $(-1)^n f^{(n+1)} \geq 0$ which shows that $(-1)^n f^{(n)}$ is increasing. If
$f(c) \leq 0$ for all $c > 0$, we use $g \equiv 0$ and $h = f$.

Definition 5.10. A C^∞ function $g : (0, \infty) \to (0, \infty)$ is said to be *logarithmically completely monotone* if

$$-(\log g)' \in \mathcal{CM}. \tag{5.4}$$

Theorem 5.11. *Suppose that* $g : (0, \infty) \to (0, \infty)$. *Then the following assertions are equivalent.*

(i) $g \in \mathcal{CM}$ *and* g *is infinitely divisible.*

(ii) $g = e^{-f}$ *where* f *is an extended Bernstein function.*

(iii) g *is logarithmically completely monotone.*

Proof. (i)$\Rightarrow$(ii) For $c > 0$ let $g_c(\lambda) := g(\lambda + c)$, $\lambda > 0$. Then

$$\lambda \mapsto \frac{g_c(\lambda)}{g_c(0+)} = \frac{g_c(\lambda)}{g(c)}$$

is again in $\mathcal{CM}$, infinitely divisible and tends to 1 as $\lambda \to 0$. By Lemma 5.8, there exists $f_c \in \mathcal{BF}$ such that $g(\lambda + c) = g_c(\lambda) = g(c)e^{-f_c(\lambda)}$ for all $\lambda > 0$. This can be written as

$$g(\lambda) = g(\lambda - c + c) = g(c)e^{-f_c(\lambda - c)} = e^{-f_c(\lambda - c) + \log g(c)}, \quad \lambda > c. \tag{5.5}$$

If $0 < b < c$ the same formula is valid with b replacing c; for $0 < b < c < \lambda$ it holds that

$$-f_c(\lambda - c) + \log g(c) = -f_b(\lambda - b) + \log g(b).$$

This implies that one can define a function $f : (0, \infty) \to \mathbb{R}$ by

$$f(\lambda) := f_c(\lambda - c) - \log g(c), \quad \lambda > c.$$

Clearly, f is C^∞ on $(0, \infty)$. For any $\lambda > 0$ find $0 < c < \lambda$. By (5.5), $g(\lambda) = e^{-f(\lambda)}$, and for $n \in \mathbb{N}$, $(-1)^{n-1} f^{(n)}(\lambda) = (-1)^{n-1} f_c^{(n)}(\lambda - c) \geq 0$. Thus, f is an extended Bernstein function.

(ii)$\Rightarrow$(iii) If $g = e^{-f}$, then $-(\log g)' = f'$ and clearly $f' \in \mathcal{CM}$.

(iii)$\Rightarrow$(i) Suppose g is logarithmically completely monotone. For $c > 0$,

$$f_c(\lambda) := \int_c^{\lambda + c} -(\log g)'(t)\, dt = -\log g(\lambda + c) + \log g(c), \quad \lambda > 0,$$

belongs to $\mathcal{BF}$. Since $g(\lambda + c) = g(c)e^{-f_c(\lambda)}$, it follows from Theorem 3.7 that $\lambda \mapsto g(\lambda + c)$ is completely monotone for all $c > 0$. Since g is continuous, we get that $g(\lambda) = \lim_{c \to 0} g(\lambda + c)$, hence $g \in \mathcal{CM}$. Note that g^t is also logarithmically completely monotone for every $t > 0$, hence by the already proven fact, $g^t \in \mathcal{CM}$. Thus, g is infinitely divisible. $\square$

Remark 5.12. (i) The concept of infinite divisibility appears naturally in the central limit problem of probability theory. This problem can be stated in the following way:

Let $(X_{n,j})_{j=1,\dots,k(n),\, n\in\mathbb{N}}$ be a (doubly indexed) family of real random variables such that the random variables in each row, $X_{n,1},\dots,X_{n,k(n)}$, are independent. When do the probability distributions

$$law\left(\frac{X_{n,1}+\cdots+X_{n,k(n)}-a_n}{b_n}\right),$$

for suitably chosen sequences of real $(a_n)_{n\in\mathbb{N}}$ and positive $(b_n)_{n\in\mathbb{N}}$ numbers, have a non-trivial weak limit as $n\to\infty$?

The classical example is, of course, the central limit theorem where a sequence of independent and identically distributed random variables $(X_j)_{j\in\mathbb{N}}$ is arranged in a triangular array of the form

$$X_1;\quad X_1,X_2;\quad X_1,X_2,X_3;\quad\dots\quad X_1,X_2,\dots,X_n;\quad\dots$$

where $k(n)=n$, $a_n=n\,\mathbb{E}X_1$ is the mean value and $b_n=\sqrt{n\operatorname{Var}X_1}$ the standard deviation of the partial sum of the variables in the nth row.

Clearly, every limiting random variable X is *infinitely divisible* in the sense that for every $n\in\mathbb{N}$ there are independent and identically distributed random variables $Y_1,\dots,Y_n$ such that X and $Y:=Y_1+\cdots+Y_n$ have the same probability distribution. The corresponding probability distributions are also called *infinitely divisible*.

It can be shown, see e.g. Gnedenko and Kolmogorov [130], that every infinitely divisible distribution can be obtained as the limiting distribution of a triangular array $(X_{n,j})_{j=1,\dots,k(n),\, n\in\mathbb{N}}$, where each row $X_{n,1},\dots,X_{n,k(n)}$ consists of independent random variables which are asymptotically negligible, i.e.

$$\lim_{n\to\infty}\max_{1\le j\le k(n)}\mathbb{P}(|X_{n,j}|\ge\epsilon)=0\quad\text{for all }\epsilon>0.$$

(ii) Specializing to one-sided distributions π supported in $[0,\infty)$ we can express the property $\pi\in\mathsf{ID}$ in terms of the Laplace transform. By Lemma 5.8, $g\in\mathcal{CM}$ with $g(0+)\le 1$ is infinitely divisible if, and only if, for every $n\in\mathbb{N}$ there is some $g_n\in\mathcal{CM}$ such that $g=g_n^n$. In probabilistic terms this means that the law $\pi\in\mathsf{ID}$ has an nth convolution root π_n such that $\mathscr{L}\pi_n=g_n$ and $\pi=\pi_n^{\star n}$. If X denotes a random variable with distribution π and Y_j, $1\le j\le n$, denote independent and identically distributed random variables with the common distribution π_n, this means that X and $Y_1+\cdots+Y_n$ have the same probability distribution.

Depending on the structure of the triangular array, the limiting distributions may have different stability properties. We will discuss this for one-sided distributions supported in $[0,\infty)$. Often it is easier to describe them in terms of the corresponding Laplace transforms.

Definition 5.13. A completely monotone function $g \in \mathcal{CM}$ with $g(0+) \leq 1$ is said to be (weakly) *stable* if

$$\text{for all } a > 0 \text{ there are } b \geq 0, \ c > 0 \text{ such that } g(\lambda)^a = g(c\lambda) \cdot e^{-b\lambda}; \quad (5.6)$$

if $b = 0$ we call g *strictly stable*.

The corresponding random variables and sub-probability distributions are again called (weakly or strictly) stable.

We will see below in Proposition 5.15 that the constant c appearing in (5.6) is necessarily of the form $c = a^{1/\alpha}$ for some $\alpha \in (0, 1]$; α is often called the *index of stability*.

It can be shown, cf. [130], that every stable π is the limiting distribution of a sequence of independent and identically distributed random variables $(X_j)_{j \in \mathbb{N}}$ and suitable centering and norming sequences $(a_n)_{n \in \mathbb{N}}$, $a_n \geq 0$, and $(b_n)_{n \in \mathbb{N}}$, $b_n > 0$:

$$\text{law}\left(\frac{1}{b_n} \left(\sum_{j=1}^{n} X_j - a_n \right) \right) \xrightarrow[n \to \infty]{\text{weak}} \pi. \quad (5.7)$$

All (weakly) stable distributions on $[0, \infty)$ have Laplace transforms which are given by $g(\lambda) = \exp(-\beta\lambda^\alpha - \gamma\lambda)$ with $\alpha \in (0, 1]$ and $\beta, \gamma \geq 0$. The distributions are (strictly) stable if $\gamma = 0$.

If the X_j are independent (but not necessarily identically distributed) and $\lim_{n \to \infty} b_n = \infty$ and $\lim_{n \to \infty} b_{n+1}/b_n = 1$, then the limit (5.7) characterizes all *self-decomposable* (also: class L) distributions. Again we focus on one-sided distributions where we can express self-decomposability conveniently in terms of Laplace transforms. We will denote the set of all self-decomposable distributions on $[0, \infty)$ by SD.

Definition 5.14. A completely monotone function $g \in \mathcal{CM}$ with $g(0+) \leq 1$ is said to be *self-decomposable*, if

$$\lambda \mapsto \frac{g(\lambda)}{g(c\lambda)} = g_c(\lambda) \text{ is completely monotone for all } c \in (0, 1). \quad (5.8)$$

We will see in Proposition 5.17 that the Laplace exponent f of a $\pi \in$ SD, i.e. $\mathscr{L}\pi = e^{-f}$, is a Bernstein function with a Lévy triplet of the form $(a, b, m(t)\, dt)$ such that $t \mapsto t\, m(t)$ is non-increasing.

The rather deep limit theorems – or the equally difficult structure results on the completely monotone functions – of Remark 5.12 show that

$$\{\text{stable laws}\} \subset \text{SD} \subset \text{ID}.$$

The following two results contain a simple, purely analytic proof for this.

Proposition 5.15. *Let* $g \in \mathcal{CM}$, $g(0+) \leq 1$, *be (weakly) stable. Then there exist* $\alpha \in (0, 1]$ *and* $\beta, \gamma \geq 0$ *such that*

$$g(\lambda) = e^{-\beta \lambda^\alpha - \gamma \lambda}.$$

In particular, $g(0+) = 1$. *In the strictly stable case,* $\gamma = 0$.

Proof. The function $\lambda \mapsto g(c\lambda)e^{-b\lambda}$ appearing in (5.6) is a product of two completely monotone functions, hence itself completely monotone. Thus, $g^a \in \mathcal{CM}$ for every $a > 0$ which means that g is infinitely divisible. By Lemma 5.8, there exists some $f \in \mathcal{BF}$ such that $g = e^{-f}$. The structure equation (5.6) becomes, in terms of f,

for all $a > 0$ there are $c > 0$, $b \geq 0$ such that $af(\lambda) = f(c\lambda) + b\lambda$.   (5.9)

This implies that $f(0+) = 0$. Differentiating this equality twice yields

$$af''(\lambda) = c^2 f''(c\lambda).$$ (5.10)

In the trivial case where $f'' \equiv 0$ we have $f(\lambda) = \gamma\lambda$ for a suitable integration constant $\gamma \geq 0$, and we are done. In all other cases, (5.10) shows that c is a function of a only and that it does not depend on b. Indeed, if both (a, c) and $(a, \tilde{c})$ satisfy (5.9), we find from (5.10) with $\lambda = 1/\tilde{c}$

$$\frac{c^2}{\tilde{c}^2} f''\left(\frac{c}{\tilde{c}}\right) = f''(1).$$

As f'' is increasing and non-trivial, it is easy to see that $c = \tilde{c}$ and $c = c(a)$. Setting $\lambda = 1$ in (5.9) we see that $b = af(1) - f(c)$ is also unique. Since the inverse f^{-1} of a non-trivial, i.e. non-constant, Bernstein function is again a continuous function, we get from (5.9)

$$c = c(a) = \frac{1}{\lambda} f^{-1}(af(\lambda) - b\lambda);$$

this proves that $c = c(a)$ depends continuously on $a > 0$. For any two $a, \tilde{a} > 0$ we can use (5.10) twice to find

$$a\tilde{a} f''(\lambda) = ac^2(\tilde{a}) f''(c(\tilde{a})\lambda) = c^2(\tilde{a})af''(c(\tilde{a})\lambda)$$
$$= c^2(\tilde{a})c^2(a) f''(c(a)c(\tilde{a})\lambda).$$

On the other hand, $a\tilde{a} f''(\lambda) = c^2(a\tilde{a}) f''(c(a\tilde{a})\lambda)$, so that $c(a\tilde{a}) = c(a)c(\tilde{a})$. Since $c(a)$ is continuous, there exists some $\alpha \in \mathbb{R} \setminus \{0\}$ such that $c(a) = a^{1/\alpha}$. Inserting this into (5.10) we have

$$f''(\lambda) = a^{\frac{2}{\alpha}-1} f''(a^{\frac{1}{\alpha}}\lambda) \quad \text{for all } a, \lambda > 0.$$

Now take $a = \lambda^{-\alpha}$ to get $f''(\lambda) = \lambda^{\alpha-2} f''(1)$ and integrate twice. Using that $f(0+) = 0$ we find

$$f(\lambda) = \frac{f''(1)}{\alpha(\alpha-1)} \lambda^\alpha + f''(1) C\lambda$$

with some integration constant $C \in \mathbb{R}$. Since $f \in \mathcal{BF}$ is non-trivial, we know that $0 < \alpha \le 1$. Finally, we can use (5.9) to work out that $f''(1) C = b/(a - a^{1/\alpha})$.   □

Corollary 5.16. *We have* {stable laws on $[0, \infty)$} $\subset$ SD $\subset$ ID.

Proof. The first inclusion follows from Proposition 5.15. If g is (weakly) stable, $f(\lambda) := -\log g(\lambda) = \beta\lambda^\alpha + \gamma\lambda$. Thus, for $0 < c < 1$,

$$f(\lambda) - f(c\lambda) = \beta(1 - c^\alpha)\lambda^\alpha + \gamma(1 - c)\lambda$$

is a Bernstein function, and we get that $g_c(\lambda) = g(\lambda)/g(c\lambda) = \exp(-(f(\lambda) - f(c\lambda)))$ is completely monotone.

If (5.8) holds for g and all $0 < c < 1$, we have

$$-\frac{g'(\lambda)}{g(\lambda)} = \lim_{c\to1-} \frac{g(c\lambda) - g(\lambda)}{(1-c)\lambda g(c\lambda)} = \lim_{c\to1-} \frac{1}{(1-c)\lambda}\left(1 - \frac{g(\lambda)}{g(c\lambda)}\right).$$

Since $g(\lambda)/g(c\lambda)$ is bounded by 1 and completely monotone, the expression inside the brackets is a Bernstein function by Proposition 3.12. Dividing by λ makes it completely monotone, cf. Corollary 3.8 (iv). Since limits of completely monotone functions are again completely monotone, cf. Corollary 1.7, we know that g is logarithmically completely monotone, hence infinitely divisible by Theorem 5.11.   □

Proposition 5.17. *Let $f \in \mathcal{BF}$ satisfy $\mathscr{L}\pi = e^{-f}$ and let μ be the Lévy measure of f. Then the following assertions are equivalent.*

 (i) μ *has a density $m(t)$ such that $t \mapsto t \cdot m(t)$ is non-increasing;*

 (ii) $\pi \in$ SD.

Proof. (ii)$\Rightarrow$(i) Assume that $\pi \in$ SD. Set $g := \mathscr{L}\pi$ and $g_c(\lambda) := g(\lambda)/g(c\lambda)$ for $c \in (0, 1)$. Since $g(\lambda) = e^{-f(\lambda)}$, we have that $g_c(\lambda) = \exp(-(f(\lambda) - f(c\lambda)))$.

We will now show that g_c is for each $c \in (0, 1)$ logarithmically completely monotone. For this we have to check that $-g_c'/g_c \in \mathcal{CM}$. A small change in the proof of Corollary 5.16 shows that

$$\phi(\lambda) := -\lambda \frac{g'(\lambda)}{g(\lambda)} = \lim_{c\to1-} \frac{g(c\lambda) - g(\lambda)}{(1-c)g(c\lambda)} = \lim_{c\to1-} \frac{1}{(1-c)}\left(1 - \frac{g(\lambda)}{g(c\lambda)}\right)$$

is a Bernstein function and, therefore, ϕ' is completely monotone. Moreover,

$$-\frac{g_c'(\lambda)}{g_c(\lambda)} = -\frac{\frac{d}{d\lambda}\frac{g(\lambda)}{g(c\lambda)}}{\frac{g(\lambda)}{g(c\lambda)}} = -\frac{g'(\lambda)g(c\lambda) - g(\lambda)cg'(c\lambda)}{g^2(c\lambda)} \cdot \frac{g(c\lambda)}{g(\lambda)}$$

$$= -\frac{g'(\lambda)}{g(\lambda)} + \frac{cg'(c\lambda)}{g(c\lambda)}$$

$$= \frac{1}{\lambda}\int_{c\lambda}^{\lambda}\phi'(s)\,ds = \int_{c}^{1}\phi'(t\lambda)\,dt.$$

Thus, $-g_c'/g_c$ is an integral mixture of completely monotone functions and as such, see Corollary 1.6, itself completely monotone.

Now define $f_c(\lambda) := f(\lambda) - f(c\lambda)$; by Theorem 5.11, $f_c \in \mathcal{BF}$. Let

$$f(\lambda) = a + b\lambda + \int_{(0,\infty)}(1 - e^{-\lambda t})\,\mu(dt) = a + b\lambda + \lambda\int_0^\infty e^{-\lambda t}M(t)\,dt,$$

where $M(t) = \mu(t,\infty)$. Then

$$f_c(\lambda) = f(\lambda) - f(c\lambda)$$

$$= \left(a + b\lambda + \int_{(0,\infty)}(1 - e^{-\lambda t})\,\mu(dt)\right)$$

$$- \left(a + bc\lambda + \int_{(0,\infty)}(1 - e^{-c\lambda t})\,\mu(dt)\right) \tag{5.11}$$

$$= b(1 - c)\lambda + \int_{(0,\infty)}(1 - e^{-\lambda t})\,\mu(dt) - \int_{(0,\infty)}(1 - e^{-\lambda t})\,\mu^{(c)}(dt),$$

where $\mu^{(c)}(\cdot) := \mu(c^{-1}\cdot)$. On the other hand, f_c being a Bernstein function, it has a representation

$$f_c(\lambda) = \tilde{b}\lambda + \int_{(0,\infty)}(1 - e^{-\lambda t})\,\mu_c(dt). \tag{5.12}$$

Hence, $\mu_c = \mu - \mu^{(c)}$, implying in particular that for every Borel set $B \subset (0,\infty)$ it holds that $\mu(B) \geq \mu(c^{-1}B)$. For $B = (s,t]$, this gives $\mu(s,t] \geq \mu(c^{-1}s, c^{-1}t]$ for every $c \in (0,1)$.

For $s \in \mathbb{R}$ define $h(s) := M(e^{-s}) = \mu(e^{-s},\infty)$. Clearly, h is non-negative and non-decreasing. Moreover, we have for all $u > 0$ and $c \in (0,1)$

$$h(s + u) - h(s) = M(e^{-s-u}) - M(e^{-s})$$

$$= \mu(e^{-s-u}, e^{-s}]$$

$$\geq \mu(e^{-s-u-\log c}, e^{-s-\log c}]$$

$$= h(s + u + \log c) - h(s + \log c).$$

This shows that for all $u > 0$ the function $s \mapsto h(s + u) - h(s)$ is non-decreasing. Therefore, h is convex. Being non-negative, non-decreasing and convex, h can be written in the form

$$h(s) = \int_{-\infty}^{s} \ell(u)\, du$$

where ℓ is non-negative and non-decreasing. Define a function $k : (0, \infty) \to \mathbb{R}$ by $k(v) := \ell(e^{-v})$. Then k is non-negative and non-increasing. For $t > 0$ a change of variables shows

$$M(t) = h(-\log t) = \int_{-\infty}^{-\log t} \ell(u)\, du = \int_{t}^{\infty} \frac{\ell(e^{-v})}{v}\, dv = \int_{t}^{\infty} \frac{k(v)}{v}\, dv,$$

and the proof is finished by defining $m(t) := k(t)/t$.

(i)$\Rightarrow$(ii) To prove the converse, let $c \in (0, 1)$ and define $f_c(\lambda) := f(\lambda) - f(c\lambda)$. Then

$$f_c(\lambda) = \left(a + b\lambda + \int_0^\infty (1 - e^{-\lambda t}) m(t)\, dt \right)$$
$$- \left(a + bc\lambda + \int_0^\infty (1 - e^{-c\lambda t}) m(t)\, dt \right)$$
$$= b(1 - c)\lambda + \int_0^\infty (1 - e^{-\lambda t})\big(m(t) - c^{-1} m(t/c)\big)\, dt.$$

Since $t \mapsto t \cdot m(t)$ is non-increasing, $n_c(t) := m(t) - c^{-1} m(t/c) \geq 0$ is non-negative for all $t \geq 0$. Hence, $f_c \in \mathcal{BF}$. Let $g := \mathscr{L}\pi = e^{-f}$. Then

$$g_c(\lambda) = \frac{g(\lambda)}{g(c\lambda)} = \frac{e^{-f(\lambda)}}{e^{-f(c\lambda)}} = e^{-f_c(\lambda)},$$

proving that g is completely monotone. Since $g(0+) \leq 1$, we see that g is self-decomposable, i.e. $\pi \in \mathsf{SD}$. $\square$

Remark 5.18. In the proof one can choose $m(t)$ so that $t \cdot m(t)$ is right-continuous. Hence, Proposition 5.17 can be stated as $\pi \in \mathsf{SD}$ if, and only if, $t \cdot m(t)$ is 1-monotone. Alternatively, $\pi \in \mathsf{SD}$ if, and only if, $t \cdot m(t) = n(\log t)$ where n is 1-monotone on $\mathbb{R}$.

The concept of self-decomposability can be extended to multiple and complete self-decomposability.

Definition 5.19. (i) A completely monotone function $g \in \mathcal{CM}$ with $g(0+) \leq 1$ is said to be *p-times self-decomposable, $p \geq 2$*, if it is self-decomposable and if

$$\lambda \mapsto \frac{g(\lambda)}{g(c\lambda)} = g_c(\lambda) \text{ is } (p-1)\text{-times self-decomposable for all } c \in (0, 1).$$

(ii) A completely monotone function $g \in \mathcal{CM}$ with $g(0+) \leq 1$ is said to be *completely self-decomposable* if it is *p*-times self-decomposable for all $p \geq 1$.

In analogy with the self-decomposable case we denote by SD_p (respectively SD_∞) the family of infinitely divisible sub-probability measures π on $[0, \infty)$ such that $\mathscr{L}\pi$ is a *p*-times self-decomposable (respectively a completely self-decomposable) completely monotone function. We have the following analogue of Proposition 5.17.

Proposition 5.20. *Let $f \in \mathcal{BF}$ satisfy $\mathscr{L}\pi = e^{-f}$ and let μ be the Lévy measure of f. Then the following assertions are equivalent.*

(i) $\pi \in \mathsf{SD}_p$ *(respectively $\pi \in \mathsf{SD}_\infty$);*

(ii) μ *has a density $m(t)$ such that $t \cdot m(t) = n(\log t)$ for some p-monotone (respectively completely monotone) function n on $\mathbb{R}$.*

Proof. We prove the result by induction in p. For $p = 1$ the claim is true by Proposition 5.17 (cf. Remark 5.18). Suppose that it is true for $p - 1$, $p \geq 2$. Let $\pi \in \mathsf{SD}_p$, define $g = \mathscr{L}\pi$, and $g_c(\lambda) = g(\lambda)/g(c\lambda)$. Then by definition of *p*-times self-decomposability, g_c is a $(p - 1)$-times self-decomposable function for all $c \in (0, 1)$. Let $g_c = e^{-f_c}$. By the induction hypothesis, the Lévy measure μ_c of f_c has a density $m_c(t)$ such that $t \cdot m_c(t) = n_c(\log t)$ where n_c is $(p - 1)$-monotone on $\mathbb{R}$. Since f corresponds to a self-decomposable distribution, its Lévy measure has a density $m(t)$. Comparing the formulae (5.11) and (5.12) for f_c in the proof of Proposition 5.17 we see that $m(t) - c^{-1}m(c^{-1}t) = m_c(t)$, hence

$$tm(t) - c^{-1}tm(c^{-1}t) = tm_c(t). \tag{5.13}$$

Define $n : \mathbb{R} \to \mathbb{R}$ by $n(s) := e^s m(e^s)$ so that $tm(t) = n(\log t)$. Then (5.13) becomes

$$n(\log t) - n(\log t - \log c) = n_c(\log t) \quad \text{for all } t \in \mathbb{R},$$

or, equivalently,

$$-\Delta_{(-\log c)}n(t) = n(t) - n(t - \log c) = n_c(t) \quad \text{for all } t > 0, \, c \in (0, 1).$$

Together with the characterization from Theorem 4.11, cf. Remark 4.13, this shows that n is *p*-monotone on $\mathbb{R}$.

Conversely, assume that $f(\lambda) = b\lambda + \int_0^\infty (1 - e^{-\lambda t}) m(t) \, dt$ and the density $m(t)$ satisfies $t \, m(t) = n(\log t)$ where n is *p*-monotone on $\mathbb{R}$. For $c \in (0, 1)$ define $n_c(t) := n(t) - n(t - \log c)$, $t \in \mathbb{R}$, and $m_c(t) := t^{-1}n_c(\log t)$. Then (5.13) is valid for all $t > 0$ and all $c \in (0, 1)$. Let $f_c(\lambda) := f(\lambda) - f(c\lambda)$. Then the computation in the proof of Proposition 5.17 shows that

$$f_c(\lambda) = b(1-c)\lambda + \int_0^\infty (1 - e^{-\lambda t}) m_c(t)\, dt.$$

Again by Theorem 4.11 we see that n_c is $(p-1)$-monotone on $\mathbb{R}$ for every $c \in (0,1)$. The part of the proposition related to SD_∞ is now clear. □

Definition 5.21. A Bernstein function f belongs to the class $\mathcal{PBF}$ if the Lévy measure μ from (3.2) has a density $m(t)$ such that $t \mapsto t \cdot m(t) = n(\log t)$ where n is completely monotone on $\mathbb{R}$, i.e.

$$f(\lambda) = a + b\lambda + \int_{(0,\infty)} (1 - e^{-\lambda t}) m(t)\, dt, \tag{5.14}$$

where $a, b \geq 0$, $\int_{(0,\infty)} (1 \wedge t) m(t)\, dt < \infty$ and $t \mapsto t \cdot m(t) = n(\log t)$ where n is completely monotone on $\mathbb{R}$.

Theorem 5.22. *For a function $f : (0,\infty) \to (0,\infty)$ the following assertions are equivalent.*

(i) $f = \mathcal{L}\pi$ *where* $\pi \in \mathsf{SD}_\infty$.

(ii) $f \in \mathcal{PBF}$.

(iii) $f \in \mathcal{BF}$ *and the Lévy measure of f has a density $m(t)$ such that for every $\gamma \geq -1$, there is a function $n = n_\gamma$ which is completely monotone on $\mathbb{R}$ such that $m(t) = t^\gamma n(\log t)$.*

(iv) f *is of the form*

$$f(\lambda) = a + b\lambda + \int_0^\infty (1 - e^{-\lambda t}) \left(\int_{(0,1)} t^{-\alpha-1}\, \nu(d\alpha) \right) dt \tag{5.15}$$

where $a, b \geq 0$ and ν is a measure on $(0,1)$ satisfying

$$\int_{(0,1)} \left(\alpha^{-1} + (1-\alpha)^{-1} \right) \nu(d\alpha) < \infty. \tag{5.16}$$

(v) f *is of the form*

$$f(\lambda) = \int_{[0,1]} \lambda^\alpha \kappa(d\alpha), \tag{5.17}$$

where κ is a finite measure on $[0,1]$.

Additionally, SD_∞ is the smallest class of sub-probability measures on $[0,\infty)$ which contains all stable laws and is closed under convolution and vague limits.

Proof. (i)⇔(ii) This is proved in Proposition 5.20.

(ii)⇒(iv) Let $m(t) = t^{-1}n(\log t)$ where n is completely monotone on $\mathbb{R}$. By (1.12), $n(t) = \int_{(0,\infty)} e^{-ts} v(ds)$ for all $t \in \mathbb{R}$ where ρ is a measure on $(0,\infty)$. In particular,

$$m(t) = t^{-1}n(\log t) = \int_{(0,\infty)} t^{-1-s} v(ds).$$

Since $m(t)$ is the density of the Lévy measure of f, it holds that $\int_0^1 t\, m(t)\, dt < \infty$ and $\int_1^\infty m(t)\, dt < \infty$. By Fubini's theorem, the first condition gives that

$$\int_{(0,\infty)} \left(\int_0^1 t^{-s}\, dt \right) v(ds) < \infty,$$

implying that $v[1,\infty) = 0$ (otherwise the inner integral would diverge), as well as $\int_{(0,1)}(1-s)^{-1} v(ds) < \infty$ (by computing the inner integral). Similarly, from the second condition on $m(t)$ we conclude that $\int_{(0,\infty)} s^{-1} v(ds) < \infty$. Therefore, $m(t) = \int_{(0,1)} t^{-s-1} v(ds)$ and $\int_{(0,1)}(s^{-1} + (1-s)^{-1}) v(ds) < \infty$.

(iv)⇒(iii) Note first that (5.16) guarantees that $m(t) := \int_{(0,1)} t^{-\alpha-1} v(d\alpha)$ is the density of a Lévy measure thus implying that $f \in \mathcal{BF}$. Fix any $\gamma \geq -1$ and let ρ be the image measure of v with respect to the mapping $\alpha \mapsto \alpha + \gamma + 1$ so that $m(t) = \int_{(\gamma+1,\gamma+2)} t^{\gamma-s} \rho(ds)$. Define $n(t) := \int_{(\gamma+1,\gamma+2)} e^{-ts} \rho(ds)$, $t \in \mathbb{R}$. Then $m(t) = t^\gamma n(\log t)$ for all $t > 0$ implying that n is well defined, and hence completely monotone on $\mathbb{R}$.

(iii)⇒(ii) This follows immediately by taking $\gamma = -1$.

(iv)⇔(v) Suppose that (v) holds and write

$$f(\lambda) = \int_{[0,1]} \lambda^\alpha \kappa(d\alpha)$$

$$= \kappa\{0\} + \kappa\{1\}\lambda + \int_{(0,1)} \lambda^\alpha \kappa(d\alpha)$$

$$= \kappa\{0\} + \kappa\{1\}\lambda + \int_{(0,1)} \left(\int_0^\infty (1-e^{-\lambda t}) \frac{\alpha}{\Gamma(1-\alpha)} t^{-\alpha-1}\, dt \right) \kappa(d\alpha)$$

$$= \kappa\{0\} + \kappa\{1\}\lambda + \int_0^\infty (1-e^{-\lambda t}) m(t)\, dt$$

with

$$m(t) = \int_{(0,1)} \frac{\alpha}{\Gamma(1-\alpha)} t^{-\alpha-1}\kappa(d\alpha).$$

Let $v(d\alpha) := \frac{\alpha}{\Gamma(1-\alpha)} \kappa(d\alpha)$. Since $\lim_{\alpha\to 1}(1-\alpha)\Gamma(1-\alpha) = 1$, it follows that

$$\int_{(0,1)} \left(\frac{1}{\alpha} + \frac{1}{1-\alpha} \right) v(d\alpha) = \int_{(0,1)} \left(\frac{1}{\alpha} + \frac{1}{1-\alpha} \right) \frac{\alpha}{\Gamma(1-\alpha)} \kappa(d\alpha) < \infty.$$

The converse direction follows by retracing the above steps.

To prove the last statement of the theorem, consider the family of all Bernstein functions f which are of the form (5.17) with a finite measure κ on $[0, 1]$. This family is a convex cone containing λ^α for all $\alpha \in [0, 1]$. We show now that it is closed under pointwise limits. Let $f_n(\lambda) = \int_{[0,1]} \lambda^\alpha \kappa_n(d\alpha)$ and assume that $f(\lambda) = \lim_{n\to\infty} f_n(\lambda)$ for all $\lambda > 0$; so, $f(1) = \lim_{n\to\infty} f_n(1) = \lim_{n\to\infty} \kappa_n[0, 1]$. Hence, the family of measures $(\kappa_n)_{n\geq 1}$ is bounded. Without loss of generality we may assume that $(\kappa_n)_{n\geq 1}$ converges weakly to a bounded measure κ. Since the function $\alpha \mapsto \lambda^\alpha$ is bounded and continuous on $[0, 1]$ for every $\lambda > 0$, it follows that $f(\lambda) = \int_{[0,1]} \lambda^\alpha \kappa(d\alpha)$. This proves that the family of Bernstein functions of the form (5.17) contains $\mathcal{PBF}$. The converse inclusion follows from the fact that every probability measure on $[0, 1]$ is a weak limit of a sequence of convex combinations of point masses. □

Remark 5.23. (i) By taking $\gamma = 0$ we see that $\mathscr{L}\pi = e^{-f}$ with $\pi \in \mathrm{SD}_\infty$ if, and only if, the Lévy measure of f has a density of the form $m(t) = n(\log t)$ where n is a completely monotone function on $\mathbb{R}$.

(ii) The equivalence (iii) and (iv) from Theorem 5.22 can be, with a small modification, extended to a slightly larger range of parameters γ.
Indeed, almost the same proof gives that f satisfies condition (iii) with $\gamma \in (-2, -1)$ if, and only if, f is of the form (5.15) where ν is a measure on $(-\gamma - 1, 1)$ satisfying $\int_{(-\gamma-1,1)} (1-\alpha)^{-1} \nu(d\alpha) < \infty$.

(iii) The representation (5.17) is called the *power representation* of f. The representing measure κ and the Lévy triplet of f are related by $a = \kappa\{0\}$, $b = \kappa\{1\}$ and

$$m(t) = \int_{(0,1)} \frac{\alpha}{\Gamma(1-\alpha)} t^{-\alpha-1} \kappa(d\alpha). \tag{5.18}$$

Now we define the potential measure of a vaguely continuous convolution semigroup $(\mu_t)_{t\geq 0}$ of sub-probability measures on $[0, \infty)$. Notice that

$$\int_0^\infty \int_{[0,\infty)} e^{-s\lambda} \mu_t(ds)\, dt = \int_0^\infty e^{-tf(\lambda)}\, dt = \frac{1}{f(\lambda)}. \tag{5.19}$$

Since every continuous function $u(s)$ with compact support in $[0, \infty)$ can be dominated by a multiple of $e^{-s\lambda}$, this implies that the vague integral $U := \int_0^\infty \mu_t\, dt$ exists, see Remark A.3. U is called the *potential measure* of the convolution semigroup $(\mu_t)_{t\geq 0}$ or, equivalently, of the corresponding killed subordinator $(S_t)_{t\geq 0}$. Moreover,

$$U(A) = \int_0^\infty \mu_t(A)\, dt = \mathbb{E} \int_0^\infty \mathbb{1}_{\{S_t \in A\}}\, dt \quad \text{for all Borel sets } A \subset [0, \infty).$$

It is a simple consequence of Fubini's theorem and (5.19) that

$$\mathscr{L}(U; \lambda) = \int_{[0,\infty)} e^{-s\lambda} \left(\int_0^\infty \mu_t(ds)\, dt \right) = \frac{1}{f(\lambda)}, \tag{5.20}$$

where $f \in \mathcal{BF}$ is the Bernstein function corresponding to the convolution semigroup $(\mu_t)_{t \geq 0}$. This shows, in particular, that for all $\lambda > 0$

$$\infty > \mathscr{L}(U; \lambda) = \int_{[0,\infty)} e^{-\lambda t} \, U(dt) \geq \int_{[0,\lambda]} e^{-\lambda t} \, U(dt) \geq e^{-\lambda^2} U[0, \lambda],$$

proving that U is finite on bounded sets.

In a similar way one defines for $\lambda > 0$ the λ-*potential measure* as the vague integral

$$U_\lambda := \int_0^\infty e^{-\lambda t} \mu_t \, dt. \tag{5.21}$$

Definition 5.24. A function $f : (0, \infty) \to (0, \infty)$ is said to be a *potential* if $f = 1/g$ where $g \in \mathcal{BF}^* = \mathcal{BF} \setminus \{0\}$. The set of all potentials will be denoted by $\mathcal{P}$.

By (5.20), $\mathcal{P}$ consists of Laplace transforms of potential measures. In particular, $\mathcal{P} \subset \mathcal{CM}$, and $\mathcal{P}$ consists of exactly those completely monotone functions f which have the property that $1/f \in \mathcal{BF}$, i.e.

$$\mathcal{P} = \left\{ \frac{1}{f} : f \in \mathcal{BF}^* \right\} = \left\{ g \in \mathcal{CM} : \frac{1}{g} \in \mathcal{BF} \right\}. \tag{5.22}$$

Proposition 5.25. *Let $g \in \mathcal{P}$ be a potential. Then g is logarithmically completely monotone and hence there exists an extended Bernstein function f such that $g = e^{-f}$.*

Proof. Let $g \in \mathcal{P}$. Then $h = 1/g \in \mathcal{BF}$ and

$$-(\log g)' = (\log h)' = h' \frac{1}{h} = h' g.$$

Since both h' and g are in $\mathcal{CM}$, it follows that their product $-(\log g)' \in \mathcal{CM}$. The second part of the statement follows from Theorem 5.11.   □

The converse does not hold. Indeed, $f(\lambda) = \lambda$ is a Bernstein function and, therefore, $g(\lambda) := e^{-f(\lambda)} = e^{-\lambda}$ is logarithmically completely monotone. Since $1/g(\lambda) = e^\lambda \notin \mathcal{BF}$, g is not a potential.

Remark 5.26. Proposition 5.25 can be restated as follows. For every $h \in \mathcal{BF}^*$, there exists an extended Bernstein function f such that $h = e^f$. In other words, $\log h$ is an extended Bernstein function.

In Corollary 3.8 (iii) we have proved that the set $\mathcal{BF}$ is closed under composition. Now we are going to give an alternative proof of this fact by explicitly producing the corresponding convolution semigroup. Suppose that $(\mu_t)_{t \geq 0}$ and $(\nu_t)_{t \geq 0}$ are two convolution semigroups on $[0, \infty)$ with the corresponding Bernstein functions

$$f(\lambda) = a + b\lambda + \int_{(0,\infty)} (1 - e^{-\lambda t}) \, \mu(dt)$$

and

$$g(\lambda) = \alpha + \beta\lambda + \int_{(0,\infty)} (1 - e^{-\lambda t}) \, v(dt).$$

Let us define a new family of measures $(\eta_t)_{t \geq 0}$ by

$$\eta_t(dr) = \int_{[0,\infty)} v_s(dr) \, \mu_t(ds) \tag{5.23}$$

where the integral is a vague integral in the sense of Remark A.3.

Theorem 5.27. *The family* $(\eta_t)_{t \geq 0}$ *is a convolution semigroup of sub-probability measures on* $[0, \infty)$ *whose corresponding Bernstein function is equal to* $f \circ g$. *Moreover,*

$$(f \circ g)(\lambda) = f(\alpha) + \beta b \lambda + \int_{(0,\infty)} (1 - e^{-\lambda t}) \, \eta(dt) \tag{5.24}$$

and the Lévy measure is given by the vague integral

$$\eta(dt) = b v(dt) + \int_{(0,\infty)} v_s(dt) \, \mu(ds). \tag{5.25}$$

Remark 5.28. (i) Once formula (5.24) is established, it follows immediately that η is a Lévy measure, i.e. $\int_{(0,\infty)} (1 \wedge t) \, \eta(dt) < \infty$.

(ii) The convolution semigroup $(\eta_t)_{t \geq 0}$ is called *subordinate* to $(v_t)_{t \geq 0}$ by $(\mu_t)_{t \geq 0}$. If $(S_t)_{t \geq 0}$ and $(T_t)_{t \geq 0}$ are independent subordinators corresponding to $(\mu_t)_{t \geq 0}$ and $(v_t)_{t \geq 0}$, respectively, then the subordinator corresponding to $(\eta_t)_{t \geq 0}$ is the process $(T(S_t))_{t \geq 0}$. This process, called a subordinate to T by the subordinator S, is a particular form of a stochastic time change: $T(S_t)(\omega) = T_{S_t(\omega)}(\omega)$.

We will extend the concept of subordination to strongly continuous semigroups on a Banach space in Section 13.2.

Proof of Theorem 5.27. We check first that each η_t is a sub-probability measure. Indeed,

$$\eta_t[0, \infty) = \int_{[0,\infty)} v_s[0, \infty) \, \mu_t(ds) \leq \int_{[0,\infty)} \mu_t(ds) \leq 1.$$

Let us compute the Laplace transform of the measure η_t:

$$\mathscr{L}(\eta_t; \lambda) = \int_{[0,\infty)} e^{-\lambda u} \, \eta_t(du) = \int_{[0,\infty)} \left(\int_{[0,\infty)} e^{-\lambda u} \, v_s(du) \right) \mu_t(ds)$$

$$= \int_{[0,\infty)} \mathscr{L}(v_s; \lambda) \, \mu_t(ds) = \int_{[0,\infty)} e^{-sg(\lambda)} \, \mu_t(ds)$$

$$= \mathscr{L}(\mu_t; g(\lambda)) = e^{-tf(g(\lambda))}.$$

This proves also that

$$\mathcal{L}(\eta_{t+s}; \lambda) = e^{-(t+s)f(g(\lambda))}$$
$$= e^{-tf(g(\lambda))} e^{-sf(g(\lambda))}$$
$$= \mathcal{L}(\eta_t; \lambda) \mathcal{L}(\eta_s; \lambda),$$

as well as vague continuity since $t \mapsto \mathcal{L}(\eta_t; \lambda) = e^{-tf(g(\lambda))}$ is continuous. In order to obtain formula (5.24), first note that $e^{-\alpha s} = \nu_s[0, \infty)$, and

$$\int_{(0,\infty)} (1 - e^{-t\lambda}) \int_{(0,\infty)} \nu_s(dt)\, \mu(ds)$$

$$= \int_{(0,\infty)} \left(\int_{(0,\infty)} (1 - e^{-t\lambda}) \nu_s(dt) \right) \mu(ds)$$

$$= \int_{(0,\infty)} (e^{-\alpha s} - \mathcal{L}(\nu_s; \lambda))\, \mu(ds)$$

$$= \int_{(0,\infty)} (e^{-s\alpha} - e^{-sg(\lambda)})\, \mu(ds)$$

$$= \int_{(0,\infty)} (1 - e^{-sg(\lambda)})\, \mu(ds) - \int_{(0,\infty)} (1 - e^{-\alpha s})\, \mu(ds)$$

$$= \int_{(0,\infty)} (1 - e^{-sg(\lambda)})\, \mu(ds) - (f(\alpha) - a - b\alpha).$$

Every function $u \in C_c(0, \infty)$ can be estimated by $|u(t)| \le c_{\lambda,u}(1 - e^{-t\lambda})$, $t > 0$, with a suitable constant $c_{\lambda,u} > 0$. Therefore, the above calculation shows that the vague integral (5.25) exists and defines a measure η on $(0, \infty)$, cf. Remark A.3. Now we have

$$f(g(\lambda)) = a + bg(\lambda) + \int_{(0,\infty)} (1 - e^{-tg(\lambda)})\, \mu(dt)$$

$$= a + b \left(\alpha + \beta\lambda + \int_{(0,\infty)} (1 - e^{-t\lambda}) \nu(dt) \right)$$

$$+ \int_{(0,\infty)} (1 - e^{-t\lambda}) \int_{(0,\infty)} \nu_s(dt)\, \mu(ds) + (f(\alpha) - a - b\alpha)$$

$$= f(\alpha) + \beta b\lambda + \int_{(0,\infty)} (1 - e^{-t\lambda}) \eta(dt). \qquad \qquad \square$$

Comments 5.29. A good treatment of convolution semigroups and their potential theory is the monograph by Berg and Forst [39]; the classic exposition by Feller, [118], covers the material from a probabilistic angle and is still readable. Subordinators are treated in most books on Lévy processes, for example in Bertoin [47] and his St.-Flour course notes [49].

Infinitely divisible and self-decomposable functions can be found in many
books on probability theory. The monograph by Steutel and van Harn [339, Sections I.5, V.2] is currently the most comprehensive study and has dedicated chapters
for one-sided laws, but Gnedenko and Kolmogorov [130], Lukacs [251], Loève [246,
vol. 1, Chapter VI], Petrov [289, pp. 82–87], [290, Chapters 3,4], Rossberg, Jesiak,
Siegel [309, Sections 7, 11–13] and Sato [314, Chapter 3] contain most of the material mentioned in Remark 5.12. That all infinitely divisible functions in $\mathcal{CM}$ are of
the form $\exp(-\mathcal{BF})$ was discovered by Schoenberg [325, Theorem 9, p. 835]. An
alternative short proof is in Horn [169, Theorem 4.4]; note that one has to have $[0, \infty]$
as support of the measure $d\mu$ and not, as stated in [169], $(0, \infty)$. The name *logarithmically completely monotone* appears in 2004 for the first time in [299] and [300],
see also the comments in [34]. Theorem 5.11 (ii) and the characterization of extended
Bernstein functions in Remark 5.9 seem to be new.

Our presentation of the stable distributions is from Sato [314, Examples 24.12,
21.7], the representation of the Laplace exponents of self-decomposable distributions
is from Steutel and van Harn [339, Theorem 2.11, p. 231]. The short proofs of Proposition 5.15 and Corollary 5.16 seem to be new. The proof of Proposition 5.17, which
is due to Lévy [243, Section VII.55], is adapted from Sato [314, p. 95].

Self-decomposable distributions were characterized by P. Lévy as limit distributions of normed partial sums of sequences of independent random variables, see [242]
and [243, pp. 192–193]; Lévy mentions that Khintchine drew his attention to this
problem in a letter in 1936. The name self-decomposable appears for the first time in
1955 in the first edition of Loève's book on probability theory [246], the notation L
goes back to Khintchine [207], see the comment in Gnedenko and Kolmogorov [130,
Section 29].

Multiple and complete self-decomposability was introduced by Urbanik [357, 358]
for distributions on $\mathbb{R}$; Sato [315] reformulated and extended it to $\mathbb{R}^d$. The standard
notation for p-times and completely self-decomposable distributions is L_{p-1} and L_∞,
respectively. In the 1980s Nguyen [277, 278, 279] obtained the continuous parameter
extension of these classes. The names *multiple* and *complete self-decomposability* are
from [276]. The multidimensional version of Proposition 5.20 is from [315] where it
is shown that almost every (w.r.t. to the spherical part of the Lévy measure) radial part
of the Lévy measure has the given representation. The analogue of the representation
(5.15) is from [315]. The last part of Theorem 5.22 appears in [357] and [315].

The best sources for potential theory of convolution semigroups are Berg and Forst
[39, Chapter III] and the original papers by F. Hirsch from the early 1970s. A more
probabilistic approach can be found in Bertoin [47, Chapter II, III]. A somewhat hard
to read but otherwise most comprehensive treatment is the monograph by Dellacherie
and Meyer [93, 94].

M. Itô [181, Section 2], see also [180], gives the following interesting characterization of potential measures and potentials: U is a potential measure on $[0, \infty)$ if, and
only if, for all vaguely continuous convolution semigroups $(\mu_t)_{t \geq 0}$ of measures on,

say $\mathbb{R}^d$, the measure $\int_0^\infty \mu_t \, U(dt)$ is of the form $\int_0^\infty \nu_t \, dt$ for some other vaguely continuous convolution semigroup $(\nu_t)_{t \geq 0}$ of measures on $\mathbb{R}^d$. Since the potential measures on $\mathbb{R}^d$ are of the form $\int_0^\infty \mu_t \, dt$, this means that $\mathcal{P}$ or the potential measures on $[0, \infty)$ are characterized by the fact that they operate on the potentials in $\mathbb{R}^d$. It is also shown in [181, pp. 118–119] that U is a potential measure on $[0, \infty)$ if, and only if, it satisfies the domination principle, i.e. if for all continuous, positive and compactly supported $v, w \in C_c^+[0, \infty)$

$$U \star v \leq U \star w \quad \text{on supp } v \quad \text{implies} \quad U \star v \leq U \star w \quad \text{on } [0, \infty).$$

Subordination was introduced by Bochner in the short note [62], see also his monograph [63, Chapter 4.4]. A rigorous functional analytic and stochastic account is in the paper by Nelson [270].

Chapter 6

Complete Bernstein functions

The class of complete Bernstein functions has been used throughout the literature in many branches of mathematics but under various names and for very different reasons, e.g. as *Pick* or *Nevanlinna functions* in (complex) interpolation theory, *Löwner* or *operator monotone functions* in functional analysis, or as *class (S)* in the Russian literature on complex function theory in a half-plane.

6.1 Representation of complete Bernstein functions

Many explicitly known Bernstein functions are actually complete Bernstein functions, which is partly due to the fact that the measure μ in the representation formula (3.2) for complete Bernstein functions has a nice density. We take this as our starting point and show that our definition coincides with other known approaches.

Definition 6.1. A Bernstein function f is said to be a *complete Bernstein function* if its Lévy measure μ in (3.2) has a completely monotone density $m(t)$ with respect to Lebesgue measure,

$$f(\lambda) = a + b\lambda + \int_{(0,\infty)} (1 - e^{-\lambda t}) m(t) \, dt. \tag{6.1}$$

We will use $\mathcal{CBF}$ to denote the collection of all complete Bernstein functions.

Denote by $\mathbb{H}^\uparrow$ and $\mathbb{H}^\downarrow$ the open upper and lower complex half-planes, i.e. the sets of all $z \in \mathbb{C}$ with $\operatorname{Im} z > 0$ and $\operatorname{Im} z < 0$, respectively.

Theorem 6.2. *Suppose that f is a non-negative function on $(0, \infty)$. Then the following conditions are equivalent.*

(i) $f \in \mathcal{CBF}$.

(ii) *The function $\lambda \mapsto f(\lambda)/\lambda$ is in $\mathcal{S}$.*

(iii) *There exists a Bernstein function g such that*

$$f(\lambda) = \lambda^2 \mathscr{L}(g; \lambda), \quad \lambda > 0.$$

(iv) *f has an analytic continuation to $\mathbb{H}^\uparrow$ such that $\operatorname{Im} f(z) \geq 0$ for all $z \in \mathbb{H}^\uparrow$ and such that the limit $f(0+) = \lim_{(0,\infty) \ni \lambda \to 0} f(\lambda)$ exists and is real.*

(v) *f has an analytic continuation to the cut complex plane* $\mathbb{C} \setminus (-\infty, 0]$ *such that* $\operatorname{Im} z \cdot \operatorname{Im} f(z) \geq 0$ *and such that the limit* $f(0+) = \lim_{(0,\infty) \ni \lambda \to 0} f(\lambda)$ *exists and is real.*

(vi) *f has an analytic continuation to* $\mathbb{H}^\uparrow$ *which is given by*

$$f(z) = a + bz + \int_{(0,\infty)} \frac{z}{z+t} \, \sigma(dt)$$

where $a, b \geq 0$ *are non-negative constants and* σ *is a measure on* $(0, \infty)$ *such that* $\int_{(0,\infty)} (1+t)^{-1} \sigma(dt) < \infty$.

Proof. (i)$\Leftrightarrow$(ii) If m is completely monotone, then we know from Theorem 1.4 that $m = \mathcal{L}\nu$ for some measure ν on $[0, \infty)$. By Fubini's theorem

$$\int_{(0,\infty)} (1 - e^{-\lambda t}) m(t) \, dt = \int_{[0,\infty)} \int_{(0,\infty)} (1 - e^{-\lambda t}) e^{-ts} \, dt \, \nu(ds)$$

$$= \int_{[0,\infty)} \left(\frac{1}{s} - \frac{1}{\lambda + s} \right) \nu(ds)$$

$$= \int_{[0,\infty)} \frac{\lambda}{\lambda + s} s^{-1} \nu(ds).$$

The expression on the right-hand side converges if, and only if, $\nu\{0\} = 0$, i.e. if ν is supported in $(0, \infty)$, and if $\int_{(0,1)} s^{-1} \nu(ds) + \int_{(1,\infty)} s^{-2} \nu(ds) < \infty$. This shows that

$$\frac{f(\lambda)}{\lambda} = \frac{a}{\lambda} + b + \int_{(0,\infty)} \frac{1}{\lambda + s} s^{-1} \nu(ds)$$

is indeed a Stieltjes function. The converse direction follows easily by retracing the above steps.

(ii)$\Rightarrow$(iii) Note that $h(\lambda) \equiv a$ has $\mathcal{L}(a; \lambda) = \frac{a}{\lambda}$ as Laplace transform. Moreover, we have $\lambda \mathcal{L}(h; \lambda) = h(0+) + \mathcal{L}(h'; \lambda)$. Thus,

$$\frac{f(\lambda)}{\lambda} = b + \frac{a}{\lambda} + \mathcal{L}^2(s^{-1} \nu(ds); \lambda) = b + \mathcal{L}\left(a + \mathcal{L}[s^{-1} \nu(ds)]; \lambda \right)$$

$$= \lambda \mathcal{L}(g; \lambda),$$

where we set $g' := a + \mathcal{L}(s^{-1} \nu(ds)) \in \mathcal{CM}$ so that $g \in \mathcal{BF}$, cf. the remark following Definition 3.1.

(iii)$\Rightarrow$(i) If $g(t) = \alpha + \beta t + \int_{(0,\infty)} (1 - e^{-ts}) \nu(ds)$, we find

$$\lambda^2 \mathcal{L}(g; \lambda) = \lambda^2 \left(\frac{\alpha}{\lambda} + \frac{\beta}{\lambda^2} + \int_0^\infty e^{-\lambda t} \int_{(0,\infty)} (1 - e^{-ts}) \nu(ds) \, dt \right)$$

$$= \alpha\lambda + \beta + \int_{(0,\infty)} \lambda^2 \left(\frac{1}{\lambda} - \frac{1}{\lambda+s}\right) v(ds)$$

$$= \alpha\lambda + \beta + \int_{(0,\infty)} s^2 \left(\frac{1}{s} - \frac{1}{\lambda+s}\right) v(ds)$$

$$= \alpha\lambda + \beta + \int_0^\infty (1 - e^{-\lambda t}) \int_{(0,\infty)} e^{-ts} s^2 v(ds)\, dt$$

and (i) follows.

(ii)$\Rightarrow$(iv) Since $f(\lambda) = a + b\lambda + \int_{(0,\infty)} \frac{\lambda}{\lambda+t}\, \sigma(dt)$, the natural way to extend f is to replace λ by $z \in \mathbb{H}^\uparrow$. If we write $z = \lambda + i\eta$, then

$$\frac{z}{z+t} = \frac{z\bar{z} + zt}{|z+t|^2} = \frac{\lambda^2 + t\lambda + \eta^2}{(\lambda+t)^2 + \eta^2} + i\, \frac{t\eta}{(\lambda+t)^2 + \eta^2}.$$

Moreover, for any fixed $z = \lambda + i\eta \in \mathbb{H}^\uparrow$ we have

$$\frac{\lambda^2 + t\lambda + \eta^2}{(\lambda+t)^2 + \eta^2} \asymp \frac{1}{1+t} \quad \text{and} \quad \frac{t\eta}{(\lambda+t)^2 + \eta^2} \asymp \frac{t}{1+t^2}$$

($\phi(t) \asymp \psi(t)$ means that there exist two absolute constants $c_2 > c_1 > 0$ such that $c_1 \phi(t) \leq \psi(t) \leq c_2 \phi(t)$) which entails that the integral

$$\int_{(0,\infty)} \frac{z}{z+t}\, \sigma(dt) = \int_{(0,\infty)} \frac{\lambda^2 + t\lambda + \eta^2}{(\lambda+t)^2 + \eta^2}\, \sigma(dt)$$

$$+ i \int_{(0,\infty)} \frac{t\eta}{(\lambda+t)^2 + \eta^2}\, \sigma(dt)$$

converges locally uniformly and defines an analytic function on $\mathbb{H}^\uparrow$ with positive imaginary part whenever $\eta = \operatorname{Im} z \geq 0$. All these remain true if we add $a + bz$.

The above estimates and the integration properties of the measure σ also show that

$$\lim_{\mathbb{H}^\uparrow \ni z \to 0} \int_{(0,\infty)} \frac{z}{z+t}\, \sigma(dt) = \lim_{(0,\infty) \ni \lambda \to 0} \int_{(0,\infty)} \frac{\lambda}{\lambda+t}\, \sigma(dt) = 0$$

so that the limit $\lim_{(0,\infty)\cup\mathbb{H}^\uparrow \ni z \to 0} f(z) = a$ exists and is real.

(iv)$\Rightarrow$(v) This is an easy consequence of the Schwarz reflection principle: since $f|_{(0,\infty)}$ is real-valued, we may use $f(\bar{z}) := \overline{f(z)}$, $z \in \mathbb{H}^\uparrow$, to extend f across $(0,\infty)$ to the lower half-plane. By the reflection principle this gives an analytic function on $\mathbb{C} \setminus (-\infty, 0] = \mathbb{H}^\uparrow \cup (0,\infty) \cup \mathbb{H}^\downarrow$. Since $\operatorname{Im} \overline{f(z)} = -\operatorname{Im} f(z) < 0$ for $z \in \mathbb{H}^\uparrow$, we get $\operatorname{Im} f(z) > 0$ or $\operatorname{Im} f(z) < 0$ for $z \in \mathbb{H}^\uparrow$ or $z \in \mathbb{H}^\downarrow$ respectively, which can be expressed by $\operatorname{Im} z \cdot \operatorname{Im} f(z) \geq 0$. Since $\lim_{\mathbb{H}^\uparrow \ni z \to 0} f(z)$ exists and is real, the above construction automatically shows that the limit $\lim_{\mathbb{C}\setminus(-\infty,0]\ni z \to 0} f(z)$ exists.

(v)⇒(vi) Note that for $p, q > 0$ the function

$$f_{p,q}(z) := \frac{p(f(z) + iq)}{p + (f(z) + iq)}$$

is analytic in $\mathbb{H}^{\uparrow}$. Write $f(z) = u(z) + iv(z)$ and observe that $v(z) \geq 0$ on $\mathbb{H}^{\uparrow}$. Then

$$f_{p,q}(z) = \frac{p(u(z) + p)u(z) + p(v(z) + q)^2}{(u(z) + p)^2 + (v(z) + q)^2} + \frac{i\, p^2(v(z) + q)}{(u(z) + p)^2 + (v(z) + q)^2}.$$

This shows that $f_{p,q}$ maps $\mathbb{H}^{\uparrow}$ into itself. Moreover, on $\mathbb{H}^{\uparrow}$

$$|f_{pq}| = \frac{p|f + iq|}{|p + (f + iq)|} \leq \frac{p|p + f + iq| + p^2}{|p + (f + iq)|}$$

$$\leq p + \frac{p^2}{|\mathrm{Im}\,(p + (f + iq))|} \leq p + \frac{p^2}{q}.$$

Now set $p = n$ and $q = 1/n, n \in \mathbb{N}$, and write

$$f_n(z) := f_{n,1/n}(z) = \frac{f(z) + \frac{i}{n}}{1 + \frac{1}{n}f(z) + \frac{i}{n^2}}.$$

Thus, f_n is analytic on $\mathbb{H}^{\uparrow}$, bounded by $2n^3$ and $\lim_{n\to\infty} f_n(z) = f(z)$. We want to apply Cauchy's theorem. For this we fix $z \in \mathbb{H}^{\uparrow}$ and consider the following integration contour Γ_R (in counterclockwise orientation) consisting of a circular arc of radius R extending from $\omega = \theta$ to $\omega = \pi - \theta$. The arc is closed by a straight line,

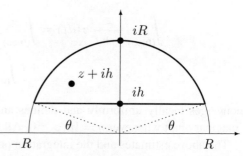

parallel to the x-axis; the distance between this line and the x-axis is $h = h(\theta, R)$. The parameters R and θ are chosen in such a way that $z + ih$ lies inside Γ_R and $\bar{z} + ih$ outside of Γ_R but not on the real axis. Using Cauchy's formula we find

$$f_n''(z + ih) = \frac{2}{2\pi i}\int_{\Gamma_R} \frac{f_n(\zeta)}{(\zeta - z - ih)^3}\, d\zeta$$

$$= \frac{1}{\pi i}\int_{-R\cos\theta}^{R\cos\theta} \frac{f_n(t + ih)}{(t - z)^3}\, dt + \frac{1}{\pi}\int_{\theta}^{\pi-\theta} \frac{Re^{i\omega}\, f_n(Re^{i\omega})}{(Re^{i\omega} - z - ih)^3}\, d\omega.$$

Since f_n is bounded, the second integral converges to 0 as $R \to \infty$. Therefore,

$$f_n''(z + ih) = \frac{1}{\pi i}\int_{-\infty}^{\infty} \frac{f_n(t + ih)}{(t - z)^3}\, dt.$$

A similar calculation where we replace $z + ih$ by $\bar{z} + ih$ which is not surrounded by Γ_R shows

$$\bar{0} = \overline{\frac{1}{\pi i} \int_{-\infty}^{\infty} \frac{f_n(t + ih)}{(t - \bar{z})^3} \, dt} = -\frac{1}{\pi i} \int_{-\infty}^{\infty} \frac{\overline{f_n(t + ih)}}{(t - z)^3} \, dt.$$

Adding these two equalities yields

$$f_n''(z + ih) = \frac{2}{\pi} \int_{-\infty}^{\infty} \frac{\operatorname{Im} f_n(t + ih)}{(t - z)^3} \, dt = \frac{2}{\pi} \int_{-\infty}^{\infty} \frac{\operatorname{Im} f_n(-s + ih)}{(s + z)^3} \, ds. \quad (6.2)$$

Since $\frac{1}{2} \frac{d^2}{dz^2} (s + z)^{-1} = (s + z)^{-3}$ we find by integrating twice that for every $\epsilon > 0$

$$f_n(i + ih) - f_n(i\epsilon + ih) = \frac{1}{\pi} \int_{-\infty}^{\infty} \frac{i(1 - \epsilon)}{(i + s)(i\epsilon + s)} \operatorname{Im} f_n(-s + ih) \, ds.$$

Therefore,

$$\operatorname{Im} f_n(i + ih) - \operatorname{Im} f_n(i\epsilon + ih)$$
$$= \frac{1}{\pi} \int_{-\infty}^{\infty} \frac{(1 - \epsilon)(s^2 - \epsilon)}{(1 + s^2)(\epsilon^2 + s^2)} \operatorname{Im} f_n(-s + ih) \, ds$$
$$= \frac{1 - \epsilon}{\pi} \left[\int_{-\infty}^{\infty} \frac{s^2 \operatorname{Im} f_n(-s + ih)}{(1 + s^2)(\epsilon^2 + s^2)} \, ds - \int_{-\infty}^{\infty} \frac{1}{(1 + t^2)} \frac{\operatorname{Im} f_n(-\epsilon t + ih)}{1 + (\epsilon t)^2} \, dt \right]$$

where we used the change of variables $s = \epsilon t$ in the last step. Using dominated convergence we get, as $\epsilon \to 0$,

$$\operatorname{Im} f_n(i + ih) - \operatorname{Im} f_n(ih)$$
$$= \frac{1}{\pi} \int_{-\infty}^{\infty} \frac{\operatorname{Im} f_n(-s + ih)}{1 + s^2} \, ds - \frac{1}{\pi} \int_{-\infty}^{\infty} \frac{\operatorname{Im} f_n(ih)}{1 + t^2} \, dt$$
$$= \frac{1}{\pi} \int_{-\infty}^{\infty} \frac{\operatorname{Im} f_n(-s + ih)}{1 + s^2} \, ds - \operatorname{Im} f_n(ih).$$

This and an application of Fatou's lemma show that

$$\operatorname{Im} f(i + ih) = \lim_{n \to \infty} \operatorname{Im} f_n(i + ih)$$
$$= \lim_{n \to \infty} \frac{1}{\pi} \int_{-\infty}^{\infty} \frac{\operatorname{Im} f_n(-s + ih)}{1 + s^2} \, ds$$
$$\geq \frac{1}{\pi} \int_{-\infty}^{\infty} \frac{\operatorname{Im} f(-s + ih)}{1 + s^2} \, ds. \quad (6.3)$$

We conclude that the family of measures $\{(1 + s^2)^{-1} \operatorname{Im} f_n(-s + ih) \, ds\}_{n \in \mathbb{N}}$ is vaguely bounded, hence relatively compact in the vague topology of measures, cf.

Theorem A.5; thus, every subsequence has a vaguely convergent subsequence. Note that $\lim_{n\to\infty} \operatorname{Im} f_n(-s+ih) = \operatorname{Im} f(-s+ih)$ boundedly as is easily seen from

$$\operatorname{Im} f_n(-s+ih) = \frac{n^2(\operatorname{Im} f(-s+ih)+n^{-1})}{(\operatorname{Re} f(-s+ih)+n)^2 + (\operatorname{Im} f(-s+ih)+n^{-1})^2}$$
$$\leq \frac{\operatorname{Im} f(-s+ih)+1}{(n^{-1}\operatorname{Re} f(-s+ih)+1)^2}$$
$$\leq C_K\,(\operatorname{Im} f(-s+ih)+1)$$

as long as s is in a compact set $K \subset \mathbb{R}$ and n is sufficiently large. This means that by dominated convergence

$$\lim_{n\to\infty} \int_{-\infty}^{\infty} \phi(s)\, \frac{\operatorname{Im} f_n(-s+ih)}{1+s^2}\,ds = \int_{-\infty}^{\infty} \phi(s)\, \frac{\operatorname{Im} f(-s+ih)}{1+s^2}\,ds$$

for all $\phi \in C_c(\mathbb{R})$. Thus, the limiting measure is $(1+s^2)^{-1}\operatorname{Im} f(-s+ih)\,ds$ and we get from (6.2)

$$f''(z+ih) = \frac{2}{\pi} \int_{-\infty}^{\infty} \frac{1}{(z+s)^3}\operatorname{Im} f(-s+ih)\,ds.$$

From (6.3) we deduce with a similar argument that $\{\operatorname{Im} f(-s+ih)\,ds\}_{h>0}$ is a vaguely relatively compact family of measures. For some subsequence $h_k \to 0$ we get

$$\frac{2}{\pi}\frac{\operatorname{Im} f(-s+ih_k)}{1+s^2}\,ds \xrightarrow{k\to\infty} \frac{\Pi(ds)}{1+s^2} \tag{6.4}$$

for a suitable limiting measure Π satisfying $\int_{-\infty}^{\infty} \frac{1}{1+s^2}\Pi(ds) \leq \operatorname{Im} f(i) < \infty$. We will see below in Corollary 6.3 that Π does actually not depend on the subsequence. Thus,

$$f''(z) = \int_{\mathbb{R}} \frac{1}{(z+s)^3}\Pi(ds).$$

Integrating twice, we find for $w,z \in \mathbb{H}^{\uparrow}$, with a suitable constant $b \in \mathbb{C}$,

$$f(z) - f(w) = b(z-w) + \frac{1}{2}\int_{\mathbb{R}} \frac{z-w}{(z+s)(w+s)}\Pi(ds),$$

and it is clear that this formula extends to all $z,w \in \mathbb{C}\setminus(-\infty,0]$. In particular, we get for $z = \lambda > 0$ and $w = \kappa > 0$

$$f(\lambda) - f(\kappa) = b(\lambda-\kappa) + \frac{1}{2}\int_{\mathbb{R}} \frac{\lambda-\kappa}{(\lambda+s)(\kappa+s)}\Pi(ds). \tag{6.5}$$

Since $\lim_{h\to0}\operatorname{Im} f(-s+ih) = 0$ locally boundedly on the negative half-axis $s < 0$, the measure Π is supported in $[0,\infty)$. If we set in (6.5) $\kappa = 1$ and let $\lambda \to \infty$, we find that $b \geq 0$. By Fatou's lemma and the finiteness of $f(0+) = \lim_{\kappa \to 0} f(\kappa)$ we conclude that

$$f(\lambda) - f(0+) \geq \frac{1}{2} \int_{[0,\infty)} \frac{\lambda}{\lambda + s} \frac{\Pi(ds)}{s}.$$

This shows that $\Pi\{0\} = 0$ and $\int_{(0,\infty)} \lambda(\lambda + s)^{-1} s^{-1} \Pi(ds) < \infty$; by the dominated convergence theorem we get

$$f(z) = a + bz + \int_{(0,\infty)} \frac{z}{z + s} \sigma(ds),$$

where $a = f(0+) \geq 0$, $b \geq 0$ and $\sigma(ds) = \frac{1}{2} s^{-1} \Pi(ds)$ is a measure on $(0, \infty)$ such that $\int_{(0,\infty)} (1 + s)^{-1} \sigma(ds) < \infty$.

(vi)$\Rightarrow$(ii) This is obvious if we take $z = \lambda \in (0, \infty)$.   □

Theorem 6.2 and its proof allow us to draw a few interesting conclusions.

Corollary 6.3. *The representation measure σ in Theorem* 6.2 (vi) *is given by*

$$\sigma(u, v] = \lim_{h \to 0+} \frac{1}{\pi} \int_{(u,v]} \frac{1}{s} \operatorname{Im} f(-s + ih) \, ds$$

$$= \lim_{h \to 0+} \frac{1}{\pi} \int_{(u,v]} \operatorname{Im} \frac{f(-s + ih)}{s - ih} \, ds \tag{6.6}$$

which holds for all positive continuity points $u, v > 0$ of the distribution function $t \mapsto \sigma(-\infty, t]$.

Proof. The proof of Theorem 6.2 (v)$\Rightarrow$(vi) shows that the limits appearing in (6.6) exist along some sequence $(h_k)_{k \in \mathbb{N}}$, $0 < h_k \to 0$. Since f is analytic in $\mathbb{H}^\uparrow$, the limit does not depend on the particular sequence and we conclude that σ (hence, ρ) is uniquely determined by f. Note that $a = \sigma\{0\}$, $a = f(0+) = \lim_{\lambda \to 0} f(\lambda)$ and $b = \lim_{\lambda \to \infty} f(\lambda)/\lambda$.   □

Note that (6.6) remains valid for $u < v < 0$ and shows that $\sigma(-\infty, 0) = 0$.

Remark 6.4. It follows from Theorem 6.2 and Corollary 6.3 that every complete Bernstein function $f \in \mathcal{CBF}$ can be uniquely represented in the following form

$$f(\lambda) = a + b\lambda + \int_{(0,\infty)} \frac{\lambda}{\lambda + t} \sigma(dt), \quad \lambda > 0, \tag{6.7}$$

where $a, b \geq 0$ are non-negative constants and σ is a measure on $(0, \infty)$ such that $\int_{(0,\infty)} (1 + t)^{-1} \sigma(dt) < \infty$. A comparison of the representation formulae (2.1) and (6.7) shows that for all $g \in \mathcal{S}$ the function $\lambda g(\lambda)$ is a complete Bernstein function.

The formula (6.7) is called the *Stieltjes representation* of the complete Bernstein function f, and the measure σ is called the *Stieltjes measure* of the complete Bernstein function f.

Corollary 6.5 (Angular limits; G. Julia et al.). *Let f be (the analytic extension of) a complete Bernstein function. Then the limit*

$$\lim_{R \to \infty} \frac{f(Re^{i\omega})}{Re^{i\omega}} = b$$

exists uniformly for all $\omega \in (\theta, \pi - \theta)$ and all fixed $\theta \in (0, \pi/2)$.

Proof. Let $z = Re^{i\omega}$ with $\theta < \omega < \pi - \theta$ for some fixed $\theta \in (0, \pi/2)$. From the formula in Theorem 6.2 (vi) we get

$$\frac{f(Re^{i\omega})}{Re^{i\omega}} = \frac{a}{Re^{i\omega}} + b + \int_{(0,\infty)} \frac{\sigma(dt)}{Re^{i\omega} + t}.$$

An elementary but somewhat tedious calculation shows that

$$\left| \frac{1}{Re^{i\omega} + t} \right| = \left| \frac{Re^{-i\omega} + t}{R^2 + 2Rt \cos \omega + t^2} \right| \leq \frac{R + t}{R^2 - 2Rt \cos \theta + t^2}$$

$$\leq \frac{5}{1 - \cos \theta} \frac{1}{R + t}.$$

From dominated convergence we conclude

$$\lim_{R \to \infty} \frac{f(Re^{i\omega})}{Re^{i\omega}} = b$$

uniformly for all $\omega \in (\theta, \pi - \theta)$. In fact, our consideration shows

$$\left| \frac{f(Re^{i\omega})}{Re^{i\omega}} - b \right| \leq \frac{a}{R} + \frac{5}{1 - \cos \theta} \int_{(0,\infty)} \frac{1}{R + t} \sigma(dt) \leq \frac{5}{1 - \cos \theta} \left[\frac{f(R)}{R} - b \right]$$

which is valid for all $\omega \in (-\pi + \theta, \pi - \theta)$.   □

Corollary 6.6. *Let f be (the analytic extension of) a complete Bernstein function. Then*

$$f(S_{[0,\theta]}) \subset S_{[0,\theta]} \quad \text{with} \quad S_{[0,\theta]} = \{z \in \mathbb{C} : 0 \leq \arg z \leq \theta\}, \ \theta \in (0, \pi). \quad (6.8)$$

Proof. This follows again from the formula in Theorem 6.2 (vi). Indeed, if we add $z = re^{i\theta}, w = Re^{i\phi} \in \mathbb{C}$, the argument of $w + z$ lies between ϕ and θ; a similar statement is true for integrals:

$$\arg \int f(t, z) \mu(dt) \in \left[\inf_t \arg f(t, z), \ \sup_t \arg f(t, z) \right].$$

Note that for all $t > 0$

$$\frac{z}{z + t} = \frac{re^{i\theta}}{re^{i\theta} + t} = \frac{re^{i\theta}}{r'e^{i\theta'}} \quad \text{where } 0 < \theta' < \theta.$$

Thus, $\arg(\frac{z}{z+t}) \leq \arg z$ and the same is true for $f(z)$ because of the formula in Theorem 6.2 (vi).   □

The proof of Theorem 6.2 (v)⟹(vi) uses essentially the fact that the function f is analytic and preserves the upper half-plane $\mathbb{H}^\uparrow$. This class of functions is well known in the literature.

Definition 6.7. A *Pick function* or *Nevanlinna function* or *Nevanlinna–Pick function* is an analytic function f defined on $\mathbb{H}^\uparrow$ or $\mathbb{C} \setminus \mathbb{R}$ such that f preserves the upper half-plane $\mathbb{H}^\uparrow$, i.e. $f(\mathbb{H}^\uparrow) \subset \mathbb{H}^\uparrow$.

If the analytic function f is only defined on $\mathbb{H}^\uparrow$, then we can extend it onto $\mathbb{C} \setminus \mathbb{R}$ by setting

$$F(z) := \begin{cases} \overline{f(\bar{z})} & \text{if } z \in \mathbb{H}^\downarrow; \\ f(z) & \text{if } z \in \mathbb{H}^\uparrow. \end{cases}$$

Although F is analytic, it is not a proper analytic extension since $\mathbb{H}^\uparrow$ and $\mathbb{H}^\downarrow$ are disjoint. If, however, f has a real-valued, continuous extension on some interval $(a, b) \subset \mathbb{R}$, then Schwarz' reflection principle applies and F is a proper analytic extension.

Corollary 6.8. *Let f be a Nevanlinna–Pick function. Then f has a unique integral representation*

$$f(z) = \alpha + \beta z + \int_{\mathbb{R}} \frac{zt - 1}{z + t}\, \rho(dt) \tag{6.9}$$

$$= \alpha + \beta z + \int_{\mathbb{R}} \left(\frac{t}{1 + t^2} - \frac{1}{z + t} \right) (1 + t^2)\, \rho(dt) \tag{6.10}$$

for all $z \in \mathbb{C} \setminus (-\infty, 0]$. Here ρ is a finite measure on $\mathbb{R}$, and α, β are constants satisfying $\beta \geq 0$ and $\alpha \in \mathbb{R}$.
If f has a continuous extension with $f : (0, \infty) \to \mathbb{R}$, then $\operatorname{supp} \rho \subset [0, \infty)$, i.e.

$$f(z) = \alpha + \beta z + \int_{[0,\infty)} \frac{zt - 1}{z + t}\, \rho(dt) \tag{6.11}$$

$$= \alpha + \beta z + \int_{[0,\infty)} \left(\frac{t}{1 + t^2} - \frac{1}{z + t} \right) (1 + t^2)\, \rho(dt). \tag{6.12}$$

If $f(0+)$ is finite, then $\rho\{0\} = 0$.

Proof. The representations (6.9) and (6.10) are clearly equivalent.

When deriving (6.5) in the proof of Theorem 6.2 (v)⟹(vi) we used only the fact that f preserves the upper half-plane. Therefore we can use (6.5) in the present context and rewrite it in the following form

$$f(\lambda) - f(\kappa) = b(\lambda - \kappa) + \frac{1}{2}\int_{\mathbb{R}} \frac{\lambda - \kappa}{(\lambda + s)(\kappa + s)}\,\Pi(ds)$$

$$= b\lambda + \frac{1}{2}\int_{\mathbb{R}}\left(\frac{s}{1 + s^2} - \frac{1}{\lambda + s}\right)\Pi(ds)$$

$$- b\kappa - \frac{1}{2}\int_{\mathbb{R}}\left(\frac{s}{1 + s^2} - \frac{1}{\kappa + s}\right)\Pi(ds).$$

Since we have

$$\left|\frac{s}{1 + s^2} - \frac{1}{\lambda + s}\right| \leq \frac{\lambda s + 1}{(1 + s^2)(\lambda + s)} \leq \frac{1}{1 + s^2}\left(\lambda + \frac{1}{\lambda}\right),$$

both integrals appearing in the formula above are convergent. This already shows that

$$f(\lambda) = c + b\lambda + \frac{1}{2}\int_{\mathbb{R}}\left(\frac{s}{1 + s^2} - \frac{1}{\lambda + s}\right)\Pi(ds)$$

for some constant $c \in \mathbb{R}$. Setting $\alpha := c$, $\beta := b$ and $\rho(dt) := \frac{1}{2}(1 + t^2)^{-1}\,\Pi(dt)$ gives (6.10).

If f has a continuous extension to $f : (0, \infty) \to \mathbb{R}$ we see, just as in the proof of Theorem 6.2 (v)$\Rightarrow$(vi), that supp $\rho \subset [0, \infty)$ and (6.12) follows. It is clear that (6.11) and (6.12) are equivalent.

Again as in the proof of Theorem 6.2 (v)$\Rightarrow$(vi) we find that $f(0+) < \infty$ implies $\rho\{0\} = 0$.   □

The following theorem shows that the family of complete Bernstein functions coincides with the family of Nevanlinna–Pick functions which are non-negative on the positive real axis and admit a right limit at the origin.

Theorem 6.9. *The set* $\mathcal{CBF}$ *consists of all Nevanlinna–Pick functions which are non-negative on* $(0, \infty)$. *In particular, the following integral representations hold for all* $f \in \mathcal{CBF}$ *and* $z \in \mathbb{C} \setminus (-\infty, 0]$

$$f(z) = \alpha + \beta z + \int_{(0,\infty)} \frac{zt - 1}{z + t}\,\rho(dt) \tag{6.13}$$

$$= \alpha + \beta z + \int_{(0,\infty)}\left(\frac{t}{1 + t^2} - \frac{1}{z + t}\right)(1 + t^2)\,\rho(dt). \tag{6.14}$$

Here ρ *is a measure on* $(0, \infty)$ *such that* $\int_{(0,1)} t^{-1}\,\rho(dt) + \int_{[1,\infty)} \rho(dt) < \infty$ *and* $\alpha, \beta \geq 0$ *are constants with* $\alpha \geq \int_{(0,\infty)} t^{-1}\,\rho(dt)$.

Proof. The equivalence of the two formulae (6.13) and (6.14) is obvious.
Assume now that f is the function given by (6.13). Then

$$f(z) - \beta z - \alpha$$
$$= \int_{(0,\infty)} \frac{zt-1}{z+t}\, \rho(dt)$$
$$= \int_{(0,\infty)} \frac{z}{z+t}\, t\, \rho(dt) - \int_{(0,\infty)} \left(\frac{1}{z+t} - \frac{1}{t}\right) \rho(dt) - \int_{(0,\infty)} \frac{1}{t}\, \rho(dt)$$
$$= \int_{(0,\infty)} \frac{z}{z+t}\left(t + \frac{1}{t}\right) \rho(dt) - \int_{(0,\infty)} \frac{1}{t}\, \rho(dt)$$

which becomes the formula in Theorem 6.2 (vi) if we set $a := \alpha - \int_{(0,\infty)} t^{-1}\rho(dt)$,
$b := \beta$ and $\sigma(dt) := (t + t^{-1})\,\rho(dt) = \frac{t^2+1}{t}\,\rho(dt)$. By assumption, $a, b \ge 0$ and

$$\int_{(0,\infty)} \frac{\sigma(dt)}{1+t} = \int_{(0,\infty)} \frac{t^2+1}{t(1+t)}\,\rho(dt) \le 2\int_{(0,1)} \frac{\rho(dt)}{t} + \int_{[1,\infty)} \rho(dt) < \infty.$$

The same calculation, run through backwards, shows that the formula in Theorem
6.2 (vi) entails (6.13), and the proof is complete.   □

Remark 6.10. (i) The (unique) representing measure ρ appearing in Corollary 6.8
and Theorem 6.9 is called the *Pick measure* or *Nevanlinna measure* or *Nevanlinna–*
Pick measure.

(ii) Just as in Corollary 6.3 we can calculate the coefficients α, β and the Nevanlinna–Pick measure ρ appearing in (6.11) and (6.13) from f. Clearly,

$$\alpha = \mathrm{Re}\, f(i), \quad \beta = \lim_{\kappa \to \infty} \frac{f(i\kappa)}{i\kappa}$$

while

$$\int_{(u,v]} (1+t^2)\,\rho(dt) = \lim_{h \to 0} \frac{1}{\pi} \int_{(u,v]} \mathrm{Im}\, f(-s+ih)\, ds \qquad (6.15)$$

holds for all $u < v$ which are continuity points of ρ. To see this, recall from the proof
of Theorem 6.9 that $t\,\sigma(dt) = (1+t^2)\,\rho(dt)$ where σ is the representing measure
from the formula in Theorem 6.2 (vi). The vague convergence in (6.15) follows now
from (6.4).

(iii) For further reference let us collect the relations between the various representing measures and densities appearing in Definition 6.1, Theorem 6.2 and Theorem 6.9:

$$m(t) = \int_{(0,\infty)} e^{-ts}\, s\, \sigma(ds) = \mathscr{L}\big(s\sigma(ds); t\big)$$

$$= \int_{(0,\infty)} e^{-ts}\, (s^2 + 1)\rho(ds) = \mathscr{L}\big((s^2 + 1)\,\rho(ds); t\big),$$

$$\sigma(dt) = \frac{t^2 + 1}{t}\, \rho(dt).$$

6.2   Extended complete Bernstein functions

In analogy with the notion of extended Bernstein functions and their characterization given in Remark 5.9, we introduce the concept of extended complete Bernstein functions.

Definition 6.11. A function $f : (0, \infty) \to \mathbb{R}$ is called an *extended complete Bernstein function* if there exist some $g \in \mathcal{CBF}$ and $h \in \mathcal{S}$ such that $f = g - h$. We write $\mathcal{CBF}_e$ for the set of all extended complete Bernstein functions.

Using the representation of Nevanlinna–Pick functions which are real on $(0, \infty)$, cf. Corollary 6.8, we can prove the following characterization of extended complete Bernstein functions.

Proposition 6.12. *The set of extended complete Bernstein functions consists of all Nevanlinna–Pick functions which are real on $(0, \infty)$. In other words, f is an extended complete Bernstein function, if $f : (0, \infty) \to \mathbb{R}$ extends analytically to $\mathbb{H}^{\uparrow}$ and preserves it. Equivalently $f \in \mathcal{CBF}_e$ if, and only if,*

$$f(z) = \alpha + \beta z + \int_{[0,\infty)} \frac{zt - 1}{z + t}\, \rho(dt) \tag{6.16}$$

$$= \alpha + \beta z + \int_{[0,\infty)} \left(\frac{t}{1 + t^2} - \frac{1}{z + t}\right)(1 + t^2)\,\rho(dt) \tag{6.17}$$

for all $z \in \mathbb{C} \setminus (-\infty, 0]$. Here ρ is a finite measure on $[0, \infty)$, and $\alpha \in \mathbb{R}$ and $\beta \geq 0$ are constants.

Proof. Suppose that f is given by (6.16). Then

$$f(\lambda) = \alpha + \beta\lambda + \int_{(0,\infty)} \frac{\lambda}{\lambda + t}\, t\rho(dt) - \int_{[0,\infty)} \frac{1}{\lambda + t}\, \rho(dt)$$

for all $\lambda > 0$. Since ρ is a finite measure, we see that

$$\int_{(0,\infty)} \frac{1}{1 + t}\, t\rho(dt) < \infty \quad \text{and} \quad \int_{[0,\infty)} \frac{1}{1 + t}\, \rho(dt) < \infty.$$

If $\alpha \geq 0$ we define

$$g(\lambda) := \alpha + \beta\lambda + \int_{(0,\infty)} \frac{\lambda}{\lambda + t}\, t\rho(dt), \quad h(\lambda) := \int_{[0,\infty)} \frac{1}{\lambda + t}\, \rho(dt),$$

while for $\alpha < 0$ we set

$$g(\lambda) := \beta\lambda + \int_{(0,\infty)} \frac{\lambda}{\lambda + t}\, t\rho(dt), \quad h(\lambda) := -\alpha + \int_{[0,\infty)} \frac{1}{\lambda + t}\, \rho(dt).$$

In both cases $g \in \mathcal{CBF}$, $h \in \mathcal{S}$ and $f = g - h$.

Conversely, suppose that $f = g - h$ with $g \in \mathcal{CBF}$ and $h \in \mathcal{S}$. Then

$$g(\lambda) = a + b\lambda + \int_{(0,\infty)} \frac{\lambda}{\lambda + t}\, \kappa(dt)$$

with $a, b \geq 0$ and a measure κ on $(0, \infty)$ such that $\int_{(0,\infty)} (1 + t)^{-1} \kappa(dt) < \infty$, and

$$h(\lambda) = c + \int_{[0,\infty)} \frac{1}{\lambda + t}\, \sigma(dt)$$

with $c \geq 0$ and a measure σ on $[0, \infty)$ such that $\int_{[0,\infty)} (1 + t)^{-1} \sigma(dt) < \infty$. Define the measure ν on $[0, \infty)$ by $\nu(dt) = t\kappa(dt) + \sigma(dt)$. As $\int_{[0,\infty)} (1+t^2)^{-1} \nu(dt) < \infty$ all integrals below are well defined. Then

$$\int_{[0,\infty)} \left(\frac{t}{1+t^2} - \frac{1}{\lambda + t} \right) \nu(dt)$$

$$= \int_{(0,\infty)} \left(\frac{t}{1+t^2} - \frac{1}{\lambda + t} \right) t\kappa(dt) + \int_{[0,\infty)} \left(\frac{t}{1+t^2} - \frac{1}{\lambda + t} \right) \sigma(dt)$$

$$= \int_{(0,\infty)} \left(\frac{t}{1+t^2} - \frac{1}{t} \right) t\kappa(dt) + \int_{(0,\infty)} \left(\frac{1}{t} - \frac{1}{\lambda + t} \right) t\kappa(dt)$$

$$+ \int_{[0,\infty)} \left(\frac{t}{1+t^2} - \frac{1}{\lambda + t} \right) \sigma(dt)$$

$$= -\int_{(0,\infty)} \frac{\kappa(dt)}{1+t^2} + \int_{(0,\infty)} \frac{\lambda}{\lambda + t}\, \kappa(dt) + \int_{[0,\infty)} \frac{t\sigma(dt)}{1+t^2}$$

$$- \int_{[0,\infty)} \frac{1}{\lambda + t}\, \sigma(dt)$$

$$= -\int_{(0,\infty)} \frac{\kappa(dt)}{1+t^2} + (g(\lambda) - a - b\lambda) + \int_{[0,\infty)} \frac{t\sigma(dt)}{1+t^2} - (h(\lambda) - c)$$

$$= f(\lambda) - \beta\lambda - \alpha,$$

where $\beta = b \geq 0$ and $\alpha = a + \int_{(0,\infty)} (1+t^2)^{-1} \kappa(dt) - \int_{[0,\infty)} t(1+t^2)^{-1} \sigma(dt) \in \mathbb{R}$. Therefore,

$$f(\lambda) = \alpha + \beta\lambda + \int_{[0,\infty)} \left(\frac{t}{1+t^2} - \frac{1}{\lambda+t} \right) (1+t^2)\rho(dt),$$

where $\rho(dt) = (1+t^2)^{-1} \nu(dt)$ is a finite measure on $[0,\infty)$. $\square$

A typical example of an extended complete Bernstein function is obtained by taking $\rho(dt) = (1+t^2)^{-1} dt$:

$$\log\lambda = \int_0^\infty \frac{\lambda t - 1}{\lambda+t} \frac{dt}{1+t^2}, \quad \lambda > 0. \tag{6.18}$$

The following results contain further connections between the classes of extended complete Bernstein functions and Nevanlinna–Pick functions.

Corollary 6.13. *Let* $f : \mathbb{H}^\uparrow \to \mathbb{H}^\uparrow$ *be a Nevanlinna–Pick function. Then the following statements are equivalent.*

(i) $f \in \mathcal{CBF}_e$.

(ii) $g(z) := \frac{z}{1+z^2}(\mathrm{Re}\, f(i) + z\mathrm{Im}\, f(i) - f(z))$ *is in* $\mathcal{CBF}$.

Proof. Let f be a Nevanlinna–Pick function. Using the representation (6.9) we find

$$f(i) = \alpha + \beta i + \int_\mathbb{R} \frac{it-1}{t+i} \rho(dt) = \alpha + i\left(\beta + \int_\mathbb{R} \rho(dt) \right).$$

Therefore,

$$g(z) = \frac{z}{1+z^2}\left(\mathrm{Re}\, f(i) + z\mathrm{Im}\, f(i) - f(z) \right)$$

$$= \frac{z}{1+z^2} \int_\mathbb{R} \left(z - \frac{tz-1}{t+z} \right) \rho(dt)$$

$$= \int_\mathbb{R} \frac{z}{t+z} \rho(dt).$$

If g is in $\mathcal{CBF}$, then $g(z) = a+bz+\int_{(0,\infty)} z/(t+z)\rho(dt)$. By the representation we have obtained above, $\lim_{|z|\to\infty} g(z)/z = 0$, so that $b = 0$. Now $a = \rho\{0\}$, implying that $\mathrm{supp}\,\rho \subset [0,\infty)$ and Proposition 6.12 shows that $f \in \mathcal{CBF}_e$.

Conversely, if $f \in \mathcal{CBF}_e$, we know from Proposition 6.12 that $\mathrm{supp}\,\rho \subset [0,\infty)$; thus g is a complete Bernstein function. $\square$

Corollary 6.14. *Let* $f : (0,\infty) \to \mathbb{R}$ *be a function. Then the following statements are equivalent.*

(i) $-f \in \mathcal{S}$;

(ii) f *has an extension which is a Nevanlinna–Pick function and $f|_{(0,\infty)}$ is continuous and non-positive.*

Proof. Assume that (i) holds. Since $f \in \mathcal{CBF}_e$, Proposition 6.12 shows that f has an extension which is a Nevanlinna–Pick function; moreover $f|_{(0,\infty)}$ is non-positive and continuous.

Conversely, by Proposition 6.12 every function satisfying (ii) is in $\mathcal{CBF}_e$. By (6.16) and the fact that $f(\lambda) \leq 0$ for all $\lambda > 0$, we get

$$\alpha + \beta\lambda + \int_{(0,\infty)} \frac{\lambda t - 1}{\lambda + t}\,\rho(dt) \leq 0.$$

Therefore, we find for every $R > 0$ and $\lambda > 0$

$$\alpha + \beta\lambda + \int_{(0,R)} \frac{\lambda t}{\lambda + t}\,\rho(dt) \leq \int_{(0,\infty)} \frac{1}{\lambda + t}\,\rho(dt).$$

Divide both sides of this inequality by λ and let $\lambda \to \infty$. Since ρ is a finite measure, we find by dominated convergence that $\beta = 0$. Taking this into account and using again dominated convergence, $\lambda \to \infty$ shows

$$-\infty < \alpha + \int_{(0,R)} t\,\rho(dt) \leq 0.$$

Since $R > 0$ is arbitrary, we see that

$$f(\lambda) = \alpha + \int_{(0,\infty)} t\,\rho(dt) + \int_{(0,\infty)} \left(\frac{\lambda t - 1}{\lambda + t} - t\right)\rho(dt)$$

$$= \alpha + \int_{(0,\infty)} t\,\rho(dt) - \int_{(0,\infty)} \frac{1}{\lambda + t}(1 + t^2)\rho(dt). \qquad \square$$

Proposition 6.15. *Let $f : \mathbb{H}^{\uparrow} \to \mathbb{H}^{\uparrow}$ be a Nevanlinna–Pick function. Then the following statements are equivalent.*

(i) *f has an extension to $\mathbb{H}^{\uparrow} \cup [0,\infty)$ such that $f|_{(0,\infty)}$ is continuous and non-positive;*

(ii) *$-zf(z)$ is a complete Bernstein function;*

(iii) *$-zf(z)$ is a Nevanlinna–Pick function.*

Proof. (i)$\Rightarrow$(ii) By Corollary 6.14, every Nevanlinna–Pick function f which is continuous and non-positive on $(0, \infty)$ is of the form $-a - g$ where $a \geq 0$ and $g \in \mathcal{S}$.

Therefore, $-zf(z) = az + zg(z)$, and this is a complete Bernstein function by (6.7), cf. Remark 6.4

(ii)$\Rightarrow$(iii) This follows from Theorem 6.9.

(iii)$\Rightarrow$(i) Since $-zf(z)$ is a Nevanlinna–Pick function, we know from (6.11) and the remark following the proof of Corollary 6.8 that

$$-zf(z) = \alpha + \beta z + \int_{\mathbb{R}} \frac{zt - 1}{z + t} \rho(dt) \tag{6.19}$$

for some constants $\alpha \in \mathbb{R}$, $\beta \geq 0$ and a finite measure ρ on $\mathbb{R}$. Since f is also a Nevanlinna–Pick function, it has a similar representation with $\hat{\alpha}, \hat{\beta}$ and $\hat{\rho}$. Therefore,

$$-zf(z) = -\hat{\alpha}z - \hat{\beta}z^2 - \int_{\mathbb{R}} \frac{zt - 1}{z + t} z\, \hat{\rho}(dt)$$

$$= -\hat{\alpha}z - \hat{\beta}z^2 - \int_{\mathbb{R}} \left(\frac{1 - zt}{z + t} t + zt - 1 \right) \hat{\rho}(dt)$$

$$= -\hat{\alpha}z - \hat{\beta}z^2 + \int_{\mathbb{R}} \hat{\rho}(dt) + \int_{\mathbb{R}} \left(\frac{zt - 1}{z + t} - z \right) t\hat{\rho}(dt).$$

Note that $(zt - 1)/(z + t) - z = -(z^2 + 1)/(z + t)$. If we subtract both representations for $-zf(z)$ we get

$$0 = \hat{\beta}z^2 + \left(\alpha - \int_{\mathbb{R}} \hat{\rho}(dt) \right) + \left(\beta + \hat{\alpha} + \int_{\mathbb{R}} \rho(dt) \right) z - \int_{\mathbb{R}} \frac{z^2 + 1}{z + t} (\rho - t\hat{\rho})(dt).$$

$$\tag{6.20}$$

(a) Divide the equality (6.20) by z^2, pick $z = iy$, $y > 0$, and let $y \to \infty$. By dominated convergence we find that $\hat{\beta} = 0$.

(b) Once we know that $\hat{\beta} = 0$, we can divide the thus modified equality (6.20) by z, set $z = iy$, $y > 0$, and let $y \to \infty$. This gives

$$0 = \left(\beta + \hat{\alpha} + \int_{\mathbb{R}} \rho(dt) \right) - \lim_{y \to \infty} \int_{\mathbb{R}} \frac{iy}{iy + t} (\rho - t\hat{\rho})(dt).$$

Since, by dominated convergence, $\lim_{y \to \infty} \int_{\mathbb{R}} iy/(iy + t)\,\rho(dt) = \int_{\mathbb{R}} \rho(dt) < \infty$, we conclude that

$$\int_{\mathbb{R}} t\hat{\rho}(dt) < \infty \quad \text{and} \quad \beta + \hat{\alpha} = -\int_{\mathbb{R}} t\hat{\rho}(dt).$$

(c) Setting $z = i$ in (6.20), we see that $\alpha = \int_{\mathbb{R}} \hat{\rho}(dt)$ and $\beta + \hat{\alpha} = -\int_{\mathbb{R}} \rho(dt)$.

If we combine (a)–(c), we get that (6.20) entails

$$\int_{\mathbb{R}} \frac{z^2 + 1}{z + t} (\rho - t\hat{\rho})(dt) = 0, \quad \text{hence,} \quad \int_{\mathbb{R}} \frac{z}{z + t} (\rho - t\hat{\rho})(dt) = 0.$$

Since the representing measure of a Nevanlinna–Pick function is unique, cf. (6.6) and Corollary 6.8, we conclude that $\rho(dt) = t\hat{\rho}(dt)$ on $\mathbb{R}$. The measures $\rho, \hat{\rho}$ are positive which means that $\operatorname{supp}\rho \subset [0, \infty)$. Using this and (a)–(c) in the representation (6.19) gives

$$f(z) = -\frac{\alpha}{z} - \beta - \int_{[0,\infty)} \left(\frac{zt-1}{z+t}\frac{1}{z} + \frac{1}{tz} - \frac{1}{tz} \right) \rho(dt)$$

$$= \frac{1}{z} \left(\int_{[0,\infty)} \frac{\rho(dt)}{t} - \alpha \right) - \beta - \int_{[0,\infty)} \frac{1}{z+t}\frac{t^2+1}{t} \rho(dt)$$

and, because of (c), the coefficient of z^{-1} vanishes. □

An analytic function Φ on a domain Ω is said to be of *bounded type* (or *bounded character*) if Φ can be written as a quotient $\Phi = G/H$ where G and H are bounded analytic functions in Ω. If Φ is of bounded type in the upper half-plane, it satisfies the *canonical* or *Nevanlinna factorization*, i.e.

$$\Phi(z) = cb(z)d(z)s(z)$$

where $|c| = 1$, $b(z)$ is a Blaschke product, see e.g. [259, Vol. 2, Section 36],

$$d(z) = \exp\left(\frac{1}{\pi i} \int_{\mathbb{R}} \left(\frac{1}{z+t} - \frac{t}{1+t^2} \right) q(t)\,dt \right)$$

with $\int |q(t)|\,dt < \infty$, and

$$g(z) = \frac{1}{2\pi i} \int_{\mathbb{R}} \frac{1-zt}{z+t} \omega(dt)$$

for some signed measure ω such that $\int_{\mathbb{R}}(1+t^2)\,|\omega|(dt) < \infty$, see e.g. Krylov [226, Theorem XX] and the comments below.

Lemma 6.16. *Let $\Phi : \mathbb{H}^{\uparrow} \to \mathbb{H}^{\uparrow}$ be an analytic function such that $\operatorname{Re}\Phi \geq 0$. Then Φ is of bounded type.*

Proof. Note that $|1 + \Phi| \geq \operatorname{Re}(1 + \Phi) \geq 1$. Therefore $\phi := \log|1 + \Phi|$ is a non-negative harmonic function on $\mathbb{H}^{\uparrow}$. Since $\mathbb{H}^{\uparrow}$ is simply connected, ϕ is the real part of an analytic function h, i.e. $\operatorname{Re}h = \phi$. Define $H := e^{-h}$ and $G := \Phi \cdot e^{-h}$. Clearly, G and H are analytic,

$$|H| = |e^{-h}| = e^{-\operatorname{Re}h} = e^{-\phi} \leq 1$$

and, since $|\Phi|^2 \leq |\Phi|^2 + 2\operatorname{Re}\Phi + 1 = |\Phi + 1|^2$,

$$\log|G| = \log|\Phi| - \log|\Phi + 1| \leq 0;$$

thus, $|G| \leq 1$. By definition $\Phi = G/H$, and the assertion follows. □

If $f \in \mathcal{CBF}$, $\Phi(z) := -if(z)$ is analytic in $\mathbb{H}^\uparrow$ and $\operatorname{Re} \Phi = \operatorname{Im} f \geq 0$. Lemma 6.16 tells us that $-if$ is of bounded type and has therefore a Nevanlinna factorization. This is the background for the next theorem. For $\mathcal{CBF}$ functions the factorization turns out to be rather easy and we can give a direct proof without resorting to the elaborate machinery of complex function theory in a half-plane.

Theorem 6.17. *For $f \in \mathcal{CBF}$, there exist a real number γ and a function η on $(0, \infty)$ taking values in $[0, 1]$ such that*

$$f(\lambda) = \exp\left(\gamma + \int_0^\infty \left(\frac{t}{1+t^2} - \frac{1}{\lambda+t}\right) \eta(t)\, dt\right), \quad \lambda > 0. \tag{6.21}$$

Conversely, any function of the form (6.21) *is in $\mathcal{CBF}$. Thus there is a one-to-one correspondence between $\mathcal{CBF}$ and the set*

$$\Gamma = \left\{(\gamma, \eta) : \gamma \in \mathbb{R} \text{ and } \eta : (0, \infty) \xrightarrow{\text{measurable}} [0, 1]\right\}. \tag{6.22}$$

Proof. Assume first that f is given by (6.21). From the structure of the integral kernel it is clear that f extends to an analytic function for all $z = \lambda + i\kappa \in \mathbb{H}^\uparrow$. Moreover,

$$\begin{aligned}
\operatorname{Im}\left(\log f(z)\right) &= \int_0^\infty \frac{\kappa}{(\lambda+t)^2 + \kappa^2} \eta(t)\, dt \\
&\leq \int_0^\infty \frac{\kappa}{(\lambda+t)^2 + \kappa^2}\, dt \\
&\leq \int_{-\infty}^\infty \frac{\kappa}{\kappa^2 + t^2}\, dt = \pi
\end{aligned}$$

shows that f preserves the upper half-plane. Since f is positive on $(0, \infty)$, $f \in \mathcal{CBF}$ by Theorem 6.2.

Conversely, assume that $f \in \mathcal{CBF}$. Using the principal branch of the logarithm we get $g(z) := \log f(z) = \log|f(z)| + i \arg f(z)$ and $\arg f(z) \in [0, \pi)$. This means that g maps $z \in \mathbb{H}^\uparrow$ to $\mathbb{H}^\uparrow$; since g is clearly analytic in $\mathbb{H}^\uparrow$ and real (but not necessarily positive) on $(0, \infty)$, it is an extended complete Bernstein function. Therefore, Proposition 6.12 applies and (6.17) shows that

$$g(z) = \alpha' + \beta' z + \int_{(0,\infty)} \left(\frac{t}{1+t^2} - \frac{1}{z+t}\right)(1+t^2)\, \rho'(dt)$$

where $\alpha' \in \mathbb{R}$, $\beta' \geq 0$ and ρ' is a finite measure on $(0, \infty)$.

We will now determine the possible values of (α', β', ρ'). Set $\gamma := \alpha'$. Since $f \in \mathcal{CBF}$ we know that $f(\lambda)$, $\lambda > 0$, grows at most linearly and $g(\lambda)$ at most logarithmically as $\lambda \to \infty$, i.e. $\beta' = 0$. Using Remark 6.10 (ii) we find that

$$\begin{aligned}
\int_{(u,v]} (1+t^2)\, \rho'(dt) &= \lim_{h \to 0+} \frac{1}{\pi} \int_{(u,v]} \operatorname{Im} g(-s + ih)\, ds \\
&= \lim_{h \to 0+} \frac{1}{\pi} \int_{(u,v]} \arg f(-s + ih)\, ds \leq v^+ - u^+,
\end{aligned}$$

where we used that ρ' is supported on $(0, \infty)$ and that $\arg f(-s + ih) \in [0, \pi)$. This means that $(1 + t^2)\, \rho'(dt)$ is absolutely continuous with respect to Lebesgue measure on $(0, \infty)$ and that the density $\eta(t)$ takes almost surely values in $[0, 1]$. □

Remark 6.18. If $f(0+) > 0$, we see that

$$\int_0^\infty \left(\frac{t}{1+t^2} - \frac{1}{t} \right) \eta(t)\, dt = \int_0^\infty \frac{-1}{t(1+t^2)}\, \eta(t)\, dt > -\infty$$

which is the same as to say that $\int_0^1 \eta(t) t^{-1}\, dt < \infty$. In this case we can rewrite the exponent in the representation (6.21) in the following form:

$$\gamma + \int_0^\infty \left(\frac{t}{1+t^2} - \frac{1}{\lambda + t} \right) \eta(t)\, dt$$

$$= \left[\gamma + \int_0^\infty \left(\frac{t}{1+t^2} - \frac{1}{t} \right) \eta(t)\, dt \right] + \int_0^\infty \left(\frac{1}{t} - \frac{1}{\lambda + t} \right) \eta(t)\, dt$$

$$= \beta + \int_0^\infty \frac{\lambda}{\lambda + t} \frac{\eta(t)}{t}\, dt;$$

this is a function of the form $\mathbb{R} + \mathcal{CBF}$, i.e. an extended (complete) Bernstein function, see Remark 5.9 and Proposition 6.12. Therefore, we get the following multiplicative representation:

$$\mathcal{CBF} \cap \{ f \, : \, f(0+) > 0 \} \subset e^{\mathbb{R} + \mathcal{CBF}} = (0, \infty) \times e^{\mathcal{CBF}}.$$

Remark 6.18 has two immediate consequences. The first result is the counterpart of Remark 5.26:

Corollary 6.19. *If f is a complete Bernstein function with $f(0+) > 0$, then $\log f$ is an extended complete Bernstein function. If $f(0+) \geq 1$, then $\log f \in \mathcal{CBF}$.*

The second result is useful if we want to construct concrete examples of complete Bernstein functions, see Section 16.11.

Corollary 6.20. *Suppose that f is a complete Bernstein function. If the Stieltjes measure of f has a density given by $\sigma(t) = \eta(t)/t$ with $0 \leq \eta(t) \leq 1$ for all $t > 0$, then e^f is a complete Bernstein function and η is the representing function appearing in the formula (6.21).*

We conclude this chapter with an application which is important in the fluctuation theory of Lévy processes.

Proposition 6.21. *If f is a complete Bernstein function, then the function*

$$g(\lambda) = \exp \left(\frac{1}{\pi} \int_0^\infty \frac{\log f(\lambda^2 \theta^2)}{1 + \theta^2}\, d\theta \right), \qquad \lambda > 0,$$

is again a complete Bernstein function. Moreover, for all $\lambda > 0$

$$e^{-\pi/2}\sqrt{f(\lambda^2)} \le g(\lambda) \le e^{\pi/2}\sqrt{f(\lambda^2)}. \qquad (6.23)$$

Proof. By Theorem 6.17, there exist some $\gamma \in \mathbb{R}$ and a function $\eta : (0, \infty) \to [0, 1]$ such that

$$\log f(\lambda) = \gamma + \int_0^\infty \left(\frac{t}{1+t^2} - \frac{1}{\lambda+t}\right) \eta(t)\,dt.$$

Thus we have

$$\log g(\lambda) = \frac{\gamma}{2} + \frac{1}{\pi} \int_0^\infty \int_0^\infty \left(\frac{t}{1+t^2} - \frac{1}{\lambda^2\theta^2 + t}\right) \eta(t)\,dt\,\frac{d\theta}{1+\theta^2}.$$

From $0 \le \eta(t) \le 1$ we get

$$\eta(t)\left|\frac{t}{1+t^2} - \frac{1}{\lambda^2\theta^2 + t}\right|\frac{1}{1+\theta^2} \le \frac{1}{1+t^2}\frac{1}{1+\theta^2}\frac{\lambda^2\theta^2 t - 1}{\lambda^2\theta^2 + t}$$

$$\le \frac{1}{1+t^2}\frac{1}{1+\theta^2}\left(\frac{1}{\lambda^2\theta^2 + t} + \frac{\lambda^2\theta^2 t}{\lambda^2\theta^2 + t}\right).$$

Since

$$\int_0^\infty \frac{1}{\lambda^2\theta^2 + t}\frac{d\theta}{1+\theta^2} \le \int_0^\infty \frac{d\theta}{\lambda^2\theta^2 + t}$$

$$= \frac{1}{t}\int_0^\infty \frac{d\theta}{\lambda^2\theta^2 t^{-1} + 1}$$

$$= \frac{1}{t}\frac{\sqrt{t}}{\lambda}\int_0^\infty \frac{d\gamma}{\gamma^2 + 1} \le \frac{\pi}{2\lambda\sqrt{t}}$$

and

$$\int_0^\infty \frac{\lambda^2\theta^2 t}{\lambda^2\theta^2 + t}\frac{d\theta}{1+\theta^2} \le \lambda^2 \int_0^\infty \frac{d\theta}{\lambda^2\theta^2 t^{-1} + 1} \le \frac{\pi\sqrt{t}}{2\lambda},$$

we can use Fubini's theorem to get

$$\log g(\lambda) = \frac{\gamma}{2} + \int_0^\infty \left(\frac{t}{2(1+t^2)} - \frac{1}{2\sqrt{t}\,(\lambda + \sqrt{t})}\right) \eta(t)\,dt$$

$$= \frac{\gamma}{2} + \int_0^\infty \left(\frac{t}{2(1+t^2)} - \frac{1}{2(1+t)}\right) \eta(t)\,dt$$

$$+ \int_0^\infty \left(\frac{1}{2(1+t)} - \frac{1}{2\sqrt{t}\,(\lambda + \sqrt{t})}\right) \eta(t)\,dt$$

$$= \gamma_1 + \int_0^\infty \left(\frac{s}{1+s^2} - \frac{1}{\lambda+s}\right) \eta(s^2)\,ds.$$

Applying Theorem 6.17 again we get that g is a complete Bernstein function.

For the second part we compute

$$\left| \log g(\lambda) - \frac{1}{2} \log f(\lambda^2) \right|$$

$$= \frac{1}{2} \left| \int_0^\infty \left[\left(\frac{t}{1+t^2} - \frac{1}{\sqrt{t}\,(\lambda + \sqrt{t})} \right) - \left(\frac{t}{1+t^2} - \frac{1}{\lambda^2 + t} \right) \right] \eta(t)\, dt \right|$$

$$\leq \frac{1}{2} \int_0^\infty \frac{\lambda(\sqrt{t} + \lambda)}{(\lambda^2 + t)\sqrt{t}\,(\lambda + \sqrt{t})}\, dt$$

$$= \frac{1}{2} \int_0^\infty \frac{\lambda}{(\lambda^2 + t)\sqrt{t}}\, dt = \frac{\pi}{2}.$$

This implies that $-\pi/2 \leq \log g(\lambda) - \frac{1}{2} \log f(\lambda^2) \leq \pi/2$ for every $\lambda > 0$, which is (6.23) □

Comments 6.22. *Section* 6.1: The notion of *complete* Bernstein function seems to appear first in the book by Prüss [297] in connection with Volterra-type integral equations. Prüss uses (iii) of Theorem 6.2 as definition of $\mathcal{CBF}$: $f(\lambda) = \lambda^2 \mathcal{L}(g; \lambda)$ where g is a Bernstein function. The specification *complete* derives from the fact that *every* Bernstein function can be written in the form $\lambda^2 \mathcal{L}(k; \lambda)$ where k is positive, non-decreasing and concave, cf. (3.4), whereas $k \in \mathcal{BF}$ if $f \in \mathcal{CBF}$. Stieltjes functions and (complete) Bernstein functions were introduced into integral equations by G. E. H. Reuter [303] who found necessary and sufficient conditions so that certain Volterra integral equations have positive solutions; the starting point for Reuter's investigations was a problem from probability theory.

The equivalent integral representations for complete Bernstein functions from Theorems 6.2 and 6.9 have a long history. This is due to the fact that the class $\mathcal{CBF}$ has been used under different names and in different contexts. One of the earliest appearances of $\mathcal{CBF}$ is in the papers by Pick [293, 294] and, independently, R. Nevanlinna [272, 273] in connection with interpolation problems in $\mathbb{C}$ and the Stieltjes moment problem. Pick gives a characterization of all analytic functions in the unit disk or the upper half-plane which attain the values w_j at the points $|z_j| \leq 1$ and $\mathrm{Im}\, z_j \geq 0$, $j = 1, 2, \ldots, n$, respectively; the class of functions on the half-plane that admit a unique solution of this interpolation problem satisfy (a discrete version of) the representation (6.13). Independently of Pick, Nevanlinna arrives in [272] at the same result for finitely and infinitely many interpolation pairs (z_j, w_j).

In [273] Nevanlinna shows that Hamburger's solution to the (extended) Stieltjes' moment problem [140] can be equivalently stated in terms of analytic functions f in the upper complex half-plane with negative imaginary part. Nevanlinna proves the equivalence of conditions (iv) and (vi) of Theorem 6.2; note that he does not assume that f has a positive continuous extension to $(0, \infty)$ (as we do). The main tool are methods from the theory of continued fractions and Schur's algorithm [327]. Later, he gives in [274] a different proof for this representation. He solves the above described interpolation problem in the unit disk for infinitely many pairs (z_j, w_j). For the case where the z_j are on the boundary of the disk or the upper half-plane, respectively, he employs earlier results of Julia [190] and Carathéodory [75] on non-tangential limits (also angular limits, *Winkelderivierte* in German). By the Cayley transform the open disk is mapped onto $\mathbb{H}^{\uparrow}$ and the z_j's are mapped into points $\lambda_j \in \mathbb{R}$. Letting the λ_j's collapse into a single point, the problem becomes the Stieltjes moment problem and one is back in the situation of [273]. Note that this approach using non-tangential limits also yields (6.6) which is nowadays called Stieltjes inversion formula.

It is interesting to note that Nevanlinna did not use Herglotz' seminal paper [156] on the representation of analytic functions in the unit disk having positive real part. Using Herglotz' result and the Cayley transform, Cauer [77] gives a short proof of the equivalence of Theorem 6.2 (iv) and Theorem 6.9, and refers to [274] for Theorem 6.2 (vi). Cauer's approach is used by most modern presentations of the material, see e.g. Akhiezer [2, Chapter 3] or Akhiezer and Glazman [3, Vol. II, Sect. 59] for particularly nice proofs. Akhiezer [2] contains many interesting notes on the Russian school of complex function theory and operator theory. In a series of papers [219, 220, 221] M. G. Kreĭn introduced the classes (S) and (S^{-1}) in connection with investigations on the theory of generalized resolvents and the theory of strings – see Chapter 15 below; a survey of Kreĭn's result is contained in Kac and Kreĭn [195]. In our notation $f \in (S)$ if $f(-z) \in \mathcal{S}$ and $f \in (S^{-1})$ if $-f(-z) \in \mathcal{CBF}$. Kreĭn's work focusses on the connection between $(S), (S^{-1})$ and the Nevanlinna–Pick functions which are denoted by (R). The paper [212] by Komatu gives a modern presentation of the material from the viewpoint of complex function theory in a half-plane.

Hirsch [162, 163, 165] introduces complete Bernstein functions – in [165] they are referred to as family $\mathscr{H}$ – through a variant of (6.7). His intention was to find a cone of functions operating on Hunt kernels and more general abstract potentials [166]. Hirsch shows, among other things, the equivalence of (6.1) and (6.7) and that $\mathcal{CBF}$ are exactly those functions which operate on all densely defined, closed operators V on a Banach space such that $(-\infty, 0)$ is in the resolvent set and $\sup_{\lambda>0} \|(\mathrm{id} + \lambda V)^{-1}\|$ is finite. This set includes the abstract potentials in the sense of Yosida [379, XIII.9], as well as all infinitesimal generators of strongly continuous contraction semigroups.

The present rather complete form of Theorems 6.2 and 6.9 appeared in [320] – this proof is reproduced in [183, Vol. 1, pp. 193–202] – see also [321] and [322] as well as the survey paper by Berg [35]. The elementary proof presented here seems to be new. The result contained in Corollary 6.5 is originally due to Julia [190], Wolff [374], Carathéodory [75] and one of the most elegant direct proofs is by Landau and Valiron [237]. Some of the above equivalences are part of the *mathematical folklore*; see e.g. Hirsch [163], Berg [32] or Nakamura [269] for corresponding remarks and statements. Corollary 6.8 and the remark following this corollary provides a new proof of the representation theorem for general Nevanlinna–Pick functions, cf. Donoghue [100, Chapter II, Theorem I and Lemma 2] for the standard proof.

The Stieltjes-type inversion formulae appearing in Corollary 6.3 and Remark 6.10 are classical and can be found, e.g. in Akhiezer [2, p. 124], de Branges [92, Chapter 1.3] or Berg [33, 35].

Section 6.2: Our definition of extended complete Bernstein function is formally inspired by the results on extended Bernstein functions, see Remark 5.9, Theorem 5.11 or Remark 5.26; all these results have their counterparts for $\mathcal{CBF}_e$ but the underlying theory is much deeper. As far as we are aware, Akhiezer [2, Notes and additions to Chapter 3] is one of the few places in the literature where $\mathcal{CBF}_e$ are (indirectly) studied as a particular subclass of Nevanlinna–Pick functions, cf. Proposition 6.12. The Corollaries 6.13 and 6.14 and Proposition 6.15 are adapted from Akhiezer [2, pp. 126–128]; Corollary 6.13 is originally due to Akhiezer and Kreĭn [4].

The proof of the canonical or Nevanlinna factorization of bounded type in $\mathbb{H}^\uparrow$ goes back to Krylov [226]. His paper is still highly readable and the formulation here is taken directly from [226, Theorem XX]. Other sources include Duren [104, Chapter 11.3], Hoffmann [168, pp. 132-3], de Branges [92, Chapter I.8] and Rosenblum and Rovnyak [308, Chapter 5]; in [308, Chapter 2] there is also a proof of Theorem 6.9 using methods from operator theory as well as a discussion of the Nevanlinna–Pick interpolation problem. Aronszajn and Donoghue [12] give a thorough discussion of exponential representations of Nevanlinna–Pick functions. The short proof of Theorem 6.17 is inspired by Kreĭn and Nudelman [225, Appendix, Theorems A.3, A.8].

The function g appearing in Proposition 6.21 is the Laplace exponent of the ladder height process of a one-dimensional subordinate Brownian motion whose subordinator has the Laplace exponent f. Proposition 6.21 is from [208], the first part is independently shown in [233], a result similar to the second part appears in [234].

Chapter 7

Properties of complete Bernstein functions

The representation results of Chapter 6 enable us to give various structural results characterising complete Bernstein and Stieltjes functions. Our first aim is to make a connection between Stieltjes and complete Bernstein functions. For this the following result is useful.

Proposition 7.1. $f \in \mathcal{CBF}$, $f \not\equiv 0$, if, and only if, the function $f^{\star}(\lambda) := \lambda/f(\lambda)$ is in $\mathcal{CBF}$.

Proof. If $f \in \mathcal{CBF}$ we may use the representation (6.7) and get

$$\frac{f(z)}{z} = \frac{a}{z} + b + \int_{(0,\infty)} \frac{1}{z+t}\, \sigma(dt), \quad z \in \mathbb{C} \setminus (-\infty, 0].$$

Since

$$\operatorname{Im} z \cdot \operatorname{Im} \frac{1}{z+t} = \operatorname{Im} z \cdot \frac{-\operatorname{Im} z}{|z+t|^2} = \frac{-(\operatorname{Im} z)^2}{|z+t|^2} \le 0$$

we see that $f(z)/z$ maps $\mathbb{H}^{\uparrow}$ into $\mathbb{H}^{\downarrow}$. As $1/z$ switches the upper and lower half-planes, $z/f(z) = (z^{-1}f(z))^{-1}$ maps $\mathbb{H}^{\uparrow}$ into itself. Further, $\lambda/f(\lambda) \in (0,\infty)$ for $\lambda > 0$, and since $a = \lim_{\lambda \to 0+} f(\lambda)$, we have

$$\lim_{\lambda \to 0+} \frac{\lambda}{f(\lambda)} = \begin{cases} 0, & \text{if } a \ne 0, \\ \dfrac{1}{\lim_{\lambda \to 0+} \frac{f(\lambda)}{\lambda}} = \dfrac{1}{b + \int_{(0,\infty)} \frac{\sigma(dt)}{t}} \in [0,\infty), & \text{if } a = 0. \end{cases}$$

Theorem 6.2 (iv) shows that $\lambda/f(\lambda)$ is in $\mathcal{CBF}$.

Conversely, if $g(\lambda) = \lambda/f(\lambda)$, $\lambda \in (0,\infty)$, is a complete Bernstein function, we can apply the just established result to this function and get

$$\mathcal{CBF} \ni \frac{\lambda}{g(\lambda)} = \frac{\lambda}{\frac{\lambda}{f(\lambda)}} = f(\lambda). \qquad \square$$

Remark 7.2. We call f, $f^{\star}$ a *conjugate pair* of complete Bernstein functions.

We can now prove a characterization of Stieltjes functions which was already announced in Chapter 2. Note that this theorem enables us to transfer all statements for complete Bernstein functions to Stieltjes functions. From a structural point of view one should compare this theorem with the definition of the cone $\mathcal{P}$, Definition 5.24.

Theorem 7.3. *A (non-trivial) function f is a complete Bernstein function if, and only if, $1/f$ is a (non-trivial) Stieltjes function. In other words*

$$\mathcal{CBF}^* = \{f : 1/f \in \mathcal{S}^*\} \quad \text{and} \quad \mathcal{S}^* = \{g : 1/g \in \mathcal{CBF}^*\}.$$

($\mathcal{CBF}^*, \mathcal{S}^*$ refer to the not identically vanishing elements of $\mathcal{CBF}$ and $\mathcal{S}$.)

Proof. If $f \in \mathcal{CBF}^*$, Proposition 7.1 shows that $\lambda \mapsto \lambda/f(\lambda)$ is a complete Bernstein function, too. As such, $z/f(z)$, $z \in \mathbb{C}\backslash(-\infty, 0]$, has a representation of the form (6.7), and dividing by z we see that $1/f(z)$ is a Stieltjes function.

Conversely, if $1/f \in \mathcal{S}^*$, it is obvious from the definition of Stieltjes functions that $z/f(z)$ has a representation of the type (6.7) which means that $z/f(z)$ is a complete Bernstein function. By Proposition 7.1 this shows that $f \in \mathcal{CBF}^*$.   □

The next Corollary is an immediate consequence of Theorem 7.3 combined with Theorem 6.2 (iv), (v).

Corollary 7.4. *Let g be a positive function on $(0, \infty)$. Then g is a Stieltjes function if, and only if, $g(0+)$ exists in $[0, \infty]$ and g extends analytically to $\mathbb{C} \setminus (-\infty, 0]$ such that $\operatorname{Im} z \cdot \operatorname{Im} g(z) \leq 0$, i.e. g maps $\mathbb{H}^\uparrow$ to $\mathbb{H}^\downarrow$ and vice versa.*

The set $\mathcal{S}$ of Stieltjes functions plays pretty much the same role for $\mathcal{CBF}$ as do the completely monotone functions $\mathcal{CM}$ for the Bernstein functions $\mathcal{BF}$. The following theorem is the $\mathcal{CBF}$-analogue of Theorem 3.7

Theorem 7.5. *Let f be a positive function on $(0, \infty)$. Then the following assertions are equivalent.*

(i) $f \in \mathcal{CBF}$.

(ii) $g \circ f \in \mathcal{S}$ *for every* $g \in \mathcal{S}$.

(iii) $\frac{1}{u+f} \in \mathcal{S}$ *for every* $u > 0$.

Proof. (i)$\Rightarrow$(ii) Assume that $f \in \mathcal{CBF}$ and $g \in \mathcal{S}$. Clearly, $g \circ f(\lambda)$ is positive for $\lambda \in (0, \infty)$ and by Theorem 6.2 (iv) and Corollary 7.4

$$g \circ f : \mathbb{H}^\uparrow \xrightarrow{\ f\ } \mathbb{H}^\uparrow \xrightarrow{\ g\ } \mathbb{H}^\downarrow.$$

Again by Corollary 7.4, $g \circ f \in \mathcal{S}$.

(ii)$\Rightarrow$(iii) This follows from the fact that $g(\lambda) := g_u(\lambda) := (u + \lambda)^{-1}$, $u > 0$, is a Stieltjes function.

(iii)$\Rightarrow$(i) Note that for $z \in \mathbb{C} \setminus (-\infty, 0]$

$$\operatorname{Im} z \cdot \operatorname{Im} \frac{1}{u + f(z)} = \operatorname{Im} z \cdot \frac{-\operatorname{Im} f(z)}{|u + f(z)|^2} \leq 0$$

so that $\operatorname{Im} z \cdot \operatorname{Im} f(z) \geq 0$. Since $f|_{(0,\infty)}$ is positive we conclude from Theorem 6.2 (v) that $f \in \mathcal{CBF}$.   □

Corollary 7.6. (i) *The set $\mathcal{CBF}$ is a convex cone: $sf_1 + tf_2 \in \mathcal{CBF}$ for all $s, t \geq 0$ and $f_1, f_2 \in \mathcal{CBF}$.*

(ii) *The set $\mathcal{CBF}$ is closed under pointwise limits: if $(f_n)_{n \in \mathbb{N}} \subset \mathcal{CBF}$ and if the limit $\lim_{n \to \infty} f_n(\lambda) = f(\lambda)$ exists for every $\lambda > 0$, then $f \in \mathcal{CBF}$ and the Stieltjes measures σ_n of f_n converge vaguely on $(0, \infty)$ to the Stieltjes measure σ of f.*

(iii) *The set $\mathcal{CBF}$ is closed under composition: if $f_1, f_2 \in \mathcal{CBF}$, then $f_1 \circ f_2 \in \mathcal{CBF}$. In particular, $\lambda \mapsto f_1(c\lambda)$ is in $\mathcal{CBF}$ for any $c > 0$.*

(iv) *$f \in \mathcal{CBF}$ is bounded on $\overrightarrow{\mathbb{H}}$ if, and only if, in (6.7) $b = 0$ and $\sigma(0, \infty) < \infty$.*

Proof. (i) This is obvious because of formula (6.7).

(ii) Without loss of generality we may assume that $f \not\equiv 0$. Then $f_n(\lambda) \to f(\lambda)$ if, and only if, $1/f_n(\lambda) \to 1/f(\lambda)$. By Theorem 7.3, $1/f_n \in \mathcal{S}$; since $\mathcal{S}$ is closed under pointwise limits, cf. Theorem 2.2, $1/f \in \mathcal{S}$ and, again by Theorem 7.3, $f \in \mathcal{CBF}$.

Let μ_n be the Lévy measure of f_n, $n \in \mathbb{N}$, and μ the Lévy measure of f. It follows by Corollary 3.9 that $\lim_{n \to \infty} \mu_n = \mu$ vaguely in $(0, \infty)$. Let m_n be the completely monotone density of μ_n, $n \in \mathbb{N}$, and m the completely monotone density of f. Since m_n and m are continuous and non-increasing, it is simple to show that $\lim_{n \to \infty} m_n(t) = m(t)$ for all $t > 0$. By Remark 6.10, $m_n(t) = \mathcal{L}(s\sigma_n(ds); t)$ and $m(t) = \mathcal{L}(s\sigma(s); t)$, and so $\lim_{n \to \infty} s\sigma_n(ds) = s\sigma(ds)$ vaguely in $[0, \infty)$, cf. the proof of Corollary 1.6. Therefore, $\lim_{n \to \infty} \sigma_n = \sigma$ vaguely in $(0, \infty)$.

(iii) Let $f_1, f_2 \in \mathcal{CBF}$. For any $g \in \mathcal{S}$ use Theorem 7.5 (i)$\Rightarrow$(ii) to get $g \circ f_1 \in \mathcal{S}$, and then $g \circ (f_1 \circ f_2) = (g \circ f_1) \circ f_2 \in \mathcal{S}$. The converse direction (ii)$\Rightarrow$(i) now shows that $f_1 \circ f_2 \in \mathcal{CBF}$.

(iv) Note that for all $z \in \overrightarrow{\mathbb{H}}$ and $t \geq 0$

$$\left| \frac{z}{z+t} \right|^2 = \frac{(\mathrm{Re}\, z)^2 + (\mathrm{Im}\, z)^2}{(\mathrm{Re}\, z + t)^2 + (\mathrm{Im}\, z)^2} \leq 1.$$

That $b = 0$ and $\sigma(0, \infty) < \infty$ imply the boundedness of $f|_{\overrightarrow{\mathbb{H}}}$ is clear from the representation (6.7). Conversely, if $f|_{\overrightarrow{\mathbb{H}}}$ is bounded, $b = 0$ follows from Remark 3.3 (iv), and $\sigma(0, \infty) < \infty$ follows from (6.7) and Fatou's Lemma for $z = \lambda \to \infty$, $\lambda \in [0, \infty)$. $\square$

Here is the $\mathcal{CBF}$-analogue of Proposition 3.12 which gives a one-to-one correspondence between bounded Stieltjes functions and bounded complete Bernstein functions.

Proposition 7.7. *If $g \in \mathcal{S}$ is bounded, then $g(0+) - g \in \mathcal{CBF}$. Conversely, if the function $f \in \mathcal{CBF}$ is bounded, there exist some constant $c > 0$ and some bounded function $g \in \mathcal{S}$, $\lim_{\lambda \to \infty} g(\lambda) = 0$, such that $f = c - g$; then $c = f(0+) + g(0+)$.*

Proof. Assume that $g(\lambda) = a\lambda^{-1} + b + \int_{(0,\infty)} (\lambda + t)^{-1}\sigma(dt)$ is bounded. This means that $a = 0$ and, by monotone convergence, $\int_{(0,\infty)} t^{-1}\sigma(dt) < \infty$. Again by monotone convergence $g(0+) = b + \int_{(0,\infty)} t^{-1}\sigma(dt) < \infty$. Hence,

$$f(\lambda) := g(0+) - g(\lambda) = \int_{(0,\infty)} \left(\frac{1}{t} - \frac{1}{\lambda + t}\right)\sigma(dt) = \int_{(0,\infty)} \frac{\lambda}{t + \lambda} \frac{\sigma(dt)}{t}$$

is a bounded complete Bernstein function, cf. Corollary 7.6 (iv).

Conversely, if $f \in \mathcal{CBF}$ is bounded, $f(\lambda) = a + \int_{(0,\infty)} \lambda (\lambda + t)^{-1}\rho(dt)$ for some bounded measure ρ, cf. Corollary 7.6. Therefore, $f(\lambda) = c - g(\lambda)$ where $g(\lambda) = \int_{(0,\infty)} t (\lambda + t)^{-1}\rho(dt)$ is a bounded Stieltjes function with $\lim_{\lambda\to\infty} g(\lambda) = 0$ and $c = a + \rho(0,\infty) = f(0+) + g(0+) > 0$. $\quad\square$

Remark 7.8. Just as in the case of Bernstein functions, see Remark 3.13, we can understand the representation formula (6.7) as a particular case of the Kreĭn–Milman or Choquet representation. The set

$$\{f \in \mathcal{CBF} : f(1) = 1\}$$

is a basis of the convex cone $\mathcal{CBF}$, and its extremal points are given by

$$e_0(\lambda) = 1, \quad e_t(\lambda) = \frac{\lambda(1 + t)}{\lambda + t}, \quad 0 < t < \infty, \quad \text{and} \quad e_\infty(\lambda) = \lambda.$$

Their extremal property follows exactly as in Remark 2.4. Thus,

$$1, \quad \lambda, \quad \frac{\lambda}{\lambda + t}, \quad (0 < t < \infty)$$

are typical functions in $\mathcal{CBF}$. Using (6.7) with $\frac{1}{\pi}\sin(\alpha\pi) t^{\alpha-1} dt$, $\mathbb{1}_{(0,1)}(t)\frac{dt}{2\sqrt{t}}$ or $\frac{1}{t}\mathbb{1}_{(1,\infty)}(t) dt$ as the Stieltjes measures $\sigma(dt)$ we see that

$$\lambda^\alpha \ (0 < \alpha < 1), \quad \sqrt{\lambda} \arctan \frac{1}{\sqrt{\lambda}}, \quad \log(1 + \lambda)$$

are also complete Bernstein functions.

The closure assertion of Corollary 7.6 says, in particular, that on the set $\mathcal{CBF}$ the notions of *pointwise convergence, locally uniform convergence*, and even *convergence in the space C^∞* coincide.

The next corollary comes in handy if we want to construct new complete Bernstein functions or Stieltjes functions from given functions in $\mathcal{CBF}$ or $\mathcal{S}$. It also shows how these classes of functions operate on each other. To simplify the exposition we use shorthand notation of the type $\mathcal{CBF} \circ \mathcal{S} \subset \mathcal{S}$ to indicate that the composition of any $f \in \mathcal{CBF}$ and $g \in \mathcal{S}$ is an element of $\mathcal{S}$.

Corollary 7.9.   (i) $\mathcal{CBF} \circ \mathcal{S} \subset \mathcal{S}$;

(ii) $\mathcal{S} \circ \mathcal{CBF} \subset \mathcal{S}$;

(iii) $\mathcal{CBF} \circ \mathcal{CBF} \subset \mathcal{CBF}$;

(iv) $\mathcal{S} \circ \mathcal{S} \subset \mathcal{CBF}$.

Proof. All assertions follow in the same way: all composite functions are positive on the positive real line $(0, \infty)$ and all that remains to be done is to track whether $\mathbb{H}^{\uparrow}$ is mapped under the composite map to $\mathbb{H}^{\uparrow}$ or $\mathbb{H}^{\downarrow}$. In the first case we get a complete Bernstein function, see Theorem 6.2 (iv), otherwise we have a Stieltjes function, see Theorem 7.3.   □

Corollary 7.9 contains many known characterizations of $\mathcal{CBF}$ and $\mathcal{S}$. Several applications of Corollary 7.9 yield

$$\lambda \mapsto f(\lambda) \in \mathcal{CBF} \quad \text{if, and only if,} \quad \lambda \mapsto \frac{1}{f(\frac{1}{\lambda})} \in \mathcal{CBF}; \tag{7.1}$$

$$\lambda \mapsto f(\lambda) \in \mathcal{S} \quad \text{if, and only if,} \quad \lambda \mapsto \frac{1}{f(\frac{1}{\lambda})} \in \mathcal{S}; \tag{7.2}$$

just note that $\lambda \mapsto 1/\lambda$ is a Stieltjes function. The same consideration applied to $\lambda \mapsto \lambda/f(\lambda)$, which is in $\mathcal{CBF}$ if, and only if, $f \in \mathcal{CBF}$, cf. Proposition 7.1, gives

$$\lambda \mapsto f(\lambda) \in \mathcal{CBF} \quad \text{if, and only if,} \quad \lambda \mapsto \lambda f\left(\frac{1}{\lambda}\right) \in \mathcal{CBF}. \tag{7.3}$$

Composing $f \in \mathcal{S}$ with the complete Bernstein function

$$\lambda \mapsto \frac{\epsilon^{-1}\lambda}{\lambda + \epsilon^{-1}} = \frac{\lambda}{\epsilon\lambda + 1}$$

shows that $f \cdot (\epsilon f + 1)^{-1} \in \mathcal{S}$. Letting $\epsilon \to 0$ brings us back to $f \in \mathcal{S}$, cf. Theorem 2.2; thus,

$$\lambda \mapsto f(\lambda) \in \mathcal{S} \quad \text{if, and only if,} \quad \lambda \mapsto \frac{f(\lambda)}{\epsilon f(\lambda) + 1} \in \mathcal{S} \quad \text{for all } \epsilon > 0. \tag{7.4}$$

Finally, if we compose $f \in \mathcal{S}$ with $1/\lambda \in \mathcal{S}$, we get $1/f \in \mathcal{CBF}$ and $1/(\lambda f(\lambda)) \in \mathcal{S}$ by Theorem 6.2 (ii). The same reasoning applied to the function $1/(\lambda f(\lambda))$ gives $f \in \mathcal{S}$, and we have shown

$$\lambda \mapsto f(\lambda) \in \mathcal{S} \quad \text{if, and only if,} \quad \lambda \mapsto \frac{1}{\lambda f(\lambda)} \in \mathcal{S}. \tag{7.5}$$

The following corollary is in the vein of Corollary 7.9 and involves additionally the class of potentials $\mathcal{P}$.

Corollary 7.10. *We have the following inclusions and identities*

(i) $\mathcal{P} \subset \mathcal{S} \circ \mathcal{BF}$;

(ii) $\dfrac{1}{\mathcal{CBF} \circ \mathcal{P}} \subset \mathcal{BF}$;

(iii) $\mathcal{CBF} \circ \mathcal{P} \subset \mathcal{P}$;

(iv) $\mathcal{S} \circ \mathcal{BF} = \mathcal{P}$.

Proof. As before, we assume implicitly that all sets used in the proof below do not include the function which is identically zero. This is clearly a slight abuse of notation.

(i) This follows from the fact that $1/\lambda$ is a Stieltjes function;

(ii) Using Theorem 7.3, (i) and Corollary 7.9 (iv), we see the following inclusions

$$\frac{1}{\mathcal{CBF} \circ \mathcal{P}} = \frac{1}{\mathcal{CBF}} \circ \mathcal{P} = \mathcal{S} \circ \mathcal{P} \subset \mathcal{S} \circ \mathcal{S} \circ \mathcal{BF} \subset \mathcal{CBF} \circ \mathcal{BF} \subset \mathcal{BF}.$$

(iii) This follows from (ii) and the very definition of $\mathcal{P}$.

(iv) The inclusion $\mathcal{S} \circ \mathcal{BF} \supset \mathcal{P}$ is just (i). For the converse we observe that, by Theorem 7.3,

$$\frac{1}{\mathcal{S} \circ \mathcal{BF}} = \frac{1}{\mathcal{S}} \circ \mathcal{BF} = \mathcal{CBF} \circ \mathcal{BF} \subset \mathcal{BF}.$$

This proves $\mathcal{S} \circ \mathcal{BF} \subset \mathcal{P}$.   □

Just as for Bernstein functions, cf. Proposition 3.5, we can characterize those complete Bernstein functions which have derivatives in $\mathcal{S}$.

Proposition 7.11. *Let $h(\lambda) = a\lambda^{-1} + b + \int_{(0,\infty)} (\lambda + t)^{-1} \tau(dt)$ be a Stieltjes function with $a, b \geq 0$ and representing measure τ, i.e. $\int_{(0,\infty)} (1+t)^{-1} \tau(dt) < \infty$.*
Then h has a primitive f which is in $\mathcal{CBF}$ if, and only if, $a = 0$ and if the representing measure satisfies

$$\int_{(0,1)} |\log t| \, \tau(dt) + \int_{[1,\infty)} \frac{1}{t} \tau(dt) < \infty. \tag{7.6}$$

If the primitive f is a complete Bernstein function, it is given by

$$f(\lambda) = c + b\lambda + \int_{(0,\infty)} \log\left(1 + \frac{\lambda}{t}\right) \tau(dt) \tag{7.7}$$

where $f(0+) = c \geq 0$ is some integration constant.
Conversely, every function of the form (7.7) with the representing measure τ satisfying (7.6) is contained in $\mathcal{CBF}$. Moreover, its derivative f' is in $\mathcal{S}$.

Proof. Let h be a Stieltjes function given by

$$h(\lambda) = \frac{a}{\lambda} + b + \int_{(0,\infty)} \frac{1}{\lambda + t}\, \tau(dt).$$

Using Fubini's theorem we find for all $\lambda > \kappa > 0$

$$\int_{\kappa}^{\lambda} h(s)\, ds = \int_{\kappa}^{\lambda} \left(\frac{a}{s} + b + \int_{(0,\infty)} \frac{1}{s + t}\, \tau(dt) \right) ds$$

$$= a \log \frac{\lambda}{\kappa} + b(\lambda - \kappa) + \int_{(0,\infty)} \int_{\kappa}^{\lambda} \frac{1}{s + t}\, ds\, \tau(dt)$$

$$= a \log \frac{\lambda}{\kappa} + b(\lambda - \kappa) + \int_{(0,\infty)} \log \frac{\left(\frac{\lambda}{t} + 1 \right)}{\left(\frac{\kappa}{t} + 1 \right)}\, \tau(dt).$$

Note that this expression has a finite limit as $\kappa \to 0$ if, and only if, $a = 0$ and if τ satisfies (7.6). Indeed, since

$$\lim_{t \to 0} \frac{\log \left(1 + \frac{\lambda}{t} \right)}{\log \left(1 + \frac{1}{t} \right)} = 1 \quad \text{and} \quad \lim_{t \to \infty} \frac{\log \left(1 + \frac{\lambda}{t} \right)}{\log \left(1 + \frac{1}{t} \right)} = \lambda,$$

the integral expression is finite, if and only if,

$$\int_{(0,\infty)} \log \left(1 + \frac{1}{t} \right) \tau(dt) < \infty.$$

Because of

$$\lim_{t \to 0} \frac{\log \left(1 + \frac{1}{t} \right)}{\log \frac{1}{t}} = 1 \quad \text{and} \quad \lim_{t \to \infty} \frac{\log \left(1 + \frac{1}{t} \right)}{\frac{1}{t}} = 1$$

we get (7.6).

If $a = 0$ and if the representing measure τ satisfies (7.6), the primitive f exists, and we have $f(0+) < \infty$. Therefore, $f(\lambda)$ is given by (7.7) and it is a complete Bernstein function. The latter follows from the fact that $\mathcal{CBF}$ is a convex cone which is closed under pointwise convergence.

Conversely, if the primitive exists and if $f \in \mathcal{CBF}$, we know that $f(0+) < \infty$ and the above calculation shows that $a = 0$ and that the measure τ satisfies (7.6). $\square$

Proposition 7.12. *If $f \in \mathcal{CBF}$, then $(-1)^{n+1} f^{(n)}(\lambda^{1/(n+1)})$, $\lambda > 0$, is a Stieltjes function.*

Proof. Differentiating the representation (6.7) n times yields

$$(-1)^{n+1} f^{(n)}(\lambda) = \begin{cases} b + \displaystyle\int_{(0,\infty)} \frac{t\,\sigma(dt)}{(t + \lambda)^2}, & \text{if } n = 1 \\[2mm] n! \displaystyle\int_{(0,\infty)} \frac{t\,\sigma(dt)}{(t + \lambda)^{n+1}}, & \text{if } n \geq 2. \end{cases}$$

By Corollary 7.15 (ii), $\lambda \mapsto (t + \lambda^{1/(n+1)})^{n+1}$ is, for every $t > 0$, a complete Bernstein function; Theorem 7.3 shows that $\lambda \mapsto 1/(t + \lambda^{1/(n+1)})^{n+1}$ is in $\mathcal{S}$, and the closedness of the convex cone $\mathcal{S}$ gives that $(-1)^{n+1} f^{(n)}(\lambda^{1/(n+1)})$ is a Stieltjes function. $\square$

Proposition 7.13. *Let $\alpha, \beta \in (0, 1)$ be such that $\alpha + \beta \leq 1$. Then*

$$f_1^\alpha \cdot f_2^\beta \in \mathcal{CBF} \quad \text{for all} \quad f_1, f_2 \in \mathcal{CBF},$$
$$g_1^\alpha \cdot g_2^\beta \in \mathcal{S} \quad \text{for all} \quad g_1, g_2 \in \mathcal{S}.$$

For $\beta = 1 - \alpha$ this means that both $\mathcal{CBF}$ and $\mathcal{S}$ are logarithmically convex cones.

Proof. Since $f \in \mathcal{CBF}$ if, and only if, $1/f \in \mathcal{S}$, we only need to show the assertion for $\mathcal{CBF}$. We give two arguments for this result:

The *first proof* is based on Theorem 6.2 (iv). Since $f \in \mathcal{CBF}$ maps $\mathbb{H}^\uparrow$ into itself, $\arg f(z) \in (0, \pi)$ for all $z \in \mathbb{H}^\uparrow$. Thus, $\arg f^\alpha(z) \in (0, \alpha\pi)$ and it is clear that $0 \leq \arg(f_1^\alpha \cdot f_2^\beta) < \alpha\pi + \beta\pi = \pi$ for $z \in \mathbb{H}^\uparrow$. This means that $f_1^\alpha \cdot f_2^\beta$ preserves the upper half-plane $\mathbb{H}^\uparrow$ and is a complete Bernstein function.

The *second proof* uses the representation of $f_1, f_2 \in \mathcal{CBF}$ from Theorem 6.17. If f_j is represented by $(\gamma_j, \eta_j) \in \Gamma$, $f_1^\alpha \cdot f_2^\beta$ corresponds to $\alpha(\gamma_1, \eta_1) + \beta(\gamma_2, \eta_2)$ which is again in Γ. $\square$

We will use the following shorthand:

$$\mathcal{CBF}^\alpha := \{f^\alpha : f \in \mathcal{CBF}\} \quad \text{and} \quad \mathcal{S}^\alpha := \{f^\alpha : f \in \mathcal{S}\}$$

where $\mathcal{CBF}^0$ and $\mathcal{S}^0$ are, by definition, the non-negative constants. Moreover, note that $\mathcal{CBF}^{\pm\alpha} = \mathcal{S}^{\mp\alpha}$.

Proposition 7.14. *For $\alpha \in [-1, 1]$, the families $\mathcal{CBF}^\alpha$ and $\mathcal{S}^\alpha$ are convex cones. In fact, for $\alpha \in [0, 1]$,*

$$\mathcal{CBF}^\alpha = \{f \in \mathcal{CBF} : \lambda^{1-\alpha} f(\lambda) \in \mathcal{CBF}\},$$
$$\mathcal{S}^\alpha = \{f \in \mathcal{S} : \lambda^{\alpha-1} f(\lambda) \in \mathcal{S}\}.$$

Proof. The convexity follows immediately from the identities given in the second half of the proposition. Since $\mathcal{S}^{-1} = \mathcal{CBF}$ and $\mathcal{CBF}^{-1} = \mathcal{S}$ it is clearly enough to consider $\alpha \in [0, 1]$ and $\mathcal{CBF}$.

When $\alpha = 1$ the assertions are trivial. If $\alpha = 0$ and $f \in \mathcal{CBF}$, then $\lambda f(\lambda)$ can only be of class $\mathcal{CBF}$ if f is a positive constant. This follows from the fact that $\lambda f(\lambda) \in \mathcal{CBF}$ implies that $\lambda/(\lambda f(\lambda)) = 1/f(\lambda) \in \mathcal{CBF}$. On the other hand, $1/f \in \mathcal{S}$ and all functions $\mathcal{CBF} \cap \mathcal{S}$ must be both non-decreasing and non-increasing, hence constant.

Fix $\alpha \in (0,1)$ and define

$$\mathscr{Q}_\alpha := \{f \in \mathcal{CBF} : \lambda^{1-\alpha} f(\lambda) \in \mathcal{CBF}\}.$$

We prove that $\mathscr{Q}_\alpha = \mathcal{CBF}^\alpha$. If $f \in \mathcal{CBF}^\alpha$, then $g = f^{1/\alpha} \in \mathcal{CBF}$. Thus by the logarithmic convexity of $\mathcal{CBF}$, we have $\lambda^{1-\alpha} f(\lambda) = \lambda^{1-\alpha} g^\alpha(\lambda) \in \mathcal{CBF}$, which implies that $\mathcal{CBF}^\alpha \subset \mathscr{Q}_\alpha$.

Conversely, suppose $f \in \mathscr{Q}_\alpha$, that is, $f \in \mathcal{CBF}$ and $h(\lambda) := \lambda^{1-\alpha} f(\lambda) \in \mathcal{CBF}$. By Theorem 6.17, there exists a pair $(\gamma, \eta) \in \Gamma$ such that

$$\log f(\lambda) = \gamma + \int_0^\infty \left(\frac{t}{1+t^2} - \frac{1}{\lambda+t}\right) \eta(t)\, dt,$$

where Γ is the set defined in (6.22). Since $\log h(\lambda) = (1-\alpha)\log\lambda + \log f(\lambda)$ and

$$\log\lambda = \int_0^\infty \left(\frac{t}{1+t^2} - \frac{1}{\lambda+t}\right) dt,$$

we have

$$\log h(\lambda) = \gamma + \int_0^\infty \left(\frac{t}{1+t^2} - \frac{1}{\lambda+t}\right) (\eta(t) + 1 - \alpha)\, dt.$$

Since $h \in \mathcal{CBF}$, we have $0 \le \eta(t) + 1 - \alpha \le 1$, which implies $0 \le \eta(t) \le \alpha$. Hence $f^{1/\alpha} \in \mathcal{CBF}$, that is, $\mathscr{Q}_\alpha \subset \mathcal{CBF}^\alpha$. □

Proposition 7.14 is very useful when it comes to compositions and products of complete Bernstein functions with fractional powers. The following corollary collects a few examples:

Corollary 7.15. *Let* $f, g \in \mathcal{CBF}$. *Then*

(i) $(f^\alpha(\lambda) + g^\alpha(\lambda))^{1/\alpha} \in \mathcal{CBF}$ *for all* $\alpha \in [-1,1]\setminus\{0\}$;

(ii) $(f(\lambda^\alpha) + g(\lambda^\alpha))^{1/\alpha} \in \mathcal{CBF}$ *for all* $\alpha \in [-1,1]\setminus\{0\}$;

(iii) $f(\lambda^\alpha)\cdot g(\lambda^{1-\alpha}) \in \mathcal{CBF}$ *for all* $\alpha \in [0,1]$.

Proof. Without loss of generality we can assume $\alpha \ne 0$ throughout. (i) follows from the fact that $\mathcal{CBF}^\alpha$ is a convex cone, cf. Proposition 7.14. (ii) can be rephrased as $h(\lambda^\alpha)^{1/\alpha} \in \mathcal{CBF}$ where $h := f + g \in \mathcal{CBF}$. Assume first that $0 < \alpha \le 1$. Note that $h \in \mathcal{CBF}$ implies that $\lambda/h(\lambda)$ is in $\mathcal{CBF}$. Therefore, $\lambda^\alpha/h(\lambda^\alpha) \in \mathcal{CBF}$ as well as $\lambda^{1-\alpha} h(\lambda^\alpha) = \lambda/(\lambda^\alpha/h(\lambda^\alpha)) \in \mathcal{CBF}$. Since $h(\lambda^\alpha)$ is in $\mathcal{CBF}$, we have $h(\lambda^\alpha) \in \mathcal{CBF}^\alpha$, and consequently $h(\lambda^\alpha)^{1/\alpha} \in \mathcal{CBF}$. If $-1 \le \alpha < 0$, set $\beta := -\alpha > 0$. Since $h(\lambda^\beta)^{1/\beta}$ is in a complete Bernstein function, (7.1) shows that also $h(\lambda^\alpha)^{1/\alpha} = 1/h(1/\lambda^\beta)^{1/\beta}$ is a complete Bernstein function.

To see (iii), observe that for $f, g \in \mathcal{CBF}$ and $\alpha \in (0,1)$ part (ii) implies that $f(\lambda^\alpha)^{1/\alpha}$ and $g(\lambda^{1-\alpha})^{1/(1-\alpha)}$ are in $\mathcal{CBF}$; because of Proposition 7.13 we get

$$f(\lambda^\alpha)g(\lambda^{1-\alpha}) = \left(f(\lambda^\alpha)^{1/\alpha}\right)^\alpha \left(g(\lambda^{1-\alpha})^{1/(1-\alpha)}\right)^{1-\alpha} \in \mathcal{CBF}. \quad □$$

The principle used in Corollary 7.15 to get new functions of type $\mathcal{CBF}$ from given ones can be generalized in the following way.

Proposition 7.16. *Let* $f, g, h \in \mathcal{CBF}$ *and* $f \not\equiv 0$. *Then*

(i) $f(\lambda) \cdot g\left(\frac{\lambda}{f(\lambda)}\right) \in \mathcal{CBF}$;

(ii) $h(f(\lambda)) \cdot g\left(\frac{\lambda}{f(\lambda)}\right) \in \mathcal{CBF}$.

Proof. (i) $f, g \in \mathcal{CBF}$ implies that $\lambda/f(\lambda)$ and $\lambda/g(\lambda)$ are in $\mathcal{CBF}$; by Corollary 7.9 (iii) the function

$$\frac{\lambda}{f(\lambda) g\left(\frac{\lambda}{f(\lambda)}\right)} = \frac{\frac{\lambda}{f(\lambda)}}{g\left(\frac{\lambda}{f(\lambda)}\right)}$$

is in $\mathcal{CBF}$, and Proposition 7.1 shows $f(\lambda)g(\lambda/f(\lambda)) \in \mathcal{CBF}$.

(ii) Since $f, g \in \mathcal{CBF}$ and f is non-zero, we know that $\lambda/f(\lambda)$, $g(\lambda/f(\lambda))$ and, by (i), $f(\lambda)g(\lambda/f(\lambda))$ are complete Bernstein functions. Thus, for every $t > 0$,

$$\frac{f(\lambda) + t}{f(\lambda)g\left(\frac{\lambda}{f(\lambda)}\right)} = \frac{1}{g\left(\frac{\lambda}{f(\lambda)}\right)} + \frac{t}{f(\lambda)g\left(\frac{\lambda}{f(\lambda)}\right)} \in \mathcal{S}$$

or, equivalently,

$$\frac{f(\lambda)}{f(\lambda) + t} g\left(\frac{\lambda}{f(\lambda)}\right) = \frac{1}{f(\lambda) + t} f(\lambda)g\left(\frac{\lambda}{f(\lambda)}\right) \in \mathcal{CBF}. \tag{7.8}$$

Since any $h \in \mathcal{CBF}$ can be written as

$$h(\lambda) = a + b\lambda + \int_{(0,\infty)} \frac{\lambda}{\lambda + t} \sigma(dt), \quad \lambda > 0,$$

we arrive at the desired conclusion by integrating (7.8) in t and adding the complete Bernstein function $ag(\lambda/f(\lambda)) + bf(\lambda)g(\lambda/f(\lambda))$. $\square$

Proposition 7.17. *If* $f \in \mathcal{CBF}$ *is non-constant, then for any* $\lambda_0 > 0$, *the functions*

$$g(\lambda) = \frac{\lambda - \lambda_0}{f(\lambda) - f(\lambda_0)} \mathbb{1}_{\{\lambda \neq \lambda_0\}} + \frac{1}{f'(\lambda_0)} \mathbb{1}_{\{\lambda = \lambda_0\}},$$

$$h(\lambda) = \frac{\lambda f(\lambda) - \lambda_0 f(\lambda_0)}{\lambda - \lambda_0} \mathbb{1}_{\{\lambda \neq \lambda_0\}} + \left(\lambda_0 f'(\lambda_0) + f(\lambda_0)\right)\mathbb{1}_{\{\lambda = \lambda_0\}},$$

$$k(\lambda) = \left(\frac{\lambda f(\lambda) - \lambda_0 f(\lambda_0)}{f(\lambda) - f(\lambda_0)}\right)^{1/2} \mathbb{1}_{\{\lambda \neq \lambda_0\}} + \left(\lambda_0 + \frac{f(\lambda_0)}{f'(\lambda_0)}\right)^{1/2} \mathbb{1}_{\{\lambda = \lambda_0\}},$$

$$\ell(\lambda) = \exp\left(\frac{\lambda \log f(\lambda) - \lambda_0 \log f(\lambda_0)}{\lambda - \lambda_0}\right) \mathbb{1}_{\{\lambda \neq \lambda_0\}}$$

$$+ \left(f(\lambda_0) + e^{\frac{\lambda_0 f'(\lambda_0)}{f(\lambda_0)}}\right) \mathbb{1}_{\{\lambda = \lambda_0\}},$$

$$p(\lambda) = \log\left(\frac{1 - \frac{\lambda}{\lambda_0}}{1 - \frac{f(\lambda)}{f(\lambda_0)}}\right) \mathbb{1}_{\{\lambda \neq \lambda_0\}} + \log\left(\frac{f(\lambda_0)}{\lambda_0 f'(\lambda_0)}\right) \mathbb{1}_{\{\lambda = \lambda_0\}},$$

$$q(\lambda) = \frac{\lambda - \lambda_0}{\log f(\lambda) - \log f(\lambda_0)} \mathbb{1}_{\{\lambda \neq \lambda_0\}} + \frac{f(\lambda_0)}{f'(\lambda_0)} \mathbb{1}_{\{\lambda = \lambda_0\}},$$

are in $\mathcal{CBF}$.

Proof. Since $f \in \mathcal{CBF}$, f has a representation of the following form, cf. Theorem 6.9,

$$f(\lambda) = \alpha + \beta\lambda + \int_{(0,\infty)} \frac{\lambda t - 1}{\lambda + t} \rho(dt)$$

for some $\alpha, \beta \geq 0$ and some measure ρ on $(0, \infty)$. For $\lambda \neq \lambda_0$ we see

$$\frac{f(\lambda) - f(\lambda_0)}{\lambda - \lambda_0} = \beta + \int_{(0,\infty)} \frac{1}{\lambda + t} \frac{1 + t^2}{\lambda_0 + t} \rho(dt).$$

The right-hand side is defined for all $\lambda > 0$ and it is clearly an element of $\mathcal{S}$. Moreover, if $\lambda = \lambda_0$, the value $f'(\lambda_0)$ extends the left-hand side continuously. This shows that the function g given in the statement of the proposition is defined for all $\lambda > 0$, it satisfies $1/g \in \mathcal{S}$ and, therefore, $g \in \mathcal{CBF}$. Further,

$$h(\lambda) = \lambda \frac{f(\lambda) - f(\lambda_0)}{\lambda - \lambda_0} + f(\lambda_0) = \frac{\lambda}{g(\lambda)} + f(\lambda_0) \in \mathcal{CBF}$$

because of Proposition 7.1. Using the log-convexity of $\mathcal{CBF}$ we see

$$k(\lambda) = \left(g(\lambda)\right)^{1/2} \left(h(\lambda)\right)^{1/2} \in \mathcal{CBF}.$$

For the function $\ell(\lambda)$ we use the representation (6.21) for f to find

$$\log \ell(\lambda) = \frac{\lambda \log f(\lambda) - \lambda_0 \log f(\lambda_0)}{\lambda - \lambda_0}$$

$$= \gamma + \int_0^\infty \left(\frac{t}{1 + t^2} - \frac{t}{(\lambda + t)(\lambda_0 + t)}\right) \eta(t)\, dt$$

$$= \gamma + \int_0^\infty \frac{\lambda_0 t\, \eta(t)}{(1 + t^2)(\lambda_0 + t)}\, dt$$

$$+ \int_0^\infty \left(\frac{t}{1 + t^2} - \frac{1}{\lambda + t}\right) \frac{t\eta(t)}{\lambda_0 + t}\, dt.$$

Since $0 \leq \eta(t) \leq 1$, we have $0 \leq t\eta(t)(\lambda_0 + t)^{-1} \leq 1$ which shows that $\ell \in \mathcal{CBF}$.

Set $\pi(\lambda) := \lambda_0^{-1} f(\lambda_0) g(\lambda)$. Since $\pi \in \mathcal{CBF}$ and $\pi(0+) = 1$, Corollary 6.19 shows that $p = \log \pi \in \mathcal{CBF}$.

For the function $q(\lambda)$ we use again the representation (6.21) for f to find

$$
\begin{aligned}
\log f(\lambda) - \log f(\lambda_0) &= \int_0^\infty \left(\frac{1}{\lambda_0 + t} - \frac{1}{\lambda + t} \right) \eta(t)\, dt \\
&= \int_0^\infty \frac{\lambda - \lambda_0}{(\lambda_0 + t)(\lambda + t)}\, \eta(t)\, dt \\
&= (\lambda - \lambda_0) \int_0^\infty \frac{1}{\lambda + t} \frac{\eta(t)}{\lambda_0 + t}\, dt.
\end{aligned}
$$

Since the function

$$
\lambda \mapsto \int_0^\infty \frac{1}{\lambda + t} \frac{\eta(t)}{\lambda_0 + t}\, dt
$$

is a Stieltjes function, the assertion follows immediately from Theorem 7.3. $\square$

Proposition 7.18. *Let* $f \in \mathcal{CBF}$ *and define* $g : (0, \infty) \to \mathbb{R}$ *by*

$$
g(\lambda) = \frac{\lambda - 1}{f(\lambda) - f(\lambda^{-1})} \mathbb{1}_{\{\lambda \neq 1\}} + \frac{1}{2 f'(1)} \mathbb{1}_{\{\lambda = 1\}}. \tag{7.9}
$$

Then $g \in \mathcal{CBF}$ *and* $g(\lambda^{-1}) = 1/g^\star(\lambda)$.

Proof. Define

$$
\phi(\lambda) := \frac{f(\lambda) - f(1)}{\lambda - 1} \mathbb{1}_{\{\lambda \neq 1\}} + f'(1)\, \mathbb{1}_{\{\lambda = 1\}}.
$$

Then

$$
\frac{1}{g(\lambda)} = \frac{f(\lambda) - f(\lambda^{-1})}{\lambda - 1} \mathbb{1}_{\{\lambda \neq 1\}} + 2 f'(1)\, \mathbb{1}_{\{\lambda = 1\}} = \phi(\lambda) + \lambda^{-1} \phi(\lambda^{-1}),
$$

and by the first entry of Proposition 7.17, $\phi \in \mathcal{S}$. By Corollary 7.9 (iv), $\phi(1/\lambda)$ is a complete Bernstein function and by Theorem 6.2 (ii), $\lambda \mapsto \lambda^{-1} \phi(1/\lambda)$ is a Stieltjes function. Now $g \in \mathcal{CBF}$ follows from Theorem 7.3. A straightforward computation gives that $g(\lambda^{-1}) = g(\lambda)/\lambda = 1/g^\star(\lambda)$. $\square$

If we apply Proposition 7.18 with $f(\lambda) = \lambda/(\lambda + t)$ and use that for every $g \in \mathcal{CBF}$ the conjugate function $\lambda/g(\lambda)$ is also in $\mathcal{CBF}$, see Proposition 7.1, we arrive at

Corollary 7.19. *For every* $t > 0$ *the functions*

$$
\lambda \mapsto \frac{(\lambda + t)(1 + \lambda t)}{\lambda + 1} \quad \text{and} \quad \lambda \mapsto \frac{\lambda}{\lambda + t} \frac{\lambda + 1}{1 + \lambda t}
$$

are in $\mathcal{CBF}$.

Recall that $f^\star(\lambda) = \lambda/f(\lambda)$. We say that $f \in \mathcal{CBF}$ is *symmetric* if it satisfies $f(\lambda^{-1}) = 1/f^\star(\lambda)$. The family of symmetric complete Bernstein function is denoted by $\mathcal{CBF}^{\text{sym}}$. This family is a convex cone, and $f \in \mathcal{CBF}^{\text{sym}}$ if, and only if, $f^\star \in \mathcal{CBF}^{\text{sym}}$. We will now derive an integral representation for symmetric complete Bernstein functions.

Let $f \in \mathcal{CBF}$ have the Stieltjes representation (6.7). Define a measure $\hat{\sigma}$ on $(0, \infty)$ by

$$\hat{\sigma}(B) = \int_{(0,\infty)} \mathbb{1}_B(t^{-1}) t^{-1} \sigma(dt), \quad B \subset (0, \infty). \tag{7.10}$$

Note that $\int_{(0,\infty)} (1+t)^{-1} \hat{\sigma}(dt) = \int_{(0,\infty)} (1+t)^{-1} \sigma(dt) < \infty$. By a change of variable,

$$\int_{(0,\infty)} \frac{\lambda}{\lambda+t} \sigma(dt) = \int_{(0,\infty)} \frac{\lambda}{1+\lambda t} \hat{\sigma}(dt).$$

Thus we have the following alternative representation of complete Bernstein functions: $f \in \mathcal{CBF}$ if, and only if, there exist non-negative constants $a, b \geq 0$ and a measure $\hat{\sigma}$ satisfying $\int_{(0,\infty)} (1+t)^{-1} \hat{\sigma}(dt) < \infty$ such that

$$f(\lambda) = a + b\lambda + \int_{(0,\infty)} \frac{\lambda}{1+\lambda t} \hat{\sigma}(dt). \tag{7.11}$$

Proposition 7.20. *Let $f \in \mathcal{CBF}$ have the representation (6.7) and let $\hat{\sigma}$ be the measure defined by (7.10). Then the following assertions are equivalent.*

(i) $f(\lambda^{-1}) = 1/f^\star(\lambda)$ *for all* $\lambda > 0$;

(ii) $a = b$ *and* $\hat{\sigma} = \sigma$;

(iii) *There exists a finite measure v on $[0, 1]$ such that*

$$f(\lambda) = \lambda \int_{[0,1]} \left(\frac{1}{\lambda+t} + \frac{1}{1+\lambda t} \right) (1+t)\, v(dt). \tag{7.12}$$

Proof. (i)$\Leftrightarrow$ (ii) Since

$$f(\lambda^{-1}) = a + b\lambda^{-1} + \int_{(0,\infty)} \frac{\lambda^{-1}}{\lambda^{-1}+t} \sigma(dt)$$

$$= a + \frac{b}{\lambda} + \int_{(0,\infty)} \frac{1}{1+\lambda t} \sigma(dt)$$

$$= a + \frac{b}{\lambda} + \int_{(0,\infty)} \frac{1}{\lambda+t} \hat{\sigma}(dt),$$

and

$$\frac{1}{f^\star(\lambda)} = \frac{f(\lambda)}{\lambda} = \frac{a}{\lambda} + b + \int_{(0,\infty)} \frac{1}{\lambda+t} \sigma(dt),$$

it follows that $f(\lambda^{-1}) = 1/f^\star(\lambda)$, $\lambda > 0$, is satisfied if, and only if, $a = b$ and $\int_{(0,\infty)}(\lambda+t)^{-1}\,\hat\sigma(dt) = \int_{(0,\infty)}(\lambda+t)^{-1}\,\sigma(dt)$. Since any Stieltjes function uniquely determines its representing measure we conclude that $\hat\sigma = \sigma$.

(ii)$\Rightarrow$(iii) By a change of variables and the equality $\hat\sigma = \sigma$ we get that

$$\int_{(1,\infty)} \frac{1}{\lambda + t}\,\sigma(dt) = \int_{(0,1)} \frac{1}{1 + \lambda t}\,\hat\sigma(dt) = \int_{(0,1)} \frac{1}{1 + \lambda t}\,\sigma(dt).$$

Therefore,

$$\frac{f(\lambda)}{\lambda} = \frac{a}{\lambda} + a + \int_{(0,1)} \frac{1}{\lambda + t}\,\sigma(dt) + \frac{\sigma\{1\}}{1 + \lambda} + \int_{(1,\infty)} \frac{1}{\lambda + t}\,\sigma(dt)$$

$$= a\left(\frac{1}{\lambda} + 1\right) + \int_{(0,1)} \left(\frac{1}{\lambda + t} + \frac{1}{1 + \lambda t}\right)\sigma(dt) + \frac{\sigma\{1\}}{1 + \lambda}$$

$$= \int_{[0,1]} \left(\frac{1}{\lambda + t} + \frac{1}{1 + \lambda t}\right)(1 + t)\,\nu(dt),$$

where $\nu\{0\} = a$ and $\nu(B) = \int_B (1 + t)^{-1}\,\sigma(dt)$ for all measurable $B \subset (0, 1]$. Since $\int_{(0,1]}(1 + t)^{-1}\,\sigma(dt) < \infty$, ν is a finite measure on $[0, 1]$.

(iii)$\Rightarrow$(i) It is straightforward to check that the representation (7.12) implies that $f(\lambda^{-1}) = 1/f^\star(\lambda)$. $\square$

For a function $f \in \mathcal{CBF}$ define its symmetrization by $f^{\mathrm{sym}}(\lambda) = g^\star(\lambda)$ where g is given by (7.9). Note that

$$f^{\mathrm{sym}}(\lambda) = \lambda\,\frac{f(\lambda) - f(\lambda^{-1})}{\lambda - 1}\,\mathbb{1}_{\{\lambda \neq 1\}} + 2f'(1)\,\mathbb{1}_{\{\lambda = 1\}}.$$

Then we can restate the conclusion of Proposition 7.18 by saying that $f^{\mathrm{sym}} \in \mathcal{CBF}^{\mathrm{sym}}$. The next corollary says that if f is already symmetric, then it is equal to its symmetrization.

Corollary 7.21. *If $f \in \mathcal{CBF}^{\mathrm{sym}}$, then $f^{\mathrm{sym}} = f$.*

Proof. By Proposition 7.20, there is a finite measure ν on $[0, 1]$ such that

$$f(\lambda) = \lambda \int_{[0,1]} \left(\frac{1}{\lambda + t} + \frac{1}{1 + \lambda t}\right)(1 + t)\,\nu(dt).$$

By a straightforward calculation,

$$f(\lambda^{-1}) = \int_{[0,1]} \left(\frac{1}{1 + \lambda t} + \frac{1}{\lambda + t}\right)(1 + t)\,\nu(dt).$$

Subtracting the last two equalities we get

$$f(\lambda) - f(\lambda^{-1}) = (\lambda - 1) \int_{[0,1]} \left(\frac{1}{\lambda + t} + \frac{1}{1 + \lambda t}\right)(1 + t)\,\nu(dt),$$

which implies that

$$f^{\text{sym}}(\lambda) = \frac{\lambda}{\lambda - 1}\left(f(\lambda) - f(\lambda^{-1})\right)$$

$$= \lambda \int_{[0,1]}\left(\frac{1}{\lambda + t} + \frac{1}{1 + \lambda t}\right)(1 + t)\,v(dt) = f(\lambda). \qquad \square$$

The following two propositions contain further results which allow us to construct complete Bernstein functions from a Bernstein function or a completely monotone function.

Proposition 7.22. *Let* $f \in \mathcal{BF}$, $\alpha \in (0, 1/2]$ *and* $\gamma \in (0, 1 - \alpha]$. *Then the function* $g(\lambda) := \lambda^\gamma f(\lambda^\alpha)$ *is in* $\mathcal{CBF}$.

Proof. Let $z = re^{i\omega} \in \mathbb{H}^\uparrow$. Since $\arg z^\alpha = \alpha\omega \in (0, \pi/2)$, i.e. $z^\alpha \in \overrightarrow{\mathbb{H}}$, Theorem 3.6 shows that $z^\gamma f(z^\alpha)$ is well defined and holomorphic for $z \in \mathbb{H}^\uparrow$. Since f preserves sectors in $\overrightarrow{\mathbb{H}}$, we find

$$\left|\arg f(z^\alpha)\right| \leq \arg z^\alpha = \alpha\omega \quad \text{and} \quad \left|\arg\left(z^\gamma f(z^\alpha)\right)\right| = (\alpha + \gamma)\omega \leq \omega < \pi.$$

Consequently, $z^\gamma f(z^\alpha) \in \mathbb{H}^\uparrow$. Since $\lambda^\gamma f(\lambda^\alpha)$ is non-negative for all $\lambda > 0$ and $\lim_{\lambda\to 0+}\lambda^\gamma f(\lambda^\alpha) = 0$, we find by Theorem 6.2 (iv) that $g(z) := z^\gamma f(z^\alpha)$ is a complete Bernstein function. $\qquad \square$

The proof of Proposition 7.22 is easily adapted to show that for every $f \in \mathcal{BF}$ and $\phi \in \mathcal{CBF}$ satisfying $\phi : \mathbb{H}^\uparrow \to \mathbb{H}^\uparrow \cap \overrightarrow{\mathbb{H}}$ the function $z \mapsto \phi(z)f(\phi(z))$ is in $\mathcal{CBF}$. Moreover,

$$\arg\left[\frac{z}{\phi(z)}f(\phi(z))\right] = \arg z + \arg f(\phi(z)) - \arg \phi(z) \leq \arg z < \pi,$$

which shows that also $z \mapsto \frac{z}{\phi(z)}f(\phi(z))$ is in $\mathcal{CBF}$. A sufficient condition such that $\phi : \mathbb{H}^\uparrow \to \mathbb{H}^\uparrow \cap \overrightarrow{\mathbb{H}}$ is, for example, that $\phi^2 \in \mathcal{CBF}$.

Proposition 7.23. *Let* $f \in \mathcal{CM}$ *such that* $f(0+) = 1$ *and let* $\alpha \in (0, 1/2]$. *Then the function* $g(\lambda) := \lambda^\alpha(1 + f(\lambda^\alpha))$ *is in* $\mathcal{CBF}$.

Proof. Fix $t > 0$ and $\alpha \in (0, 1/2]$. For every $z \in \mathbb{H}^\uparrow$ we have $z^\alpha \in \mathbb{H}^\uparrow \cap \overrightarrow{\mathbb{H}}$, i.e. $z^\alpha = x + iy$ with $x = x(\alpha) > 0$ and $y = y(\alpha) > 0$. Therefore,

$$z^\alpha\left(1 + e^{-tz^\alpha}\right) = (x + iy)\left(1 + e^{-tx}\cos(ty) - ie^{-tx}\sin(ty)\right).$$

Using the elementary inequalities $e^{tx} - 1 \geq tx$ and $\sin(ty) \leq ty$ we see

$$\text{Im}\left(z^\alpha(1 + e^{-tz^\alpha})\right) = e^{-tx}\left(ye^{tx} + y\cos(ty) - x\sin(ty)\right)$$
$$\geq e^{-tx}\left(y(e^{tx} - 1) - x\sin(ty)\right)$$
$$\geq e^{-tx}\left(ytx - xty\right) = 0.$$

This means that the function $z \mapsto z^\alpha (1 + e^{-tz^\alpha})$ is holomorphic and preserves the upper half-plane. Since $\lambda^\alpha (1 + e^{-t\lambda^\alpha})$, $\lambda > 0$, is real and tends to 0 as $\lambda \to 0$, it is a complete Bernstein function, cf. Theorem 6.2 (iv).

From $f(0+) = 1$ and Bernstein's theorem, Theorem 1.4, we conclude that there exists a probability measure π on $[0, \infty)$ such that

$$f(\lambda) = \mathscr{L}(\pi; \lambda) = \int_{[0,\infty)} e^{-t\lambda}\, \pi(dt).$$

Then

$$g(\lambda) = \lambda^\alpha (1 + f(\lambda^\alpha)) = \int_{[0,\infty)} \lambda^\alpha \left(1 + e^{-t\lambda^\alpha}\right) \pi(dt),$$

and the claim follows because g is a mixture of $\mathcal{CBF}$ functions, cf. Corollary 7.6. □

Comments 7.24. Proposition 7.1 appears in [269, 35] and, independently, in [322]. The second equality of Theorem 7.3 is already mentioned in [269] and it is implicitly contained in [162, 163]. Stability properties of $\mathcal{CBF}$ and $\mathcal{S}$ under particular compositions have been studied by Hirsch [162, 163], see also [33, 35]: the 'if' directions of (7.1)–(7.4) appear in [162], (7.5) is due to Reuter [303] and (7.2) and the Stieltjes version of (7.3) can be found in van Herk [361, Theorems 7.49, 7.50]. The general statement of Corollary 7.9 and the characterizations of $\mathcal{CBF}$ and $\mathcal{S}$ in terms of their stability properties, Theorem 7.5, seem to be new.

The calculations for the examples in Remark 7.8 are the same as those of the corresponding Remark 2.4 for Stieltjes functions. An easier way to arrive there, without duplicating all calculations, would be to use the examples of Remark 2.4 and to observe that $f \in \mathcal{CBF}$ if, and only if, $f(\lambda)/\lambda \in \mathcal{S}$, see Theorem 6.2 (ii).

Corollary 7.10 is an extension of Teke and Deshmukh [346, Theorem 2.1] which contains the assertion 7.10 (iii). The case $n = 1$ of Proposition 7.12 is due to [354], while the case $n > 1$ seems to be new. The log-convexity of $\mathcal{CBF}$ and $\mathcal{S}$, Proposition 7.13, was originally shown by Berg [32] using a rather complicated function-theoretic proof. Our first argument is from Nakamura [269] while the second is due to Berg, see [33].

The cones $\mathcal{CBF}^\alpha$ and $\mathcal{S}^\alpha$ were introduced by Nakamura [269] for $\alpha \neq 0$, and Proposition 7.14 along with Corollary 7.15 can be found there for positive $\alpha > 0$. Since $\lim_{\alpha \to 0}(\frac{1}{2} f^\alpha + \frac{1}{2} g^\alpha)^{1/\alpha} = \sqrt{fg}$ (use, e.g. de l'Hospital's rule), the limiting case of Corollary 7.15 (i) contains yet another proof of the log-convexity of $\mathcal{CBF}$. Part (ii) of Corollary 7.15 seems to be due to Ando [9]; for general $|\alpha| \leq 1$ it appears, with a different proof using the mapping property of $\mathcal{CBF}$, in [320, 321]. Proposition 7.16 is from Uchiyama [353, Lemma 2.1], but his proof of part (ii) has a gap. Proposition 7.17 is mainly from [269], the function p appears in [233, Lemma 2] and q is from [355, Theorem 2.1], both with different proofs.

Proposition 7.18 is from [132], the given proof is different. The concept of symmetric complete Bernstein functions can be traced back to [227], the equivalence of (i) and (iii) in Proposition 7.20 is from [127], our proof is new. It is also shown in [127] that $\Lambda f(\lambda) := f(0)^{-1}(1+\lambda) - f(\lambda)^{-1}(\lambda-1)^2$ is a bijection from $\mathcal{CBF}^{\mathrm{sym}} \setminus \mathcal{CBF}_0^{\mathrm{sym}}$ to $\mathcal{CBF}_0^{\mathrm{sym}}$.

The statement of Proposition 7.22 was communicated to us by Wissem Jedidi and Sonia Fourati [119], see also [120] for a different proof, the related Proposition 7.23 seems to be new. In some sense, they are generalizations of the entries 18 and 19 in the Tables 16.3, page 312.

Complete Bernstein functions can also be characterized in the following way: a continuously differentiable function $f : (0, \infty) \to \mathbb{R}$ is in $\mathcal{CBF}$ if, and only if, the kernel

$$K(s,t) = \begin{cases} \dfrac{f(s) - f(t)}{s - t}, & \text{if } s, t > 0, \ s \neq t, \\[2mm] f'(t), & \text{if } s, t > 0, \ s = t \end{cases}$$

is positive definite in the sense that

$$\int_0^\infty \int_0^\infty K(s,t)\,\phi(s)\overline{\phi(t)}\,ds\,dt \geq 0$$

for all compactly supported continuous functions $\phi : (0, \infty) \to \mathbb{C}$, cf. also the Comments 4.16. A proof can be found in Rosenblum and Rovnyak [308, Chapter 2.12]. The discrete version requiring for all $n \in \mathbb{N}$, $s_1, \dots, s_n \in \mathbb{R}$ and $c_1, \dots, c_n \in \mathbb{C}$

$$\sum_{j,k=1}^n K(s_j, s_k) c_j \overline{c}_k \geq 0$$

can be found in Korányi [214] and Korányi and Nagy [215]. These are variants of Löwner's original result that a C^1-function $f : (0, \infty) \to \mathbb{R}$ is in $\mathcal{CBF}$ if, and only if, the Pick matrix

$$\left(\left(f(z_j) - \overline{f(z_k)} \right) / \left(z_j - \bar{z}_k \right) \right)_{j,k=1,\dots n}, \qquad z_1, \dots, z_n \in \mathbb{H}^\uparrow$$

is non-negative hermitian, cf. [248], see also Comments 12.18.

Chapter 8

Thorin–Bernstein functions

Proposition 7.11 contains a characterization of those complete Bernstein functions f whose derivative f' is a Stieltjes function. For probabilists this subclass of $\mathcal{CBF}$ is quite important, see Proposition 9.11 and Remark 9.18. We begin with a definition which reminds of the definition of complete Bernstein functions, Definition 6.1, based on the representation result for general Bernstein functions, cf. Theorem 3.2.

Definition 8.1. A Bernstein function f is called a *Thorin–Bernstein function* if the Lévy measure μ from (3.2) has a density $m(t)$ such that $t \mapsto t \cdot m(t)$ is completely monotone, i.e.

$$f(\lambda) = a + b\lambda + \int_{(0,\infty)} (1 - e^{-\lambda t}) m(t) \, dt, \tag{8.1}$$

where $a, b \geq 0$, $\int_{(0,\infty)} (1 \wedge t) m(t) \, dt < \infty$ and $t \mapsto t \cdot m(t)$ is completely monotone. We will use $\mathcal{TBF}$ to denote the family of all Thorin–Bernstein functions.

Observe that the complete monotonicity of $t \mapsto t \cdot m(t)$ implies immediately that $m(t) = t^{-1} \cdot (t \cdot m(t))$ is completely monotone. Therefore, $\mathcal{TBF} \subset \mathcal{CBF}$.

Theorem 8.2. *For a function $f : (0, \infty) \to (0, \infty)$ the following assertions are equivalent.*

(i) $f \in \mathcal{TBF}$.

(ii) $f' \in \mathcal{S}$ and $f(0+) = \lim_{\lambda \to 0+} f(\lambda)$ exists.

(iii) f is of the form

$$f(\lambda) = a + b\lambda + \int_{(0,\infty)} \log\left(1 + \frac{\lambda}{t}\right) \tau(dt), \tag{8.2}$$

where $a, b \geq 0$ and with an unique representing measure τ on $(0, \infty)$ satisfying $\int_{(0,1)} |\log t| \, \tau(dt) + \int_{[1,\infty)} t^{-1} \tau(dt) < \infty$.

(iv) $f \in \mathcal{CBF}$ and $f' \in \mathcal{S}$.

(v) f is of the form

$$f(\lambda) = a + b\lambda + \int_0^\infty \frac{\lambda}{\lambda + t} \frac{w(t)}{t} \, dt \tag{8.3}$$

where $a, b \geq 0$ are positive constants and $w : (0, \infty) \to [0, \infty)$ is a non-decreasing function such that $\int_0^\infty (1 + t)^{-1} t^{-1} w(t)\, dt < \infty$. In order to assure the uniqueness of w we can, and will, assume that w is left-continuous.

Proof. (i)$\Rightarrow$(ii) Let $f \in \mathcal{TBF}$. Then $f(0+)$ exists and is finite. Moreover,

$$f(\lambda) = a + b\lambda + \int_0^\infty (1 - e^{-\lambda t}) m(t)\, dt,$$

where $t \cdot m(t)$ is completely monotone. Write $t \cdot m(t) = \mathscr{L}(v; t)$ with a suitable measure v. Differentiation gives

$$f'(\lambda) = b + \int_0^\infty e^{-\lambda t} t\, m(t)\, dt = b + \mathscr{L}\big(\mathscr{L}(v; t)\, dt; \lambda\big),$$

and by Theorem 2.2 (i) we find $f' \in \mathcal{S}$.

(ii)$\Rightarrow$(iii) By assumption,

$$f'(\lambda) = \frac{\alpha}{\lambda} + \beta + \int_{(0,\infty)} \frac{1}{\lambda + t}\, \tau(dt).$$

Since the primitive $f(\lambda)$ exists with $f(0+)$ finite, we see that necessarily $\alpha = 0$ as well as

$$f(\lambda) = a + \beta\lambda + \int_{(0,\infty)} \big(\log(\lambda + t) - \log t\big)\, \tau(dt)$$

$$= a + \beta\lambda + \int_{(0,\infty)} \log\left(1 + \frac{\lambda}{t}\right) \tau(dt).$$

Since f is non-negative, the integration constant a has to be non-negative and if we set $b = \beta$ we get (8.2). Since $f(1)$ is finite, τ integrates $t \mapsto \log(1 + t^{-1})$; equivalently, see Proposition 7.11, $\int_{(0,1)} |\log t|\, \tau(dt) + \int_{[1,\infty)} t^{-1}\, \tau(dt) < \infty$. The uniqueness of τ follows immediately from the uniqueness of the representing measure of Stieltjes functions, see the comment following Definition 2.1.

(iii)$\Rightarrow$(iv) This is the converse direction in Proposition 7.11.

(iv)$\Rightarrow$(v) By Proposition 7.11 every $f \in \mathcal{CBF}$ with $f' \in \mathcal{S}$ is of the form (7.7). Using Fubini's theorem we get

$$f(\lambda) = f(0+) + b\lambda + \int_{(0,\infty)} \int_t^\infty (-1)\frac{d}{ds} \log\left(1 + \frac{\lambda}{s}\right) ds\, \tau(dt)$$

$$= f(0+) + b\lambda + \int_{(0,\infty)} \int_t^\infty \frac{1}{s} \frac{\lambda}{\lambda + s}\, ds\, \tau(dt)$$

$$= f(0+) + b\lambda + \int_0^\infty \frac{\lambda}{\lambda + s} \frac{1}{s} \int_{(0,s)} \tau(dt)\, ds.$$

This shows (8.3) with the left-continuous density $w(t) = \tau(0, t)$ and $a = f(0+)$; the integrability condition on $w(t)$ is a consequence of Fubini's theorem and the fact that $f(1) < \infty$.

(v)$\Rightarrow$(i) We may use Fubini's theorem to get

$$f(\lambda) = a + b\lambda + \int_0^\infty \frac{\lambda}{\lambda + s} \frac{w(s)}{s} \, ds$$

$$= a + b\lambda + \int_0^\infty \left(\frac{1}{s} - \frac{1}{\lambda + s} \right) w(s) \, ds$$

$$= a + b\lambda + \int_0^\infty \int_0^\infty \left(e^{-st} - e^{-(\lambda + s)t} \right) w(s) \, dt \, ds$$

$$= a + b\lambda + \int_0^\infty (1 - e^{-\lambda t}) \left(\int_0^\infty e^{-ts} w(s) \, ds \right) dt.$$

Since w is non-decreasing, we have

$$\int_0^\infty \frac{1}{1+s} \frac{w(0+)}{s} \, ds \leq \int_0^\infty \frac{1}{1+s} \frac{w(s)}{s} \, ds < \infty$$

which is only possible if $w(0+) = 0$. Thus, $w(s) = \rho(0, s)$ for a suitable measure ρ defined on $(0, \infty)$, and thus

$$m(t) := \int_0^\infty e^{-ts} w(s) \, ds = \int_0^\infty e^{-ts} \int_{(0,s)} \rho(du) \, ds$$

$$= \int_{(0,\infty)} \int_u^\infty e^{-ts} \, ds \, \rho(du)$$

$$= \frac{1}{t} \int_{(0,\infty)} e^{-tu} \rho(du).$$

Hence, $t \cdot m(t) = \int_{(0,\infty)} e^{-tu} \rho(du)$ is completely monotone, and (i) follows. A comparison with the formulae in the second half of the proof of Proposition 7.11 allows us, because of the uniqueness of the measure in (8.2), to identify $\rho(ds)$ with $\tau(ds)$. $\square$

Remark 8.3. (i) (8.2) is called the Thorin representation of $f \in \mathcal{TBF}$ and the (unique) representing measure τ appearing in (8.2) is called the *Thorin measure* of f.

(ii) For further reference let us collect the relations between the various representing measures and densities from Definition 8.1 and Theorem 8.2:

$$m(t) = \int_{(0,\infty)} e^{-ts} w(s) \, ds = \mathcal{L}(w; t);$$

$$t \cdot m(t) = \int_{(0,\infty)} e^{-ts} \, dw(s) = \int_{(0,\infty)} e^{-ts} \tau(ds) = \mathcal{L}(\tau; t);$$

$$w(t) = \tau(0,t);$$

$$\sigma(dt) = \frac{w(t)}{t}\, dt.$$

(iii) The Thorin measure τ of $f \in \mathcal{TBF}$ is the Stieltjes measure of the complete Bernstein function $\lambda f'(\lambda)$. This useful observation can be seen in the following way: Let $f \in \mathcal{TBF}$ be given by

$$f(\lambda) = a + b\lambda + \int_{(0,\infty)} \log\left(1 + \frac{\lambda}{t}\right) \tau(dt).$$

Then $f' \in \mathcal{S}$ and

$$f'(\lambda) = b + \int_{(0,\infty)} \frac{1}{1 + \frac{\lambda}{t}} \frac{1}{t}\, \tau(dt).$$

Thus,

$$\lambda f'(\lambda) = b\lambda + \int_{(0,\infty)} \frac{\lambda}{\lambda + t}\, \tau(dt).$$

This shows that whenever we have an explicit representation for $f \in \mathcal{TBF}$ in terms of the Thorin measure, then we have an explicit representation of the complete Bernstein function $\lambda f'(\lambda)$ in terms of the Stieltjes measure.

(iv) With formula (8.3) it is easy to find examples of complete Bernstein functions which do not have derivatives in $\mathcal{S}$. For example,

$$f(\lambda) := \sqrt{\lambda}\arctan\frac{1}{\sqrt{\lambda}} = \int_0^1 \frac{\lambda}{\lambda + t}\frac{dt}{2\sqrt{t}} = \frac{1}{2}\int_0^1 \frac{\lambda}{\lambda + t}\frac{\sqrt{t}}{t}\, dt,$$

cf. Remark 7.8; but the density $\frac{1}{2}\frac{\sqrt{t}}{t}\mathbb{1}_{(0,1)}(t)$ is clearly not of the form $\frac{1}{t}\int_{(0,t)}\sigma(ds)$ which shows that f is of class $\mathcal{CBF}$ having a derivative which is not a Stieltjes function; in other words, $f \in \mathcal{CBF}\setminus\mathcal{TBF}$.

The next two theorems describe the structure of the family $\mathcal{TBF}$.

Theorem 8.4. *Let $f \in \mathcal{TBF}$. The following statements are equivalent.*

(i) $g \circ f \in \mathcal{TBF}$ *for every* $g \in \mathcal{TBF}$.

(ii) $f'/f \in \mathcal{S}$.

Proof. (ii)$\Rightarrow$(i) Assume that $f \in \mathcal{TBF}$ and $f'/f \in \mathcal{S}$. By Theorem 8.2, $f' \in \mathcal{S}$. From $1/\mathcal{S} = \mathcal{CBF}$ we conclude that

$$\frac{1}{\frac{f'(\lambda)}{f(\lambda)+t}} = \frac{f(\lambda)+t}{f'(\lambda)} = \frac{f(\lambda)}{f'(\lambda)} + \frac{t}{f'(\lambda)} = \frac{1}{\frac{f'(\lambda)}{f(\lambda)}} + \frac{t}{f'(\lambda)}$$

is a complete Bernstein function. This means that $f'(\lambda)/(f(\lambda) + t)$ is a Stieltjes function. For all $g \in \mathcal{TBF}$ we know that $g' \in \mathcal{S}$ and

$$g'(\lambda) = b + \int_{(0,\infty)} \frac{1}{\lambda + t}\, \tau(dt).$$

Therefore

$$(g \circ f)'(\lambda) = g'\big(f(\lambda)\big) f'(\lambda) = b f'(\lambda) + \int_{(0,\infty)} \frac{f'(\lambda)}{f(\lambda) + t}\, \tau(dt).$$

The integrand is a Stieltjes function, thus $(g \circ f)' \in \mathcal{S}$. Since $(g \circ f)(0+) < \infty$, Theorem 8.2 shows that $g \circ f \in \mathcal{TBF}$.

(i)$\Rightarrow$(ii) As $g(\lambda) = \log(1 + \lambda/t) \in \mathcal{TBF}$ for all $t > 0$, we see that $\log(1 + f(\lambda)/t)$ is a Thorin–Bernstein function for all $t > 0$. This implies that

$$\lambda \mapsto \frac{d}{d\lambda} \log\left(1 + \frac{f(\lambda)}{t}\right) = \frac{f'(\lambda)}{t + f(\lambda)} \in \mathcal{S}$$

for all $t > 0$. By letting $t \to 0$, we get that $f'/f \in \mathcal{S}$. $\qquad\square$

Theorem 8.5. *The set $\mathcal{TBF}$ is a convex cone which is closed under pointwise limits.*

Proof. That $\mathcal{TBF}$ is a convex cone follows immediately from the representation (8.1). Suppose that $(f_n)_{n \in \mathbb{N}}$ is a sequence of functions in $\mathcal{TBF}$ such that $\lim_{n \to \infty} f_n = f$ pointwise. It follows from Corollary 7.6 (iii) that $f \in \mathcal{CBF}$ and by Corollary 3.9 we find that $\lim_{n \to \infty} f'_n = f'$. Since $f'_n \in \mathcal{S}$, we get from Theorem 2.2 that $f' \in \mathcal{S}$. Hence, $f \in \mathcal{TBF}$, cf. Theorem 8.2. $\qquad\square$

Corollary 8.6. *If $(f_n)_{n \in \mathbb{N}} \subset \mathcal{TBF}$ with Thorin measures $(\tau_n)_{n \in \mathbb{N}}$, and $f = \lim_n f_n$ pointwise, then $\lim_{n \to \infty} \tau_n = \tau$ vaguely in $(0, \infty)$, where τ is the Thorin measure of f.*

Proof. This is proved similarly to the proof of Corollary 7.6 (ii). $\qquad\square$

The following assertion should be compared with Theorem 5.11.

Proposition 8.7. *Suppose that $g : (0, \infty) \to (0, \infty)$. Then the following statements are equivalent.*

(i) *$g = e^{-f}$ where $f \in \mathcal{TBF}$.*

(ii) *$-(\log g)' \in \mathcal{S}$ and $g(0+) \leq 1$.*

Proof. Suppose that $g = e^{-f}$ where $f \in \mathcal{TBF}$. Then $g = e^{-f}$ where $f' \in \mathcal{S}$, and since $f = -\log g$ we get (ii). The converse assertion follows by retracing the steps backwards. $\qquad\square$

Remark 8.8. Just as in the case of (complete) Bernstein functions, see Remarks 3.13 and 7.8, we can understand the representation formula (8.2) as a particular case of the Kreĭn–Milman or Choquet representation.

For this we rewrite (8.2) as

$$f(\lambda) = a + b\lambda + \int_{(0,\infty)} \frac{\log\left(1 + \frac{\lambda}{t}\right)}{\log\left(1 + \frac{1}{t}\right)} \log\left(1 + \frac{1}{t}\right) \tau(dt)$$

$$= \int_{[0,\infty]} \frac{\log\left(1 + \frac{\lambda}{t}\right)}{\log\left(1 + \frac{1}{t}\right)} \bar{\tau}(dt),$$

where $\bar{\tau} = a\delta_0 + \log(1 + t^{-1})\,\tau + b\delta_\infty$ is a finite measure on $[0, \infty]$. Here we used, cf. the proof of Proposition 7.11, that $\int_{(0,\infty)} \log\left(1 + t^{-1}\right)\tau(dt) < \infty$ and that

$$\lim_{t\to 0} \frac{\log\left(1 + \frac{\lambda}{t}\right)}{\log\left(1 + \frac{1}{t}\right)} = 1 \quad \text{and} \quad \lim_{t\to\infty} \frac{\log\left(1 + \frac{\lambda}{t}\right)}{\log\left(1 + \frac{1}{t}\right)} = \lambda.$$

The set

$$\{f \in \mathcal{TBF} : f(1) = 1\}$$

is a basis of the convex cone $\mathcal{TBF}$, and its extremal points are given by

$$e_0(\lambda) = 1, \quad e_t(\lambda) = \frac{\log\left(1 + \frac{\lambda}{t}\right)}{\log\left(1 + \frac{1}{t}\right)}, \quad 0 < t < \infty, \quad \text{and} \quad e_\infty(\lambda) = \lambda.$$

Their extremal property follows exactly as in Remark 2.4. Thus,

$$1, \quad \lambda, \quad \frac{\log\left(1 + \frac{\lambda}{t}\right)}{\log\left(1 + \frac{1}{t}\right)}, \quad (0 < t < \infty),$$

are typical functions in $\mathcal{TBF}$. Note that these are the logarithms of the Laplace transforms of the following (degenerate) Gamma distributions

$$\gamma_{0,0}, \quad \gamma_{\infty,\infty}, \quad \text{and} \quad \gamma_{1/\log(1+\frac{1}{t}),t}, \quad (0 < t < \infty), \quad \text{respectively.}$$

As usual,

$$\gamma_{\alpha,\beta}(x) = \frac{\beta^\alpha}{\Gamma(\alpha)} x^{\alpha-1} e^{-\beta x}, \quad x > 0.$$

The following generalization of the class $\mathcal{TBF}$ is due to Bondesson.

Definition 8.9. A Bernstein function f is said to be in $\mathcal{TBF}_\gamma$ for some $\gamma > 0$ if the Lévy measure μ from (3.2) has a density $m(t)$ such that $t \mapsto t^{2-\gamma} \cdot m(t)$ is completely monotone, i.e.

$$f(\lambda) = a + b\lambda + \int_{(0,\infty)} (1 - e^{-\lambda t}) m(t)\, dt, \tag{8.4}$$

where $a, b \geq 0$, $\int_{(0,\infty)} (1 \wedge t) m(t)\, dt < \infty$ and $t \mapsto t^{2-\gamma} \cdot m(t)$ is completely monotone.

Obviously, $\mathcal{TBF} = \mathcal{TBF}_1$ and $\mathcal{CBF} = \mathcal{TBF}_2$. For $\gamma, \epsilon > 0$ we see that

$$t^{2-\gamma-\epsilon} m(t) = t^{2-\gamma} m(t) \cdot t^{-\epsilon};$$

since $t^{-\epsilon}$ is completely monotone and since the product of two completely monotone functions is again in $\mathcal{CM}$ we conclude that $\mathcal{TBF}_\gamma \subset \mathcal{TBF}_{\gamma+\epsilon}$ for all $\gamma > 0$ and $\epsilon > 0$.

Proposition 8.10. *Let $\gamma > 1$. A function $f : (0, \infty) \to [0, \infty)$ is in $\mathcal{TBF}_\gamma$ if, and only if, f is differentiable, $f(0+)$ exists, and*

$$f'(\lambda) = b + \int_{(0,\infty)} \frac{1}{(\lambda + t)^\gamma} \sigma(dt) \tag{8.5}$$

for some $b \geq 0$ and a measure σ on $(0, \infty)$ such that $\int_{(0,\infty)} (1+t)^{-\gamma} \sigma(dt) < \infty$.

Proof. Assume that f is such that f' is of the form (8.5). By Fubini's theorem we can calculate the primitive

$$f(\lambda) = a + b\lambda + \int_{(0,\infty)} \left(\frac{1}{t^{\gamma-1}} - \frac{1}{(\lambda + t)^{\gamma-1}} \right) \sigma(dt)$$

$$= a + b\lambda + \int_{(0,\infty)} \left(\int_0^\infty y^{\gamma-2} e^{-yt} \, dy - \int_0^\infty y^{\gamma-2} e^{-y(t+\lambda)} \, dy \right) \frac{\sigma(dt)}{\Gamma(\gamma - 1)}$$

$$= a + b\lambda + \int_0^\infty (1 - e^{-y\lambda}) y^{\gamma-2} \int_{(0,\infty)} e^{-yt} \sigma(dt) \, dy.$$

In the second equality we used the elementary identity

$$t^{-\alpha} = \frac{1}{\Gamma(\alpha)} \int_0^\infty y^{\alpha-1} e^{-yt} \, dy.$$

Since f is non-negative, $a \geq 0$; setting $m(y) := \frac{y^{\gamma-2}}{\Gamma(\gamma-1)} \int_{(0,\infty)} e^{-yt} \sigma(dt)$ shows that $y^{2-\gamma} m(y)$ is, up to a constant, the Laplace transform of the measure σ, hence completely monotone.

Conversely, if $f \in \mathcal{TBF}_\gamma$ we can differentiate (8.9) and use that $t^{\gamma-2} m(t)$ is the Laplace transform of a measure $\tilde{\sigma}$:

$$f'(\lambda) = b + \int_0^\infty e^{-t\lambda} t \, m(t) \, dt$$

$$= b + \int_0^\infty e^{-t\lambda} t^{1-\gamma} \int_{[0,\infty)} e^{-yt} \tilde{\sigma}(dy) \, dt$$

$$= b + \int_{[0,\infty)} \int_0^\infty t^{1-\gamma} e^{-t(\lambda+y)} \, dt \, \tilde{\sigma}(dy)$$

$$= b + \int_{[0,\infty)} \Gamma(\gamma - 1) \frac{1}{(\lambda + y)^\gamma} \tilde{\sigma}(dy).$$

Since f' has a primitive with $f(0+) < \infty$, we see that $\int_{[0,\infty)} (1+t)^{-\gamma} \tilde{\sigma}(dt) < \infty$ and $\tilde{\sigma}\{0\} = 0$. This proves (8.5). $\qquad\square$

Comments 8.11. The class $\mathcal{TBF}$ appears for the first time in Thorin's 1977 papers [347, 348] as the family of Laplace exponents of a class of probability distributions, the so-called *generalized Gamma convolutions*. This class of distributions will be discussed in the next chapter, Chapter 9. Nowadays the standard reference are the comprehensive lecture notes [67] by Bondesson and the monograph [339] by Steutel and van Harn.

The exposition in Bondesson [67] focusses on probability distributions which are generalized Gamma convolutions, GGC for short. Bondesson calls, in honour of O. Thorin, the family of (sometimes: Laplace exponents of) distributions from GGC the *Thorin class*, $\mathcal{T}$. It is easy to see that for $f \in \mathcal{T}$ the reflected function $\lambda \mapsto f(-\lambda)$ is a Thorin–Bernstein function $\mathcal{TBF}$ in the sense of Definition 8.1.

Taking this reflection into account, the class $\mathcal{TBF}_\gamma$ can be found in [66, 67] where the assertion of Proposition 8.10 is used as the definition of $\mathcal{TBF}_\gamma$. The integral appearing in (8.5) is sometimes called generalized Stieltjes transform, see e.g. Hirschman and Widder [167]. Bondesson shows in [67, §9.5, pp. 150–1] that the closure of the set $\bigcup_\gamma \mathcal{TBF}_\gamma$ are the Bernstein functions.

The above mentioned reflection in the argument of the log-Laplace exponent makes Bondesson sometimes difficult to read. The presentation in Steutel and van Harn [339] uses our convention but does not always contain (complete) proofs.

Theorem 8.2 is, in less comprehensive form, essentially contained in [66], see also [67]. Note that Thorin [347] uses Theorem 8.2 (iii) as *definition* of GGC or $\mathcal{TBF}$, respectively. Since we want to emphasize the '(complete) Bernstein' nature of $\mathcal{TBF}$, we prefer (8.1) as the basic definition. Part (iv) of Theorem 8.2 is the *main characterization theorem* of GGC in [67]. It appears in [347] and in the present form in [66]; both references use the connection with complex analysis that complete Bernstein (respectively, Stieltjes) functions preserve (respectively, switch) the upper and lower half-planes, see Theorem 6.2. The stability result of Theorem 8.4 is the counterpart of [67, Theorem 3.3.1], the closure result Theorem 8.5 is due to Thorin [348]; among probabilists it is often referred to as Thorin's continuity theorem. Theorem 9.13 was the starting point of Thorin's investigations [347, 348]. Proposition 8.7 is due to Bondesson [66, Remark 3.2]. It provides, in particular, another proof for the fact that all probability distributions in GGC are infinitely divisible; this will again be proved in Proposition 9.11 of the next chapter. The interpretation of (8.2) in the context of Choquet theory seems to be new.

Chapter 9

A second probabilistic intermezzo

In Chapter 5 we discussed the relation between (extended) Bernstein functions $\mathcal{BF}$ and the infinitely divisible elements in $\mathcal{CM}$, see Lemma 5.8 and Theorem 5.11. This allowed us to give a probabilistic characterization of the cone $\mathcal{BF}$. A similar characterization is possible for complete Bernstein functions. To do this we use Lemma 5.8 as starting point.

Definition 9.1. A measure π on $[0, \infty)$ belongs to the *Bondesson class* if

$$\mathscr{L}(\pi; \lambda) = e^{-f(\lambda)}$$

for some $f \in \mathcal{CBF}$. We write $\pi \in$ BO.

Note that $\pi \in$ BO is a sub-probability measure: $\pi[0, \infty) = \mathscr{L}(\pi; 0+) \leq 1$.
A probabilistic characterization of the class BO will be given in Theorem 9.7 below. Let us first collect a few elementary properties of the Bondesson class BO.

Lemma 9.2. (i) *A sub-probability measure π on $[0, \infty)$ is in* BO *if, and only if, there exist a measure v on $(0, \infty)$ with $\int_{(0,1)} s^{-1} v(ds) + \int_{(1,\infty)} s^{-2} v(ds) < \infty$ and constants $a, b \geq 0$ such that*

$$\mathscr{L}(\pi; \lambda) = \exp\left(-a - b\lambda - \int_{(0,\infty)} \left(\frac{1}{s} - \frac{1}{\lambda + s}\right) v(ds)\right). \qquad (9.1)$$

(ii) *Every $\pi \in$ BO is infinitely divisible.*

(iii) BO *is closed under convolutions: if $\pi, \rho \in$ BO, then $\pi \star \rho \in$ BO.*

(iv) BO *is closed under vague convergence.*

Proof. (i) Because of Theorem 6.2, see also Remark 6.4, every $f \in \mathcal{CBF}$ has a representation

$$f(\lambda) = a + b\lambda + \int_{(0,\infty)} \left(\frac{1}{s} - \frac{1}{\lambda + s}\right) v(ds),$$

with a, b and v as claimed.
  (ii) Since $\mathcal{CBF} \subset \mathcal{BF}$, Lemma 5.8 tells us that each $\pi \in$ BO is infinitely divisible.

(iii) Since $\mathcal{CBF}$ is a convex cone, this follows immediately from the convolution theorem for the Laplace transform: $\mathcal{L}(\pi \star \rho) = \mathcal{L}\pi \cdot \mathcal{L}\rho$.

(iv) Let $(\pi_n)_{n\in\mathbb{N}} \subset$ BO be a sequence of sub-probability measures and denote by $(f_n)_{n\in\mathbb{N}} \subset \mathcal{CBF}$ the complete Bernstein functions such that $\mathcal{L}(\pi_n; \lambda) = e^{-f_n(\lambda)}$. If $\lim_{n\to\infty} \pi_n = \pi$ vaguely, then $\lim_{n\to\infty} \mathcal{L}(\pi_n; \lambda) = \mathcal{L}(\pi; \lambda)$ for every $\lambda > 0$, and the claim follows from the fact that $\mathcal{CBF}$ is closed under pointwise limits. $\qquad\square$

Since $\mathcal{CBF} \subset \mathcal{BF}$, Theorem 5.2 guarantees the existence of a unique semigroup of sub-probability measures $(\mu_t)_{t\geq 0}$ on $[0, \infty)$ such that $\mathcal{L}(\mu_t; \lambda) = e^{-tf(\lambda)}$, $t \geq 0$; in particular, $\mu_t \in$ BO. Clearly, the converse is also true and we arrive at the analogue of Theorem 5.2.

Theorem 9.3. *Let $(\mu_t)_{t\geq 0} \subset$ BO be a convolution semigroup of sub-probability measures on $[0, \infty)$. Then there exists a unique $f \in \mathcal{CBF}$ such that the Laplace transform of μ_t is given by*

$$\mathcal{L}\mu_t = e^{-tf} \quad \text{for all } t \geq 0. \tag{9.2}$$

Conversely, given $f \in \mathcal{CBF}$, there exists a unique convolution semigroup of sub-probability measures $(\mu_t)_{t\geq 0} \subset$ BO such that (9.2) holds true.

Note that the set $e^{-\mathcal{BF}} = \{g : \exists f \in \mathcal{BF}, \ g = e^{-f}\}$ is a relatively simple subset of $\mathcal{CM}$: it consists of all infinitely divisible functions from $\mathcal{CM} \cap \{g : g(0+) \leq 1\}$, cf. Lemma 5.8. The structure of $e^{-\mathcal{CBF}} \subset \mathcal{CM}$ is more complicated. If we restrict our attention to $g \in \mathcal{CBF}$ such that $g(0+) \geq 1$, we find that

$$f = \log g = \log\big((g - 1) + 1\big) \in \mathcal{CBF}$$

and $e^{-f} = 1/g \in \mathcal{S}$ with $1/g(0+) \leq 1$. This motivates the following definition.

Definition 9.4. A measure π on $[0, \infty)$ is a *mixture of exponential distributions* if

$$\mathcal{L}\pi \in \mathcal{S} \quad \text{and} \quad \mathcal{L}(\pi; 0+) \leq 1. \tag{9.3}$$

We write $\pi \in$ ME.

Note that $\pi \in$ ME is a sub-probability measure: $\pi[0, \infty) = \mathcal{L}(\pi; 0+) \leq 1$.

The reason for the name *mixture of exponential distributions* is given in part (iii) of the following theorem.

Theorem 9.5. *Let π be a measure on $[0, \infty)$. Then the following conditions are equivalent.*

(i) $\pi \in$ ME.

(ii) *There exist $\beta \geq 0$ and a measurable function $\eta : (0,\infty) \to [0,1]$ satisfying $\int_0^1 \eta(t) t^{-1} dt < \infty$, such that*

$$\mathscr{L}(\pi;\lambda) = \exp\left(-\beta - \int_0^\infty \left(\frac{1}{t} - \frac{1}{\lambda+t}\right)\eta(t)\,dt\right). \tag{9.4}$$

(iii) *There exists a sub-probability measure ρ on $(0,\infty]$ such that*

$$\pi[0,t] = \int_{(0,\infty]} (1 - e^{-ts})\,\rho(ds). \tag{9.5}$$

Proof. (i)⇒(ii) Since $\mathscr{L}(\pi;\lambda) = g(\lambda)$ with $g \in \mathcal{S}$ and $g(0+) \leq 1$ we know from Theorem 7.3 that $f := 1/g \in \mathcal{CBF}$ and $f(\lambda) \geq f(0+) \geq 1$. From the representation (6.21) of Theorem 6.17, we find that

$$\log f(\lambda) = \gamma + \int_0^\infty \left(\frac{t}{1+t^2} - \frac{1}{\lambda+t}\right)\eta(t)\,dt$$

for some $\gamma \in \mathbb{R}$ and a measurable function $\eta : (0,\infty) \to [0,1]$. As $\infty > f(0+) \geq 1$, we see that

$$\gamma + \int_0^\infty \frac{1}{t(1+t^2)}\,\eta(t)\,dt < \infty$$

which entails that $\int_0^1 \eta(t) t^{-1} dt < \infty$. Because of this integrability condition, we have

$$\log f(\lambda) = \gamma + \int_0^\infty \left(\frac{t}{1+t^2} - \frac{1}{t}\right)\eta(t)\,dt - \int_0^\infty \left(\frac{1}{\lambda+t} - \frac{1}{t}\right)\eta(t)\,dt$$

$$= \beta + \int_0^\infty \left(\frac{1}{t} - \frac{1}{\lambda+t}\right)\eta(t)\,dt.$$

Since $f(0+) \geq 1$, we conclude that $\beta \geq 0$.

(ii)⇒(iii) In view of Theorem 6.17, formula (9.4) means that $\mathscr{L}(\pi;\lambda)$ is of the form $1/f$ where $f \in \mathcal{CBF}$ and $f(0+) \geq 1$; hence, $h(\lambda) = \mathscr{L}(\pi;\lambda)$ is a Stieltjes function satisfying $h(0+) \leq 1$. But then $h(\lambda) = b + \int_{(0,\infty)}(\lambda+s)^{-1}\sigma(ds)$ and $b + \int_{(0,\infty)} s^{-1}\sigma(ds) = h(0+) \leq 1$. Define a sub-probability measure ρ on $(0,\infty]$ by $\rho(ds) = s^{-1}\sigma(ds)$ on $(0,\infty)$, and $\rho\{\infty\} = b$. By Fubini's theorem

$$\mathscr{L}(\pi;\lambda) = b + \int_{(0,\infty)} \frac{s}{\lambda+s}\,\rho(ds)$$

$$= b + \int_{(0,\infty)}\left(\int_0^\infty e^{-(\lambda+s)t}\,dt\right)s\rho(ds)$$

$$= \int_{[0,\infty)} e^{-\lambda t}\,b\delta_0(dt) + \int_{[0,\infty)} e^{-\lambda t}\int_{(0,\infty)} se^{-ts}\,\rho(ds)\,dt$$

$$= \int_{[0,\infty)} e^{-\lambda t}\left(b\delta_0(dt) + \int_{(0,\infty)} se^{-ts}\,\rho(ds)\,dt\right)$$

which is equivalent to saying that $\pi[0,t] = \int_{(0,\infty]}(1 - e^{-ts})\,\rho(ds)$.

(iii)$\Rightarrow$(i) If $\pi[0,t] = \int_{(0,\infty]}(1 - e^{-ts})\,\rho(ds)$ we can perform the calculation of the previous step in reverse direction and arrive at

$$\mathscr{L}(\pi;\lambda) = b + \int_{(0,\infty)} \frac{s}{\lambda+s}\,\rho(ds).$$

Set $\sigma(ds) = s\rho(ds)$. Since $\int_{(0,\infty)}(1+s)^{-1}\sigma(ds) = \int_{(0,\infty)} s(1+s)^{-1}\rho(ds) < \infty$, we see that $\mathscr{L}\pi \in \mathcal{S}$. Moreover, $\mathscr{L}(\pi;0+) = b + \rho(0,\infty) = \rho(0,\infty] \leq 1$. $\quad\square$

Corollary 9.6. *We have* ME $\subset$ BO. *Moreover,* ME *is closed under vague limits.*

Proof. Let $\pi \in$ ME. By Theorem 9.5, the Laplace transform $\mathscr{L}\pi$ is of the form e^{-f} where $f \in \mathcal{CBF} \subset \mathcal{BF}$. This shows that $\pi \in$ BO.

If $(\pi_n)_{n\in\mathbb{N}} \subset$ ME converges vaguely to π, we see for the Laplace transforms $\lim_{n\to\infty}\mathscr{L}(\pi_n;\lambda) = \mathscr{L}(\pi;\lambda)$. Since Stieltjes functions are closed under pointwise limits, cf. Theorem 2.2 (iii), $\mathscr{L}\pi \in \mathcal{S}$ is a Stieltjes function. Moreover, we have $\mathscr{L}(\pi;0+) = \lim_{\lambda\to0+}\mathscr{L}(\pi;\lambda) = \lim_{\lambda\to0+}\lim_{n\to\infty}\mathscr{L}(\pi_n,\lambda) \leq 1$. Thus, π is a mixture of exponential distributions. $\quad\square$

We are now ready for the probabilistic interpretation of the class BO.

Theorem 9.7. BO *is the vague closure of the set*

$$\text{ME}^\star := \{\pi_1 \star \cdots \star \pi_k \,:\, \pi_j \in \text{ME},\ j = 1,2,\ldots,k,\ k \in \mathbb{N}\},$$

i.e. BO *is the smallest class of sub-probability measures on* $[0,\infty)$ *which contains* ME *and which is closed under convolutions and vague limits.*

Proof. We have already seen in Corollary 9.6 that ME $\subset$ BO. That BO is vaguely closed and stable under convolutions has been established in Lemma 9.2. It is therefore enough to show that the extremal elements of $\mathcal{CBF}$,

$$\lambda \mapsto 1, \quad \lambda \mapsto \frac{\lambda}{\lambda+s}, \quad s > 0, \quad \lambda \mapsto \lambda,$$

can be represented as pointwise limits of Laplace exponents of measures in $\text{ME}^\star$. For $\lambda \mapsto 1$ there is nothing to show. Note that

$$\lim_{r\to\infty} \frac{1}{\log\frac{r+1}{r}} \int_r^\infty \frac{\lambda}{\lambda+t}\frac{dt}{t} = \lambda.$$

If we set $\eta_r(t) := \mathbb{1}_{[r,\infty)}(t)$ and pick a sequence of r such that $(\log\frac{r+1}{r})^{-1} \in \mathbb{N}$, we see that $\lambda \mapsto \lambda$ is the limit of Laplace exponents of convolutions of measures from ME. Further,

$$\frac{\lambda}{\lambda+s} = \lim_{r\to\infty} r \int_{s-1/r}^{s+1/r} \frac{\lambda}{\lambda+t}\frac{\eta_{r,s}(t)}{t}\,dt,$$

where $r \in \mathbb{N}$, $t^{-1}\eta_{r,s}(t)$ is a tent-function centered at s with basis $[s - r^{-1}, s + r^{-1}]$ and height 1, i.e. $\eta_{r,s}(t) = rt[(t-s+r^{-1})^+ \wedge (s+r^{-1}-t)^+]$. Therefore $\lambda\,(\lambda+s)^{-1}$ is the limit of Laplace exponents of convolutions of measures from ME. This proves that BO $\subset$ *vague-closure*(ME*). $\square$

Remark 9.8. The proof of Theorem 9.7 uses implicitly the representation of $\mathcal{CBF}$ in terms of its extremal points $e_s(\lambda) = \lambda\,(1 + s)/(\lambda + s)$, $0 \le s \le \infty$, cf. Remark 7.8. It shows that we can approximate the corresponding representing measures δ_s as vague limits of probability measures of the form $c\,(1+t)^{-1}t^{-1}\eta(t)\,dt$ with a function $\eta : [0, \infty) \to [0, 1]$.

Remark 9.9. Suppose that $\mathcal{L}\pi = e^{-f}$ where $\pi \in \mathrm{SD}_\infty$ and f is given by the representation (5.17). From the Stieltjes representation of λ^α we see that

$$f(\lambda) = \kappa\{0\} + \kappa\{1\}\lambda + \int_0^\infty \frac{\lambda}{\lambda + t} \left(\int_{(0,1)} \frac{\sin(\alpha\pi)}{\pi} t^\alpha \kappa(d\alpha) \right) \frac{dt}{t}.$$

If also $\pi \in$ ME, then by Theorem 9.5

$$f(\lambda) = \beta + \int_0^\infty \frac{\lambda}{\lambda + t} \frac{\eta(t)}{t}\,dt$$

where $\beta \ge 0$ and $\eta : (0, \infty) \to [0, 1]$. By comparing the last two expressions for f, we see that $\pi \in \mathrm{SD}_\infty \cap$ ME if, and only if, $\beta = \kappa(0)$ and $\kappa(0, 1] = 0$. Thus, $\mathrm{SD}_\infty \cap$ ME $= \{e^{-\beta}\delta_0 : \beta \ge 0\}$.

Let us introduce a class of sub-probability measures which is related to the family $\mathcal{TBF}$ in the same way as BO is related to $\mathcal{CBF}$, see Definition 9.1.

Definition 9.10. A measure π on $[0, \infty)$ is a *generalized Gamma convolution* if

$$\mathcal{L}(\pi; \lambda) = e^{-f(\lambda)}$$

for some $f \in \mathcal{TBF}$. We write $\pi \in$ GGC.

Note that $\pi \in$ GGC is a sub-probability measure: $\pi[0, \infty) = \mathcal{L}(\pi; 0+) \le 1$.

A probabilistic characterization of the class GGC will be given below. Let us first collect a few elementary properties of this class.

Proposition 9.11. (i) *A sub-probability measure π is in GGC if, and only if, there exist a measure τ on $(0, \infty)$ with $\int_{(0,1)} |\log t|\,\tau(dt) + \int_{(1,\infty)} t^{-1}\tau(dt) < \infty$ and constants $a, b \ge 0$ such that*

$$\mathcal{L}(\pi; \lambda) = \exp\left(-a - b\lambda - \int_{(0,\infty)} \log\left(1 + \frac{\lambda}{t}\right) \tau(dt) \right),$$

or a non-decreasing function $w(t)$ on $(0, \infty)$ with $\int_0^\infty (1+t)^{-1} t^{-1} w(t)\, dt < \infty$ such that

$$\mathcal{L}(\pi; \lambda) = \exp\left(-a - b\lambda - \int_{(0,\infty)} \frac{\lambda}{\lambda + t} \frac{w(t)}{t}\, dt\right).$$

(ii) $\mathsf{SD}_\infty \subset \mathsf{GGC} \subset \mathsf{SD} \subset \mathsf{ID}$.

(iii) $\mathsf{GGC} \subset \mathsf{BO} \subset \mathsf{ID}$.

(iv) GGC *contains all Gamma distributions* $\gamma_{\alpha,\beta}$, $\alpha, \beta > 0$.

(v) GGC *is closed under convolutions: if* $\pi, \rho \in \mathsf{GGC}$, *then* $\pi \star \rho \in \mathsf{GGC}$.

(vi) GGC *is closed under vague convergence.*

Proof. Since $\mathcal{TBF} \subset \mathcal{CBF} \subset \mathcal{BF}$, and since $\mathcal{TBF}$ is a convex cone which is closed under pointwise limits, (iii), (v) and (vi) follow as in Lemma 9.2. To see (ii), note that the Lévy measure of a Thorin–Bernstein function has a density $m(t)$ such that $t \cdot m(t)$ is completely monotone, cf. Definition 8.1; as such, $t \cdot m(t)$ is non-increasing and the inclusion $\mathsf{GGC} \subset \mathsf{SD}$ follows from the structure result for Laplace exponents of SD distributions, cf. Proposition 5.17. Since the Laplace exponent of any SD_∞ distribution has a Lévy density m such that $t \cdot m(t) = n(\log t)$ with n completely monotone on $\mathbb{R}$, cf. Proposition 5.20, $t \cdot m(t)$ is completely monotone by Proposition 1.16 proving that $\mathsf{SD}_\infty \subset \mathsf{GGC}$. For (i) we have to use the representations (8.2) and (8.3) of Thorin–Bernstein functions. Property (iv) can be directly verified: suppose that $\gamma_{\alpha,\beta}$ is a Gamma distribution with parameters $\alpha, \beta > 0$, then

$$\mathcal{L}(\gamma_{\alpha,\beta}; \lambda) = \left(1 + \frac{\lambda}{\beta}\right)^{-\alpha} = e^{-\alpha \log(1+\frac{\lambda}{\beta})} = e^{-f(\lambda)}.$$

The Thorin measure τ of f is equal to $\alpha \delta_\beta$. Thus, GGC contains all Gamma distributions. □

Since $\mathcal{TBF} \subset \mathcal{BF}$, Theorem 5.2 guarantees the existence of a unique semigroup of sub-probability measures $(\mu_t)_{t\geq 0}$ on $[0, \infty)$ such that $\mathcal{L}(\mu_t; \lambda) = e^{-tf(\lambda)}$, $t \geq 0$; in particular, $\mu_t \in \mathsf{GGC}$. Clearly, the converse is also true and we arrive at the analogue of Theorem 5.2:

Theorem 9.12. *Let* $(\mu_t)_{t\geq 0} \subset \mathsf{GGC}$ *be a convolution semigroup of sub-probability measures on* $[0, \infty)$. *Then there exists a unique* $f \in \mathcal{TBF}$ *such that the Laplace transform of* μ_t *is given by*

$$\mathcal{L}\mu_t = e^{-tf} \quad \text{for all } t \geq 0. \tag{9.6}$$

Conversely, given $f \in \mathcal{TBF}$, *there exists a unique convolution semigroup of sub-probability measures* $(\mu_t)_{t\geq 0} \subset \mathsf{GGC}$ *such that (9.6) holds true.*

We are now ready for the probabilistic interpretation of the class GGC.

Theorem 9.13. GGC *is the vague closure of the set* $\{\gamma_{\alpha,\beta} \,:\, \alpha, \beta > 0\}^\star$ *of all finite convolutions of Gamma distributions. In particular,* GGC *is the smallest class of sub-probability measures on* $[0, \infty)$ *which contains all Gamma distributions and which is closed under convolutions and vague limits.*

Proof. We have seen that $\gamma_{\alpha,\beta} \in$ GGC in Proposition 9.11. Moreover,

$$\mathcal{L}(\gamma_{\alpha,\beta}; \lambda) = e^{-\alpha \log(1+\frac{\lambda}{\beta})}$$

which shows that the extreme points $e_t(\lambda) = \log(1 + \frac{\lambda}{t})/\log(1 + \frac{1}{t})$, $0 < t < \infty$, of $\mathcal{TBF}$ are the Laplace exponents of $\gamma_{1/\log(1+\frac{1}{t}),t}$, $0 < t < \infty$.

Since $\mathcal{TBF}$ consists of integral mixtures of its extreme points, cf. Remark 8.8, we conclude that GGC is the vague closure of finite convolutions of Gamma distributions. $\square$

There seems to be no deeper relation between the classes ME and GGC. Indeed, the Laplace exponent of $\pi \in$ ME is of the form

$$\beta + \int_0^\infty \frac{\lambda}{\lambda + t} \frac{\eta(t)}{t}\, dt$$

with η taking values in $[0, 1]$; on the other hand, the Laplace exponent of a distribution in GGC looks like

$$a + b\lambda + \int_0^\infty \frac{\lambda}{\lambda + t} \frac{w(t)}{t}\, dt,$$

where w is a non-decreasing function. The intersection GGC $\cap$ ME is characterized in Proposition 9.14 (iii).

We have seen in Proposition 9.11 that GGC distributions are self-decomposable, i.e. they satisfy (5.8). In fact, GGC contains all distributions π that are $\mathcal{S}$-*self decomposable* in the following sense: $g = \mathcal{L}\pi$, $g(0+) \le 1$ and

$$\lambda \mapsto \frac{g(\lambda)}{g(c\lambda)} \quad \text{is a Stieltjes function for all } c \in (0, 1). \tag{9.7}$$

Proposition 9.14. *Let* π *be a sub-probability measure and* $g = \mathcal{L}\pi$*. The condition* (9.7) *is equivalent to any of the following assertions.*

(i) $\pi \in$ GGC *with Laplace exponent* $f \in \mathcal{TBF}$ *having no linear part, i.e.* $b = 0$ *in* (8.3)*, and with a bounded density* $w(t) \in [0, 1]$*.*

(ii) $\pi \in$ GGC *with Laplace exponent* $f \in \mathcal{TBF}$ *having no linear part, i.e.* $b = 0$ *in* (8.1) *and such that* $t \cdot m(t) = \mathcal{L}(\tau; t)$ *for some sub-probability measure* τ *on* $(0, \infty)$*.*

(iii) $\pi \in \mathsf{GGC} \cap \mathsf{ME}$.

Proof. The equivalence of (i) and (ii) follows immediately from Remark 8.3 (ii).

Assume that (9.7) holds true; we want to show (i). Let us begin with the proof that $\pi \in \mathsf{GGC}$. By (9.7), we get

$$-\frac{g'(\lambda)}{g(\lambda)} = \lim_{c \to 1-} \frac{g(c\lambda) - g(\lambda)}{(1-c)\lambda g(c\lambda)} = \lim_{c \to 1-} \frac{1}{(1-c)\lambda} \left(1 - \frac{g(\lambda)}{g(c\lambda)} \right).$$

Since $g(\lambda)/g(c\lambda)$ is bounded by 1 and since it is a Stieltjes function, the expression inside the brackets is a complete Bernstein function by Proposition 7.7; dividing by λ transforms it into a Stieltjes function. Since limits of Stieltjes functions are again Stieltjes functions, cf. Theorem 2.2, we conclude that $-g'/g \in \mathsf{S}$ and so $\pi \in \mathsf{GGC}$ by Proposition 8.7.

Consequently, $\mathscr{L}(\pi; \lambda) = e^{-f(\lambda)}$ where $f \in \mathsf{TBF}$; since f can be written as

$$f(\lambda) = a + b\lambda + \int_0^\infty \frac{\lambda}{t+\lambda} \frac{w(t)}{t} \, dt,$$

we see that

$$\frac{g(c\lambda)}{g(\lambda)} = \exp\left(b(1-c)\lambda + \int_0^\infty \frac{\lambda}{t+\lambda} \frac{w(t) - w(ct)}{t} \, dt \right).$$

By assumption, $\lambda \mapsto g(\lambda)/g(c\lambda)$ is a Stieltjes function so that, by Theorem 7.3, the function $\lambda \mapsto g(c\lambda)/g(\lambda)$ is in CBF. In view of the uniqueness of the representation in Theorem 6.17 and Remark 6.18 this is only possible, if in the above representation $b = 0$ and $0 \leq w(t) - w(ct) \leq 1$ for all $t \geq 0$ and $c \in (0, 1)$; letting $c \to 0$ proves $w(t) \in [0, 1]$. Here we used that $w(0+) = 0$, which follows from the integrability condition $\int_0^\infty (1+t)^{-1} t^{-1} w(t) \, dt < \infty$.

Conversely, assume that $g = e^{-f}$ where $f \in \mathsf{TBF}$ can be represented by (8.3) where $b = 0$ and $w(t)$ takes values in $[0, 1]$ for all $t \geq 0$. For any $0 < c < 1$ we find

$$f(\lambda) - f(c\lambda) = \int_0^\infty \frac{\lambda}{t+\lambda} \frac{w(t)}{t} \, dt - \int_0^\infty \frac{c\lambda}{t+c\lambda} \frac{w(t)}{t} \, dt$$

$$= \int_0^\infty \frac{\lambda}{t+\lambda} \frac{w(t) - w(ct)}{t} \, dt.$$

Since $w(t)$ is non-decreasing and bounded by 1, we get $0 \leq w(t) - w(ct) \leq 1$. Thus,

$$\frac{g(c\lambda)}{g(\lambda)} = \exp\left(f(\lambda) - f(c\lambda) \right) = \exp\left(\int_0^\infty \frac{\lambda}{t+\lambda} \frac{w(t) - w(ct)}{t} \, dt \right),$$

and we conclude from Theorem 6.17 and Remark 6.18 that $\lambda \mapsto g(c\lambda)/g(\lambda)$ is for all $c \in (0, 1)$ a complete Bernstein function. Together with Theorem 7.3 this proves (9.7).

Since S is closed under pointwise limits, cf. Theorem 2.2, and since $g(0+) > 0$, (iii) follows from (9.7) if we let $c \to 0$.

On the other hand, we infer (i) from (iii) by Theorem 6.17 and Remark 6.18. $\square$

We will now discuss another class of probability measures on $[0, \infty)$. This class is related to the exponential distributions in the same way as GGC is related to the Gamma distributions.

Definition 9.15. Let CE be the smallest class of sub-probability measures on $[0, \infty)$ which contains all exponential laws and which is closed under convolutions and weak limits.

The class CE is called the class of *convolutions of exponential distributions*. It is straightforward to see that $\pi \in$ CE if, and only if, $\pi = \delta_c$ for some $c \geq 0$, or if there exist $c \geq 0$ and a sequence $(b_n)_{n\in\mathbb{N}}$ satisfying $\sum_n b_n^{-1} < \infty$ such that

$$\mathscr{L}(\pi; \lambda) = e^{-c\lambda} \prod_n \left(1 + \frac{\lambda}{b_n}\right)^{-1}. \tag{9.8}$$

The sequence $(b_n)_{n\in\mathbb{N}}$ will be called the representing sequence of the measure π.

Corollary 9.16. (i) CE $\subset$ GGC.

(ii) CE $\cap$ ME $=$ Exp, *where* Exp *denotes the family of all exponential distributions. Here the point mass at zero δ_0 is interpreted as an exponential distribution with mean zero.*

(iii) CE $\cap$ SD$_\infty = \{\delta_c : c \geq 0\}$

Proof. (i) Let $\pi \in$ CE. It follows from the representation (9.8) that $\mathscr{L}\pi = e^{-f}$ with

$$f(\lambda) = c\lambda + \sum_n \log\left(1 + \frac{\lambda}{b_n}\right) = c\lambda + \int_{(0,\infty)} \log\left(1 + \frac{\lambda}{t}\right) \tau(dt), \tag{9.9}$$

where $\tau = \sum_n \delta_{b_n}$. Thus $f \in \mathcal{TBF}$, and hence $\pi \in$ GGC.
    (ii) Assume that $\pi \in$ CE $\cap$ ME and $\pi \neq \delta_0$. Let $\mathscr{L}\pi = e^{-f}$ where f has the representation (9.9). Then the Stieltjes measure σ of f is given by

$$\sigma(dt) = \frac{\tau(0, t)}{t} \, dt = \frac{\#\{n : b_n < t\}}{t} \, dt,$$

cf. Remark 8.3 (ii). On the other hand, comparing this representation with (9.4), it follows that $t \mapsto \tau(0, t)$ is bounded by 1. Hence, $\tau = \delta_b$ for some $b > 0$. Again by comparing (9.4) and (9.9), it follows that $c = 0$ and $\beta = 0$. Hence π is an exponential distribution.

(iii) Assume that $\pi \in CE \cap SD_\infty$ and let $\mathscr{L}\pi = e^{-f}$ where f is given by (9.9). Since $\pi \in SD_\infty$, the Lévy measure of f has a density m of the form (5.18). Therefore,

$$
\begin{aligned}
t\, m(t) &= \int_{(0,1)} \frac{\alpha}{\Gamma(1-\alpha)} t^{-\alpha} \kappa(d\alpha) \\
&= \int_0^\infty e^{-ts} \left(\int_{(0,1)} \frac{\alpha}{\Gamma(1-\alpha)\Gamma(\alpha)} s^{\alpha-1} \kappa(d\alpha) \right) ds.
\end{aligned}
$$

By Remark 8.3 (ii), the expression in parentheses is the density of the Thorin measure of f. Comparing with (9.9) it follows that the Thorin measure is equal to zero. Hence $f(\lambda) = c\lambda$ proving the claim. □

Suppose that $0 < b < a \le \infty$ and define a measure ρ on $(0, \infty]$ by

$$
\rho = \left(1 - \frac{b}{a}\right) \delta_b + \frac{b}{a} \delta_\infty,
$$

and a measure $\pi \in ME$ by

$$
\pi[0, t] = \int_{(0,\infty]} (1 - e^{-ts}) \rho(ds).
$$

Note that π is the mixture of a point mass at 0 and the exponential distribution with parameter b. By integration it follows that

$$
\mathscr{L}(\pi; \lambda) = \left(1 + \frac{\lambda}{a}\right) \left(1 + \frac{\lambda}{b}\right)^{-1}.
$$

In the case when $a = \infty$, π is the exponential law with parameter b.

Corollary 9.17. *Let $(a_n)_{n\in\mathbb{N}}$ and $(b_n)_{n\in\mathbb{N}}$ be two sequences of non-negative real numbers such that $0 < b_n < a_n \le \infty$ for all $n \in \mathbb{N}$ and $\sum_n b_n^{-1} < \infty$. Then the function*

$$
g(\lambda) := \prod_n \left(1 + \frac{\lambda}{a_n}\right) \left(1 + \frac{\lambda}{b_n}\right)^{-1}, \quad \lambda > 0,
$$

is well defined, strictly positive, and it is the Laplace transform of a probability measure from the Bondesson class.

If all $a_n = \infty$, then g is the Laplace transform of a probability measure from CE.

Proof. By the assumptions on the sequences $(a_n)_{n\in\mathbb{N}}$ and $(b_n)_{n\in\mathbb{N}}$ the product converges to a strictly positive function. Since the factors are Laplace transforms of probability measures in BO, the claim follows from Theorem 9.7. The case when $a_n = \infty$ for all $n \in \mathbb{N}$ follows from the definition of CE. □

Table 9.1. Overview of the classes of distributions, Laplace transforms and exponents.

measure	Laplace transform	Laplace exponent	references
ID	log-$\mathcal{CM}$	extended $\mathcal{BF}$	5.6, 5.10, 5.11
ID, $\pi[0,\infty) \leq 1$	log-$\mathcal{CM}$, $f(0+) \leq 1$	$\mathcal{BF}$	5.8
SD	—	$\mathcal{BF}, \mu(dt) = m(t)\,dt$, $t \cdot m(t)$ non-increasing	5.17
SD$_\infty$	—	$\mathcal{PBF}$	5.21, 5.22
BO	—	$\mathcal{CBF}$	9.1, 9.2
ME	$\mathcal{S}$, $f(0+) \leq 1$	$\mathcal{CBF}, \sigma(dt) = \frac{\eta(t)}{t}\,dt$, $0 \leq \eta(t) \leq 1$	9.4, 9.5
GGC	log-$\mathcal{S}$, $f(0+) \leq 1$	$\mathcal{TBF}$	8.7, 9.10, 9.11
CE	$e^{-c\lambda} \prod_n \left(1 + \frac{\lambda}{b_n}\right)^{-1}$	$\mathcal{TBF}$, $\tau = \sum_n \delta_{b_n}$	(9.8), 9.16
Exp	$\left(1 + \frac{\lambda}{b}\right)^{-1}$	$\mathcal{TBF}$, $\tau = \delta_b$	9.16

Another way to prove the corollary is to notice that

$$\lambda \mapsto \log\left(1 + \frac{\lambda}{a_n}\right) - \log\left(1 + \frac{\lambda}{b_n}\right) \in \mathcal{CBF}$$

and to apply the closure property of $\mathcal{CBF}$, Corollary 7.6.

Remark 9.18. Let us briefly give an overview of all classes of distributions and their Laplace exponents we have encountered so far. The numbers n.m in the 'references' column of Table 9.1 refer to the corresponding statements in this tract. We use log-$\mathcal{CM}$ as a shorthand for *logarithmically completely monotone* (see Definition 5.10) and, analogously, log-$\mathcal{S}$ as a shorthand for *logarithmically Stieltjes* in the sense that the condition in Proposition 8.7 (ii) holds.

We follow Bondesson [67] and use Venn diagrams to illustrate the relations among the classes of distributions and their Laplace exponents. The non-inclusions shown in Figures 9.1 and 9.2 are easily verified by the following examples:

SD $\not\subset$ BO, in particular SD $\not\subset$ ME. Consider the Lévy measure with density

$$m(t) = \frac{1}{t}\mathbb{1}_{(0,1)}(t), \quad t > 0.$$

The corresponding Bernstein function is $f(\lambda) = \gamma + \log(\lambda) - \mathrm{Ei}(\lambda)$ where $\gamma = 0.5772\ldots$ is Euler's constant and $\mathrm{Ei}(\lambda) = -\text{p.v.}\int_{-\lambda}^{\infty} t^{-1} e^{-t}\,dt$ stands

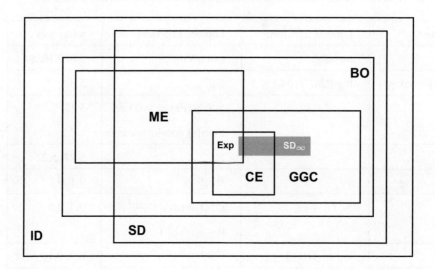

Figure 9.1. Relations between various classes of distributions...

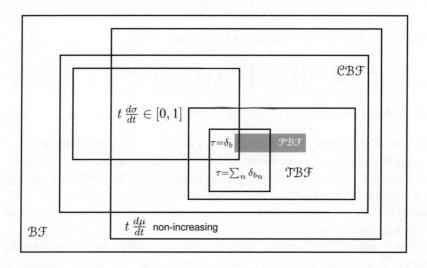

Figure 9.2. ...and their Laplace exponents.

for the exponential integral. Clearly, $t \cdot m(t)$ is non-increasing, but $m(t)$ is not completely monotone.

ME $\not\subset$ SD, in particular BO $\not\subset$ SD. This is shown by Entry 3 in the Tables 16.2. The corresponding Bernstein function is $f(\lambda) = 1 - (1 + \lambda)^{\alpha - 1}$, $0 < \alpha < 1$.

GGC $\not\subset$ ME. This follows from Entry 1 in the Tables 16.2. The corresponding Bernstein function is $f(\lambda) = \lambda^{\alpha}, 0 < \alpha < 1$.

ME $\not\subset$ GGC. This follows from Entry 3 in the Tables 16.2. The corresponding Bernstein function is $f(\lambda) = 1 - (1 + \lambda)^{\alpha-1}, 0 < \alpha < 1$.

GGC $\not\subset$ SD$_\infty$. This follows from Entry 26 in the Tables 16.4 and Remark 1.17. Indeed, the density of the Lévy measure is $m(t) = t^{-1}e^{-at}$ so that $t \cdot m(t)$ is completely monotone. If $t \cdot m(t) = n(\log t)$, then $n(t) = e^{e^{-at}}, t \in \mathbb{R}$. But this function is not completely monotone on $\mathbb{R}$.

Comments 9.19. The class of Bondesson distributions, BO, appears first in Bondesson's 1981 paper [66] under the name *g.c.m.e.d.* distributions. In his lecture notes [67] he also calls them class $\mathcal{T}_2$ distributions. The best source for the history is the following quote from Bondesson [67, p. 151] (with references pointing to our bibliography): *The $\mathcal{T}_2$ [here: BO] distributions were introduced in [66] for the purpose of finding a class of ID densities containing the MED's [here: ME] as well as the $\mathcal{B}$-densities [$\mathcal{B}$ denotes the hyperbolically completely monotone functions]. The attempt was not successful and I [i.e.: L. Bondesson] have often considered the class as a failure. Others have been much more interested in it and my name was even attached to it by Kent [205]. Unfortunately, some people have also called it B, e.g. Küchler & Lauritzen [232]. Earlier, Ito & McKean [178, pp. 214–217] encountered the class in connection with first passage times for diffusion processes* [see Chapter 15].

Thorin introduced the notion of *generalized Gamma convolutions* in his two 1977 papers [347, 348] in order to show the infinite divisibility of certain probability distributions. Subsequently, a large number of papers appeared on the subject, many of them in the Scandinavian Actuarial Journal. Nowadays the standard texts in the field are the survey paper [66] and the comprehensive lecture notes [67] by Bondesson. Chapter VI of the monograph [339] by Steutel and van Harn contains a digest of this material.

Most of the results of this section are already in the original papers by Thorin [347, 348] and Bondesson [66]. New is the presentation, which emphasizes the analogy with the material of Chapter 5, in particular with infinite divisibility and self-decomposability. Theorem 9.3 is from [66], Theorem 9.5 and its Corollary 9.6 are due to Steutel [338]. Our proof of Theorem 9.7 seems to be new. Alternatively, it is possible to construct $\pi \in vague\text{-}closure(\text{ME}^{\star})$ such that $\mathscr{L}(\pi; \lambda) = \exp(-\lambda/(\lambda + t))$ resp. $\mathscr{L}(\pi; \lambda) = \exp(-b\lambda)$. Typically, such distributions π are vague limits of convolutions of elementary measures of the form $\epsilon(t + \epsilon)^{-1}\delta_t + t(t + \epsilon)^{-1}\delta_\infty$. This is the usual way to prove Theorem 9.7, see e.g. Steutel and van Harn [339, Chapter VI.3] and Bondesson [67, p. 141]. The first part of Proposition 9.11 appears as Theorem 5.2 in [66], while all other assertions seem to be general knowledge. Theorems 9.12 and 9.13 are due to Thorin [347, 348], while Proposition 9.14 and the notion of $\mathcal{S}$-*self decomposability* seem to be new. Probabilistically, (9.7) can be interpreted in the following way: (the distribution of) a random variable X is $\mathcal{S}$-self decomposable if, and only if, X has a GGC distribution and if for every $c \in (0, 1)$ there is a further random variable $X^{(c)}$ with distribution in GGC such that $X = cX + X^{(c)}$.

The class of convolutions of exponential distributions, CE, is introduced by Yamazato in [376], see also [377], in the context of first passage distributions. The notion itself appears in the same context much earlier, see Keilson [201] and Kent [203, 204]. Corollary 9.17 is taken from Kent [203, 204]; he refers to this class of distributions as *infinite convolutions of elementary MEDs*, i.e. mixtures of exponential distributions.

Figures 9.1 and 9.2 appearing in Remark 9.18 owe much to the presentation in Bondesson [67, p. 4].

Chapter 10

Transformations of Bernstein functions

Suppose that $S = (S_t)_{t \geq 0}$ is a subordinator and $\theta : (0, \infty) \to [0, \infty)$ a non-negative function. The integral $\int_{(0,\infty)} \theta(t) \, dS_t$ defines a non-negative, possibly infinite, random variable. If this random variable is finite, its distribution is infinitely divisible. Various subfamilies of infinitely divisible sub-probability measures on $[0, \infty)$ can be obtained by integrating the same function θ against all subordinators. In this chapter we will consider general properties of such transformations and also the properties of three concrete families of transformations. We will identify the domains and ranges of these transformations and their iterations. It turns out that familiar classes of infinitely divisible distributions on $[0, \infty)$ such as BO and GGC can be obtained as ranges of particular transformations or their iterates. In the spirit of the monograph we will focus on transformations of the corresponding Laplace exponents, i.e. Bernstein functions.

To be more specific, let $\theta : (0, A) \to (0, B)$, $0 < A, B \leq \infty$, be a strictly decreasing function, and let $S = (S_t)_{t \geq 0}$ be a (possibly killed) subordinator defined on a probability space $(\Omega, \mathscr{F}, \mathbb{P})$ with Laplace exponent $f \in \mathcal{BF}$. We define a random variable X by

$$X := \int_{(0,A)} \theta(t) \, dS_t. \tag{10.1}$$

Then X is well defined and takes values in $[0, \infty]$. We start by computing the Laplace transform of X.

Lemma 10.1. *Let X be defined by* (10.1). *Then*

$$\mathbb{E}\left[e^{-\lambda X}\right] = \exp\left\{-\int_0^A f(\lambda \theta(t)) \, dt\right\}, \quad \lambda > 0. \tag{10.2}$$

Proof. Assume first that $A < \infty$. Since θ is decreasing we have by the monotone convergence theorem that

$$\int_{(0,A)} \theta(t) \, dS_t = \lim_{n \to \infty} \sum_{j=1}^{2^n - 1} \theta\left(\frac{j}{2^n} A\right) \left(S_{\frac{jA}{2^n}} - S_{\frac{(j-1)A}{2^n}}\right).$$

By the dominated convergence theorem, the independence and stationarity of the increments of S, and then the monotone convergence theorem, we get that

$$
\mathbb{E}\left[e^{-\lambda X}\right] = \mathbb{E}\left[\exp\left\{-\lambda \lim_{n\to\infty} \sum_{j=1}^{2^n-1} \theta\left(\frac{j}{2^n}A\right)\left(S_{\frac{jA}{2^n}} - S_{\frac{(j-1)A}{2^n}}\right)\right\}\right]
$$

$$
= \lim_{n\to\infty} \prod_{j=1}^{2^n-1} \mathbb{E}\left[\exp\left\{-\lambda\theta\left(\frac{j}{2^n}A\right)\left(S_{\frac{jA}{2^n}} - S_{\frac{(j-1)A}{2^n}}\right)\right\}\right]
$$

$$
= \lim_{n\to\infty} \prod_{j=1}^{2^n-1} \exp\left\{-\frac{A}{2^n} f\left(\lambda\theta\left(\frac{j}{2^n}A\right)\right)\right\}
$$

$$
= \lim_{n\to\infty} \exp\left\{-\sum_{j=1}^{2^n-1} f\left(\lambda\theta\left(\frac{j}{2^n}A\right)\right)\frac{A}{2^n}\right\}
$$

$$
= \exp\left\{-\int_0^A f(\lambda\theta(t))\,dt\right\}.
$$

The case of the infinite interval follows by letting $A \to \infty$.   □

Remark 10.2. Lemma 10.1 remains valid for general non-negative measurable functions θ. To see this we use that θ is an increasing limit of step functions and the fact that each subordinator generates a Poisson random measure on $[0,\infty) \times (0,\infty)$.

Motivated by the expression on the right-hand side of (10.2), for a function $f \in \mathcal{BF}$ we define

$$
\Theta f(\lambda) := \int_0^A f(\lambda\theta(t))\,dt, \quad \lambda > 0. \tag{10.3}
$$

Clearly, $\Theta f : (0,\infty) \to (0,\infty]$. From now on we will assume that θ is a bijection from $(0, A)$ onto $(0, B)$, which is equivalent to assuming that θ is continuous. Let $\theta^{-1} : (0, B) \to (0, A)$ be the inverse function of θ. The function θ^{-1} is strictly decreasing, hence, almost everywhere differentiable. Define $\vartheta : (0, B) \to (0,\infty)$ by $\vartheta(t) := -(\theta^{-1})'(t)$ at all points t where θ^{-1} is differentiable, and $\vartheta(t) = 0$ elsewhere. With this definition we have

$$
\theta^{-1}(t) = \int_t^B \vartheta(s)\,ds, \quad 0 < t < B. \tag{10.4}
$$

A change of variables in (10.3) yields an alternative representation of Θf:

$$
\Theta f(\lambda) = \int_0^B f(\lambda t)\vartheta(t)\,dt, \quad \lambda > 0. \tag{10.5}
$$

Alternatively, we could start with a measurable function $\vartheta : (0, B) \to (0,\infty)$ such that the integral in (10.4) is finite for all $t \in (0, B)$, and define a continuous, strictly decreasing function θ as the inverse of that integral. In fact, this is the most common way of prescribing the mapping Θ.

For a given mapping Θ we denote by $\mathfrak{D}(\Theta) := \{f \in \mathcal{BF} : \Theta f \in \mathcal{BF}\}$ its *domain* and by $\mathfrak{R}(\Theta) := \{\Theta f : f \in \mathfrak{D}(\Theta)\}$ its *range*. Since the prescription (10.3) defining Θ is linear, we see that $\mathfrak{D}(\Theta)$ is a convex cone.

Recall that there is a one-to-one correspondence between the family $\mathcal{BF}$ of all Bernstein functions and the family ID of all infinitely divisible sub-probability measures on $[0, \infty)$ which is realized as follows: For $f \in \mathcal{BF}$, let $(\mu_t)_{t \geq 0}$ be the unique convolution semigroup of sub-probability measures on $[0, \infty)$ such that $\mathscr{L}\mu_t = e^{-tf}$ for all $t \geq 0$. This correspondence allows to define an analogue of the mapping Θ on ID which we will denote again by Θ. Let $\mu_1 \in$ ID and let $f \in \mathcal{BF}$ be the unique function such that $\mathscr{L}\mu_1 = e^{-f}$. If $\Theta f \in \mathcal{BF}$, there exists a unique $\nu_1 \in$ ID such that $\mathscr{L}\nu_1 = e^{-\Theta f}$. In this case we define $\Theta\mu_1 := \nu_1$. Note that ν_1 is the distribution of the random variable X defined by (10.1). If $\Theta f = \infty$, then we define $\Theta\mu_1$ to be the zero measure on $[0, \infty)$. In this case $X = +\infty$ a.s. We use the same notation $\mathfrak{D}(\Theta)$ to denote the family of all $\mu_1 \in$ ID such that $\Theta\mu_1$ is not the zero measure – the domain of Θ acting on measures.

We will focus on the three families of mappings Θ which are defined by the following two-parameter functions, where $\alpha \in (-\infty, 1)$, $\beta, p > 0$:

$$\vartheta_{\alpha,\beta} : (0, \infty) \to (0, \infty), \quad \vartheta_{\alpha,\beta}(t) = c_{\alpha,\beta} e^{-t^\beta} t^{-\alpha-1},$$
$$\widehat{\vartheta}_{\alpha,p} : (0, 1) \to (0, \infty), \quad \widehat{\vartheta}_{\alpha,p}(t) = \widehat{c}_{\alpha,p}(1 - t)^{p-1} t^{-\alpha-1}, \tag{10.6}$$
$$\overline{\vartheta}_{\alpha,p} : (0, 1) \to (0, \infty), \quad \overline{\vartheta}_{\alpha,p}(t) = \overline{c}_{\alpha,p}(-\log t)^{p-1} t^{-\alpha-1}.$$

The constants $c_{\alpha,\beta}$, $\widehat{c}_{\alpha,p}$ and $\overline{c}_{\alpha,p}$ are either chosen to be 1 (if ϑ is not integrable) or $1 / \int \vartheta(t) \, dt$ (if ϑ is integrable). Thus

$$c_{\alpha,\beta} = \frac{\beta}{\Gamma(-\alpha/\beta)}, \quad \widehat{c}_{\alpha,p} = \frac{1}{B(-\alpha, p)}, \quad \overline{c}_{\alpha,p} = \frac{(-\alpha)^p}{\Gamma(p)}, \quad -\infty < \alpha < 0,$$
$$\tag{10.7}$$

and $c_{\alpha,\beta} = \widehat{c}_{\alpha,p} = \overline{c}_{\alpha,p} = 1$ if $0 \leq \alpha < 1$. The corresponding mappings Θ will be denoted by $\Theta_{\alpha,\beta}$, $\widehat{\Theta}_{\alpha,p}$ and $\overline{\Theta}_{\alpha,p}$ respectively. Note that we have $\widehat{\Theta}_{\alpha,1} = \overline{\Theta}_{\alpha,1}$ for all $\alpha \in (-\infty, 1)$.

In order to describe the domains of these mappings, we need to introduce subfamilies of Bernstein functions whose Lévy measures have additional integrability properties at infinity. For $p > 0$ and $\alpha > 0$, let

$$\mathcal{BF}_{\log^p} := \left\{ f \in \mathcal{BF} : \int_{(1,\infty)} (\log t)^p \, \mu(dt) < \infty \right\},$$

$$\mathcal{BF}_\alpha := \left\{ f \in \mathcal{BF} : \int_{(1,\infty)} t^\alpha \, \mu(dt) < \infty \right\},$$

$$\mathcal{BF}_{\alpha,\log^{p-1}} := \left\{ f \in \mathcal{BF} : \int_{(1,\infty)} (\log t)^{p-1} t^\alpha \, \mu(dt) < \infty \right\}.$$

Table 10.1. Domains and ranges of $\Theta_{\alpha,\beta}$, $\widehat{\Theta}_{\alpha,p}$ and $\overline{\Theta}_{\alpha,p}$.

ϑ	$\alpha \in$	$\mathfrak{D}(\Theta)$	$\mathfrak{R}(\Theta)$	
$\vartheta_{\alpha,\beta}$	$(-\infty,0)$	$\mathcal{BF}$	$t^{(\alpha+1)/\beta} \cdot \widetilde{m}(t^{1/\beta})$	completely monotone
	$\{0\}$	$\mathcal{BF}_{\log,0}$	$t^{1/\beta} \cdot \widetilde{m}(t^{1/\beta})$	completely monotone
	$(0,1)$	$\mathcal{BF}_{\alpha,0}$	$t^{(\alpha+1)/\beta} \cdot \widetilde{m}(t^{1/\beta})$	completely monotone and vanishes at ∞
$\widehat{\vartheta}_{\alpha,p}$	$(-\infty,0)$	$\mathcal{BF}$	$t^{\alpha+1} \cdot \widetilde{m}(t)$	p-monotone
	$\{0\}$	$\mathcal{BF}_{\log,0}$	$t \cdot \widetilde{m}(t)$	p-monotone
	$(0,1)$	$\mathcal{BF}_{\alpha,0}$	$t^{\alpha+1} \cdot \widetilde{m}(t)$	p-monotone and vanishes at ∞
$\overline{\vartheta}_{\alpha,p}$	$(-\infty,0)$	$\mathcal{BF}$	$t^{\alpha+1} \cdot \widetilde{m}(t) = n(\log t)$	n p-monotone on $\mathbb{R}$
	$\{0\}$	$\mathcal{BF}_{\log^p,0}$	$t \cdot \widetilde{m}(t) = n(\log t)$	n p-monotone on $\mathbb{R}$
	$(0,1)$	$\mathcal{BF}_{\alpha,\log^{p-1},0}$	$t^{\alpha+1} \cdot \widetilde{m}(t) = n(\log t)$	n p-monotone on $\mathbb{R}$ and vanishes at ∞

For a family of Bernstein functions, the subscript 0 denotes the subfamily consisting of those functions f satisfying $f(0+) = 0$, e.g. $\mathcal{BF}_{\log,0} = \{f \in \mathcal{BF}_{\log} : f(0+) = 0\}$.

Table 10.1 summarizes some of the main results of this chapter. It gives the domains and ranges of the mappings $\Theta_{\alpha,\beta}$, $\widehat{\Theta}_{\alpha,p}$ and $\overline{\Theta}_{\alpha,p}$. In each case, the Lévy measure of the Bernstein function in the range has a density denoted by $\widetilde{m}$. The ranges are completely characterized by monotonicity properties of $\widetilde{m}$.

Example 10.3. In this example we illustrate the results stated in Table 10.1 by considering eight particular choices of the mappings $\Theta_{\alpha,\beta}$, $\widehat{\Theta}_{\alpha,p}$ and $\overline{\Theta}_{\alpha,p}$, and identify their ranges with some of the known subfamilies of Bernstein functions.

(i) $\widehat{\Theta}_{-1,1}$. Then $\widehat{\vartheta}_{-1,1}(t) = 1$ and $\widehat{\theta}_{-1,1}(t) = 1 - t$. By the fourth line of Table 10.1, the range $\mathfrak{R}(\widehat{\Theta}_{-1,1})$ consists of all Bernstein functions whose Lévy measure has a non-increasing density. The multidimensional version of the mapping $\widehat{\Theta}_{-1,1}$ is known in the literature as $\mathfrak{U}$, while the analogue of $\widehat{\Theta}_{-1,1}(\mathsf{ID})$ is known as the *Jurek class* or the *class of s-self-decomposable distributions*.

(ii) $\Theta_{-1,1}$. Then $\vartheta_{-1,1}(t) = e^{-t}$ and $\theta_{-1,1}(t) = \log \frac{1}{t}$. By Definition 6.1 and the first line of Table 10.1, we have that $\Theta_{-1,1}(\mathcal{BF}) = \mathcal{CBF}$ and $\Theta_{-1,1}(\mathsf{ID}) = \mathsf{BO}$. Let $\mathcal{BF}_{\mathsf{SD}}$ denote the family of Bernstein functions f such that the corresponding sub-probability π is self-decomposable. By Proposition 5.17, $f \in \mathcal{BF}_{\mathsf{SD}}$ if, and only if, the Lévy measure μ of f has a density $m(t)$ such that $t \mapsto t \cdot m(t)$ is non-increasing.

It follows from Definition 8.1 and Corollary 10.13 that $\Theta_{-1,1}(\mathcal{BF}_{\mathrm{SD}}) = \mathcal{TBF}$, and consequently $\Theta_{-1,1}(\mathrm{SD}) = \mathrm{GGC}$. The mapping $\Theta_{-1,1}$ is known in the literature as the Υ *(upsilon) transformation*; it is introduced in [21, 22] in the one-dimensional case, and [20] in the multidimensional case.

(iii) $\Theta_{-1,2}$. Then $\vartheta_{-1,2}(t) = \sqrt{2/\pi}\, e^{-t^2}$ and $\theta_{-1,2} : (0,1) \to (0,\infty)$ is the inverse function of $t \mapsto \int_t^\infty \sqrt{2/\pi}\, e^{-s^2}\, ds$. By the first line of Table 10.1, the range of $\Theta_{-1,2}$ consists of all Bernstein functions whose Lévy measure has a density $\widetilde{m}$ such that $t \mapsto \widetilde{m}(\sqrt{t})$ is completely monotone. This class of Lévy measures is related to subordinate Brownian motions: A rotationally invariant Lévy process in $\mathbb{R}^d$ is a subordinate Brownian motion if, and only if, the radial part of its Lévy measure has a density n such that $t \mapsto n(\sqrt{t})$ is completely monotone, cf. Bochner [63, Theorem 4.3.3] or Example 13.16 below. The multidimensional version of the mapping $\Theta_{-1,2}$ is known in the literature as $\mathcal{G}$, while the analogue of $\Theta_{-1,2}(\mathrm{ID})$ is denoted by G.

(iv) $\widehat{\Theta}_{0,1}$. Then $\widehat{\vartheta}_{0,1}(t) = t^{-1}$ and $\widehat{\theta}_{0,1}(t) = e^{-t}$. By the fifth line of Table 10.1 and Proposition 5.17, $\widehat{\Theta}_{0,1}(\mathcal{BF}_{\log,0}) = \mathcal{BF}_{\mathrm{SD},0}$. The multidimensional version of the mapping $\widehat{\Theta}_{0,1}$ is known in the literature as Φ, while the analogue of $\widehat{\Theta}_{0,1}(\mathrm{ID})$ is the class of all self-decomposable distributions.

(v) $\Theta_{0,1}$. Then $\vartheta_{0,1}(t) = t^{-1}e^{-t}$ and $\theta_{0,1} : (0,\infty) \to (0,\infty)$ is the inverse function of $t \mapsto \int_t^\infty s^{-1} e^{-s}\, ds$. By the second line of Table 10.1 and Definition 8.1 it follows that $\Theta_{0,1}(\mathcal{BF}_{\log,0}) = \mathcal{TBF}_0$. This can also be deduced from (ii) and (iv) by using that $\Theta_{0,1} = \Theta_{-1,1} \circ \widehat{\Theta}_{0,1}$, cf. Proposition 10.11. The multidimensional version of the mapping $\Theta_{0,1}$ is known in the literature as Ψ.

(vi) $\Theta_{0,2}$. Then $\vartheta_{0,2}(t) = t^{-1}e^{-t^2}$ and $\theta_{0,2} : (0,\infty) \to (0,\infty)$ is the inverse function of $t \mapsto \int_t^\infty s^{-1}e^{-s^2}\, ds$. By the second line of Table 10.1, the family $\Theta_{0,2}(\mathcal{BF}_{\log,0})$ consists of all functions in $\mathcal{BF}_0$ whose Lévy measure has a density $\widetilde{m}$ such that $t \mapsto t \cdot \widetilde{m}(\sqrt{t})$ is completely monotone. The multidimensional version of the mapping $\Theta_{0,2}$ is known in the literature as $\mathcal{M}$.

(vii) $\overline{\Theta}_{0,p}$. Then $\overline{\vartheta}_{0,p}(t) = (-\log t)^{p-1} t^{-1}$ and $\theta_{0,p}(t) = \exp(-(pt)^{1/p})$. Let $\mathcal{BF}_{\mathrm{SD}_p}$ denote the family of Bernstein functions f such that the corresponding subprobability measure is in SD_p. By Proposition 5.20, $f \in \mathcal{BF}_{\mathrm{SD}_p}$ if, and only if, the Lévy measure μ of f has a density $m(t)$ such that $t \mapsto t \cdot m(t) = n(\log t)$ where n is p-monotone on $\mathbb{R}$. It follows from the eighth line in Table 10.1 that $\overline{\Theta}_{0,p}(\mathcal{BF}_{\log^p,0}) = \mathcal{BF}_{\mathrm{SD}_p,0}$.

(viii) $\Theta_{1-\gamma,1}$. Then $\vartheta_{1-\gamma,1}(t) = c_{1-\gamma,1}e^{-t}t^{\gamma-2}$. By the first three lines of Table 10.1 we see that $\mathfrak{R}(\Theta_{1-\gamma,1})$ is equal to $\mathcal{TBF}_\gamma$ for $\gamma > 1$ and $\mathcal{TBF}_{\gamma,0}$ for $\gamma \in (0,1]$, see Definition 8.9.

Table 10.2. Overview of the classical cases.

Θ		$\mathfrak{D}(\Theta)$	$\mathfrak{R}(\Theta)$	$\theta(t)$	$\vartheta(t)$	$\widetilde{m}(t)$
$\widehat{\Theta}_{-1,1}$	$\mathcal{U}$	$\mathcal{BF}$		$1-t$	1	$\int_t^\infty \frac{\mu(ds)}{s}$
$\Theta_{-1,1}$	Υ	$\mathcal{BF}$	$\mathcal{CBF}$	$\log\frac{1}{t}$	e^{-t}	$\int_0^\infty e^{-ts^{-1}}\frac{\mu(ds)}{s}$
$\Theta_{-1,1}$	Υ	$\mathcal{BF}_{SD}$	$\mathcal{TBF}$	$\log\frac{1}{t}$	e^{-t}	$\int_0^\infty e^{-ts^{-1}}\frac{\mu(ds)}{s}$
$\Theta_{-1,2}$	$\mathcal{G}$	$\mathcal{BF}$		$\sqrt{\frac{2}{\pi}}e^{-t^2}$		$\sqrt{\frac{2}{\pi}}\int_0^\infty e^{-t^2 s^{-2}}\frac{\mu(ds)}{s}$
$\widehat{\Theta}_{0,1}$	Φ	$\mathcal{BF}_{\log,0}$	$\mathcal{BF}_{SD,0}$	e^{-t}	$\frac{1}{t}$	$\frac{1}{t}\mu(t,\infty)$
$\Theta_{0,1}$	Ψ	$\mathcal{BF}_{\log,0}$	$\mathcal{TBF}_0$		$\frac{1}{t}e^{-t}$	$\frac{1}{t}\int_0^\infty e^{-ts^{-1}}\mu(ds)$
$\Theta_{0,2}$	$\mathcal{M}$	$\mathcal{BF}_{\log,0}$			$\frac{1}{t}e^{-t^2}$	$\frac{1}{t}\int_0^\infty e^{-t^2 s^{-2}}\mu(ds)$
$\overline{\Theta}_{0,p}$		$\mathcal{BF}_{\log^p,0}$	$\mathcal{BF}_{SD_p,0}$	$e^{-(pt)^{1/p}}$	$\frac{1}{t}(-\log t)^p$	$\frac{1}{t}\int_{(t,\infty)}(\log\frac{s}{t})^{p-1}\mu(ds)$

Table 10.2 summarizes the discussion in Example 10.3. The last column, the form of the Lévy measure of Θf, follows from (10.12). The notation $\mathcal{BF}_{SD,0}$ and $\mathcal{BF}_{SD_p,0}$ is explained in Example 10.3 (ii) and (vii), respectively.

Most of the remaining part of this chapter is devoted to proving the results listed in Table 10.1. We start with a discussion of the properties of the general mapping Θ.

Proposition 10.4. *Let $f \in \mathcal{BF}$ be given by*

$$f(\lambda) = a + b\lambda + \int_{(0,\infty)} (1 - e^{-\lambda t})\,\mu(dt). \qquad (10.8)$$

(i) *If Θf is defined by (10.3), then*

$$\Theta f(\lambda) = \tilde{a} + \tilde{b}\lambda + \int_0^\infty (1 - e^{-\lambda t})\,\widetilde{m}(t)\,dt, \qquad (10.9)$$

where

$$\tilde{a} = a \left(\int_0^B \vartheta(t)\, dt \right), \tag{10.10}$$

$$\tilde{b} = b \left(\int_0^B t\vartheta(t)\, dt \right), \tag{10.11}$$

$$\widetilde{m}(t) = \int_{(t/B,\infty)} \vartheta(ts^{-1})\, s^{-1}\, \mu(ds), \tag{10.12}$$

with the convention that $t/B = 0$ if $B = \infty$, and $0 \cdot \infty = 0$.

(ii) *If $\Theta f(1) < \infty$, then $\Theta f \in \mathcal{BF}$. In particular, $\tilde{a} < \infty$, $\tilde{b} < \infty$ and $\widetilde{m}$ is the density of a Lévy measure.*

(iii) *$\Theta f \in \mathcal{BF}$ if, and only if, the following three conditions hold:*

$$a = 0 \quad or \quad \int_0^B \vartheta(t)\, dt < \infty, \tag{10.13}$$

$$b = 0 \quad or \quad \int_0^B t\vartheta(t)\, dt < \infty, \tag{10.14}$$

$$\int_{1/B}^\infty \left(\int_0^{1/t} s\vartheta(s)\, ds \right) \mu(t,\infty)\, dt < \infty. \tag{10.15}$$

Proof. (i) From (10.5) we get that

$$\Theta f(\lambda) = \int_0^B \left(a + b\lambda t + \int_{(0,\infty)} (1 - e^{-\lambda ts})\, \mu(ds) \right) \vartheta(t)\, dt$$

$$= a \left(\int_0^B \vartheta(t)\, dt \right) + b \left(\int_0^B t\vartheta(t)\, dt \right) \lambda$$

$$+ \int_{(0,\infty)} \left(\int_0^B (1 - e^{-\lambda ts})\vartheta(t)\, dt \right) \mu(ds)$$

$$= \tilde{a} + \tilde{b}\lambda + \int_{(0,\infty)} \left(\int_0^{Bs} (1 - e^{-\lambda t})\vartheta(ts^{-1})s^{-1}\, dt \right) \mu(ds)$$

$$= \tilde{a} + \tilde{b}\lambda + \int_0^\infty (1 - e^{-\lambda t}) \left(\int_{(t/B,\infty)} \vartheta(ts^{-1})s^{-1}\, \mu(ds) \right) dt$$

$$= \tilde{a} + \tilde{b}\lambda + \int_0^\infty (1 - e^{-\lambda t})\widetilde{m}(t)\, dt,$$

with $\tilde{a}$, $\tilde{b}$ and $\widetilde{m}$ as in the statement.

(ii) If $\Theta f(1) < \infty$, then clearly $\tilde{a} < \infty$, $\tilde{b} < \infty$, and $\int_0^\infty (1 - e^{-t})\widetilde{m}(t)\, dt < \infty$. Remark 3.3 (iii) shows that the measure $\widetilde{m}(t)\, dt$ is a Lévy measure.

(iii) It is clear that $\tilde{a} < \infty$ is equivalent to (10.13), and $\tilde{b} < \infty$ is equivalent to (10.14). Since Θ is additive we may from now on assume that $a = b = 0$. Note that $\Theta f \in \mathcal{BF}$ is equivalent to

$$\Theta f(1) = \int_0^B f(s)\vartheta(s)\,ds < \infty.$$

By Lemma 3.4

$$\frac{e}{e-1}\,tI_\mu(1/t) \leq f(t) \leq tI_\mu(1/t) \quad \text{for all } t > 0,$$

where $I_\mu(s) = \int_0^s \mu(t,\infty)\,dt$. We write this fact as $f(s) \asymp sI_\mu(1/s)$. Assume that $B < \infty$. Then by Fubini's theorem

$$\Theta f(1) \asymp \int_0^B sI_\mu(1/s)\vartheta(s)\,ds$$

$$= \int_0^B s\vartheta(s)\left[\int_0^{1/s} \mu(t,\infty)\,dt\right]ds$$

$$= \left[\int_0^{1/B} \mu(t,\infty)\,dt\right]\left[\int_0^B s\vartheta(s)\,ds\right]$$

$$+ \int_{1/B}^\infty \left[\int_0^{1/t} s\vartheta(s)\,ds\right]\mu(t,\infty)\,dt.$$

If $\Theta f(1) < \infty$, then clearly (10.15) holds. Conversely, suppose that the integral in (10.15) is finite. Then $s \mapsto s\vartheta(s)$ is Lebesgue integrable near zero implying that $\int_0^B s\vartheta(s)\,ds < \infty$. Since $\int_0^{1/B} \mu(t,\infty)\,dt = \int_{(0,\infty)}((1/B) \wedge t)\,\mu(dt) < \infty$ we see that $\Theta f(1) < \infty$. The case when $B = \infty$ is proved analogously.   □

Remark 10.5. (i) We have given necessary and sufficient conditions for $\Theta f \in \mathcal{BF}$ in terms of the function ϑ. Alternatively, we can give conditions using the function θ:

$$a = 0 \quad \text{or} \quad A < \infty, \tag{10.16}$$

$$b = 0 \quad \text{or} \quad \int_0^A \theta(t)\,dt < \infty, \tag{10.17}$$

$$\int_0^A \left(\int_{(0,\infty)} (t\theta(s) \wedge 1)\,\mu(dt)\right)ds < \infty. \tag{10.18}$$

To see (10.18) we assume that $a = b = 0$, and note that $\Theta f(1) < \infty$ if, and only if,

$$\int_0^A \left(\int_{(0,\infty)} (1 - e^{-t\theta(s)}) \mu(dt) \right) ds < \infty.$$

We conclude by invoking the inequality $\frac{1}{2}(1 \wedge t) \le 1 - e^{-t} \le 1 \wedge t$ valid for all $t > 0$.

(ii) Let μ be the Lévy measure of $f \in \mathcal{BF}$ and define the measure $\widetilde{\mu}$ on $(0, \infty)$ by

$$\widetilde{\mu}(C) := \int_0^A \left(\int_{(0,\infty)} \mathbb{1}_C (t\theta(s)) \mu(dt) \right) ds \tag{10.19}$$

$$= \int_0^B \left(\int_{(0,\infty)} \mathbb{1}_C (tu) \mu(dt) \right) \vartheta(u)\, du, \tag{10.20}$$

for any Borel set $C \subset (0, \infty)$. Then

$$\int_{(0,\infty)} (1 - e^{-\lambda t}) \widetilde{\mu}(dt) = \int_0^A \left(\int_{(0,\infty)} (1 - e^{-\lambda t\theta(s)}) \mu(dt) \right) ds,$$

showing that $\widetilde{\mu}$ is the Lévy measure of Θf. Suppose now that μ is any measure on $(0, \infty)$ and define $\widetilde{\mu}$ by (10.19). If $\widetilde{\mu}$ is a Lévy measure, then so is μ. Indeed, there exist $\gamma \in (0, 1]$ and $C \subset (0, \infty)$ with $0 < \mathrm{Leb}(C) < \infty$ such that $\theta(s) \ge \gamma$ for $s \in C$. Then

$$\int_{(0,\infty)} (t \wedge 1) \widetilde{\mu}(dt) = \int_0^\infty \left(\int_{(0,\infty)} (t\theta(s) \wedge 1) \mu(dt) \right) ds$$

$$\ge \int_C \left(\int_{(0,\infty)} (t\gamma \wedge 1) \mu(dt) \right) ds$$

$$\ge \gamma\, \mathrm{Leb}(C) \int_{(0,\infty)} (t \wedge 1) \mu(dt).$$

(iii) Suppose that $\alpha \ge 1$ and $\beta > 0$. If we define $\vartheta_{\alpha,\beta}(t) = t^{-\alpha-1} \exp(-t^\beta)$, then $\int_0^{1/t} s\, \vartheta_{\alpha,\beta}(s)\, ds = \infty$. Hence, for every Lévy measure μ, the integral in (10.15) will be infinite and the domain of $\Theta_{\alpha,\beta}$ consists only of the function identically equal to zero. The same argument works for $t^{-\alpha-1}(1-t)^{p-1}$ and $t^{-\alpha-1}(-\log t)^{p-1}$. This explains the range of the parameter α in (10.6).

The family $\mathcal{PBF}$ of Laplace exponents of completely self-decomposable sub-probability measures on $[0, \infty)$ plays a distinguished role in studying the mappings Θ. In the next proposition we prove that it is invariant under all mappings Θ. We will later see that it naturally appears as the intersection of iterates of certain families of mappings Θ, cf. Theorem 10.22. Recall that every $f \in \mathcal{PBF}$ admits the power representation (5.17), $f(\lambda) = \int_{[0,1]} \lambda^\alpha \kappa(d\alpha)$. For any measurable set $E \subset [0, 1]$ let

$$\mathcal{PBF}^E := \left\{ f \in \mathcal{BF} : f(\lambda) = \int_{[0,1]} \lambda^\alpha \kappa(d\alpha), \quad \kappa([0,1] \setminus E) = 0 \right\}. \tag{10.21}$$

Proposition 10.6. *For $0 \le \alpha \le 1$, let $c(\alpha) := \int_0^B t^\alpha \vartheta(t)\, dt \in (0, \infty]$.*

(i) *If $f(\lambda) = \lambda^\alpha$, then $\Theta f = c(\alpha) f$.*

(ii) *If $E \subset [0, 1]$ and $f \in \mathcal{PBF}^E \cap \mathfrak{D}(\Theta)$, then $\Theta f \in \mathcal{PBF}^E$.*

(iii) *If $E \subset [0, 1]$ is a measurable set and $c(\alpha) \in (0, \infty)$ for all $\alpha \in E$, then $\Theta(\mathcal{PBF}^E \cap \mathfrak{D}(\Theta)) = \mathcal{PBF}^E$.*

Proof. (i) $\Theta f(\lambda) = \int_0^B f(\lambda t)\vartheta(t)\, dt = \left(\int_0^B t^\alpha \vartheta(t)\, dt \right) \lambda^\alpha = c(\alpha) f(\lambda)$.

(ii) By (10.21), there exists a finite measure κ on $[0, 1]$ such that $\kappa([0, 1] \setminus E) = 0$ and $f(\lambda) = \int_{[0,1]} \lambda^\alpha \kappa(d\alpha)$. Therefore

$$\Theta f(\lambda) = \int_0^B f(\lambda t)\vartheta(t)\, dt = \int_0^B \left(\int_{[0,1]} (\lambda t)^\alpha \kappa(d\alpha) \right) \vartheta(t)\, dt$$

$$= \int_{[0,1]} \left(\int_0^B t^\alpha \vartheta(t)\, dt \right) \lambda^\alpha \kappa(d\alpha).$$

Let $\widetilde{\kappa}(d\alpha) := c(\alpha)\kappa(d\alpha)$. Then $\Theta f(\lambda) = \int_{[0,1]} \lambda^\alpha \widetilde{\kappa}(d\alpha)$. Since $\Theta f(1) < \infty$ by the assumption, it follows that $\widetilde{\kappa}$ is a finite measure on $[0, 1]$. Moreover, $\widetilde{\kappa}([0, 1] \setminus E) = 0$; thus $\Theta f \in \mathcal{PBF}^E$.

(iii) Let $\tilde{f}(\lambda) = \int_{[0,1]} \lambda^\alpha \widetilde{\kappa}(d\alpha) \in \mathcal{PBF}^E$, and define a finite measure κ on $[0, 1]$ by $\kappa(d\alpha) = c(\alpha)^{-1} \widetilde{\kappa}(d\alpha)$. Since $c(\alpha) \ge \int_0^{B \wedge 1} t^\alpha \vartheta(t)\, dt \ge \int_0^{B \wedge 1} t \vartheta(t)\, dt > 0$, we have that $c(\alpha)^{-1}$ is a bounded function on $[0, 1]$. Thus κ is a finite measure and $\kappa([0, 1] \setminus E) = 0$. Let $f(\lambda) := \int_{[0,1]} \lambda^\alpha \kappa(d\alpha) \in \mathcal{PBF}^E$. The computation in (ii) shows that $\Theta f = \tilde{f}$.   □

The next proposition states that, if ϑ satisfies certain integrability conditions, $\mathfrak{D}(\Theta)$ and $\mathfrak{R}(\Theta)$ are closed under pointwise limits, and Θ is continuous with respect to pointwise convergence. In Remark 10.8 we will show that continuity need not be true without the integrability conditions.

Proposition 10.7. *Assume that $\int_0^B \vartheta(t)\, dt < \infty$ if $B < \infty$, and that $\int_0^1 \vartheta(t)\, dt < \infty$ and $\int_1^\infty t\vartheta(t)\, dt < \infty$ if $B = \infty$.*

(i) *If $(f_n)_{n\in\mathbb{N}} \subset \mathfrak{D}(\Theta)$ is such that $\lim_{n\to\infty} f_n(\lambda) = f(\lambda)$ for all $\lambda \in (0, \infty)$, and $f(1) < \infty$, then $f \in \mathfrak{D}(\Theta)$ and $\lim_{n\to\infty} \Theta f_n(\lambda) = \Theta f(\lambda)$ for all $\lambda \in (0, \infty)$.*

(ii) *If $(f_n)_{n\in\mathbb{N}} \subset \mathfrak{D}(\Theta)$ is such that $\lim_{n\to\infty} \Theta f_n(\lambda) = \tilde{f}(\lambda) \in (0, \infty)$ for all $\lambda \in (0, \infty)$, then there exists $f \in \mathcal{BF}$ such that $\tilde{f} = \Theta f \in \mathcal{BF}$. Moreover, if Θ is one-to-one, then $\lim_{n\to\infty} f_n = f$.*

Proof. (i) Since $f(1) < \infty$, we have that $f \in \mathcal{BF}$. Suppose first that $B < \infty$. Then

$$\Theta f(\lambda) = \int_0^B f(\lambda t)\vartheta(t)\, dt \le f(\lambda B) \int_0^B \vartheta(t)\, dt < \infty,$$

implying that $f \in \mathfrak{D}(\Theta)$. Let $\lambda > 0$. Since $\lim_{n\to\infty} f_n(\lambda B) = f(\lambda B)$, there exists $n_0 \in \mathbb{N}$ such that $f_n(\lambda t) \leq f_n(\lambda B) \leq 2f(\lambda B)$ for all $t \in (0, B]$ and all $n \geq n_0$. By the dominated convergence theorem it follows that

$$\lim_{n\to\infty} \Theta f_n(\lambda) = \lim_{n\to\infty} \int_0^B f_n(\lambda t)\vartheta(t)\, dt = \int_0^B f(\lambda t)\vartheta(t)\, dt = \Theta f(\lambda).$$

If $B = \infty$, then the same argument as above shows that

$$\lim_{n\to\infty} \int_0^1 f_n(\lambda t)\vartheta(t)\, dt = \int_0^1 f(\lambda t)\vartheta(t)\, dt < \infty.$$

Fix $\lambda > 0$ and let $t \geq 1$. Then

$$f(\lambda t) = a + b\lambda t + \lambda t \int_0^\infty e^{-\lambda ts}\mu(s, \infty)\, ds$$

$$\leq ta + t(b\lambda) + t\lambda \int_0^\infty e^{-\lambda s}\mu(s, \infty)\, ds = tf(\lambda).$$

Hence, there exists some $n_0 \in \mathbb{N}$ such that $f_n(\lambda t) \leq f_n(\lambda)t \leq 2f(\lambda)t$ for all $n \geq 0$ and all $t \geq 1$. Again by the dominated convergence theorem we conclude that

$$\lim_{n\to\infty} \int_1^\infty f_n(\lambda t)\vartheta(t)\, dt = \int_1^\infty f(\lambda t)\vartheta(t)\, dt < \infty.$$

(ii) Because of Remark 3.11 we can find a subsequence $(n_k)_{k\in\mathbb{N}}$ and a function $f : (0, \infty) \to [0, \infty]$ such that $\lim_{k\to\infty} f_{n_k}(\lambda) = f(\lambda)$ for all $\lambda \in (0, \infty)$. By Fatou's lemma

$$\int_0^B f(\lambda t)\vartheta(t)\, dt \leq \liminf_{k\to\infty} \int_0^B f_{n_k}(\lambda t)\vartheta(t)\, dt = \lim_{k\to\infty} \Theta f_{n_k}(\lambda) = \tilde{f}(\lambda) < \infty,$$

implying that $f(\lambda) < \infty$ for all $\lambda \in (0, \infty)$. Thus $f \in \mathcal{BF}$. From (i) we conclude that $\Theta f = \tilde{f}$. If Θ is one-to-one, then all subsequential limits are equal to f, hence the whole sequence $(f_n)_{n\in\mathbb{N}}$ converges to f. □

Remark 10.8. Let $\alpha \in [0, 1)$ and suppose that $c(\alpha) = \int_0^B t^\alpha \vartheta(t)\, dt = +\infty$, but $c(\gamma) = \int_0^B t^\gamma \vartheta(t)\, dt < \infty$ for all $\gamma \in (\alpha, 1]$. Examples that satisfy these conditions are $\vartheta_{\alpha,\beta}$, $\hat{\vartheta}_{\alpha,p}$ and $\overline{\vartheta}_{\alpha,p}$.

Consider the sequence $(f_n)_{n\geq 0}$ where $f_n(\lambda) := \lambda^{\alpha+1/n}$. By Proposition 10.6 (i), $\Theta f_n = c(\alpha + 1/n) f_n$, hence $f_n \in \mathfrak{D}(\Theta)$. On the other hand, with $f(\lambda) = \lambda^\alpha$ we have that $\lim_{n\to\infty} f_n(\lambda) = f(\lambda)$ for all $\lambda > 0$. Since $\Theta f = c(\alpha)f \equiv +\infty$ we conclude that $f \notin \mathfrak{D}(\Theta)$. Therefore, $\mathfrak{D}(\Theta)$ is not closed with respect to pointwise convergence.

Let $g_n(\lambda) = c(\alpha + 1/n)^{-1}\lambda^{\alpha+1/n}$, $n \geq 1$. Then $g_n \in \mathfrak{D}(\Theta)$, $\Theta g_n(\lambda) = \lambda^{\alpha+1/n}$, and $\lim_{n\to\infty} \Theta g_n(\lambda) = \lambda^\alpha$ for all $\lambda > 0$. Setting $g(\lambda) \equiv 0$ we see that $g \in \mathfrak{D}(\Theta)$, $\Theta g(\lambda) = 0$, and $\lim_{n\to\infty} g_n(\lambda) = g(\lambda)$ for all $\lambda > 0$. This shows that Θ is not continuous with respect to pointwise convergence.

The next example shows that the mapping Θ need not be one-to-one.

Example 10.9. Let $\alpha > 0$ and define $\vartheta : (0,\infty) \to (0,\infty)$ by $\vartheta(t) = \alpha t^{-\alpha-1}$. Then

$$\int_t^\infty \vartheta(s)\,ds = t^{-\alpha},$$

implying that $\theta : (0,\infty) \to (0,\infty)$ is given by $\theta(t) = t^{-1/\alpha}$. Let $f \in \mathcal{BF}$ with the generating triplet (a,b,μ). If $a > 0$ (respectively $b > 0$), then $\tilde{a} = \infty$ (respectively $\tilde{b} = \infty$). Assume that $a = b = 0$. By using (10.12) we get that

$$\widetilde{m}(t) = t^{-\alpha-1} \int_{(0,\infty)} s^\alpha \,\mu(ds), \quad t > 0.$$

If $\int_{(0,\infty)} s^\alpha \,\mu(ds) = \infty$, then $\widetilde{m}$ is identically infinite, hence $\Theta f = \infty$. On the other hand, if $\int_{(0,\infty)} s^\alpha \,\mu(ds) < \infty$, $\widetilde{m}$ is proportional to $t^{-\alpha-1}$. If $\alpha \geq 1$, then since $\int_0^\infty (t \wedge 1) t^{-\alpha-1} = \infty$, $\widetilde{m}(t)\,dt$ is not a Lévy measure and again, $\Theta f = \infty$. Thus, for $\alpha \geq 1$, $\mathfrak{D}(\Theta) = \{0\}$. On the other hand, if $0 < \alpha < 1$, then $\int_0^\infty (t \wedge 1) t^{-\alpha-1} < \infty$, and $\widetilde{m}$ is a Lévy measure. Hence

$$\mathfrak{D}(\Theta) = \left\{ f \in \mathcal{BF} : a = b = 0, \int_{(0,\infty)} s^\alpha \,\mu(ds) < \infty \right\}$$

(this can be also computed by using (10.15)). By using the formula for $\widetilde{m}$ we obtain that $\Theta f(\lambda) = c_{\mu,\alpha} \lambda^\alpha \in \mathcal{BF}$, where $c_{\mu,\alpha} = \int_{(0,\infty)} s^\alpha \,\mu(ds)$. There exist different Lévy measures μ and ν such that $\int_{(0,\infty)} s^\alpha \,\mu(ds) = \int_{(0,\infty)} s^\alpha \,\nu(ds) < \infty$. Hence, Θ is not one-to-one.

We show now that any two operators Θ_1 and Θ_2 commute and give the formula for their composition.

Proposition 10.10. *Let $\vartheta_j(t) = -(\theta_j^{-1})'(t)$, $j = 1, 2$, define*

$$\vartheta(t) = \int_{t/B_1}^{B_2} \vartheta_1(ts^{-1})\vartheta_2(s)s^{-1}\,ds = \int_{t/B_2}^{B_1} \vartheta_2(ts^{-1})\vartheta_1(s)s^{-1}\,ds, \qquad (10.22)$$

and let θ be the inverse of $t \mapsto \int_t^{B_1 B_2} \vartheta(s)\,ds$. Then $\Theta = \Theta_2 \circ \Theta_1 = \Theta_1 \circ \Theta_2$. In particular, Θ_1 and Θ_2 commute.

Proof. By a change of variables and Fubini's theorem we have that

$$\Theta_2(\Theta_1 f)(\lambda) = \int_0^{B_2} \Theta_1 f(\lambda s)\vartheta_2(s)\,ds$$

$$= \int_0^{B_2}\left(\int_0^{B_1} f(\lambda s u)\vartheta_1(u)\,du\right)\vartheta_2(s)\,ds$$

$$= \int_0^{B_2}\left(\int_0^{B_1 s} f(\lambda t)\vartheta_1(ts^{-1})s^{-1}\,dt\right)\vartheta_2(s)\,ds$$

$$= \int_0^{B_1 B_2} f(\lambda t)\left(\int_{t/B_1}^{B_2}\vartheta_1(ts^{-1})\vartheta_2(s)s^{-1}\,ds\right)dt$$

$$= \int_0^{B_1 B_2} f(\lambda t)\vartheta(t)\,dt,$$

proving the first equality in (10.22). The second equality follows by another change
of variables.   □

Proposition 10.10 applied to the three two-parameter families of mappings defined
in (10.6) reveals certain relations among those families.

Proposition 10.11.   (i) *Assume that $p, q > 0$ and $-\infty < \alpha < 1$. Then*

$$\widehat{\Theta}_{\alpha,p} \circ \widehat{\Theta}_{\alpha-p,q} = \widehat{c}(\alpha, p, q)\,\widehat{\Theta}_{\alpha,p+q}, \tag{10.23}$$

where $\widehat{c}(\alpha, p, q) > 0$ is a positive constant. If $-\infty < \alpha \le 0$, $\widehat{c}(\alpha, p, q) = 1$.

(ii) *Assume that $\beta > 0$, $-\infty < \alpha < 0$ and $\alpha + \beta < 1$. Then*

$$\widehat{\Theta}_{\alpha+\beta,1} \circ \widehat{\Theta}_{\alpha,\beta} = c(\alpha, \beta)\,\widehat{\Theta}_{\alpha+\beta,\beta}, \tag{10.24}$$

where $c(\alpha, \beta) > 0$ is a positive constant. If $\alpha + \beta \le 0$, $c(\alpha, \beta) = 1$.

(ii') *Assume that $\alpha, \beta < 1$ and $\alpha > \beta$. Then*

$$\widehat{\Theta}_{\alpha,\alpha-\beta} \circ \widehat{\Theta}_{\beta,1} = c'(\alpha, \beta)\,\widehat{\Theta}_{\alpha,1}. \tag{10.25}$$

where $c'(\alpha, \beta) > 0$ is a positive constant. If $\alpha \le 0$, then $c'(\alpha, \beta) = 1$.

(iii) *Assume that $p, q > 0$ and $-\infty < \alpha < 1$. Then*

$$\overline{\Theta}_{\alpha,p} \circ \overline{\Theta}_{\alpha,q} = \overline{c}(\alpha, p, q)\,\overline{\Theta}_{\alpha,p+q}, \tag{10.26}$$

where $\overline{c}(\alpha, p, q) > 0$ is a positive constant. If $-\infty < \alpha < 0$, $\overline{c}(\alpha, p, q) = 1$.

Proof. (i) We use Proposition 10.10 to compute the function ϑ corresponding to the composition $\widehat{\Theta}_{\alpha,p} \circ \widehat{\Theta}_{\alpha-p,q}$:

$$
\begin{aligned}
\vartheta(t) &= \int_t^1 \widehat{\vartheta}_{\alpha,p}(ts^{-1})\widehat{\vartheta}_{\alpha-p,q}(s)s^{-1}\,ds \\
&= \widehat{c}_{\alpha,p}\,\widehat{c}_{\alpha-p,q}\int_t^1 \left(1 - \frac{t}{s}\right)^{p-1}\left(\frac{t}{s}\right)^{-\alpha-1}(1-s)^{q-1}s^{-\alpha+p-1}s^{-1}\,ds \\
&= \widehat{c}_{\alpha,p}\,\widehat{c}_{\alpha-p,q}t^{-\alpha-1}\int_t^1 (s-t)^{p-1}(1-s)^{q-1}\,ds \\
&= \widehat{c}_{\alpha,p}\,\widehat{c}_{\alpha-p,q}\frac{\Gamma(p)\Gamma(q)}{\Gamma(p+q)}(1-t)^{p+q-1}t^{-\alpha-1}.
\end{aligned}
$$

It remains to check that $\widehat{c}_{\alpha,p}\,\widehat{c}_{\alpha-p,q}\Gamma(p)\Gamma(q)/\Gamma(p+q) = \widehat{c}_{\alpha,p+q}$ when $\alpha \le 0$. If $\alpha < 0$, then

$$
\begin{aligned}
\widehat{c}_{\alpha,p}\,\widehat{c}_{\alpha-p,q}\frac{\Gamma(p)\Gamma(q)}{\Gamma(p+q)} &= \frac{\Gamma(-\alpha+p)}{\Gamma(-\alpha)\Gamma(p)}\frac{\Gamma(-\alpha+p+q)}{\Gamma(-\alpha+p)\Gamma(q)}\frac{\Gamma(p)\Gamma(q)}{\Gamma(p+q)} \\
&= \frac{\Gamma(-\alpha+p+q)}{\Gamma(-\alpha)\Gamma(p+q)} = \frac{1}{B(-\alpha, p+q)} = \widehat{c}_{\alpha,p+q}.
\end{aligned}
$$

If $\alpha = 0$, then

$$
\widehat{c}_{0,p}\,\widehat{c}_{-p,q}\frac{\Gamma(p)\Gamma(q)}{\Gamma(p+q)} = \frac{\Gamma(p+q)}{\Gamma(p)\Gamma(q)}\frac{\Gamma(p)\Gamma(q)}{\Gamma(p+q)} = 1 = \widehat{c}_{0,p+q}.
$$

(ii) Again, by Proposition 10.10, the function ϑ corresponding to the composition $\widehat{\Theta}_{\alpha+\beta,1} \circ \Theta_{\alpha,\beta}$ is equal to

$$
\begin{aligned}
\vartheta(t) &= \int_t^\infty \widehat{\vartheta}_{\alpha+\beta,1}(ts^{-1})\vartheta_{\alpha,\beta}(s)s^{-1}\,ds \\
&= \widehat{c}_{\alpha+\beta,1}\,c_{\alpha,\beta}\int_t^\infty \left(\frac{t}{s}\right)^{-\alpha-\beta-1}s^{-\alpha-1}\exp\left(-s^\beta\right)s^{-1}\,ds \\
&= \widehat{c}_{\alpha+\beta,1}\,c_{\alpha,\beta}\,t^{-\alpha-\beta-1}\int_t^\infty s^{\beta-1}\exp\left(-s^\beta\right)\,ds \\
&= \frac{1}{\beta}\widehat{c}_{\alpha+\beta,1}\,c_{\alpha,\beta}\,t^{-\alpha-\beta-1}\int_{t^\beta}^\infty e^{-u}\,du \\
&= \frac{1}{\beta}\widehat{c}_{\alpha+\beta,1}\,c_{\alpha,\beta}\,t^{-\alpha-\beta-1}e^{-t^\beta}.
\end{aligned}
$$

It remains to check that $\beta^{-1}\widehat{c}_{\alpha+\beta,1}\,c_{\alpha,\beta} = c_{\alpha+\beta,\beta}$ when $\alpha+\beta \le 0$. If $\alpha+\beta < 0$,

$$
\frac{1}{\beta}\widehat{c}_{\alpha+\beta,1}\,c_{\alpha,\beta} = \frac{-\alpha-\beta}{\Gamma\left(-\frac{\alpha}{\beta}\right)} = \frac{\beta}{\Gamma\left(-\frac{\alpha+\beta}{\beta}\right)} = c_{\alpha+\beta,\beta}.
$$

If $\alpha + \beta = 0$, $\alpha^{-1}\widehat{c}_{0,1} c_{-\alpha,\alpha} = \Gamma(1) = 1 = c_{0,\alpha}$.

(ii') Similarly as in (ii) we have

$$\vartheta(t) = \int_t^\infty \widehat{\vartheta}_{\alpha,\alpha-\beta}(ts^{-1})\vartheta_{\beta,1}(s)s^{-1}\,ds$$

$$= \widehat{c}_{\alpha,\alpha-\beta}\, c_{\beta,1} \int_t^\infty \left(1 - \frac{t}{s}\right)^{\alpha-\beta-1} \left(\frac{t}{s}\right)^{-\alpha-1} s^{-\beta-1}e^{-s}s^{-1}\,ds$$

$$= \widehat{c}_{\alpha,\alpha-\beta}\, c_{\beta,1}\, t^{-\alpha-1} \int_t^\infty (s-t)^{\alpha-\beta-1}e^{-s}\,ds$$

$$= \widehat{c}_{\alpha,\alpha-\beta}\, c_{\beta,1}\Gamma(\alpha-\beta)\, t^{-\alpha-1}e^{-t}.$$

It remains to check that $\widehat{c}_{\alpha,\alpha-\beta}\, c_{\beta,1}\Gamma(\alpha-\beta) = c_{\alpha,1}$ for $\alpha \leq 0$. If $\alpha < 0$, then

$$\widehat{c}_{\alpha,\alpha-\beta}\, c_{\beta,1}\Gamma(\alpha-\beta) = \frac{1}{B(-\alpha,\alpha,-\beta)}\frac{1}{\Gamma(-\beta)}\Gamma(\alpha-\beta) = \frac{1}{\Gamma(-\alpha)} = c_{\alpha,1}.$$

If $\alpha = 0$, then $\widehat{c}_{0,-\beta}\, c_{\beta,1}\Gamma(-\beta) = 1 = c_{0,1}$.

(iii) We use Proposition 10.10 to compute the function ϑ corresponding to the composition $\overline{\Theta}_{\alpha,p} \circ \overline{\Theta}_{\alpha,q}$. In the line marked by $*$ we change variables according to $u = -\log s/\log t + 1$:

$$\vartheta(t) = \int_t^1 \overline{\vartheta}_{\alpha,p}(ts^{-1})\overline{\vartheta}_{\alpha,q}(s)s^{-1}\,ds$$

$$= \overline{c}_{\alpha,p}\,\overline{c}_{\alpha,q} \int_t^1 \left(-\log\frac{t}{s}\right)^{p-1}\left(\frac{t}{s}\right)^{-\alpha-1}(-\log s)^{q-1}s^{-\alpha-1}s^{-1}\,ds$$

$$= \overline{c}_{\alpha,p}\,\overline{c}_{\alpha,q}\,t^{-\alpha-1} \int_t^1 (-\log t + \log s)^{p-1}(-\log s)^{q-1}s^{-1}\,ds$$

$$\overset{*}{=} \overline{c}_{\alpha,p}\,\overline{c}_{\alpha,q}\,t^{-\alpha-1}(-\log t)^{p+q-1}\int_0^1 u^{p-1}(1-u)^{q-1}\,du$$

$$= \overline{c}_{\alpha,p}\,\overline{c}_{\alpha,q}\,t^{-\alpha-1}(-\log t)^{p+q-1}B(p,q)$$

$$= c(\alpha,p,q)\,\overline{\vartheta}_{\alpha,p+q}$$

where

$$c(\alpha,p,q) = \frac{\overline{c}_{\alpha,p}\,\overline{c}_{\alpha,q}\,\Gamma(p)\Gamma(q)}{\overline{c}_{\alpha,p+q}\,\Gamma(p+q)}.$$

It is straightforward to check that $c(\alpha,p,q) = 1$ if $-\infty < \alpha < 0$. $\qquad\qquad\square$

By using (10.10), (10.11) and (10.6) we see that each of the functions $\vartheta_{\alpha,\beta}$, $\widehat{\vartheta}_{\alpha,p}$ and $\overline{\vartheta}_{\alpha,p}$ satisfies

$$\tilde{a} = \left(\int_0^\infty \vartheta(t)\, dt \right) a = \begin{cases} a, & \alpha < 0 \\ \infty \cdot a, & 0 \le \alpha < 1, \end{cases}$$

$$\tilde{b} = \left(\int_0^\infty t\vartheta(t)\, dt \right) b < \infty.$$

For $f \in \mathcal{BF}$ given by (10.8), let $m_{\alpha,\beta}$, $\widehat{m}_{\alpha,p}$ and $\overline{m}_{\alpha,p}$ denote the densities of the Lévy measures of $\Theta_{\alpha,\beta} f$, $\widehat{\Theta}_{\alpha,p} f$ and $\overline{\Theta}_{\alpha,p} f$, respectively. By (10.12) we have

$$m_{\alpha,\beta}(t) = c_{\alpha,\beta}\, t^{-\alpha-1} \int_{(0,\infty)} \exp\left(-t^\beta s^{-\beta} \right) s^\alpha\, \mu(ds), \qquad (10.27)$$

$$\widehat{m}_{\alpha,p}(t) = \widehat{c}_{\alpha,p}\, t^{-\alpha-1} \int_{(t,\infty)} (s-t)^{p-1} s^{\alpha+1-p}\, \mu(ds), \qquad (10.28)$$

$$\overline{m}_{\alpha,p}(t) = \overline{c}_{\alpha,p}\, t^{-\alpha-1} \int_{(t,\infty)} (\log s - \log t)^{p-1} s^\alpha\, \mu(ds). \qquad (10.29)$$

If $0 < \alpha < 1$, $\widehat{m}_{\alpha,p}$ is well defined only if $f \in \mathcal{BF}_{\alpha,0}$, while for $0 \le \alpha < 1$, $\overline{m}_{\alpha,p}$ is well defined only if $f \in \mathcal{BF}_{\alpha,\log^{p-1}}$.

We are now going to state and prove three theorems which provide a complete description of the domains and ranges of the mappings $\Theta_{\alpha,\beta}$, $\widehat{\Theta}_{\alpha,p}$ and $\overline{\Theta}_{\alpha,p}$, cf. Table 10.1; we will also show that these mappings are one-to-one and, in case $\alpha < 0$, continuous with respect to pointwise convergence, cf. Remark 10.8.

Theorem 10.12.   (i) *Let $\beta > 0$ and $-\infty < \alpha < 0$. Then $\mathfrak{D}(\Theta_{\alpha,\beta}) = \mathcal{BF}$ and $\Theta_{\alpha,\beta} : \mathcal{BF} \to \mathcal{BF}$ is one-to-one and continuous with respect to pointwise convergence. Moreover, $\mathfrak{R}(\Theta_{\alpha,\beta})$ consists of all Bernstein functions whose Lévy measure has a density $\widetilde{m}$ such that $t \mapsto t^{(\alpha+1)/\beta} \cdot \widetilde{m}(t^{1/\beta})$ is a completely monotone function.*

(ii) *Let $\beta > 0$ and $\alpha = 0$. Then $\mathfrak{D}(\Theta_{0,\beta}) = \mathcal{BF}_{\log,0}$ and $\Theta_{0,\beta} : \mathcal{BF}_{\log,0} \to \mathcal{BF}_0$ is one-to-one. Moreover, $\mathfrak{R}(\Theta_{0,\beta})$ consists of all Bernstein functions in $\mathcal{BF}_0$ whose Lévy measure has a density $\widetilde{m}$ such that $t \mapsto t^{1/\beta} \cdot \widetilde{m}(t^{1/\beta})$ is a completely monotone function.*

(iii) *Let $\beta > 0$ and $0 < \alpha < 1$. Then $\mathfrak{D}(\Theta_{0,\beta}) = \mathcal{BF}_{\alpha,0}$ and $\Theta_{\alpha,\beta} : \mathcal{BF}_{\alpha,0} \to \mathcal{BF}_0$ is one-to-one. Moreover, $\mathfrak{R}(\Theta_{\alpha,\beta})$ consists of all Bernstein functions in $\mathcal{BF}_0$ whose Lévy measure has a density $\widetilde{m}$ such that $t \mapsto t^{(\alpha+1)/\beta} \cdot \widetilde{m}(t^{1/\beta})$ is a completely monotone function vanishing at infinity.*

Proof. Let v be the image measure of $s^{\alpha-1} \mu(ds)$ under the mapping $\phi(s) := s^{-\beta}$. Then

$$m_{\alpha,\beta}(t) = c_{\alpha,\beta} \, t^{-\alpha-1} \int_{(0,\infty)} e^{-t^\beta s} \, v(ds) = c_{\alpha,\beta} \, t^{-\alpha-1} \mathscr{L}(v; t^\beta).$$

This shows that $t \mapsto t^{(\alpha+1)/\beta} \cdot m_{\alpha,\beta}(t^{1/\beta}) = c_{\alpha,\beta} \mathscr{L}(v; t)$ is a completely monotone function vanishing at infinity (since $v\{0\} = 0$), and that the measure μ is uniquely determined by $m_{\alpha,\beta}$.

(i) We check that $\Theta_{\alpha,\beta} f \in \mathcal{BF}$ for every $f \in \mathcal{BF}$. Clearly, $\int_0^\infty \vartheta_{\alpha,\beta}(t) \, dt < \infty$ and $\int_0^\infty t \vartheta_{\alpha,\beta}(t) \, dt < \infty$, showing that (10.13) and (10.14) are satisfied. Further,

$$\int_0^\infty \left(\int_0^{1/t} s \vartheta_{\alpha,\beta}(s) \, ds \right) \mu(t,\infty) \, dt$$

$$\leq c_{\alpha,\beta} \int_0^1 \left(\int_0^\infty s^{-\alpha} e^{-s^\beta} \, ds \right) \mu(t,\infty) \, dt$$

$$+ c_{\alpha,\beta} \int_1^\infty \left(\int_0^{1/t} s^{-\alpha} \, ds \right) \mu(t,\infty) \, dt$$

$$\leq c_{\alpha,\beta} \left(\int_0^1 \mu(t,\infty) \, dt \right) \left(\int_0^\infty s^{-\alpha} e^{-s^\beta} \, ds \right) dt$$

$$+ \frac{c_{\alpha,\beta}}{1-\alpha} \int_1^\infty t^{\alpha-1} \mu(t,\infty) \, dt.$$

Since this is finite, (10.15) holds as well. Further, $\tilde{a}$, $\tilde{b}$ and $m_{\alpha,\beta}$ are uniquely determined by the triplet (a, b, μ), i.e. $\Theta_{\alpha,\beta}$ is one-to-one from $\mathcal{BF}$ onto $\Theta_{\alpha,\beta}(\mathcal{BF})$. Since $\vartheta_{\alpha,\beta}$ satisfies the integrability conditions of Proposition 10.7, $\Theta_{\alpha,\beta}$ is continuous with respect to pointwise convergence.

We have already seen that every $\tilde{f} \in \mathfrak{R}(\Theta_{\alpha,\beta})$ has a Lévy measure with density $\tilde{m}$ such that $t \mapsto t^{(\alpha+1)/\beta} \cdot \tilde{m}(t^{1/\beta})$ is completely monotone. Conversely, suppose that $\tilde{f} \in \mathcal{BF}$ with the generating triplet $(\tilde{a}, \tilde{b}, \tilde{\mu})$, where $\tilde{\mu}(dt) = \tilde{m}(t) \, dt$ such that $t \mapsto t^{(\alpha+1)/\beta} \cdot \tilde{m}(t^{1/\beta}) \in \mathcal{CM}$. Since $\int_1^\infty \tilde{m}(t) \, dt < \infty$, it follows that $\lim_{t\to\infty} t^{(\alpha+1)/\beta} \tilde{m}(t^{1/\beta}) = \lim_{s\to\infty} s^{\alpha+1} \tilde{m}(s) = 0$. Therefore, there exists a measure v on $(0,\infty)$ such that $t^{(\alpha+1)/\beta} \cdot \tilde{m}(t^{1/\beta}) = \mathscr{L}(v; t)$. Define a measure μ on $(0,\infty)$ by $\mu(ds) := s^{-\alpha+1}(v \circ \phi)(ds)$ (recall that $\phi(s) = s^{-\beta}$). Then

$$t^{-\alpha-1} \int_{(0,\infty)} \exp\left(-t^\beta s^{-\beta}\right) s^{\alpha-1} \mu(ds) = t^{-\alpha-1} \mathscr{L}(v; t^\beta) = \tilde{m}(t).$$

Let $a = \tilde{a}$ and $b = \left(\int_0^\infty t \vartheta_{\alpha,\beta}(t) \, dt\right)^{-1} \tilde{b}$. If f is given by (10.8), then $\Theta_{\alpha,\beta} f = \tilde{f}$.

(ii) We check that $\Theta_{0,\beta} f$ is a Bernstein function if, and only if, $f \in \mathcal{BF}_{\log,0}$. Since $\int_0^\infty \vartheta_{0,\beta}(t) \, dt = \infty$, the condition (10.13) holds if, and only if, $a = 0$; (10.14) is always satisfied because $\int_0^\infty t\vartheta_{0,\beta}(t)\,dt < \infty$. Finally,

$$\int_0^\infty \left[\int_0^{1/t} s\vartheta_{0,\beta}(s)\,ds\right] \mu(t,\infty)\,dt$$

$$= \int_0^1 \left[\int_0^{1/t} \exp\left(-s^\beta\right)ds\right] \mu(t,\infty)\,dt$$

$$+ \int_1^\infty \left[\int_0^{1/t} \exp\left(-s^\beta\right)ds\right] \mu(t,\infty)\,dt.$$

The first term is always finite. In the second term, we use that $e^{-1} \le \exp(-s^\beta) \le 1$ for $s \in (0,1)$. This shows that the second term is comparable to

$$\int_1^\infty t^{-1}\mu(t,\infty)\,dt = \int_{(1,\infty)} \log t\, \mu(dt).$$

Thus, $\Theta_{0,\beta} f \in \mathcal{BF}$ if, and only if, $f \in \mathcal{BF}_{\log,0}$. The rest follows as in (i).

(iii) The proof is similar to the proof of (ii). Note that for the converse we have to *assume* that $t \mapsto t^{(\alpha+1)/\beta} \cdot \widetilde{m}(t^{1/\beta})$ vanishes at infinity, because for $\alpha \in (0,1)$ this does not come for free. $\qquad\square$

Let us record two consequences of Theorem 10.12.

Corollary 10.13. *Let $\beta > 0$ and $-\infty < \alpha < 1$. Suppose that the Lévy measure μ has a density m. Then, under the transformation $\Theta_{\alpha,\beta}$,*

$$t \mapsto t^{(\alpha+1)/\beta+1} \cdot m(t^{1/\beta}) \quad \text{is decreasing}$$

if, and only if,

$$t \mapsto t^{(\alpha+1)/\beta+1} \cdot \widetilde{m}(t^{1/\beta}) \quad \text{is completely monotone.}$$

Proof. Assume that $t \mapsto t^{(\alpha+1)/\beta+1} \cdot m(t^{1/\beta})$ is decreasing. By a change of variable $u = s^{-\beta}$ in (10.27), we get that

$$t^{(\alpha+1)/\beta}\, \widetilde{m}(t^{1/\beta}) = \frac{c_{\alpha,\beta}}{\beta} \int_0^\infty e^{-tu} u^{(-\alpha-1)/\beta-1} m(u^{-1/\beta})\,du.$$

Let $n(u) := u^{(-\alpha-1)/\beta-1} m(u^{-1/\beta})$, $u > 0$. By the assumption, n is an increasing function. Denote by N the measure corresponding to the increasing function n. Then

$$t^{(\alpha+1)/\beta+1}\, \widetilde{m}(t^{1/\beta}) = \frac{c_{\alpha,\beta}}{\beta} \int_0^\infty te^{-tu} n(u)\,du$$

$$= \frac{c_{\alpha,\beta}}{\beta} \int_{(0,\infty)} e^{-tu}\, N(du) = \frac{c_{\alpha,\beta}}{\beta}\, \mathscr{L}(N;t).$$

The converse follows by retracing the steps backwards. $\qquad\square$

Corollary 10.14. (i) *Let $\beta > 0$ and $-\infty < \alpha_1 < \alpha_2 < 0$. Then*

$$\mathfrak{R}(\Theta_{0,\beta}) \subset \mathfrak{R}(\Theta_{\alpha_2,\beta}) \subset \mathfrak{R}(\Theta_{\alpha_1,\beta}).$$

(ii) *Let $0 < \beta_1 < \beta_2$ and $\alpha \leq 0$. Then $\mathfrak{R}(\Theta_{\alpha,\beta_1}) \subset \mathfrak{R}(\Theta_{\alpha,\beta_2})$.*

Proof. (i) We show the second inclusion, the first follows similarly. If $f \in \mathfrak{R}(\Theta_{\alpha_2,\beta})$, then the Lévy measure of f has a density n such that $t \mapsto t^{(\alpha_2+1)/\beta} \cdot n(t^{1/\beta})$ is completely monotone. Since $t \mapsto t^{(\alpha_1-\alpha_2)/\beta}$ is also completely monotone, the product $t \mapsto t^{(\alpha_1+1)/\beta} \cdot n(t^{1/\beta})$ is completely monotone. Therefore, $f \in \mathfrak{R}(\Theta_{\alpha_1,\beta})$.

(ii) Let $\alpha \leq 0$. If $f \in \mathfrak{R}(\Theta_{\alpha,\beta_1})$, then the Lévy measure of f has a density n such that $t \mapsto t^{(\alpha+1)/\beta_1} \cdot n(t^{1/\beta_1})$ is completely monotone. The function $t \mapsto t^{\beta_1/\beta_2}$ is a Bernstein function. Since by Theorem 3.7 $\mathcal{CM} \circ \mathcal{BF} \subset \mathcal{CM}$, we conclude that $t \mapsto t^{(\alpha+1)/\beta_2} \cdot n(t^{1/\beta_2})$ is a completely monotone function. Thus $f \in \mathfrak{R}(\Theta_{\alpha,\beta_2})$. $\square$

Theorem 10.15. *Let $p \in \mathbb{N}$.*

(i) *Let $-\infty < \alpha < 0$. Then $\mathfrak{D}(\widehat{\Theta}_{\alpha,p}) = \mathcal{BF}$, $\widehat{\Theta}_{\alpha,p} : \mathcal{BF} \to \mathcal{BF}$ is one-to-one and continuous with respect to pointwise convergence. Moreover, $\mathfrak{R}(\widehat{\Theta}_{\alpha,p})$ consists of all Bernstein functions whose Lévy measure has a density $\widetilde{m}$ such that $t \mapsto t^{\alpha+1} \cdot \widetilde{m}(t)$ is p-monotone.*

(ii) *Let $\alpha = 0$. Then $\mathfrak{D}(\widehat{\Theta}_{0,p}) = \mathcal{BF}_{\log,0}$ and $\widehat{\Theta}_{0,p} : \mathcal{BF}_{\log,0} \to \mathcal{BF}_0$ is one-to-one. Moreover, $\mathfrak{R}(\widehat{\Theta}_{0,p})$ consists of all Bernstein functions in $\mathcal{BF}_0$ whose Lévy measure has a density $\widetilde{m}$ such that $t \mapsto t \cdot \widetilde{m}(t)$ is p-monotone.*

(iii) *Let $0 < \alpha < 1$. Then $\mathfrak{D}(\widehat{\Theta}_{\alpha,p}) = \mathcal{BF}_{\alpha,0}$ and $\widehat{\Theta}_{\alpha,p} : \mathcal{BF}_{\alpha,0} \to \mathcal{BF}_0$ is one-to-one. Further, $\mathfrak{R}(\widehat{\Theta}_{\alpha,p})$ consists of all Bernstein functions in $\mathcal{BF}_0$ whose Lévy measure has a density $\widetilde{m}$ such that $t \mapsto t^{\alpha+1} \cdot \widetilde{m}(t)$ is p-monotone and vanishes at infinity.*

Proof. We first compute the domains by checking (10.15). Since

$$\int_1^2 \left(\int_0^{1/t} (1-s)^{p-1} s^{-\alpha} \, ds \right) \mu(t,\infty) \, dt \leq B(p, 1-\alpha) \int_1^2 \mu(t,\infty) \, dt < \infty,$$

it suffices to check that

$$\int_2^\infty \left(\int_0^{1/t} (1-s)^{p-1} s^{-\alpha} \, ds \right) \mu(t,\infty) \, dt < \infty.$$

For $s \in (0, 1/2)$ we have $2^{1-p} \leq (1-s)^{p-1} \leq 1$ if $p \geq 1$, and $1 \leq (1-s)^{p-1} \leq 2^{1-p}$ if $0 < p < 1$. Thus, (10.15) is valid if, and only if,

$$\int_2^\infty \left(\int_0^{1/t} s^{-\alpha} \, ds \right) \mu(t,\infty) \, dt < \infty.$$

If $-\infty < \alpha < 0$ we see by a straightforward calculation and Fubini's theorem

$$\int_2^\infty \left(\int_0^{1/t} s^{-\alpha}\, ds \right) \mu(t,\infty)\, dt = \frac{-\alpha}{1-\alpha} \int_2^\infty t^{\alpha-1} \mu(t,\infty)\, dt$$

$$\leq \frac{-\alpha}{1-\alpha} \int_1^\infty t^{\alpha-1} \mu(t,\infty)\, dt$$

$$= \frac{1}{1-\alpha} \int_{(1,\infty)} (1-s^\alpha)\, \mu(ds) < \infty.$$

If $\alpha = 0$, then

$$\int_2^\infty \left(\int_0^{1/t} ds \right) \mu(t,\infty)\, dt \leq \int_1^\infty t^{-1} \mu(t,\infty)\, dt = \int_{(1,\infty)} \log s\, \mu(ds) < \infty.$$

Finally, let $0 < \alpha < 1$. Then

$$\int_2^\infty \left(\int_0^{1/t} s^{-\alpha}\, ds \right) \mu(t,\infty)\, dt \leq \frac{1}{1-\alpha} \int_1^\infty t^{\alpha-1} \mu(t,\infty)\, dt$$

$$= \frac{1}{\alpha(1-\alpha)} \int_{(1,\infty)} (s^\alpha - 1)\, \mu(ds).$$

Therefore, $\mathfrak{D}(\widehat{\Theta}_{\alpha,p}) = \mathcal{BF}$ for $-\infty < \alpha < 0$, $\mathfrak{D}(\widehat{\Theta}_{\alpha,p}) = \mathcal{BF}_{\alpha,0}$ for $0 < \alpha < 1$, and $\mathfrak{D}(\widehat{\Theta}_{0,p}) = \mathcal{BF}_{\log,0}$.

Now we prove that the mappings are one-to-one on their domains. If $p = 1$, $t^{\alpha+1}\widehat{m}_{\alpha,1}(t) = \widehat{c}_{\alpha,1} \int_{(t,\infty)} s^\alpha \, \mu(ds)$ for every $t > 0$, implying that the measure $s^\alpha \mu(ds)$ is uniquely determined by $\widehat{m}_{\alpha,p}$. Hence, the transformation $\widehat{\Theta}_{\alpha,1}$ is one-to-one for all $\alpha \in (-\infty, 1)$.

Let $p \geq 2$ and suppose that $\widehat{\Theta}_{\alpha,p-1}$ is one-to-one for all $\alpha \in (-\infty, 1)$. From Proposition 10.11 (i) we know that $\widehat{\Theta}_{\alpha,p} = \widehat{c}(\alpha, p-1, 1)^{-1} \widehat{\Theta}_{\alpha,p-1} \circ \widehat{\Theta}_{\alpha+1-p,1}$. Therefore, $\widehat{\Theta}_{\alpha,p}$ is also one-to-one for all $\alpha \in (-\infty, 1)$. From Proposition 10.7 it follows that $\widehat{\Theta}_{\alpha,p}$ is continuous with respect to pointwise convergence for $\alpha < 0$. Finally, from (10.28) we see that $t^{\alpha+1}\widehat{m}_{\alpha,p}(t) \leq \widehat{c}_{\alpha,p} \int_{(t,\infty)} s^\alpha \, \mu(ds)$, hence it vanishes at infinity.

(i) It is clear from (10.28) and Theorem 1.11 that $t \mapsto t^{\alpha+1} \cdot \widehat{m}_{\alpha,p}(t)$ is p-monotone. Conversely, suppose that $\tilde{f}$ is a Bernstein function with the generating triplet $(\tilde{a}, \tilde{b}, \tilde{\mu})$, where $\tilde{\mu}(dt) = \widetilde{m}(t)\, dt$ and $t \mapsto t^{\alpha+1} \cdot \widetilde{m}(t)$ is p-monotone. Moreover, since $\int_1^\infty \widetilde{m}(t)\, dt < \infty$, we have that $\lim_{t\to\infty} t^{\alpha+1} \widetilde{m}(t) = 0$. Hence, there exists a measure ν on $(0, \infty)$ such that

$$t^{\alpha+1}\widetilde{m}(t) = \widehat{c}_{\alpha,p} \int_{(t,\infty)} (s-t)^{p-1} \nu(ds).$$

Let μ be the measure on $(0, \infty)$ defined by $\mu(dt) := t^{-\alpha+p-1}\nu(dt)$. Then for $u > 0$,

$$
\begin{aligned}
\int_u^\infty \widetilde{m}(t)\,dt &= \widehat{c}_{\alpha,p} \int_u^\infty t^{-\alpha-1}\left(\int_{(t,\infty)} (s-t)^{p-1}s^{\alpha+1-p}\,\mu(ds)\right)dt \\
&= \widehat{c}_{\alpha,p} \int_{(u,\infty)} s^{\alpha+1-p}\left(\int_u^s (s-t)^{p-1}t^{-\alpha-1}\,dt\right)\mu(ds) \\
&= \widehat{c}_{\alpha,p} \int_{(u,\infty)} \left(\int_{u/s}^1 (1-v)^{p-1}v^{-\alpha-1}\,dv\right)\mu(ds) \\
&= \int_{(0,\infty)} \left(\int_{1\wedge(u/s)}^1 \widehat{\vartheta}_{\alpha,p}(v)\,dv\right)\mu(ds) \\
&= \int_{(0,\infty)} \left(\int_0^1 \mathbb{1}_{(u,\infty)}(sv)\widehat{\vartheta}_{\alpha,p}(v)\,dv\right)\mu(ds) \\
&= \int_0^1 \left(\int_{(0,\infty)} \mathbb{1}_{(u,\infty)}(sv)\widehat{\vartheta}_{\alpha,p}(v)\,\mu(ds)\right)dv.
\end{aligned}
\tag{10.30}
$$

This shows that the measure with density $\widetilde{m}$ is of the form (10.20). Since by assumption this is a Lévy measure, it follows by Remark 10.5 (ii) that μ is also a Lévy measure. Finally, let $a = \tilde{a}$ and $b = \tilde{b}/\int_0^1 t\,\widehat{\vartheta}_{\alpha,p}(t)\,dt$. If f is given by (10.8), then $\widehat{\Theta}_{\alpha,p}f = \tilde{f}$. This shows that $\widehat{\Theta}_{\alpha,p}$ maps $\mathcal{BF}$ onto the family of all Bernstein functions whose Lévy measure has a density $\widetilde{m}$ such that $t \mapsto t^{\alpha+1}\cdot\widetilde{m}(t)$ is p-monotone.

(ii) The proof is analogous to (i). The only additional ingredient is the part of the converse claiming that the Lévy measure μ constructed in the proof satisfies $\int_{(1,\infty)} \log t\,\mu(dt) < \infty$. To prove this we use (10.30) with $u = 1$:

$$
\begin{aligned}
\infty > \int_1^\infty \widetilde{m}(t)\,dt &= \int_0^1 \left(\int_{(0,\infty)} \mathbb{1}_{(1,\infty)}(sv)\widehat{\vartheta}_{0,p}(v)\,\mu(ds)\right)dv \\
&= \int_{(0,\infty)} \left(\int_{1\wedge s^{-1}}^1 (1-u)^{p-1}u^{-1}\,du\right)\mu(ds) \\
&= \int_{(1,\infty)} \left(\int_{1/s}^1 (1-u)^{p-1}u^{-1}\,du\right)\mu(ds) \\
&\geq \int_{(2,\infty)} \left(\int_{1/s}^{1/2} (1-u)^{p-1}u^{-1}\,du\right)\mu(ds) \\
&\geq (2^{1-p}\wedge 1)\int_{(2,\infty)} \left(\int_{1/s}^{1/2} u^{-1}\,du\right)\mu(ds) \\
&= (2^{1-p}\wedge 1)\int_{(2,\infty)} (\log s - \log 2)\,\mu(ds).
\end{aligned}
$$

Therefore, $\int_{(2,\infty)} \log t\,\mu(dt) < \infty$ which is equivalent to $\int_{(1,\infty)} \log t\,\mu(dt) < \infty$.

(iii) Again, the proof is analogous to the proof of (i). We only need to check that $\int_{(1,\infty)} t^\alpha \, \mu(dt) < \infty$. The same calculation as in (ii) yields

$$\infty > \int_1^\infty \widetilde{m}(t) \, dt = \int_0^1 \left(\int_{(0,\infty)} \mathbb{1}_{(1,\infty)}(sv)\widehat{\vartheta}_{\alpha,p}(v) \, \mu(ds) \right) dv$$

$$\geq \frac{2^{1-p} \wedge 1}{\alpha} \int_{(2,\infty)} (s^\alpha - 2^\alpha) \, \mu(ds),$$

proving the claim.   □

Remark 10.16. Although the theorem is stated for integer values of p, it is true for all $p > 0$ (see [317, Theorem 2.10]). The only place in the proof where we used the assumption that p is an integer was to show that the mappings $\widehat{\Theta}_{\alpha,p}$ are one-to-one, more precisely that $\widehat{\Theta}_{\alpha,1}$ is one-to-one. It is shown in [317] that $\widehat{\Theta}_{\alpha,p}$ is one-to-one for all $p \in (0,1)$ which suffices to prove Theorem 10.15 for all values $p > 0$.

Corollary 10.17.   (i) *Let* $-\infty < \alpha < 1$ *and* $0 < p < q$. *Then*

$$\mathfrak{R}(\widehat{\Theta}_{\alpha,q}) \subset \mathfrak{R}(\widehat{\Theta}_{\alpha,p}).$$

(ii) *Let* $-\infty < \alpha < 1$. *Then* $\bigcap_{p>0} \mathfrak{R}(\widehat{\Theta}_{\alpha,p}) = \mathfrak{R}(\Theta_{\alpha,1})$.

Proof. (i) This is a direct consequence of Proposition 10.11 (i).

   (ii) First note that because of (i), $\bigcap_{p>0} \mathfrak{R}(\widehat{\Theta}_{\alpha,p}) = \bigcap_{p\in\mathbb{N}} \mathfrak{R}(\widehat{\Theta}_{\alpha,p})$. Suppose that $f \in \bigcap_{p\in\mathbb{N}} \mathfrak{R}(\widehat{\Theta}_{\alpha,p})$. By Theorem 10.15, for every $p \in \mathbb{N}$, the Lévy measure of of f has a density $\widetilde{m}$ such that $t \mapsto t^{\alpha+1} \cdot \widetilde{m}(t)$ is p-monotone. Hence $t \mapsto t^{\alpha+1} \cdot \widetilde{m}(t)$ is completely monotone. By Theorem 10.12, $f \in \mathfrak{R}(\Theta_{\alpha,1})$. The other direction is proved by retracing the step backwards.   □

Theorem 10.18. *Let* $p \in \mathbb{N}$.

   (i) *Let* $-\infty < \alpha < 0$. *Then* $\mathfrak{D}(\overline{\Theta}_{\alpha,p}) = \mathcal{BF}$, $\overline{\Theta}_{\alpha,p} : \mathcal{BF} \to \mathcal{BF}$ *is one-to-one and continuous with respect to pointwise convergence. Moreover,* $\mathfrak{R}(\overline{\Theta}_{\alpha,p})$ *consists of all Bernstein functions whose Lévy measure has a density* $\widetilde{m}$ *of the form* $t^{-\alpha-1}n(\log t)$ *where* n *is* p-*monotone on* $\mathbb{R}$.

   (ii) *Let* $\alpha = 0$. *Then* $\mathfrak{D}(\overline{\Theta}_{0,p}) = \mathcal{BF}_{\log^p,0}$ *and* $\overline{\Theta}_{0,p} : \mathcal{BF}_{\log^p,0} \to \mathcal{BF}_0$ *is one-to-one. Moreover,* $\mathfrak{R}(\overline{\Theta}_{0,p})$ *consists of all Bernstein functions whose Lévy measure has a density* $\widetilde{m}$ *of the form* $t^{-1}n(\log t)$ *where* n *is* p-*monotone on* $\mathbb{R}$.

   (iii) *Let* $0 < \alpha < 1$. *Then* $\mathfrak{D}(\overline{\Theta}_{\alpha,p}) = \mathcal{BF}_{\log^{p-1},\alpha,0}$ *and* $\overline{\Theta}_{0,p} : \mathcal{BF}_{\log^{p-1},\alpha,0} \to \mathcal{BF}_0$ *is one-to-one. Moreover,* $\mathfrak{R}(\overline{\Theta}_{\alpha,p})$ *consists of all Bernstein functions whose Lévy measure has a density* $\widetilde{m}$ *of the form* $t^{-\alpha-1} \, n(\log t)$ *where* n *is* p-*monotone on* $\mathbb{R}$ *and vanishes at infinity.*

Proof. We first determine the domains by checking (10.15). For $\alpha < 0$, let $\epsilon > 0$ be such that $\alpha + \epsilon < 0$. There is a constant $c > 0$ such that $(-\log s)^{p-1} s^{-\alpha} \leq c s^{-\alpha-\epsilon}$ for all $s \in (0, 1)$. Hence,

$$\int_1^\infty \left(\int_0^{1/t} \overline{c}_{\alpha,p} (-\log s)^{p-1} s^{-\alpha} \, ds \right) \mu(t, \infty) \, dt$$

$$\leq c \, \overline{c}_{\alpha,p} \int_{(1,\infty)} \left(\int_0^{1/t} s^{-\alpha-\epsilon} \, ds \right) \mu(t, \infty) \, dt < \infty.$$

This shows that $\mathfrak{D}(\overline{\Theta}_{\alpha,p}) = \mathcal{BF}$ for $-\infty < \alpha < 0$. If $\alpha = 0$, after the change of variables $u = \log(1/s)$, we arrive at

$$\int_1^\infty \left(\int_0^{1/t} (-\log s)^{p-1} s^{-\alpha} \, ds \right) \mu(t, \infty) \, dt$$

$$= \int_{(1,\infty)} \left(\int_{\log t}^\infty u^{p-1} e^{-u} \, du \right) \mu(t, \infty) \, dt$$

$$= \int_{(1,\infty)} \Gamma(p, \log t) \, \mu(t, \infty) \, dt.$$

($\Gamma(p, x)$ is the incomplete Gamma function, cf. Section 16.1.)

As $\lim_{x \to \infty} \Gamma(p, x)/(x^{p-1} e^{-x}) = 1$, the integral $\int_{(2,\infty)} \Gamma(p, \log t) \mu(t, \infty) \, dt$ is comparable with

$$\int_{(2,\infty)} t^{-1} (\log t)^{p-1} \mu(t, \infty) \, dt = \int_{(2,\infty)} \left(\int_1^s t^{-1} (\log t)^{p-1} \, dt \right) \mu(ds)$$

$$= \int_{(2,\infty)} (\log s)^p \, \mu(ds).$$

Thus, $\mathfrak{D}(\overline{\Theta}_{0,p}) = \mathcal{BF}_{\log^p, 0}$. A similar analysis shows that

$$\mathfrak{D}(\overline{\Theta}_{\alpha,p}) = \mathcal{BF}_{\alpha, \log^{p-1}, 0}$$

for $0 < \alpha < 1$.

We prove that the mappings are one-to-one on their domains. As $\overline{\Theta}_{\alpha,1} = \widehat{\Theta}_{\alpha,1}$, it is one-to-one, cf. Theorem 10.15. For integers $p \geq 2$, the proof is the same as the proof of Theorem 10.15, the only difference being that instead of Proposition 10.11 (i) we use 10.11 (iii). Let ν be the image measure of μ under the map $s \mapsto \log s$, and define $n : \mathbb{R} \to (0, \infty)$ by

$$n(t) := \overline{c}_{\alpha,p} \int_{(t,\infty)} (s-t)^{p-1} e^{\alpha s} \, \nu(ds).$$

Then n is p-monotone on $\mathbb{R}$, vanishes at infinity, and

$$t^{\alpha+1} \overline{m}_{\alpha,p}(t) = \overline{c}_{\alpha,p} \int_{(t,\infty)} (\log s - \log t)^{p-1} s^\alpha \mu(ds) = n(\log t).$$

(i) Let $\tilde{f} \in \mathcal{BF}$ with the generating triplet $(\tilde{a}, \tilde{b}, \tilde{\mu})$, where $\tilde{\mu}(dt) = \widetilde{m}(t)\,dt$ and $t^{\alpha+1}\widetilde{m}(t) = n(\log t)$ where n is p-monotone on $\mathbb{R}$. Since $\int_1^\infty \widetilde{m}(t)\,dt < \infty$, we have that $\lim_{t\to\infty} t^{\alpha+1}\widetilde{m}(t) = 0$. Hence, there exists a measure ν on $(0,\infty)$ such that

$$t^{\alpha+1}\widetilde{m}(t) = \overline{c}_{\alpha,p} \int_{(\log t,\infty)} (s - \log t)^{p-1}\,\nu(ds).$$

Let ρ be the image measure of ν under the map $s \mapsto e^s$, and $\mu(ds) = s^{-\alpha}\rho(ds)$. Then for $u > 0$,

$$
\begin{aligned}
\int_u^\infty \widetilde{m}(t)\,dt &= \overline{c}_{\alpha,p} \int_u^\infty t^{-\alpha-1} \left(\int_{(t,\infty)} (\log s - \log t)^{p-1}\,\rho(ds) \right) dt \\
&= \overline{c}_{\alpha,p} \int_{(u,\infty)} \left(\int_u^s (-\log(t/s))^{p-1} t^{-\alpha-1}\,dt \right) \rho(ds) \\
&= \overline{c}_{\alpha,p} \int_{(u,\infty)} \left(\int_{u/s}^1 (-\log v)^{p-1} v^{-\alpha-1}\,dv \right) s^{-\alpha}\,\rho(ds) \\
&= \int_{(0,\infty)} \left(\int_{1\wedge(u/s)}^1 \overline{\vartheta}_{\alpha,p}(v)\,dv \right) \mu(ds) \\
&= \int_{(0,\infty)} \left(\int_0^1 \mathbb{1}_{(u,\infty)}(sv)\,\overline{\vartheta}_{\alpha,p}(v)\,dv \right) \mu(ds) \\
&= \int_0^1 \left(\int_{(0,\infty)} \mathbb{1}_{(u,\infty)}(sv)\,\overline{\vartheta}_{\alpha,p}(v)\,\mu(ds) \right) dv.
\end{aligned}
\tag{10.31}
$$

This shows that the measure with density $\widetilde{m}$ is of the form (10.20). By our assumption, this is a Lévy measure, and Remark 10.5 (ii) shows that μ is also a Lévy measure. Finally, let $a = \tilde{a}$ and $b = \tilde{b}/\int_0^1 t\,\vartheta_{\alpha,p}(t)\,dt$. If f is given by (10.8), then $\overline{\Theta}_{\alpha,p} f = \tilde{f}$. This shows that $\overline{\Theta}_{\alpha,p}$ maps $\mathcal{BF}$ onto the family of all Bernstein functions whose Lévy measure has a density $\widetilde{m}$ such that $t^{\alpha+1}\widetilde{m}(t) = n(\log t)$ where n is p-monotone on $\mathbb{R}$.

(ii) The proof is analogous to (i). The only additional ingredient is the part of the converse claiming that the Lévy measure μ constructed in the proof satisfies $\int_{(1,\infty)} (\log t)^p\,\mu(dt) < \infty$. To prove this we use (10.31) with $u = 1$:

$$
\begin{aligned}
\infty > \int_1^\infty \widetilde{m}(t)\,dt &= \int_0^1 \left(\int_{(0,\infty)} \mathbb{1}_{(1,\infty)}(sv)\,\overline{\vartheta}_{0,p}(v)\,\mu(ds) \right) dv \\
&= \int_{(0,\infty)} \left(\int_{1\wedge s^{-1}}^1 (-\log u)^{p-1} u^{-1}\,du \right) \mu(ds) \\
&= \int_{(1,\infty)} \left(\int_{1/s}^1 (-\log u)^{p-1} u^{-1}\,du \right) \mu(ds)
\end{aligned}
$$

$$= \int_{(1,\infty)} \left(\int_1^{\log s} t^{p-1}\, dt \right) \mu(ds)$$

$$= \frac{1}{p} \int_{(1,\infty)} \left((\log s)^p - (\log 2)^p \right) \mu(ds).$$

Therefore, $\int_{(1,\infty)} (\log t)^p\, \mu(dt) < \infty$.

(iii) Again, the proof is similar to the proof of Theorem 10.15. $\qquad\square$

Corollary 10.19. (i) *Let* $-\infty < \alpha < 1$ *and* $0 < p < q$. *Then* $\Re(\overline{\Theta}_{\alpha,q}) \subset \Re(\overline{\Theta}_{\alpha,p})$.

(ii) *Let* $-\infty < \alpha < 1$ *and let* $p \in \mathbb{N}$. *Then* $\Re(\overline{\Theta}_{\alpha,p}) \subset \Re(\widehat{\Theta}_{\alpha,p})$.

(iii) *Let* $-\infty < \alpha < 1$. *Then* $\bigcap_{p>0} \Re(\overline{\Theta}_{\alpha,p}) \subset \bigcap_{p>0} \Re(\widehat{\Theta}_{\alpha,p})$.

Proof. (i) This is a direct consequence of Proposition 10.11 (iii).

(ii) Suppose $f \in \Re(\overline{\Theta}_{\alpha,p})$. By Theorem 10.18 the Lévy measure of f has a density $\widetilde{m}$ such that $t^{\alpha+1}\widetilde{m}(t) = n(\log t)$ where n is p-monotone on $\mathbb{R}$. By Proposition 1.16, $t \mapsto n(\log t)$ is p-monotone on $(0,\infty)$. The claim follows from Theorem 10.15.

(iii) This follows immediately from (i), (ii) and Corollary 10.17 (i). $\qquad\square$

Remark 10.20. The inclusions in (ii) and (iii) are strict, see Remark 1.17.

Theorem 10.21. (i) *Let* $-\infty < \alpha < 0$. *Then* $\bigcap_{p>0} \Re(\overline{\Theta}_{\alpha,p}) = \mathcal{PBF}$.

(ii) *Let* $\alpha = 0$. *Then* $\bigcap_{p>0} \Re(\overline{\Theta}_{0,p}) = \mathcal{PBF}_0$.

(iii) *Let* $0 < \alpha < 1$. *Then* $\bigcap_{p>0} \Re(\overline{\Theta}_{0,p}) = \mathcal{PBF}_0^{(\alpha,1]}$.

Proof. (i) Let $f \in \bigcap_{p>0} \Re(\overline{\Theta}_{\alpha,p}) = \bigcap_{p\in\mathbb{N}} \Re(\overline{\Theta}_{\alpha,p})$. By Theorem 10.18, the Lévy measure of f has a density $\widetilde{m}$ such that $\widetilde{m}(t) = t^{-\alpha-1} n(\log t)$ where n is p-monotone on $\mathbb{R}$ for every $p \in \mathbb{N}$, hence completely monotone on $\mathbb{R}$. Since $-\alpha - 1 \geq -1$, Theorem 5.22 implies that $f \in \mathcal{PBF}$.

Conversely, if $f \in \mathcal{PBF}$, then again by Theorem 5.22, the Lévy measure of f has a density $\widetilde{m}$ such that $\widetilde{m}(t) = t^{-\alpha-1} n(\log t)$ for some function n which is completely monotone on $\mathbb{R}$ and, in particular, p-monotone on $\mathbb{R}$ for every $p \in \mathbb{N}$. By Theorem 10.18, $f \in \bigcap_{p\in\mathbb{N}} \Re(\overline{\Theta}_{\alpha,p})$.

(ii) This is proved in the same way as (i).

(iii) Let $f \in \bigcap_{p>0} \Re(\overline{\Theta}_{\alpha,p}) = \bigcap_{p\in\mathbb{N}} \Re(\overline{\Theta}_{\alpha,p})$. By Theorem 10.18, the Lévy measure of f has a density $\widetilde{m}$ such that $\widetilde{m}(t) = t^{-\alpha-1} n(\log t)$ where n is p-monotone on $\mathbb{R}$ for every $p \in \mathbb{N}$, hence completely monotone on $\mathbb{R}$. By Remark 5.23 (ii), $f \in \mathcal{PBF}$ with a representing measure κ such that $\kappa[0,\alpha] = 0$. Thus $f \in \mathcal{PBF}_0^{(\alpha,1]}$. The converse is proved similarly as in (i) by using Remark 5.23 (ii) instead of Theorem 5.22. $\qquad\square$

For a mapping Θ and $m \in \mathbb{N}$, let $\Theta^m = \Theta \circ \cdots \circ \Theta$ be the m-fold composition of Θ with itself.

Theorem 10.22.   (i) *Let* $-\infty < \alpha < 0$. *Then* $\bigcap_{m=1}^{\infty} \mathfrak{R}(\overline{\Theta}_{\alpha,1}^m) = \mathcal{PBF}$.

(ii) *Let* $-\infty < \alpha < 0$. *Then* $\bigcap_{m=1}^{\infty} \mathfrak{R}(\widehat{\Theta}_{\alpha,1}^m) = \mathcal{PBF}$.

(iii) *Let* $-\infty < \alpha < 0$ *and* $\beta > 0$. *Then* $\bigcap_{m=1}^{\infty} \mathfrak{R}(\Theta_{\alpha,\beta}^m) = \mathcal{PBF}$.

Proof. (i) By Proposition 10.11 (iii), $\overline{\Theta}_{\alpha,1}^m = \overline{\Theta}_{\alpha,m}$, and the claim follows from Theorem 10.21.

(ii) This immediately follows from (i) since $\widehat{\Theta}_{\alpha,1} = \overline{\Theta}_{\alpha,1}$.

(iii) Recall from Proposition 10.11 (ii) that $\Theta_{\alpha,\beta} = \widehat{\Theta}_{\alpha,1} \circ \Theta_{\alpha-\beta,\beta}$. Since the various transformations Θ commute, we find for every $m \in \mathbb{N}$,

$$\Theta_{\alpha,\beta}^m = \widehat{\Theta}_{\alpha,1}^m \circ \Theta_{\alpha-\beta,\beta}^m = \Theta_{\alpha-\beta,\beta}^m \circ \widehat{\Theta}_{\alpha,1}^m,$$

implying that $\mathfrak{R}(\Theta_{\alpha,\beta}^m) \subset \mathfrak{R}(\widehat{\Theta}_{\alpha,1}^m)$. Hence,

$$\bigcap_{m=1}^{\infty} \mathfrak{R}(\Theta_{\alpha,\beta}^m) \subset \bigcap_{m=1}^{\infty} \mathfrak{R}(\widehat{\Theta}_{\alpha,1}^m) = \mathcal{PBF}.$$

The other direction is simpler. For $0 \le s \le 1$, we see that

$$c(s) = \int_0^{\infty} t^s \vartheta_{\alpha,\beta}(t)\, dt = c_{\alpha,\beta} \int_0^{\infty} t^{s-\alpha-1} e^{-t^\beta}\, dt < \infty.$$

Hence, by Proposition 10.6 (iii), $\Theta_{\alpha,\beta}(\mathcal{PBF}) = \mathcal{PBF}$. Since $\mathcal{BF} \supset \mathcal{PBF}$, it follows that $\mathfrak{R}(\Theta_{\alpha,\beta}) = \Theta_{\alpha,\beta}(\mathcal{BF}) \supset \Theta_{\alpha,\beta}(\mathcal{PBF}) = \mathcal{PBF}$. By induction we get that $\mathfrak{R}(\Theta_{\alpha,\beta}^m) \supset \mathcal{PBF}$ for every $m \in \mathbb{N}$. Therefore,

$$\bigcap_{m=1}^{\infty} \mathfrak{R}(\Theta_{\alpha,\beta}^m) \supset \mathcal{PBF}. \qquad \square$$

Remark 10.23. Theorem 10.22 remains valid for $\alpha = 0$ if we replace $\mathcal{PBF}$ by the family $\mathcal{PBF}_0$.

Comments 10.24. The study of representations of infinitely divisible distributions in $\mathbb{R}^d$ by means of stochastic integrals of deterministic functions was started by Wolfe in [373]. He considered an Ornstein–Uhlenbeck stochastic differential equation $dX_t = -\gamma X_t\, dY_t$ driven by a centered one-dimensional Lévy process $Y = (Y_t)_{t \ge 0}$ and showed that $X_t \xrightarrow{t \to \infty} X$ (in law) if, and only if, $\mathbb{E}[\log^+ |Y_1|] < \infty$, in which case X has self-decomposable distribution on $\mathbb{R}$. In this way he could determine the

domain and the range of the transformation $\mathrm{law}(Y_1) \mapsto \mathrm{law}(\int_0^\infty e^{-\gamma t}\, dY_t)$. This relation of self-decomposable distributions to Ornstein–Uhlenbeck processes and stochas-
tic integrals driven by a Lévy process was discovered around the same time (or slightly
later) by Jurek and Vervaat, [193], for Banach space valued random variables, Sato
and Yamazato, [319], for an Ornstein–Uhlenbeck process in $\mathbb{R}^d$, and Gravereaux
[135], cf. [305] for a more detailed historical account. Jurek [191] introduced another
integral transformation, $\int_0^1 t\, dY_t$ (the *Jurek transformation*), showed that its domain
is the family of all infinitely divisible distributions and that its range is the class $\mathcal{U}$ of
s-self-decomposable distributions which had been introduced slightly earlier. The interest in integral representations of various subfamilies of ID distributions was revived
in the last ten years, starting in [20] and culminating with the definitive [317]. The
focus was to determine the domains, ranges and limits of ranges of nested subfamilies.

The present chapter is an exposition of these results for the case of distributions on
$[0, \infty)$. Although in many respects simpler, the presentation encompasses most of the
main features of the multivariate case. This comes from the fact that the ranges of
integral transformations are described in terms of the radial part of the d-dimensional
Lévy measure, cf. [20, Lemma 2.1] for the polar decomposition of the Lévy measure.

Lemma 10.1 and Remark 10.2 are quite standard. In a similar context they ap-
pear already in Lukacs [250]. The families of transformations $\Theta_{\alpha,\beta}$, $\widehat{\Theta}_{\alpha,p}$ and $\overline{\Theta}_{\alpha,p}$
have been introduced over the years for certain ranges of parameters and with vary-
ing notation. Transformations $\Theta_{-1,1}$ (known as Υ, notation introduced in [22]) and
$\Theta_{0,1}$ (denoted by Ψ in [316]) are from [20], $\Theta_{-1,2}$ (denoted by $\mathcal{G}$) from [254, 255],
and $\Theta_{0,2}$ (denoted by $\mathcal{M}$) from [14]. The family $\Theta_{\alpha,1}$ is from [316] (denoted by Ψ_α),
$\Theta_{-\alpha,\alpha}$ from [13] (denoted by $\mathcal{E}_\alpha$), and the general $\Theta_{\alpha,\beta}$ from [253] (denoted by $\mathcal{J}_{\alpha,\beta}$).
The transformation $\widehat{\Theta}_{0,1}$ is the earliest, cf. [135, 193, 319, 373], and is known as Φ
(the notation already used in [305]), $\widehat{\Theta}_{-1,1}$ is from [191], now known as $\mathcal{U}$. The family $\widehat{\Theta}_{\alpha,p}$ is introduced in [316] with different parametrization and denoted by $\Phi_{\beta,\alpha}$
($\beta = \alpha - p$); in [317] the notation $\overline{\Phi}_{p,\alpha}$ is used. Finally, $\overline{\Theta}_{\alpha,p}$ is from [317], denoted
by $\Lambda_{p,\alpha}$.

Propositions 10.4 and 10.7 – in the context of Bernstein functions and in this generality – seem to be new; various versions for infinitely divisible distributions on $\mathbb{R}^d$ are
known. The question of the domain of Θ is particularly simple here. Various versions
of Proposition 10.6 have been proved over the years, cf. [193, 191] for Φ, [191] for
$\mathcal{U}$, [20] for Υ. If Θ is one-to-one, then Proposition 10.6 (i) has an obvious converse:
if $\Theta f(\lambda) = \lambda^\alpha$, then $f(\lambda) = c(\alpha)^{-1}\lambda^\alpha$.

The fact that any two transformations commute, cf. Proposition 10.10, is particularly simple in the context of Bernstein functions, the formula for the composition seems to be new. Concrete examples have appeared in the literature several
times: for instance, $\Theta_{-1,1} \circ \widehat{\Theta}_{0,1} = \widehat{\Theta}_{0,1} \circ \Theta_{-1,1} = \Theta_{0,1}$ was proved in [20],
and $\Theta_{-2,2} \circ \widehat{\Theta}_{0,1} = \widehat{\Theta}_{0,1} \circ \Theta_{-2,2} = \Theta_{0,2}$ appeared in [13], (10.25) is from [316],
(10.23) and (10.26) from [317], while (10.24) seems to be new.

Theorem 10.12 appears in [20] for $\Theta_{-1,1}$ and $\Theta_{0,1}$, in [255] for $\Theta_{-1,2}$, in [14] for $\Theta_{0,2}$, in [316] for $\Theta_{\alpha,1}$ and in [253] for the general $\Theta_{\alpha,\beta}$. Theorem 10.15 has a long history – for $\widehat{\Theta}_{0,1}$ see [373, 193], for $\widehat{\Theta}_{-1,1}$ cf. [191], the general case is discussed in [316, 317]. Theorem 10.18 is from [317], Corollaries 10.14, 10.17 and 10.19 are simple consequences of the corresponding theorems and can be found in the papers mentioned earlier. Corollary 10.13 is contained in [20] for $\alpha = -1$, $\beta = 1$, in [316] for $\beta = 1$ and general α, and in [253] in the general case.

The first systematic study of the limits of nested subfamilies of infinitely divisible distributions seems to be [256] where Theorem 10.22 is proved for $\widehat{\Theta}_{-1,1}$, $\widehat{\Theta}_{0,1}$, $\Theta_{-1,1}$, $\Theta_{-1,2}$ and $\Theta_{0,1}$. For earlier results cf. [20, 192, 255]. The general result is from [318] for $\Theta_{\alpha,1}$, $\widehat{\Theta}_{\alpha,p}$ and $\overline{\Theta}_{\alpha,p}$, and from [257] for $\Theta_{\alpha,\beta}$. For Theorem 10.21 we refer to [317].

Chapter 11

Special Bernstein functions and potentials

In the first section of this chapter we will focus our attention on a less known family of Bernstein functions, the special Bernstein functions. This family, denoted by $\mathcal{SBF}$, contains $\mathcal{CBF}$ and inherits some of its good properties. The main interest in special Bernstein functions lies in the fact that the potential measure, restricted to $(0, \infty)$, of the convolution semigroup corresponding to a special Bernstein function admits a non-increasing density. On the other hand, a non-increasing function on $(0, \infty)$ need not be a potential density of any convolution semigroup. The best known sufficient condition for this to be true is that the function is non-increasing and log-convex. The Laplace transforms of such functions form the so-called Hirsch class of functions which is between the families $\mathcal{S}$ and $\mathcal{CM}$. The Hirsch class is studied in the second part of this chapter.

11.1 Special Bernstein functions

It was shown in Proposition 7.1 that $f \not\equiv 0$ belongs to $\mathcal{CBF}$ if, and only if, the function $\lambda \mapsto \lambda/f(\lambda)$ belongs to $\mathcal{CBF}$. Generalizing this property leads to the larger class of special Bernstein functions which enjoy certain desirable properties. On the other hand, the class itself does not have good structural properties.

Definition 11.1. A function $f \in \mathcal{BF}$ is said to be a *special Bernstein function* if the function $f^\star(\lambda) := \lambda/f(\lambda)$ is again a Bernstein function, i.e. $f^\star \in \mathcal{BF}$. We will use $\mathcal{SBF}$ to denote the collection of all special Bernstein functions.

A subordinator S whose Laplace exponent f belongs to $\mathcal{SBF}$ will be called a *special subordinator*.

Note that if $f \in \mathcal{SBF}$, then also $f^\star \in \mathcal{SBF}$, where $f^\star(\lambda) := \lambda/f(\lambda)$. We call f and $f^\star$ a *conjugate pair* of special Bernstein functions. It is clear from Proposition 7.1 that $\mathcal{CBF} \subset \mathcal{SBF}$. We will show later that $\mathcal{CBF} \subsetneq \mathcal{SBF} \subsetneq \mathcal{BF}$, cf. Propositions 11.16, 11.17 and Example 11.18.

Remark 11.2. Suppose that $f \in \mathcal{BF}$. Then $\lambda/f(\lambda)$ is a Bernstein function if, and only if, $f(\lambda)/\lambda \in \mathcal{P}$. Hence $f \in \mathcal{SBF}$ if, and only if, $f \in \mathcal{BF}$ and $f(\lambda)/\lambda \in \mathcal{P}$.

Let

$$f(\lambda) = a + b\lambda + \int_{(0,\infty)} (1 - e^{-\lambda t})\, \mu(dt), \tag{11.1}$$

$$f^\star(\lambda) = a^\star + b^\star \lambda + \int_{(0,\infty)} (1 - e^{-\lambda t})\, \mu^\star(dt) \tag{11.2}$$

be representations of f and its conjugate $f^\star$. Then, using the convention $\frac{1}{\infty} = 0$,

$$a^\star = \lim_{\lambda \to 0} f^\star(\lambda) = \lim_{\lambda \to 0} \frac{\lambda}{f(\lambda)} = \begin{cases} 0, & a > 0, \\ \dfrac{1}{b + \int_{(0,\infty)} t\, \mu(dt)}, & a = 0, \end{cases} \tag{11.3}$$

$$b^\star = \lim_{\lambda \to \infty} \frac{f^\star(\lambda)}{\lambda} = \lim_{\lambda \to \infty} \frac{1}{f(\lambda)} = \begin{cases} 0, & b > 0, \\ \dfrac{1}{a + \mu(0,\infty)}, & b = 0. \end{cases} \tag{11.4}$$

The following theorem gives a characterization of a special subordinator in terms of its potential measure. Roughly speaking, it says that a subordinator is special if, and only if, its potential measure restricted to $(0, \infty)$ has a non-increasing density.

Theorem 11.3. *Let S be a subordinator with potential measure U. Then S is special if, and only if,*

$$U(dt) = c\delta_0(dt) + u(t)\, dt \tag{11.5}$$

for some $c \geq 0$ and some non-increasing function $u : (0, \infty) \to (0, \infty)$ satisfying $\int_0^1 u(t)\, dt < \infty$.

Remark 11.4. (i) In general, the measure U given by (11.5) – with $c \geq 0$ and a non-increasing function u satisfying $\int_0^1 u(t)\, dt < \infty$ – need not be the potential measure of any subordinator.

For example, if $U(dt) = u(t)\, dt$ with $u(t) = 1 \wedge e^{1-t}$, then

$$\mathscr{L}(U; \lambda) = \frac{1 + \lambda - e^{-\lambda}}{\lambda(1 + \lambda)}.$$

If U were the potential measure of a subordinator, then $f(\lambda) := 1/\mathscr{L}(U; \lambda)$ would be a Bernstein function, see (5.22). A direct calculation shows that, for example, $f''(1) > 0$, hence $f \notin \mathcal{BF}$.

(ii) Let $f \in \mathcal{BF}$ be the Laplace exponent of the subordinator S. Theorem 11.3 can be equivalently stated in the following way: $f \in \mathcal{SBF}$ if, and only if,

$$\frac{1}{f(\lambda)} = c + \int_0^\infty e^{-\lambda t} u(t)\, dt \tag{11.6}$$

for some $c \geq 0$ and some non-increasing function $u : (0, \infty) \to (0, \infty)$ satisfying $\int_0^1 u(t)\, dt < \infty$.

Proof of Theorem 11.3. Suppose that S is a special subordinator, and let f be its Laplace exponent. Then f and its conjugate $f^\star$ have representations (11.1) and (11.2) where the coefficients $a^\star$ and $b^\star$ are given by (11.3) and (11.4). Define

$$u(t) := a^\star + \mu^\star(t, \infty), \quad t > 0, \tag{11.7}$$

and note that

$$\mathcal{L}(U; \lambda) = \frac{1}{f(\lambda)} = \frac{f^\star(\lambda)}{\lambda} = \frac{a^\star}{\lambda} + b^\star + \int_0^\infty e^{-\lambda t} \mu^\star(t, \infty)\, dt$$

$$= b^\star + \int_0^\infty e^{-\lambda t} u(t)\, dt.$$

This shows that $U(dt) = b^\star \delta_0(dt) + u(t)\, dt$, with u given by (11.7). It is clear that u is non-increasing and that $\int_0^1 u(t)\, dt < \infty$.

Conversely, suppose that (11.5) holds with a density $u : (0, \infty) \to (0, \infty)$ which is non-increasing and satisfies $\int_0^1 u(t)\, dt < \infty$. Then

$$\frac{1}{f(\lambda)} = \mathcal{L}(U; \lambda) = c + \int_0^\infty e^{-\lambda t} u(t)\, dt.$$

Let γ be the measure on $(0, \infty)$ defined by $\gamma(t, \infty) = u(t) - u(\infty)$, where we define $u(\infty) = \lim_{t \to \infty} u(t)$. Then

$$\int_{(0,1]} s\, \gamma(ds) = \int_{(0,1]} \int_0^s dt\, \gamma(ds) = \int_0^1 \int_{(t,1]} \gamma(ds)\, dt$$

$$= \int_0^1 \big(u(t) - u(1)\big)\, dt < \infty$$

by the assumption on u. Hence, $\int_{(0,\infty)} (1 \wedge s)\, \gamma(ds) < \infty$. It follows from (11.6) by Fubini's theorem that

$$\frac{\lambda}{f(\lambda)} = c\lambda + \int_0^\infty \lambda e^{-\lambda t} u(\infty)\, dt + \int_0^\infty \lambda e^{-\lambda t} \big(u(t) - u(\infty)\big)\, dt$$

$$= c\lambda + u(\infty) + \int_0^\infty \lambda e^{-\lambda t} \gamma(t, \infty)\, dt$$

$$= c\lambda + u(\infty) + \int_{(0,\infty)} (1 - e^{-\lambda t})\, \gamma(ds). \tag{11.8}$$

Therefore, $\lambda/f(\lambda)$ is a Bernstein function and S is a special subordinator.   □

Remark 11.5. In case $c = 0$, we will call u the potential density of the subordinator S or of the Laplace exponent f.

The following remark provides yet another characterization of complete Bernstein functions, this time in terms of the corresponding potential measures.

Remark 11.6. Let S be a subordinator with Laplace exponent f and potential measure U. Then f is a complete Bernstein function if, and only if, U restricted to $(0, \infty)$ has a completely monotone density u. This follows immediately from Theorem 7.3 and Theorem 2.2.

If we compare the expressions (11.2) and (11.8) for $\lambda/f(\lambda)$, and use formulae (11.3) and (11.4), we get immediately

$$c = b^\star = \begin{cases} 0, & b > 0, \\ \frac{1}{a+\mu(0,\infty)}, & b = 0, \end{cases}$$

$$u(\infty) = a^\star = \begin{cases} 0, & a > 0, \\ \frac{1}{b+\int_{(0,\infty)} t\,\mu(dt)}, & a = 0, \end{cases}$$

$$u(t) = a^\star + \mu^\star(t, \infty). \tag{11.9}$$

In particular, it cannot happen that both a and $a^\star$ are positive, or that both b and $b^\star$ are positive. Moreover, it is clear from the definition of $a^\star$ and $b^\star$ that $a^\star > 0$ if, and only if, $a = 0$ and $\int_{(0,\infty)} t\,\mu(dt) < \infty$, and $b^\star > 0$ if, and only if, $b = 0$ and $\mu(0,\infty) < \infty$.

The following two corollaries follow immediately, if we combine Theorem 11.3 and Remark 3.3 (v) with the formulae for $a^\star, b^\star$ and (11.9).

Corollary 11.7. *Suppose that* $S = (S_t)_{t \geq 0}$ *is a subordinator whose Laplace exponent*

$$f(\lambda) = a + b\lambda + \int_{(0,\infty)} (1 - e^{-\lambda t})\,\mu(dt)$$

is a special Bernstein function with $b > 0$ *or* $\mu(0,\infty) = \infty$. *Then the potential measure* U *of* S *has a non-increasing density* u *satisfying*

$$\lim_{t \to 0} tu(t) = 0 \quad \text{and} \quad \lim_{t \to 0} \int_0^t s\,du(s) = 0. \tag{11.10}$$

Corollary 11.8. *Suppose that* $S = (S_t)_{t \geq 0}$ *is a special subordinator with the Laplace exponent given by*

$$f(\lambda) = a + \int_{(0,\infty)} (1 - e^{-\lambda t})\,\mu(dt)$$

where μ *satisfies* $\mu(0,\infty) = \infty$. *Then*

$$f^\star(\lambda) := \frac{\lambda}{f(\lambda)} = a^\star + \int_{(0,\infty)} (1 - e^{-\lambda t})\,\mu^\star(dt) \tag{11.11}$$

where the Lévy measure $\mu^\star$ *satisfies* $\mu^\star(0,\infty) = \infty$.

Let T be a subordinator with Laplace exponent $f^\star$. If u and v denote the potential densities of S and T, respectively, then

$$v(t) = a + \mu(t, \infty). \tag{11.12}$$

In particular, $a = v(\infty)$ and $a^\star = u(\infty)$.

Assume that f is a special Bernstein function of the form (11.1) where $b > 0$ or $\mu(0, \infty) = \infty$. Let S be a subordinator with Laplace exponent f, and let U denote its potential measure. By Corollary 11.7, the measure U has a non-increasing density $u : (0, \infty) \to (0, \infty)$.

Let T be a subordinator with Laplace exponent $f^\star(\lambda) = \lambda/f(\lambda)$ and let V denote its potential measure. Then $V(dt) = b\delta_0(dt) + v(t)\,dt$ where $v : (0, \infty) \to (0, \infty)$ is a non-increasing function. If $b > 0$, the potential measure V has an atom at zero, and hence the subordinator T is a compound Poisson process. If $b = 0$, we require that $\mu(0, \infty) = \infty$, and then, by Corollary 11.8, $f^\star$ has the same structure as f, namely $b^\star = 0$ and $\mu^\star(0, \infty) = \infty$. In this case, the subordinators S and T play symmetric roles.

The defining property of special Bernstein functions says that the identity function is factorized into the product of two Bernstein functions. The following theorem is the counterpart of this property in terms of potential densities.

Theorem 11.9. *Let f be a special Bernstein function with representation (11.1) satisfying $b > 0$ or $\mu(0, \infty) = \infty$. Then*

$$b\,u(t) + \int_0^t u(s)v(t-s)\,ds = b\,u(t) + \int_0^t v(s)u(t-s)\,ds = 1, \quad t > 0. \tag{11.13}$$

Proof. Since for all $\lambda > 0$ we have

$$\frac{1}{f(\lambda)} = \mathscr{L}(u; \lambda), \quad \frac{f(\lambda)}{\lambda} = b + \mathscr{L}(v; \lambda),$$

we get after multiplying

$$\frac{1}{\lambda} = b\mathscr{L}(u; \lambda) + \mathscr{L}(u; \lambda)\mathscr{L}(v; \lambda)$$
$$= b\mathscr{L}(u; \lambda) + \mathscr{L}(u \star v; \lambda).$$

By Laplace inversion we see

$$1 = b\,u(t) + \int_0^t u(s)v(t-s)\,ds, \quad t > 0. \qquad \square$$

Our next goal is to describe a sufficient condition on the Lévy measure of the subordinator which guarantees that the subordinator is special. Since logarithmic convexity will play a major role, we recall the definition.

Definition 11.10. (i) A function $f : (0, \infty) \to (0, \infty)$ is said to be *logarithmically convex* (or *log-convex*) if $\log f$ is convex.

(ii) A sequence $(a_n)_{n \geq 0}$ of non-negative real numbers is *logarithmically convex* (or *log-convex*) if $a_n^2 \leq a_{n-1} a_{n+1}$ for all $n \in \mathbb{N}$.

Let f be a Bernstein function with the representation (11.1). It will be convenient to extend the measure μ onto $(0, \infty]$ by defining $\mu\{\infty\} = a$. Let $\bar{\mu}(x) := \mu(x, \infty]$, $x > 0$, be the tail of μ. Then one can write, as in Remark 3.3 (ii),

$$f(\lambda) = b\lambda + \lambda \int_0^\infty e^{-\lambda t} \bar{\mu}(t)\, dt. \tag{11.14}$$

Theorem 11.11. *Suppose that f is given by* (11.14) *and that $x \mapsto \bar{\mu}(x)$ is log-convex on $(0, \infty)$.*

(i) *If $b > 0$ or $\mu(0, \infty) = \infty$, then the potential measure U has a non-increasing density u.*

(ii) *If $b = 0$ and $\mu(0, \infty) < \infty$, then the restriction $U|_{(0,\infty)}$ has a non-increasing density u.*

In both cases, $f \in \mathcal{SBF}$.

Before we give the proof of the theorem we note some immediate consequences.

Corollary 11.12. *Suppose that $u : (0, \infty) \to (0, \infty)$ is non-increasing, log-convex and satisfies $\int_0^1 u(t)\, dt < \infty$, and let $c \geq 0$. Then the measure U defined by*

$$U(dt) = c\delta_0(dt) + u(t)\, dt$$

is the potential measure of a special subordinator $S = (S_t)_{t \geq 0}$.

Proof. Define a measure $\mu^\star$ on $(0, \infty)$ by $\mu^\star(t, \infty) := u(t) - a^\star$ with $a^\star := u(+\infty)$. Then $\mu^\star$ is a Lévy measure, i.e. $\int (1 \wedge t)\, \mu^\star(dt) < \infty$, and the tail $\mu^\star(t, \infty] = u(t)$ is log-convex.

Define $f^\star(\lambda) := a^\star + b^\star \lambda + \int_{(0,\infty)} (1 - e^{-\lambda t})\, \mu^\star(dt)$, where $b^\star = c$. By Theorem 11.11, $f^\star$ is a special Bernstein function. Let $f(\lambda) := \lambda / f^\star(\lambda)$ with corresponding special subordinator $S = (S_t)_{t \geq 0}$. It follows from Theorem 11.3 and (11.9) that the potential measure of S is given by

$$b^\star \delta_0(dt) + a^\star + \mu^\star(t, \infty)\, dt = c\delta_0(dt) + u(t)\, dt = U(dt).$$

Hence, U is the potential measure of the special subordinator S. $\qquad \square$

Remark 11.13. As a consequence of the above corollary, we can conclude that the Laplace transform $\mathscr{L}U \in \mathcal{P}$. In this form, Corollary 11.12 is due to [164]. See also [39, Corollary 14.9], and [180, 181]. Corollary 11.12 will be strengthened in the next section, see Theorem 11.24.

In order to prove Theorem 11.11 we need two preliminary lemmas.

Lemma 11.14. *Let $(v_n)_{n\geq0}$ be a sequence satisfying $v_0 = 1$ and $0 < v_n \leq 1$, $n \in \mathbb{N}$. Assume that $(v_n)_{n\geq0}$ is a log-convex sequence. Then there exists a non-increasing sequence $(r_n)_{n\geq0}$ such that $r_n \geq 0$ for all $n \geq 0$, and*

$$1 = \sum_{j=0}^{n} r_j\, v_{n-j} \quad \text{for all } n \geq 0. \tag{11.15}$$

Proof. Note that the inequality $v_n^2 \leq v_{n-1}v_{n+1}$ is equivalent to $v_n/v_{n-1} \leq v_{n+1}/v_n$. Therefore, the sequence $(v_n/v_{n-1})_{n\in\mathbb{N}}$ is non-decreasing.

Define $f_1 := v_1$ and recursively

$$f_n := v_n - \sum_{j=1}^{n-1} f_j\, v_{n-j}, \quad n = 2, 3, \dots \tag{11.16}$$

We claim that $f_n \geq 0$ for all $n \in \mathbb{N}$. This is clear for f_1. Assume that $f_k \geq 0$ for $1 \leq k \leq n$. Then

$$
\begin{aligned}
f_{n+1} &= \frac{v_{n+1}}{v_n} v_n - \sum_{j=1}^{n} f_j \frac{v_{n+1-j}}{v_{n-j}} v_{n-j} \\
&\geq \frac{v_{n+1}}{v_n} \left(v_n - \sum_{j=1}^{n} f_j\, v_{n-j} \right) \\
&= 0,
\end{aligned}
$$

where the inequality follows since $(v_n/v_{n-1})_{n\in\mathbb{N}}$ is non-decreasing; the last equality comes from (11.16) and the fact that $v_0 = 1$.

Let $V(z) := \sum_{n=0}^{\infty} v_n z^n$ and $F(z) := \sum_{n=1}^{\infty} f_n z^n$. Since $(v_n)_{n\geq0}$ is a bounded sequence and $f_n \leq v_n$ for all $n \in \mathbb{N}$, we get that $V(z)$ and $F(z)$ are well defined for $z \in (0,1)$. The renewal equation $v_n = \sum_{j=1}^{n} f_j\, v_{n-j}$ implies $V(z) = 1 + F(z)V(z)$, hence

$$F(z) = \frac{V(z) - 1}{V(z)} \leq 1.$$

By letting $z \to 1$, it follows that $\sum_{n=1}^{\infty} f_n \leq 1$.

Define the sequence $(r_n)_{n \geq 0}$ by $r_0 := 1$ and $r_n := 1 - \sum_{j=1}^{n} f_k$, $n \geq 1$. Clearly, $(r_n)_{n \geq 0}$ is a non-increasing sequence of non-negative numbers. Since $r_{n-1} - r_n = f_n$ for all $n \geq 1$, we see

$$
\begin{aligned}
v_n = \sum_{j=1}^{n} f_j \, v_{n-j} &= \sum_{j=1}^{n} (r_{j-1} - r_j) \, v_{n-j} \\
&= \sum_{j=1}^{n} r_{j-1} \, v_{n-j} - \sum_{j=1}^{n} r_j \, v_{n-j} \\
&= \sum_{j=0}^{n-1} r_j \, v_{n-1-j} - \sum_{j=1}^{n} r_j \, v_{n-j}.
\end{aligned}
$$

This implies that $\sum_{j=0}^{n} r_j \, v_{n-j} = \sum_{j=0}^{n-1} r_j \, v_{n-1-j}$ for all $n \geq 1$. But for $n = 1$ we have that $\sum_{j=0}^{n-1} r_j \, v_{n-1-j} = r_0 v_0 = 1$. This proves that $\sum_{j=0}^{n} r_j \, v_{n-j} = 1$ for all $n \geq 0$. $\square$

Lemma 11.15. *Suppose that $x \mapsto \bar{\mu}(x)$ is absolutely continuous on $(0, \infty)$. If the total mass $\mu(0, \infty) = \infty$ or if $b > 0$, then the potential measure U is absolutely continuous. If the total mass $\mu(0, \infty) < \infty$ and $b = 0$, then $U|_{(0,\infty)}$ is absolutely continuous.*

Proof. Assume first that $a = 0$. If $b > 0$, then it is well known that U is absolutely continuous, see e.g. [47, Section III.2, Theorem 5]. Assume that $\mu(0, \infty) = \infty$. Since $\bar{\mu}(x)$ is absolutely continuous, by [314, Theorem 27.7] the transition probabilities of S are absolutely continuous and therefore U is absolutely continuous.

If $a > 0$, then the potential measure of the killed subordinator is equal to the a-potential measure of the (non-killed) subordinator, hence again absolutely continuous, see e.g. [314, Remark 41.12].

Assume now that $\mu(0, \infty) < \infty$ and $b = 0$. Since $x \mapsto \bar{\mu}(x)$ is absolutely continuous, we have $\mu(dx) = m(x)\, dx$ for some function m. Let $c := \mu(0, \infty)$. The transition probability at time t of the non-killed subordinator is given by

$$
\mu_t = \sum_{k=0}^{\infty} e^{-tc} \frac{t^k}{k!} \mu^{\star k}.
$$

Here and in the rest of this proof $\mu^{\star k}$ stands for the k-fold convolution of μ, and $\mu^{\star 0} = \delta_0$. Therefore, the potential measure of the killed subordinator is equal to

$$
\begin{aligned}
U &= \int_0^{\infty} e^{-at} \mu_t \, dt \\
&= \int_0^{\infty} e^{-at} \left(\sum_{k=0}^{\infty} e^{-tc} \frac{t^k}{k!} \mu^{\star k} \right) dt
\end{aligned}
$$

$$= \sum_{k=0}^{\infty} \frac{\mu^{\star k}}{k!} \int_0^{\infty} e^{-(a+c)t}\, t^k\, dt$$

$$= \frac{1}{a+c}\,\delta_0 + \frac{1}{a+c} \sum_{k=1}^{\infty} \left(\frac{\mu}{a+c}\right)^{\star k}.$$

This shows that $U|_{(0,\infty)}$ is absolutely continuous with the density

$$u(x) = \frac{1}{a+c} \sum_{k=1}^{\infty} \left(\frac{m}{a+c}\right)^{\star k}(x). \qquad \square$$

Proof of Theorem 11.11. The log-convexity of $\bar{\mu}(x)$ implies that $\bar{\mu}(x)$ is absolutely continuous on $(0,\infty)$. By Lemma 11.15 we know that the density of U exists if $\mu(0,\infty) = \infty$ or $b > 0$. We choose a version of u such that

$$\limsup_{h\to 0} \frac{U(x, x+h)}{h} = u(x) \quad \text{for all } x > 0. \qquad (11.17)$$

Because of the log-convexity, $\bar{\mu}$ is strictly positive everywhere. This excludes the case where $a + \mu(x,\infty) = 0$ for some $x > 0$. Fix $c > 0$ and define a sequence $(v_n(c))_{n\geq 0}$ by

$$v_0(c) := \frac{b/c + \bar{\mu}(c)}{b/c + \bar{\mu}(c)} = 1, \quad v_n(c) := \frac{\bar{\mu}(nc+c)}{b/c + \bar{\mu}(c)}, \quad n \geq 1.$$

Clearly $0 < v_n(c) \leq 1$ for all $n \geq 0$. Moreover, $v_n(c)^2 \leq v_{n-1}(c)v_{n+1}(c)$ for all $n \geq 1$. Indeed, for $n \geq 2$ this is equivalent to

$$\bar{\mu}(nc+c)^2 \leq \bar{\mu}\big((n-1)c+c\big) \cdot \bar{\mu}\big((n+1)c+c\big)$$

which is a consequence of the log-convexity of $\bar{\mu}$. For $n = 1$ we have

$$\bar{\mu}(2c) \leq \bar{\mu}(c)\bar{\mu}(3c) \leq \big(b/c + \bar{\mu}(c)\big)\bar{\mu}(3c).$$

By Lemma 11.14, there exists a non-increasing sequence $(r_n(c))_{n\geq 0}$ such that

$$\sum_{j=0}^{n} r_j(c)v_{n-j}(c) = 1 \quad \text{for all } n \geq 0. \qquad (11.18)$$

Define

$$u_n(c) := \frac{r_n(c)}{b/c + \bar{\mu}(c)}, \quad n \geq 0.$$

Inserting this expression into (11.18) we get for all $n \geq 0$

$$\frac{b}{c} u_n(c) + \sum_{j=0}^{n} u_j(c)\bar{\mu}\big((n-j)c+c\big) = 1. \qquad (11.19)$$

Multiplying (11.19) by $c\lambda e^{-(n+1)c\lambda}$ and summing over all $n \geq 0$, we obtain

$$b\lambda \sum_{n=0}^{\infty} e^{-(n+1)c\lambda} u_n(c) + \sum_{n=0}^{\infty} \sum_{j=0}^{n} c\lambda e^{-(n+1)c\lambda} u_j(c) \bar{\mu}\big((n-j)c + c\big)$$

$$= \sum_{n=0}^{\infty} c\lambda e^{-(n+1)c\lambda}.$$

This can be simplified to

$$b\lambda e^{-c\lambda} \sum_{n=0}^{\infty} e^{-nc\lambda} u_n(c) + \left(\sum_{n=0}^{\infty} e^{-nc\lambda} u_n(c) \right) \left(c\lambda \sum_{n=1}^{\infty} e^{-nc\lambda} \bar{\mu}(nc) \right)$$

$$= \frac{c\lambda e^{-c\lambda}}{1 - e^{-c\lambda}}.$$

Define a measure U_c on $(0, \infty)$ by $U_c := \sum_{n=0}^{\infty} u_n(c) \delta_{nc}$. Then the above equation becomes

$$\left(\int_0^{\infty} e^{-\lambda t} \, dU_c(t) \right) \left(b\lambda e^{-c\lambda} + \sum_{n=1}^{\infty} c\lambda e^{-nc\lambda} \bar{\mu}(nc) \right) = \frac{c\lambda e^{-c\lambda}}{1 - e^{-c\lambda}}.$$

Let $c \downarrow 0$. The right-hand side converges to 1, while

$$\lim_{c \downarrow 0} \left(b\lambda e^{-c\lambda} + \sum_{n=1}^{\infty} c\lambda e^{-nc\lambda} \bar{\mu}(nc) \right) = b\lambda + \int_0^{\infty} \lambda e^{-\lambda t} \bar{\mu}(t) \, dt = f(\lambda).$$

Therefore

$$\lim_{c \downarrow 0} \int_{[0,\infty)} e^{-\lambda t} \, dU_c(t) = \frac{1}{f(\lambda)} = \int_{[0,\infty)} e^{-\lambda t} \, dU(t)$$

which means that U_c converges vaguely to U. Since U is absolutely continuous, this implies that for all $x > 0$ and all $h > 0$,

$$\lim_{c \downarrow 0} U_c(x, x + h) = \int_x^{x+h} u(t) \, dt.$$

Now suppose that $0 < x < y$ and choose $h > 0$ such that $x < x + h < y$. Moreover, let c be such that none of the points $x, x + h, y, y + h$ is an integer multiple of c. By the monotonicity of $(u_n(c))_{n \geq 0}$, it follows that

$$U_c(y, y + h) \leq U_c(x, x + h).$$

Let c go to zero along values such that the points $x, x + h, y, y + h$ are not integer multiples of c. It follows that

$$U(y, y + h) \leq U(x, x + h)$$

and (11.17) guarantees that $u(y) \leq u(x)$ for all $x < y$. $\qquad\qquad\square$

We give now two sufficient conditions such that a Bernstein function is *not* special.

Proposition 11.16. *Let* $f(\lambda) = b\lambda + \int_{(0,\infty)} (1 - e^{-\lambda t}) \mu(dt)$, $b \geq 0$, *be a Bernstein function. If* μ *has bounded support, then* f *cannot be special.*

Proof. Assume that $f \in \mathcal{SBF}$ and $\mu(t_0, \infty) = 0$ for some $t_0 > 0$. Let $T = (T_t)_{t \geq 0}$ be the subordinator with Laplace exponent $f^\star(\lambda) := \lambda/f(\lambda)$. Denote by V the potential measure of T and by v the density of $V|_{(0,\infty)}$; then $v(t) = \mu(t, \infty) = 0$ for all $t \geq t_0$. But this implies $V(t_0, \infty) = 0$ which is impossible.   □

Proposition 11.17. *Let* $f(\lambda) = a + \int_{(0,\infty)} (1 - e^{-\lambda t}) \mu(dt)$, $a \geq 0$, *be a Bernstein function. Assume that the Lévy measure* μ *is nontrivial and that* $\mu(0, t_0] = 0$ *for some* $t_0 > 0$. *Then* f *is not special.*

Proof. Suppose, on the contrary, that f is a special Bernstein function. Then the conjugate function $f^\star(\lambda) = \lambda/f(\lambda)$ is of the form (11.2). The potential density of the corresponding subordinator T is given by the formula $v(t) = a + \mu(t, \infty)$. In particular, $v(t) = a + \mu(t_0, \infty) =: \gamma$ for $0 < t < t_0$. It follows from (11.4) that $b^\star = 1/\gamma > 0$. On the other hand, from Theorem 11.9 (with the roles of u and v reversed), we have that

$$b^\star v(t) + \int_0^t u(s)v(t-s)\,ds = 1 \quad \text{for all } t > 0.$$

For $0 < t < t_0$ we have $v(t) = \gamma$, so

$$1 + \gamma \int_0^t u(s)\,ds = 1, \quad 0 < t < t_0.$$

It follows that $\int_0^t u(s)\,ds = 0$ for all $0 < t < t_0$, and since u is non-increasing we have $u \equiv 0$. Thus $U(dt) = b^\star \delta_0$, implying that $f = 1/b^\star$. But this contradicts the assumption that the Lévy measure μ is not trivial.   □

The last two propositions clearly show that $\mathcal{SBF}$ is strictly smaller than $\mathcal{BF}$. A typical example of a Bernstein function which is not special is $f(\lambda) = 1 - e^{-\gamma\lambda}$. The corresponding subordinator is a Poisson process with rate $\gamma > 0$. The next example shows that the family $\mathcal{SBF}$ is strictly larger than $\mathcal{CBF}$.

Example 11.18. (i) Define

$$v(t) := \begin{cases} t^{-\alpha}, & 0 < t < 1, \\ t^{-\beta}, & 1 \leq t < \infty \end{cases}$$

where $0 < \beta < \alpha < 1$. Obviously, v is non-increasing, log-convex and satisfies $\int_0^1 v(t)\,dt < \infty$. By Corollary 11.12, v is the potential density of a special subordinator. Since v is not completely monotone – it is clearly not C^∞ –, the corresponding Laplace exponent is not a complete Bernstein function.

(ii) For $0 < \alpha < 1$ define

$$v(t) := \begin{cases} t^{-\alpha}, & 0 < t < 1, \\ 1, & 1 \le t < \infty. \end{cases}$$

Again, v is non-increasing, log-convex and satisfies $\int_0^1 v(t)\,dt < \infty$. Hence, there exists a special subordinator $T = (T_t)_{t \ge 0}$ with potential measure V such that v is the density of V. Let $f^\star$ be the Laplace exponent of T, and define $f(\lambda) := \lambda/f^\star(\lambda)$. Since $v(\infty) = 1$, we have that $f(\lambda) = 1 + \int_{(0,\infty)}(1-e^{-\lambda t})\,\mu(dt)$ with Lévy measure $\mu(dt) = m(t)\,dt$, where

$$m(t) := \begin{cases} \alpha\, t^{-\alpha-1}, & 0 < t < 1, \\ 0, & 1 \le t < \infty. \end{cases}$$

Note that the Lévy measure μ has bounded support. This does not contradict Proposition 11.16 since we have a non-zero killing term of f equal to 1.

This example shows that for $f \in \mathcal{BF}$ it may happen that $\lambda \mapsto a + f(\lambda) \in \mathcal{SBF}$ for some $a > 0$, while $f \notin \mathcal{SBF}$.

(iii) This example is similar to the previous one, but the Lévy measure is finite. Let

$$v(t) := \begin{cases} e^{1-t}, & 0 < t < 1, \\ 1, & 1 \le t < \infty. \end{cases}$$

Again, v is non-increasing, log-convex and satisfies $\int_0^1 v(t)\,dt < \infty$. Hence, there exists a special subordinator $T = (T_t)_{t \ge 0}$ with potential measure V such that v is the density of V. Let $f^\star$ be the Laplace exponent of T, and define $f(\lambda) := \lambda/f^\star(\lambda)$. Then $f(\lambda) = 1 + \int_{(0,\infty)}(1 - e^{-\lambda t})\,\mu(dt)$ with Lévy measure $\mu(dt) = m(t)\,dt$, where

$$m(t) := \begin{cases} e^{1-t}, & 0 < t < 1, \\ 0, & 1 \le t < \infty. \end{cases}$$

Remark 11.19. In the third example above, the density of the Lévy measure $\mu(x)$ is discontinuous at $x = 1$. Put

$$f(x) := \begin{cases} e^{-x}, & 0 < x < 1, \\ 0, & 1 \le x < \infty. \end{cases}$$

Clearly,

$$f^{\star 2}(x) = \begin{cases} xe^{-x}, & 0 < x < 1, \\ (2-x)e^{-x}, & 1 \le x < 2, \\ 0, & 2 \le x < \infty, \end{cases}$$

and all convolutions $f^{\star n}$, $n \ge 2$, are continuous everywhere. It is easy to check that, for $x \in (0,1)$,

$$f^{\star k}(x) = \frac{x^{k-1}}{(k-1)!}\, e^{-x}, \quad k \ge 1.$$

Using the formula above, one can check that for $x \in \left[1, \frac{5}{4}\right)$,

$$f^{\star k}(x) \le \frac{5}{8}\left(\frac{1}{2}\right)^{k-3} e^{-x}, \quad k \ge 3.$$

Therefore, the series $e^{-1}\sum_{k=1}^{\infty} f^{\star k}(x)$ is uniformly convergent for $x \in \left(0, \frac{5}{4}\right)$. By the last formula in the proof of Lemma 11.15, p. 167, we get that $u(x)$ is discontinuous at $x = 1$. On the other hand, $u(x) = \tilde{a} + \nu(x, \infty)$, implying that $x \mapsto \nu(x, \infty)$ has a discontinuity at $x = 1$. Hence, ν has an atom at 1. This shows that the Lévy measure of a special subordinator may have atoms. As a consequence, the tail is not log-convex. This resolves the conjecture made in [72] in the negative.

As already pointed out, the class $\mathcal{SBF}$ does not have good structural properties. Here are a few positive results.

Proposition 11.20.   (i) $\mathcal{SBF}$ is a cone closed under pointwise convergence.

(ii) $f \in \mathcal{SBF}$ if, and only if, $\lambda \mapsto b\lambda + f(\lambda) \in \mathcal{SBF}$ for all $b > 0$.

(iii) $f \in \mathcal{SBF}$ if, and only if, $\frac{tf}{t+f} \in \mathcal{SBF}$ for all $t > 0$.

(iv) $f \in \mathcal{SBF}$ if, and only if, $\lambda \mapsto f(\lambda + c) \in \mathcal{SBF}$ for all $c > 0$.

Proof. (i) follows immediately from the fact that $\mathcal{BF}$ is a cone which is closed under pointwise convergence.
   (ii) Let $f \in \mathcal{SBF}$ and $b > 0$. Define

$$h(\lambda) := \frac{\lambda}{1 + b\lambda}.$$

Then $h \in \mathcal{BF}$, and by the composition result for Bernstein functions, cf. Corollary 3.8 or Theorem 5.27, it follows for the conjugate $f^{\star}(\lambda) = \lambda/f(\lambda)$ that

$$h \circ f^{\star}(\lambda) = \frac{f^{\star}(\lambda)}{1 + bf^{\star}(\lambda)} = \frac{\lambda}{b\lambda + f(\lambda)} \in \mathcal{BF}.$$

Since $\lambda \mapsto b\lambda + f(\lambda) \in \mathcal{BF}$, we know that $\lambda \mapsto b\lambda + f(\lambda) \in \mathcal{SBF}$, too. The converse statement follows from (i) by letting $b \to 0$.

(iii) Let $f \in \mathcal{SBF}$ and $t > 0$. As in (ii), $tf/(t+f) \in \mathcal{BF}$. Further,

$$\frac{\lambda}{\frac{tf(\lambda)}{t+f(\lambda)}} = \frac{t\lambda + \lambda f(\lambda)}{tf(\lambda)} = \frac{\lambda}{f(\lambda)} + \frac{\lambda}{t} \in \mathcal{BF},$$

implying that $tf/(t+f) \in \mathcal{SBF}$. The converse follows from (i) by letting $t \to \infty$.

(iv) First note that if $f \in \mathcal{BF}$, $g \in \mathcal{CM}$ and $f - g \geq 0$, then $f - g \in \mathcal{BF}$ which is easily seen from the definition of $\mathcal{BF}$ and $\mathcal{CM}$. Let now $f \in \mathcal{SBF}$ and $c > 0$. Then

$$\frac{\lambda}{f(\lambda + c)} = \frac{\lambda + c}{f(\lambda + c)} - \frac{c}{f(\lambda + c)} \geq 0.$$

Since $\lambda \mapsto (\lambda + c)/f(\lambda + c) \in \mathcal{BF}$ and $\lambda \mapsto c/f(\lambda + c) \in \mathcal{CM}$, we conclude that $\lambda/f(\lambda + c)$ is a Bernstein function which proves the claim. The converse follows by letting $c \to 0$.   □

Remark 11.21. We do not know whether $\mathcal{SBF}$ is a convex cone. It is even unclear whether $a + f \in \mathcal{SBF}$, $a > 0$, for $f \in \mathcal{SBF}$. Let $f \in \mathcal{SBF}$. Then $a + f \in \mathcal{SBF}$ if, and only if, $\lambda \mapsto (a + f(\lambda))/\lambda \in \mathcal{P}$. We can also use potential densities to characterize when $a + f \in \mathcal{SBF}$. Assume, for simplicity, that $f \in \mathcal{SBF}$ is of the form

$$f(\lambda) = \lambda \int_0^\infty e^{-\lambda t} \bar{\mu}(t)\, dt.$$

Then $v(t) = \bar{\mu}(t)$ is the potential density of the conjugate function $f^\star(\lambda) = \lambda/f(\lambda)$. For $a > 0$,

$$\frac{a + f(\lambda)}{\lambda} = \int_0^\infty e^{-\lambda t}(v(t) + a)\, dt,$$

showing that $a + f \in \mathcal{SBF}$ if, and only if, $v(t) + a$ is a potential density.

11.2   Hirsch's class

In this section we describe another family of completely monotone functions which was introduced by F. Hirsch in [164]. It turns out that this class, $\mathcal{H}$, is a rich intermediate class between the Stieltjes functions $\mathcal{S}$ on the one side, and the completely monotone functions of the form $1/f$ where $f \in \mathcal{SBF}$ on the other side. In some sense $\mathcal{H}$ is the largest known family of concrete examples for potentials.

Definition 11.22. We call a function $f \in \mathcal{CM}$ a *Hirsch function* if for all $\lambda > 0$ the sequence

$$\left(\frac{(-1)^n}{n!} f^{(n)}(\lambda) \right)_{n \geq 0}$$

is log-convex in the sense of Definition 11.10. The set of all Hirsch functions will be denoted by $\mathcal{H}$.

Proposition 11.23. *The set $\mathcal{H}$ is a convex cone closed under pointwise convergence.*

Proof. The convexity of $\mathcal{H}$ follows from the fact that log-convex sequences form a convex cone. Indeed, let $(a_n)_{n\geq 0}$ and $(b_n)_{n\geq 0}$ be two log-convex sequences. Then it holds for all $n \geq 1$ that $a_n^2 \leq a_{n-1}a_{n+1}$ and $b_n^2 \leq b_{n-1}b_{n+1}$. Therefore

$$(a_n + b_n)^2 = a_n^2 + b_n^2 + 2a_n b_n$$
$$\leq a_{n-1}a_{n+1} + b_{n-1}b_{n+1} + 2\sqrt{a_{n-1}a_{n+1}b_{n-1}b_{n+1}}$$
$$\leq a_{n-1}a_{n+1} + b_{n-1}b_{n+1} + a_{n-1}b_{n+1} + a_{n+1}b_{n-1}$$
$$= (a_{n-1} + b_{n-1})(a_{n+1} + b_{n+1}).$$

Since complete monotonicity and log-convexity are preserved under pointwise convergence, $\mathcal{H}$ is closed under pointwise convergence.   □

Theorem 11.24. *Suppose that $u : (0, \infty) \to (0, \infty)$ is non-increasing, log-convex and satisfies $\int_0^1 u(t)\,dt < \infty$, and let $c \geq 0$. Define a measure U by*

$$U(dt) = c\delta_0(dt) + u(t)\,dt. \qquad (11.20)$$

Then $\mathscr{L}U \in \mathcal{H}$. Conversely, every $f \in \mathcal{H}$ is the Laplace transform of a measure U of the form (11.20).

Proof. Suppose that U is given by (11.20) and that the function u is twice continuously differentiable. For every $\lambda > 0$, the function g_λ defined by

$$g_\lambda(s) = e^{-\lambda s}u(s), \quad s > 0,$$

is also logarithmically convex and of class C^2, hence

$$(g_\lambda')^2 \leq g_\lambda g_\lambda''.$$

Let

$$f(\lambda) := \mathscr{L}(u; \lambda) = \int_0^\infty e^{-\lambda s}u(s)\,ds, \quad \lambda > 0.$$

Then for $n \geq 1$,

$$\frac{(-1)^n}{n!}f^{(n)}(\lambda) = \int_0^\infty \frac{s^n}{n!}g_\lambda(s)\,ds = -\int_0^\infty \frac{s^{n+1}}{(n+1)!}g_\lambda'(s)\,ds.$$

As $g'_\lambda \le \sqrt{g_\lambda g''_\lambda}$, we find by the Cauchy–Schwarz inequality that

$$
\left(\frac{(-1)^n}{n!} f^{(n)}(\lambda) \right)^2 \le \int_0^\infty \frac{s^{n+1}}{(n+1)!} g_\lambda(s)\, ds \int_0^\infty \frac{s^{n+1}}{(n+1)!} g''_\lambda(s)\, ds
$$

$$
= \int_0^\infty \frac{s^{n+1}}{(n+1)!} g_\lambda(s)\, ds \int_0^\infty \frac{s^{n-1}}{(n-1)!} g_\lambda(s)\, ds
$$

$$
= \left(\frac{(-1)^{n+1}}{(n+1)!} f^{(n+1)}(\lambda) \right) \left(\frac{(-1)^{n-1}}{(n-1)!} f^{(n-1)}(\lambda) \right),
$$

showing that $f \in \mathcal{H}$. Every non-increasing log-convex function $u : (0,\infty) \to (0,\infty)$ is the supremum of a sequence of C^2 functions $u_n : (0,\infty) \to (0,\infty)$ which are non-increasing and log-convex. For such functions we have proved that $\mathscr{L}u_n \in \mathcal{H}$. By the monotone convergence theorem it follows that $\mathscr{L}u_n$ increases pointwise to $\mathscr{L}u$. By Proposition 11.23 it follows that $\mathscr{L}u \in \mathcal{H}$. Since positive constants are in $\mathcal{H}$, and since $\mathcal{H}$ is a convex cone, we see that $\mathscr{L}U \in \mathcal{H}$.

Now we prove the converse statement. Let $f \in \mathcal{H} \subset \mathcal{CM}$ and define

$$
f_n(\lambda) := \frac{(-1)^n}{n!} \left(\frac{n}{\lambda} \right)^{n+1} f^{(n)}\left(\frac{n}{\lambda} \right), \quad \lambda > 0,
$$

(compare with (1.8)). Then f_n is C^∞ and strictly positive. We are going to prove that it is log-convex, by showing that $(\log f_n)''(\lambda) > 0$. We first compute

$$
(\log f)''(\lambda) = \frac{n+1}{\lambda^2} + \frac{2n}{\lambda^3} \frac{f^{(n+1)}\left(\frac{n}{\lambda} \right)}{f^{(n)}\left(\frac{n}{\lambda} \right)}
$$

$$
+ \left(\frac{n}{\lambda^2} \right)^2 \frac{f^{(n+2)}\left(\frac{n}{\lambda} \right)}{f^{(n)}\left(\frac{n}{\lambda} \right)} - \left(\frac{n}{\lambda^2} \frac{f^{(n+1)}\left(\frac{n}{\lambda} \right)}{f^{(n)}\left(\frac{n}{\lambda} \right)} \right)^2.
$$

By the inequality

$$
f^{(n)}\left(\frac{n}{\lambda} \right) f^{(n+2)}\left(\frac{n}{\lambda} \right) \ge \frac{n+2}{n+1} \left(f^{(n+1)}\left(\frac{n}{\lambda} \right) \right)^2,
$$

we obtain

$$
(\log f)''(\lambda) \ge \frac{n+1}{\lambda^2} + \frac{2n}{\lambda^3} \frac{f^{(n+1)}\left(\frac{n}{\lambda} \right)}{f^{(n)}\left(\frac{n}{\lambda} \right)} + \frac{1}{n+1} \left(\frac{n}{\lambda^2} \frac{f^{(n+1)}\left(\frac{n}{\lambda} \right)}{f^{(n)}\left(\frac{n}{\lambda} \right)} \right)^2
$$

$$
= \left(\frac{\sqrt{n+1}}{\lambda} + \frac{f^{(n+1)}\left(\frac{n}{\lambda} \right)}{f^{(n)}\left(\frac{n}{\lambda} \right)} \frac{n}{\lambda^2 \sqrt{n+1}} \right)^2 \ge 0.
$$

The next step is to show that f_n is non-increasing. Since $f \in \mathcal{CM}$, there exists a measure U on $[0, \infty)$ such that $f = \mathcal{L}U$. Assume that U is bounded. Differentiating n times under the integral sign yields

$$f^{(n)}(\lambda) = \int_{[0,\infty)} (-1)^n \, e^{-\lambda t} \, U(dt).$$

Using the elementary inequality $e^{-\lambda t} (\lambda t)^n \le (n/e)^n$, $\lambda, t \ge 0$, and dominated convergence we see

$$\lim_{\lambda \to 0} f^{(n)}(\lambda) \lambda^{n+1} = 0.$$

Hence

$$\lim_{\lambda \to \infty} f_n(\lambda) = \lim_{\lambda \to \infty} \frac{(-1)^n}{n!} f^{(n)} \left(\frac{n}{\lambda}\right) \left(\frac{n}{\lambda}\right)^{n+1} = 0.$$

Since f_n is convex, this implies that it is non-increasing. For a general measure U, pick $c > 0$ and set $f_c(\lambda) := f(\lambda + c) = \int_{[0,\infty)} e^{-\lambda t} e^{-ct} U(dt)$. Then $f_c \in \mathcal{H}$ with a bounded measure $e^{-ct} U(dt)$. By the proof above, $f_{c,n}$ is non-increasing. As $f_n(\lambda) = \lim_{c \to 0} f_{c,n}(\lambda)$, $\lambda \ge 0$, it follows that f_n is non-increasing.

Let $a := \lim_{\lambda \to \infty} f(\lambda)$. Then $U = a\delta_0 + U_0$ where $U_0 = U|_{(0,\infty)}$. Moreover, if $\tilde{f} := f - a$, then the proof of Theorem 1.4 shows that $\tilde{f}_n(x) \, dx$ converges weakly to $U_0(dx)$ as $n \to \infty$. Since $f_n = \tilde{f}_n$, $n \ge 1$, we get that $f_n(x) \, dx$ converges weakly to $U_0(dx)$. Since $f_n' \le 0$ and $f_n'' \ge 0$, it follows that $DU_0 \le 0$ and $D^2 U_0 \ge 0$ where we take derivatives in the distributional sense. Hence, there exists a convex non-increasing function $u : (0, \infty) \to (0, \infty)$ such that $U_0(dx) = u(x) \, dx$. In particular, u is continuous. By the result in [370, p. 288] it follows that $\lim_{n \to \infty} f_n(x) = u(x)$ for all $x > 0$. Since the pointwise limit of log-convex functions is again log-convex, this finishes the proof.   □

We record several simple consequences of the preceding theorem.

Corollary 11.25. (i) $\frac{1}{\mathcal{H}^*} \subset \mathcal{SBF}$ *and* $\mathcal{H}^* \subset \mathcal{P}$.

(ii) *If* $h \in \mathcal{H}$, *then* $\lambda h(\lambda) \in \mathcal{SBF}$.

(iii) *If* $f \in \mathcal{BF}$ *and* $\bar{\mu}$ *is log-convex, then* $f(\lambda)/\lambda \in \mathcal{H}$.

(iv) $\mathcal{S} \subset \mathcal{H}$.

Proof. (i) Let $g \in \mathcal{H}^*$. By Theorem 11.24, $g = \mathcal{L}U$ for a measure U of the form (11.20). By Corollary 11.12, U is the potential measure of a special subordinator, i.e. $\mathcal{L}U = 1/f$ for $f \in \mathcal{SBF}$. Thus, $1/g = f \in \mathcal{SBF}$ and, in particular, $g \in \mathcal{P}$.
   (ii) Let $f := 1/h$. By (i), $f \in \mathcal{SBF}$, and therefore $\lambda h(\lambda) = \lambda/f(\lambda) \in \mathcal{SBF}$.

(iii) By (11.14) we have

$$f(\lambda) = b\lambda + \lambda \int_0^\infty e^{-\lambda t}\, \bar{\mu}(t)\, dt = \lambda\, \mathscr{L}\big(b\delta_0(dt) + \bar{\mu}(t)\, dt; \lambda\big)$$

which proves the claim.

(iv) Let $g \in \mathcal{S}$. By Theorem 7.3, $g = 1/f$ where $f \in \mathcal{CBF}$. By Remark 11.6, the potential measure U of f is of the form $U(dt) = c\delta_0(dt) + u(t)\, dt$ with $u \in \mathcal{CM}$. Since completely monotone functions are logarithmically convex, Theorem 11.24 yields $\mathscr{L}U \in \mathcal{H}$. But $\mathscr{L}U = g$. □

Corollary 11.26. $\mathcal{CBF} \circ \mathcal{H} \subset \mathcal{CM}$

Proof. Let $h \in \mathcal{H}$. We know from Corollary 11.25 (i) that $h = \frac{1}{f}$ for some $f \in \mathcal{BF}$. Consider for every $t > 0$

$$\frac{h}{t+h} = \frac{\frac{1}{f}}{t + \frac{1}{f}} = \frac{1}{tf+1}.$$

Since $\lambda \mapsto tf(\lambda) + 1$ is a Bernstein function, we find

$$\frac{h}{t+h} \in \frac{1}{\mathcal{BF}} \subset \mathcal{CM}.$$

Since $\mathcal{CM}$ is a closed convex cone, we conclude that

$$\int_{(0,\infty)} \frac{h}{t+h}\, \sigma(dt) \in \mathcal{CM}$$

whenever the integral exists. Using the Stieltjes representation (6.7) of a complete Bernstein function, we find that $g \circ h \in \mathcal{CM}$ for all $g \in \mathcal{CBF}$. □

Corollary 11.27. *Assume that $\pi(dx) := c\delta_0(dx) + p(x)\, dx$ is a probability measure where $c \geq 0$ and $p(x)$ is a non-increasing, log-convex density. Then $\pi \in \mathsf{ID}$.*

Proof. By Theorem 11.24 we see that $\mathscr{L}\pi \in \mathcal{H}$. By Corollary 11.26 we get that $(\mathscr{L}\pi)^\alpha \in \mathcal{CM}$ for all $0 < \alpha < 1$ and this means that $\mathscr{L}\pi$ is infinitely divisible, i.e. $\pi \in \mathsf{ID}$, see Definition 5.6 and Remark 5.7. □

Remark 11.28. In [39] there is a proof of Corollary 11.25 (i), $\mathcal{H}^* \subset \mathcal{P}$, which does not rely on Theorems 11.11 and 11.24. By using this fact and Theorem 11.24 one can give an alternative proof of Theorem 11.11.

Indeed, suppose that $f \in \mathcal{BF}$, $f(\lambda) = a + b\lambda + \int_{(0,\infty)}(1 - e^{-\lambda t})\, \mu(dt)$ and that $\bar{\mu}$ is log-convex. Define the measure $V(dt) := b\delta_0(dt) + v(t)\, dt$ where $v(t) = \bar{\mu}(t)$. Then $\mathscr{L}V \in \mathcal{H}^* \subset \mathcal{P}$, and $f(\lambda) = \lambda\mathscr{L}V(\lambda)$. Therefore, $\lambda/f(\lambda) = 1/\mathscr{L}V(\lambda) \in \mathcal{BF}$, proving that $f \in \mathcal{SBF}$.

Remark 11.29. (i) It is shown in Remark 11.19 that there exists a special Bernstein function f such that the density u of $U|_{(0,\infty)}$ is not continuous. This shows that $1/\mathcal{H}^*$ is strictly contained in $\mathcal{SBF}$.

(ii) Suppose that $f \in \mathcal{SBF}$ such that $\frac{f}{\lambda} \in \mathcal{H}$. Let $a > 0$. Then

$$\frac{f(\lambda) + a}{\lambda} = \frac{f(\lambda)}{\lambda} + \frac{a}{\lambda} \in \mathcal{H} \subset \mathcal{P},$$

implying that $\lambda/(f(\lambda) + a) \in \mathcal{BF}$, i.e. $f + a \in \mathcal{SBF}$. Compare with Remark 11.21.

At the end of this section we point out a unified way to describe some of the considered classes of functions. Consider the family

$$\{f : f = \mathcal{L}U, \ U(dt) = c\delta_0 + u(t)\,dt\} \subset \mathcal{CM},$$

where $c \geq 0$ and $u : (0,\infty) \to (0,\infty)$ such that $\int_0^1 u(t)\,dt < \infty$. Different choices of u result in different subfamilies of completely monotone functions. More precisely,

 (i) $u \in \mathcal{CM}$ if, and only if, $\mathcal{L}U \in \mathcal{S}$;

 (ii) u is non-increasing and log-convex if, and only if, $\mathcal{L}U \in \mathcal{H}$;

 (iii) u is non-increasing and is a potential density if, and only if, $\mathcal{L}U \in \frac{1}{\mathcal{SBF}}$;

 (iv) u is non-increasing if, and only if, $\lambda\mathcal{L}(U;\lambda) \in \mathcal{BF}$.

Statements (i), (ii) and (iii) are proved, respectively, in Remark 11.6, Theorem 11.24 and Theorem 11.3. For the last statement note that if u is non-increasing, then it is the tail of a measure v, $u(t) = \bar{v}(t)$. Thus

$$\lambda\mathcal{L}(U;\lambda) = c\lambda + \lambda \int_0^\infty e^{-\lambda t} u(t)\,dt = c\lambda + \lambda \int_0^\infty e^{-\lambda t} \bar{v}(t)\,dt \in \mathcal{BF}.$$

The converse is proved similarly. Hence,

$$\mathcal{S} \subset \mathcal{H} \subset \frac{1}{\mathcal{SBF}} \quad
\begin{array}{c}
\subset \ \dfrac{\mathcal{BF}}{\lambda} \ \supset \\[2ex]
\subset \ \dfrac{1}{\mathcal{BF}} \ \supset
\end{array}
\quad \mathcal{CM},$$

and all inclusions are strict. For the second inclusion see Remark 11.29 (i), and for the third notice that $\mathcal{SBF}$ is strictly contained in $\lambda/\mathcal{BF}$ and $\mathcal{BF}$, respectively.

Comments 11.30. *Section* 11.1*:* The term *special Bernstein function* was introduced in [332], although the notion itself is encountered in the literature earlier. Theorem 11.3 is essentially contained in Bertoin [48, Corollaries 1 and 2] who attributes the result to Van Harn and Steutel [360]. Our presentation follows [332] where the result was rediscovered. Theorem 11.9 also comes from [332].

Special Bernstein functions were used in [235] to characterize the scale functions of one-dimensional Lévy processes having only negative jumps, and in [236] to study de Finetti's optimal dividend problem.

Log-convexity of the tail $\bar{\mu}$ of the Lévy measure was studied by J. Hawkes in [153, Theorem 2.1]. Theorem 11.11, cf. [333], is a slight extension as it allows for the killing term, the drift term, and finite Lévy measure. Lemma 11.14 appears also in [153] with a proof having a minor gap (it works only for the case $\sum_{j=1}^{\infty} f_j = 1$). The conclusion of the lemma, equation (11.15), is a discrete version of Chung's equation, see [86]. Examples 11.18 and Propositions 11.16 and 11.17 are from [333], the proof of the latter is new. One direction of Proposition 11.20 (iv) is proved in [235] where also the corresponding Lévy measure is identified.

Section 11.2*:* The class $\mathcal{H}$ was introduced by F. Hirsch in [163] and [164]. The starting point of his investigations is the problem to find a structurally well-behaved family of functions which operate on the (abstract) potential operators in the sense of Yosida [379, Chapter XIII.9]; these are densely defined, injective operators V such that $A = -V^{-1}$ is the generator of a strongly continuous contraction semigroup, cf. Chapter 13 below. Although the Laplace transform induces a bijection between Hunt kernels on the half-line $[0, \infty)$ (leading to potential operators) and the family of potentials $\mathcal{P}$, see e.g. [162, page 175], the set $\mathcal{P}$ lacks nice structural properties. Hirsch shows that the family $\mathcal{H} \subset \mathcal{P}$ is a convex cone which allows to define a symbolic calculus on (abstract) potential operators along the lines of Faraut's symbolic calculus [112]. See also the Comments 2.5, 5.29 and 6.22. Theorem 11.24 is from [164, Théorème 2], our proof follows the presentation in Berg and Forst [39, Section 14]. Corollary 11.27 appears, without point mass at the origin, for the first time in Steutel [338, Theorem 4.2.6], see also [339, Theorem III.10.2, p. 117].

Chapter 12

The spectral theorem and operator monotonicity

In this chapter we look at applications of complete Bernstein functions in operator theory on Hilbert spaces. First we give a proof of the spectral theorem for self-adjoint operators on a Hilbert space which uses the Stieltjes representation of complete Bernstein functions. Then we prove Löwner's theorem which says that the family of complete Bernstein functions coincides with the family of non-negative operator monotone functions on $(0, \infty)$.

12.1 The spectral theorem

A surprising consequence of the representation formulae for complete Bernstein functions is a simple proof of the spectral theorem for self-adjoint operators on Hilbert spaces. This was first observed by Doob and Koopman [101] and Lengyel [240]. The paper by Nevanlinna and Nieminen [275] contains a full account of this approach and its extensions. Our exposition follows these lines; the monograph by Kato [200] will be our standard reference for spectral analysis in Hilbert spaces.

Let $\mathfrak{H}$ be a complex Hilbert space with scalar product $\langle \cdot, \cdot \rangle$ and norm $\| \cdot \|$. We assume that the scalar product is linear in the first, and skew-linear in the second argument.

A linear operator $A : \mathfrak{D}(A) \to \mathfrak{H}$ is said to be *self-adjoint* if $\mathfrak{D}(A) \subset \mathfrak{H}$ is dense and $(A, \mathfrak{D}(A)) = (A^*, \mathfrak{D}(A^*))$. The *resolvent set* ϱ_A consists of all $\zeta \in \mathbb{C}$ such that the operator $\zeta - A := \zeta \operatorname{id} - A$ has a bounded inverse $R_\zeta := (\zeta - A)^{-1}$; the operators $(R_\zeta)_{\zeta \in \varrho_A}$ are the *resolvent* of A. It is well known that the resolvent set ϱ_A is open. The set $\sigma_A := \mathbb{C} \setminus \varrho_A$ is called the *spectrum* of A.

All self-adjoint operators A have real spectrum $\sigma_A \subset \mathbb{R}$ and their resolvents satisfy

$$\| R_\zeta u \| \leq \frac{1}{\operatorname{dist}(\zeta, \sigma_A)} \| u \| \leq \frac{1}{|\operatorname{Im} \zeta|} \| u \|, \quad \zeta \in \varrho_A, \ u \in \mathfrak{H}. \tag{12.1}$$

For $\zeta, z \in \varrho_A$ the following *resolvent equation* holds

$$R_\zeta - R_z = (z - \zeta) R_z R_\zeta \tag{12.2}$$

which is easily seen from $R_\zeta = R_\zeta (z - A) R_z = R_\zeta ((z - \zeta) + (\zeta - A)) R_z$. Since $\zeta \mapsto \langle R_\zeta u, v \rangle$ is continuous, the resolvent equation tells us

$$\lim_{z \to \zeta} \left\langle \frac{R_\zeta - R_z}{\zeta - z} u, v \right\rangle = - \lim_{z \to \zeta} \langle R_z R_\zeta u, v \rangle = -\langle R_\zeta R_\zeta u, v \rangle. \tag{12.3}$$

This means that $\langle R_\zeta u, v \rangle$, $u, v \in \mathfrak{H}$, is complex differentiable for all $\zeta \in \varrho_A$, hence it is analytic.

Recall that a self-adjoint operator $(A, \mathfrak{D}(A))$ is *dissipative* if

$$\langle Au, u \rangle \le 0 \quad \text{for all } u \in \mathfrak{D}(A). \tag{12.4}$$

Lemma 12.1. *Let $(A, \mathfrak{D}(A))$ be a self-adjoint operator on $\mathfrak{H}$. Then A is dissipative if, and only if, the following holds*

$$\|\lambda u - Au\| \ge \lambda \|u\| \quad \text{for all } u \in \mathfrak{D}(A), \ \lambda > 0. \tag{12.5}$$

Proof. Clearly,

$$\|(\lambda - A)u\|^2 = \lambda^2 \|u\|^2 - 2\lambda \langle Au, u \rangle + \|Au\|^2.$$

If $\langle Au, u \rangle \le 0$ we see that $\|(\lambda - A)u\|^2 \ge \lambda^2 \|u\|^2$ and (12.5) follows. Conversely, under (12.5) we have

$$\lambda^2 \|u\|^2 - 2\lambda \langle Au, u \rangle + \|Au\|^2 \ge \lambda^2 \|u\|^2$$

or $2\lambda \langle Au, u \rangle \le \|Au\|^2$. Since λ can be arbitrarily large, this is only possible if $\langle Au, u \rangle \le 0$ and (12.4) follows.   □

Proposition 12.2. *Let $(A, \mathfrak{D}(A))$ be a self-adjoint operator on $\mathfrak{H}$. Then A is dissipative if, and only if, the spectrum σ_A is contained in $(-\infty, 0]$.*

Proof. First we suppose that $\sigma_A \subset (-\infty, 0]$. Since $\lambda > 0$ is in the resolvent set, we may use (12.1) with $\zeta = \lambda$ and $u = (\lambda - A)w$, $w \in \mathfrak{D}(A)$, to get

$$\|w\| = \|R_\lambda(\lambda - A)w\| \le \frac{1}{\text{dist}(\lambda, \sigma_A)} \|\lambda w - Aw\| \le \frac{1}{\lambda} \|\lambda w - Aw\|.$$

This proves dissipativity.

Now we suppose that A is dissipative and we want to show that $\sigma_A \subset (-\infty, 0]$. For any $\lambda > 0$, let $v \in ((A - \lambda)(\mathfrak{D}(A)))^\perp$. Then $v \in \mathfrak{D}(A)$ and $Av = \lambda v$ since A is self-adjoint. Hence $\lambda \|v\|^2 = \langle Av, v \rangle \le 0$, which gives $v = 0$. Thus $(A - \lambda)(\mathfrak{D}(A)) = \mathfrak{H}$ and consequently $\sigma_A \subset (-\infty, 0]$.   □

From now on we will assume that A is a dissipative self-adjoint operator.

Lemma 12.3. *Let $(A, \mathfrak{D}(A))$ be a dissipative self-adjoint operator on $\mathfrak{H}$. For every $u \in \mathfrak{H}$ the function $f(\zeta) := \langle \zeta R_\zeta u, u \rangle$, $\zeta \in \mathbb{C} \setminus (-\infty, 0]$, is (the extension of) a complete Bernstein function such that*

$$f(\lambda) = \langle \lambda R_\lambda u, u \rangle \le \|u\|^2 \quad \text{for all } \lambda > 0, \ u \in \mathfrak{H}.$$

Moreover,

$$\lim_{\lambda \to \infty} \langle \lambda R_\lambda u, v \rangle = \langle u, v \rangle, \quad u, v \in \mathfrak{H}, \tag{12.6}$$

and there exists a uniquely determined finite measure $E = E_{u,u}$ supported in $(-\infty, 0]$ such that $E_{u,u}(-\infty, 0] = \|u\|^2$ and

$$\langle R_\zeta u, u \rangle = \int_{(-\infty,0]} \frac{1}{\zeta - t} E_{u,u}(dt), \quad \zeta \in \mathbb{C} \setminus (-\infty, 0]. \tag{12.7}$$

Proof. It is clear that $f(\zeta)$ is analytic on ϱ_A. Note that

$$f(\zeta) = \langle (\zeta - A + A) R_\zeta u, u \rangle = \langle u, u \rangle + \langle A R_\zeta u, u \rangle. \tag{12.8}$$

Setting $u = (\zeta - A)\phi$ for a suitable $\phi = R_\zeta u \in \mathfrak{D}(A)$ and $\zeta = \lambda + i\kappa$, we conclude that

$$f(\zeta) = \langle \zeta R_\zeta (\zeta - A)\phi, (\zeta - A)\phi \rangle = \langle \zeta \phi, (\zeta - A)\phi \rangle = \zeta \bar{\zeta} \langle \phi, \phi \rangle - \zeta \langle \phi, A\phi \rangle.$$

Thus $\operatorname{Im} f(\zeta) = -\operatorname{Im} \zeta \langle \phi, A\phi \rangle \geq 0$ for $\operatorname{Im} \zeta \geq 0$, and for $\zeta = \lambda > 0$ we see

$$f(\lambda) = \lambda^2 \langle \phi, \phi \rangle - \lambda \langle \phi, A\phi \rangle \geq 0.$$

From (12.3) and (12.8) we get

$$\frac{d}{d\lambda} f(\lambda) = \frac{d}{d\lambda} \langle A R_\lambda u, u \rangle = \frac{d}{d\lambda} \langle R_\lambda u, Au \rangle = -\langle R_\lambda R_\lambda u, Au \rangle$$
$$= -\langle R_\lambda u, A R_\lambda u \rangle \geq 0.$$

Therefore, $f(0+) = \inf_{\lambda > 0} f(\lambda) < \infty$ exists, and by Theorem 6.2 (iv), $f \in \mathcal{CBF}$; in particular,

$$f(\lambda) = a + b\lambda + \int_{(0,\infty)} \frac{\lambda}{\lambda + t} \sigma(dt)$$

where $a, b \geq 0$ and σ is a measure on $(0, \infty)$ such that $\int_{(0,\infty)} (1 + t)^{-1} \sigma(dt) < \infty$.

The Cauchy–Schwarz inequality and the resolvent estimate (12.1) show for any self-adjoint operator $(A, \mathfrak{D}(A))$ on $\mathfrak{H}$ and $u \in \mathfrak{D}(A)$

$$\lim_{\kappa \to \infty} |\langle A R_{i\kappa} u, u \rangle| \leq \lim_{\kappa \to \infty} \|R_{i\kappa} Au\| \cdot \|u\| \leq \lim_{\kappa \to \infty} \left(\frac{1}{\kappa} \|Au\| \cdot \|u\| \right) = 0.$$

Thus,

$$\lim_{\kappa \to \infty} f(i\kappa) = \langle u, u \rangle + \lim_{\kappa \to \infty} \langle A R_{i\kappa} u, u \rangle = \|u\|^2$$

and we find from the integral representation of $f \in \mathcal{CBF}$ that

$$\operatorname{Re} f(i\kappa) = a + \int_{(0,\infty)} \frac{\kappa^2}{\kappa^2 + t^2} \sigma(dt).$$

Monotone convergence now shows that

$$\|u\|^2 = \lim_{\kappa \to \infty} f(i\kappa) = a + \sigma(0, \infty).$$

On the other hand,

$$\operatorname{Im} f(i\kappa) = b\kappa + \int_{(0,\infty)} \frac{\kappa t}{\kappa^2 + t^2} \sigma(dt),$$

so that $\lim_{\kappa \to \infty} f(i\kappa) = \|u\|^2$ combined with the dominated convergence theorem proves that $b = 0$.

This shows

$$f(\zeta) = a + \int_{(0,\infty)} \frac{\zeta}{\zeta + t} \sigma(dt), \quad a + \sigma(0, \infty) = \|u\|^2.$$

Moreover,

$$f(\lambda) = a + \int_{(0,\infty)} \frac{\lambda}{\lambda + t} \sigma(dt) \le a + \sigma(0, \infty) = \|u\|^2,$$

and we can use monotone convergence to get

$$\lim_{\lambda \to \infty} \langle \lambda R_\lambda u, u \rangle = \lim_{\lambda \to \infty} f(\lambda) = \|u\|^2, \quad u \in \mathfrak{H}.$$

A standard polarization argument proves then (12.6).

To see (12.7), set $E_{u,u}(I) := \delta_0(I) + \sigma(-I)$ for all Borel sets $I \subset (-\infty, 0]$. The uniqueness of the measure $E_{u,u}$ follows from the uniqueness of the measure σ in the Stieltjes representation of complete Bernstein functions.   □

With a simple polarization argument we can extend (12.7) to

$$\langle R_\zeta u, v \rangle = \int_{(-\infty, 0]} \frac{1}{\zeta - t} E_{u,v}(dt), \quad \zeta \in \mathbb{C} \setminus (-\infty, 0], \ u, v \in \mathfrak{H}.$$

Since $E_{u,u}$ is uniquely determined by u, the complex-valued measure $E_{u,v}$ is uniquely determined by u and v, and $E_{u,v}$ depends linearly on u and skew-linearly on v; moreover, $E_{u,v} = \overline{E_{v,u}}$ is skew-symmetric and $E_{u,v}$ has total variation less or equal than $\|u\| \cdot \|v\|$.

Using the inversion formula (6.6) we can actually work out the measure $E_{u,v}$. For $\alpha < \beta \le 0$ we get the *Hellinger–Stone formula*

$$
\begin{aligned}
E_{u,v}[\alpha, \beta] &= \sigma(-\beta, -\alpha] \\
&= \lim_{h \to 0+} \frac{1}{\pi} \int_{(-\beta, -\alpha]} \operatorname{Im} \left(\frac{\langle (s - ih) R_{-s+ih} u, v \rangle}{s - ih} \right) ds \\
&= - \lim_{h \to 0+} \frac{1}{\pi} \int_{[\alpha, \beta)} \operatorname{Im} \langle R_{t+ih} u, v \rangle \, dt
\end{aligned}
\tag{12.9}
$$

at all continuity points α, β of $\lambda \mapsto E_{u,v}(-\infty, \lambda)$. Note that this formula is still valid for $\alpha, \beta > 0$ with $E_{u,v}[\alpha, \beta) = 0$; this allows to extend $E_{u,v}[\alpha, \beta)$ trivially by 0 onto the positive real line.

If $I \subset (-\infty, 0]$ is a Borel set, this shows that $(u, v) \mapsto E_{u,v}(I)$ is a skew symmetric sesquilinear form in the Hilbert space $\mathfrak{H}$. By a standard argument, see Lengyel [240] or Lax [239, Chapter 32], we deduce that there exists a bounded self-adjoint operator $E(I)$ such that

$$E_{u,v}(I) = \langle E(I)u, v \rangle, \quad u, v \in \mathfrak{H}, \ I \subset (-\infty, 0] \text{ Borel.}$$

This shows that (12.7) can be written as

$$\langle R_\zeta u, v \rangle = \int_{(-\infty, 0]} \frac{1}{\zeta - t} \langle E(dt)u, v \rangle, \quad \zeta \in \mathbb{C} \setminus (-\infty, 0], \ u, v \in \mathfrak{H}.$$

From this point onwards we can argue as usual: one concludes that $(E(-\infty, \lambda))_{\lambda \in \mathbb{R}}$ is a spectral resolution of the operator A and proves the spectral theorem for (dissipative) self-adjoint operators.

Theorem 12.4. *Let $(A, \mathfrak{D}(A))$ be a dissipative self-adjoint operator on $\mathfrak{H}$. Then there exists an orthogonal projection-valued measure E on the Borel sets of $\mathbb{R}$ with support σ_A such that for all Borel sets $I, J \subset \mathbb{R}$*

(i) $E(\emptyset) = 0, \ E(\mathbb{R}) = \text{id};$

(ii) $E(I \cap J) = E(I)E(J);$

(iii) $E(I) : \mathfrak{D}(A) \to \mathfrak{D}(A)$ *and* $AE(I) = E(I)A;$

(iv) $Au = \int_{\sigma_A} \lambda\, E(d\lambda)u$ *for* $u \in \mathfrak{D}(A) = \{u \in \mathfrak{H} : \int_{\sigma_A} \lambda^2 \langle E(d\lambda)u, u \rangle < \infty\}.$

Note that $\|Au\|^2 = \int_{\sigma_A} \lambda^2 \langle E(d\lambda)u, u \rangle$. For $\Phi : (-\infty, 0] \to \mathbb{R}$ one can define the function $\Phi(A)$ of the operator A by

$$\Phi(A)u := \int_{(-\infty, 0]} \Phi(\lambda)\, E(d\lambda)u, \tag{12.10}$$

$$\mathfrak{D}(\Phi(A)) := \left\{ u \in \mathfrak{H} : \|\Phi(A)u\|^2 = \int_{(-\infty, 0]} |\Phi(\lambda)|^2 \langle E(d\lambda)u, u \rangle < \infty \right\}. \tag{12.11}$$

Example 12.5. Let $(A, \mathfrak{D}(A))$ be a dissipative self-adjoint operator on $\mathfrak{H}$. Then A generates a semigroup of linear operators $(T_t)_{t \geq 0}$ which is strongly continuous, i.e. $t \mapsto T_t u$ is continuous from $[0, \infty)$ to $\mathfrak{H}$, and contractive, i.e. $\|T_t u\| \leq \|u\|$.

Indeed: Let $E(d\lambda)$ be the spectral resolution of the operator A. Then

$$T_t u = e^{tA} u = \int_{(-\infty,0]} e^{t\lambda} E(d\lambda)u, \quad u \in \mathfrak{H}, \ t > 0,$$

and by Theorem 12.4 we get for all $u, v \in \mathfrak{H}$ and all $s, t > 0$

$$\langle T_t T_s u, v \rangle = \left\langle T_t \int_{(-\infty,0]} e^{s\lambda} E(d\lambda)u, v \right\rangle$$

$$= \int_{(-\infty,0]} \int_{(-\infty,0]} e^{t\kappa} e^{s\lambda} \langle E(d\lambda)u, E(d\kappa)v \rangle$$

$$= \int_{(-\infty,0]} e^{(t+s)\lambda} \langle E(d\lambda)u, v \rangle$$

$$= \langle T_{s+t} u, v \rangle$$

which shows that $(T_t)_{t\geq 0}$ is a semigroup. That $t \mapsto T_t u$, $u \in \mathfrak{H}$, is continuous in the Hilbert space topology follows directly from the continuity of $t \mapsto e^{t\lambda}$, $\lambda \in (-\infty, 0]$, and the fact that

$$\|T_t u - u\|^2 = \int_{(-\infty,0]} (e^{t\lambda} - 1)^2 \langle E(d\lambda)u, u \rangle.$$

In a similar way we deduce from the fact that $e^{t\lambda} \leq 1$, $t > 0$ and $\lambda \in (-\infty, 0]$, that $\|T_t u\| \leq \|u\|$ which proves that T_t is a contraction operator.

Example 12.6. Let $(A, \mathfrak{D}(A))$ be a dissipative self-adjoint operator on $\mathfrak{H}$ and denote by $(T_t)_{t\geq 0}$ the strongly continuous contraction semigroup generated by A. Let f be a Bernstein function given by the Lévy–Khintchine representation (3.2),

$$f(\lambda) = a + b\lambda + \int_{(0,\infty)} (1 - e^{-t\lambda}) \mu(dt),$$

where $a, b \geq 0$ and $\int_{(0,\infty)} (1 \wedge t) \mu(dt) < \infty$. Then $\mathfrak{D}(A) \subset \mathfrak{D}(f(-A))$ and

$$-f(-A)u = -au + bAu + \int_{(0,\infty)} (T_t u - u) \mu(dt), \quad u \in \mathfrak{D}(A). \tag{12.12}$$

Indeed: Let $E(d\lambda)$ be the spectral resolution of the operator A. The elementary inequality $1 \wedge (t\lambda) \leq (1 + \lambda)(1 \wedge t)$, $\lambda, t \geq 0$, shows that

$$f(\lambda) \leq a + b\lambda + (1 + \lambda) \int_{(0,\infty)} (1 \wedge t) \mu(dt) \leq c (1 + \lambda)$$

for some suitable constant $c > 0$. Therefore

$$\int_{(-\infty,0]} |f(-\lambda)|^2 \langle E(d\lambda)u, u \rangle \leq c \int_{(-\infty,0]} (1 - \lambda)^2 \langle E(d\lambda)u, u \rangle$$

which means that $\mathfrak{D}(A) \subset \mathfrak{D}(f(-A))$. The representation (12.12) follows now directly from the integral formula (3.2) for f. For $u \in \mathfrak{D}(A)$ we get from Fubini's theorem

$$
\begin{aligned}
f(-A)u &= \int_{(-\infty,0]} f(-\lambda)\, E(d\lambda)u \\
&= \int_{(-\infty,0]} \left(a - b\lambda + \int_{(0,\infty)} (1 - e^{t\lambda})\, \mu(dt) \right) E(d\lambda)u \\
&= au - bAu + \int_{(0,\infty)} \int_{(-\infty,0]} (1 - e^{t\lambda})\, E(d\lambda)u\, \mu(dt) \\
&= au - bAu + \int_{(0,\infty)} (u - T_t u)\, \mu(dt).
\end{aligned}
$$

Example 12.7. Let $(A, \mathfrak{D}(A))$ be a dissipative self-adjoint operator on $\mathfrak{H}$, denote by $(R_\zeta)_{\zeta \in \varrho_A}$ its resolvent, and let f be a complete Bernstein function. We use the Stieltjes representation (6.7)

$$
f(\lambda) = a + b\lambda + \int_{(0,\infty)} \frac{\lambda}{\lambda + t}\, \sigma(dt),
$$

where $a, b \geq 0$ and $\int_{(0,\infty)} (1 + t)^{-1}\, \sigma(dt) < \infty$. Then $\mathfrak{D}(A) \subset \mathfrak{D}(f(-A))$ and

$$
-f(-A)u = -au + bAu + \int_{(0,\infty)} AR_t u\, \sigma(dt), \quad u \in \mathfrak{D}(A). \tag{12.13}
$$

Indeed: Since $\mathcal{CBF} \subset \mathcal{BF}$, the inclusion $\mathfrak{D}(A) \subset \mathfrak{D}(f(-A))$ follows from Example 12.6. Let $E(d\lambda)$ be the spectral resolution of the operator A. Then

$$
-AR_t u = \int_{(-\infty,0]} \frac{-\lambda}{t - \lambda}\, E(d\lambda)u, \quad u \in \mathfrak{H}, \ t > 0,
$$

defines a bounded operator. Since $(-\lambda)/(t - \lambda) \leq 1 \wedge (-\lambda/t)$ for $\lambda \leq 0$ and $t > 0$, we conclude from this that

$$
\|AR_t u\| \leq \min \left\{ \|u\|, \frac{1}{t} \|Au\| \right\}, \quad t > 0.
$$

From the elementary inequality $\min\{1, s\} \leq 2s/(1 + s)$, $s \geq 0$, we get

$$
\|AR_t u\| \leq 2 \frac{\frac{1}{t} \frac{\|Au\|}{\|u\|}}{1 + \frac{1}{t} \frac{\|Au\|}{\|u\|}} \|u\| = 2 \frac{\frac{\|Au\|}{\|u\|}}{t + \frac{\|Au\|}{\|u\|}} \|u\|. \tag{12.14}
$$

Let us now determine $f(-A)$.

$$\langle -f(-A)u, u\rangle = \int_{(-\infty,0]} -f(-\lambda)\langle E(d\lambda)u, u\rangle$$

$$= -a\langle u, u\rangle + b\langle Au, u\rangle + \int_{(-\infty,0]}\int_{(0,\infty)} \frac{\lambda}{t-\lambda}\,\sigma(dt)\langle E(d\lambda)u, u\rangle$$

$$= -a\langle u, u\rangle + b\langle Au, u\rangle + \int_{(0,\infty)}\int_{(-\infty,0]} \frac{\lambda}{t-\lambda}\,\langle E(d\lambda)u, u\rangle\,\sigma(dt)$$

$$= -a\langle u, u\rangle + b\langle Au, u\rangle + \int_{(0,\infty)} \langle AR_t u, u\rangle\,\sigma(dt).$$

The change in the order of integration is possible since all integrands are non-negative.
The estimate (12.14) tells us that for $u \in \mathfrak{D}(A)$

$$\|f(-A)u\| \leq a\|u\| + b\|Au\| + \int_{(0,\infty)} \|AR_t u\|\,\sigma(dt)$$

$$\leq a\|u\| + b\|Au\| + 2\int_{(0,\infty)} \frac{\frac{\|Au\|}{\|u\|}}{t + \frac{\|Au\|}{\|u\|}}\,\sigma(dt)\,\|u\|$$

$$\leq 2f\left(\frac{\|Au\|}{\|u\|}\right)\|u\|.$$

Remark 12.8. Let $(A, \mathfrak{D}(A))$ be a dissipative self-adjoint operator on $\mathfrak{H}$ with resolvent $(R_\lambda)_{\lambda>0}$. By Lemma 12.3 the function $\lambda \mapsto \langle \lambda R_\lambda u, u\rangle$ is for all $u \in \mathfrak{H}$ a complete Bernstein function. Therefore, $f(\lambda) := \langle R_\lambda u, u\rangle$ is a Stieltjes function.

Consider the special situation where $\mathfrak{H} = L^2(\mathbb{R}^d, dx)$ and where resolvent operators R_λ are integral operators given by kernels $R_\lambda(x, y)$, i.e.

$$R_\lambda u(x) = \int_{\mathbb{R}^d} R_\lambda(x, y)u(y)\,dy, \quad u \in L^2.$$

If $x \mapsto R_\lambda(x, x)$ is continuous at $x = x_0$, then $f(\lambda) := R_\lambda(x_0, x_0)$ is a Stieltjes function.

Without loss of generality we can assume that $x_0 = 0$. Consider a sequence of type delta, i.e. $u_n \in C_c^\infty(\mathbb{R}^d)$ such that $u_n \xrightarrow{n\to\infty} \delta_0$ weakly, then

$$\langle R_\lambda u_n, u_n\rangle = \iint R_\lambda(x, y)\,u_n(x)\,u_n(y)\,dx\,dy \xrightarrow{n\to\infty} R_\lambda(0, 0)$$

since R_λ is continuous at $x = 0$. The assertion follows since $\mathcal{S}$ is stable under pointwise limits, cf. Theorem 2.2.

We have seen in Example 12.5 that A generates a strongly continuous contraction semigroup $(T_t)_{t\geq0}$. Obviously

$$s \mapsto \langle T_s u, u\rangle = \int_{(-\infty,0]} e^{s\lambda}\langle E(d\lambda)u, u\rangle, \quad u \in L^2,$$

is completely monotone. If T_s is an integral operator with kernel $p_s(x, y)$, we can show as above that

$$s \mapsto p_s(x, x) \text{ is completely monotone}$$

for all x such that $x \mapsto p_s(x, x)$ is continuous.

A particularly interesting example is given by strongly continuous convolution semigroups or, equivalently, Lévy processes; for details we refer to Jacob [183, Vol. 1, Section 3.6] or [184]. In this case, $A = -\psi(D)$ is a pseudo differential operator with symbol ψ which is negative definite in the sense of Schoenberg, cf. (4.8), and

$$T_s u(x) = \int e^{-s\psi(\xi)} \, \widehat{u}(\xi) \, e^{ix\xi} \, d\xi = \int u(x + y) \, p_s(dy), \quad u \in C_c^\infty(\mathbb{R}^d),$$

where $\widehat{u}$ denotes the Fourier transform and $\widehat{p}_s = e^{-s\psi}$. The resolvent operators are given by

$$R_\lambda u(x) = \int \frac{1}{\lambda + \psi(\xi)} \widehat{u}(\xi) \, e^{ix\xi} \, d\xi, \quad u \in C_c^\infty(\mathbb{R}^d).$$

Since $\widehat{p}_s(\xi) = e^{-s\psi(\xi)}$, it is clear that the measure $p_s(dy)$ is for all $s > 0$ absolutely continuous with density $p_s \in L^\infty$ if, and only if, $e^{-s\psi} \in L^1$ for all $s > 0$. In this case, the Riemann–Lebesgue lemma applies and shows that $y \mapsto p_s(y)$ is continuous. Sufficiency is obvious, necessity follows from the fact that $p_{s/2} \in L^\infty \cap L^1$ is already an L^2-function. Therefore,

$$e^{-\frac{1}{2}s\psi} = \widehat{p}_{s/2} \in L^2$$

or $e^{-s\psi} \in L^1$. This proves that the semigroup and resolvent densities exist and that

$$\left[t \mapsto p_t(0) \right] \in \mathcal{CM} \quad \text{and} \quad \left[\lambda \mapsto R_\lambda(0) \right] \in \mathcal{S}$$

whenever $e^{-s\psi} \in L^1$ for all $s > 0$.

12.2 Operator monotone functions

Let $(\mathfrak{H}, \langle \cdot, \cdot \rangle)$ be a finite or infinite-dimensional real Hilbert space and let A and B be bounded self-adjoint operators on $\mathfrak{H}$. If $\dim \mathfrak{H} = n$, we will identify $\mathfrak{H}$ with $\mathbb{R}^n$ and A, B with symmetric $n \times n$ matrices. We write $B \leq A$ if $B - A$ is dissipative. This is equivalent to saying that

$$\langle Bu, u \rangle \leq \langle Au, u \rangle \quad \text{for all } u \in \mathfrak{H}. \tag{12.15}$$

As usual we define functions of A by the spectral theorem (in finite or infinite dimension).

Definition 12.9. A function $f : (0, \infty) \to \mathbb{R}$ is said to be *matrix monotone of order n* if for all symmetric matrices $A, B \in \mathbb{R}^{n \times n}$ with $\sigma_A, \sigma_B \subset (0, \infty)$

$$B \le A \quad \text{implies} \quad f(B) \le f(A).$$

We write P_n for the set of matrix monotone functions of order n. If f is matrix monotone of all orders $n \in \mathbb{N}$, we call f *operator monotone*.

In the remaining part of this section we will assume that the operators $(A, \mathfrak{D}(A))$ and $(B, \mathfrak{D}(B))$ are dissipative self-adjoint operators on $\mathfrak{H}$ which are not necessarily bounded. It may happen that the domain $\mathfrak{D}(A) \cap \mathfrak{D}(B)$ of the difference $B - A$ of two unbounded operators is not dense in $\mathfrak{H}$; this means that (12.15) might not be well defined. To circumvent this, we use quadratic forms. Denote by $(-A)^{1/2}$ the unique positive square root of $(-A)$ defined via the spectral theorem. Then

$$Q_A(u, u) := \begin{cases} \langle (-A)^{1/2} u, (-A)^{1/2} u \rangle, & \text{if } u \in \mathfrak{D}(Q_A) = \mathfrak{D}((-A)^{1/2}); \\ +\infty, & \text{for all other } u \in \mathfrak{H} \end{cases}$$

defines the quadratic form generated by the operator A. If $u \in \mathfrak{D}(A)$, then we have $Q_A(u, u) = -\langle Au, u \rangle$; this is, in particular, the case for bounded operators. Therefore,

$$0 \le Q_A(u, u) \le Q_B(u, u) \le +\infty, \quad u \in \mathfrak{H}, \tag{12.16}$$

is the proper generalization of (12.15) for arbitrary dissipative operators.

Definition 12.10. Let A and B be dissipative self-adjoint operators on $\mathfrak{H}$. If (12.16) holds for the quadratic forms, we say that $(-B)$ *dominates* $(-A)$ *in quadratic form sense* and write $(-A) \le_{\mathrm{qf}} (-B)$.

Note that $(-A) \le_{\mathrm{qf}} (-B)$ entails

$$\mathfrak{D}((-B)^{1/2}) = \mathfrak{D}(Q_B) \subset \mathfrak{D}(Q_A) = \mathfrak{D}((-A)^{1/2})$$

and that $0 \le (-A)$ amounts to saying that A is dissipative.

Obviously, the condition for $f : (0, \infty) \to \mathbb{R}$ that $f(-A) \le_{\mathrm{qf}} f(-B)$ whenever $(-A) \le_{\mathrm{qf}} (-B)$ seems to be stronger than $f \in \bigcap_{n \in \mathbb{N}} P_n$. But we will see later in Theorem 12.17 that these properties are equivalent. A hint in this direction is given by the following lemma.

Lemma 12.11. *Let $(A, \mathfrak{D}(A))$ and $(B, \mathfrak{D}(B))$ be two dissipative self-adjoint operators on $\mathfrak{H}$ and denote by $(R_\zeta^A)_{\zeta \in \varrho_A}$ and $(R_\zeta^B)_{\zeta \in \varrho_B}$ their resolvents. Then the following assertions are equivalent*

(i) $0 \le_{\mathrm{qf}} (-A) \le_{\mathrm{qf}} (-B)$;

(ii) $0 \leq_{\mathrm{qf}} R_t^B \leq_{\mathrm{qf}} R_t^A$ for all $t > 0$;

(iii) $0 \leq_{\mathrm{qf}} (-A)R_t^A \leq_{\mathrm{qf}} (-B)R_t^B$ for all $t > 0$.

Proof. (i)$\Rightarrow$(ii) Fix $t > 0$. For any $u \in \mathfrak{H}$ set

$$\phi := (t - A)^{-1}u \in \mathfrak{D}(A) \subset \mathfrak{D}(Q_A) \subset \mathfrak{D}(Q_B),$$
$$\psi := (t - B)^{-1}u \in \mathfrak{D}(B) \subset \mathfrak{D}(Q_B).$$

Since $A - t$ is dissipative, we can use the spectral theorem, Theorem 12.4, to define the square root $(t - A)^{1/2}$; this is again a self-adjoint operator and

$$\mathfrak{D}(A) \subset \mathfrak{D}((-A)^{1/2}) = \mathfrak{D}((t - A)^{1/2}).$$

Since by assumption $(t - A) \leq_{\mathrm{qf}} (t - B)$, we get by the Cauchy–Schwarz inequality

$$
\begin{aligned}
\left((t - B)^{-1}u, u\right)^2 &= \left((t - B)^{-1}u, (t - A)\phi\right)^2 \\
&= \left((t - A)^{1/2}(t - B)^{-1}u, (t - A)^{1/2}\phi\right)^2 \\
&\leq \left((t - A)^{1/2}\psi, (t - A)^{1/2}\psi\right) \cdot \left(\phi, (t - A)\phi\right) \\
&\leq \left((t - B)^{1/2}\psi, (t - B)^{1/2}\psi\right) \cdot \left((t - A)^{-1}u, u\right) \\
&= \left((t - B)^{-1}u, u\right)\left((t - A)^{-1}u, u\right).
\end{aligned}
$$

This proves $(t - B)^{-1} \leq_{\mathrm{qf}} (t - A)^{-1}$.

(ii)$\Rightarrow$(iii) We have for all $t > 0$ and $u \in \mathfrak{H}$

$$
\begin{aligned}
\left((-A)R_t^A u, u\right) &= \left((t - A)R_t^A u - tR_t^A u, u\right) \\
&= \langle u, u \rangle - \langle tR_t^A u, u \rangle \\
&\leq \langle u, u \rangle - \langle tR_t^B u, u \rangle \\
&= \left((-B)R_t^B u, u\right).
\end{aligned}
$$

(iii)$\Rightarrow$(i) Lemma 12.3 shows that

$$
\begin{aligned}
\lim_{t \to \infty} \left(tR_t^A(-A)u, u\right) &= \lim_{t \to \infty} \left(tR_t^A(-A)^{1/2}u, (-A)^{1/2}u\right) \\
&= \left((-A)^{1/2}u, (-A)^{1/2}u\right)
\end{aligned}
$$

for all $u \in \mathfrak{D}(Q_A)$. Using a similar argument for $tR_t^B(-B)$ yields

$$\left((-A)^{1/2}u, (-A)^{1/2}u\right) \leq \left((-B)^{1/2}u, (-B)^{1/2}u\right). \qquad \square$$

Theorem 12.12. *Let $f \in \mathcal{CBF}$ and let $(A, \mathfrak{D}(A))$ and $(B, \mathfrak{D}(B))$ be two dissipative self-adjoint operators on $\mathfrak{H}$; denote by $(R_\zeta^A)_{\zeta \in \varrho_A}$ and $(R_\zeta^B)_{\zeta \in \varrho_B}$ the corresponding resolvents. Then*

$$(-A) \leq_{\mathrm{qf}} (-B) \quad \text{implies} \quad f(-A) \leq_{\mathrm{qf}} f(-B). \tag{12.17}$$

In particular, f is operator monotone in the sense of Definition 12.9.

Proof. We have seen in Example 12.7 that

$$\langle f(-A)u, u \rangle = a \langle u, u \rangle + b \langle (-A)u, u \rangle + \int_{(0,\infty)} \langle (-A) R_t^A u, u \rangle \sigma(dt)$$

holds for all $u \in \mathfrak{D}(A)$ – and even $u \in \mathfrak{H}$ if we understand the expression on the right as a quadratic form which is $+\infty$ if $u \notin \mathfrak{D}(Q_{f(-A)})$. The operators $t R_t A$ are bounded, see Example 12.7, and $\sigma(0, n) < \infty$ for all $n \in \mathbb{N}$. Therefore,

$$\int_{(0,n)} \langle A R_t^A u, u \rangle \sigma(dt) < \infty \quad \text{for all } u \in \mathfrak{H}, \ n \in \mathbb{N}.$$

Since $\lim_{n \to \infty} \langle n R_n^A A u, u \rangle = \langle A u, u \rangle$ on $\mathfrak{D}(A)$, we have for $u \in \mathfrak{D}(A)$

$$\langle f(-A)u, u \rangle = \lim_{n \to \infty} \left(a \langle u, u \rangle + b \langle n R_n^A(-A)u, u \rangle + \int_{(0,n)} \langle (-A) R_t^A u, u \rangle \sigma(dt) \right)$$

where the expression inside the bracket defines a bounded operator. If we understand this in the sense of quadratic forms, the right-hand side converges to a finite value if, and only if, $u \in \mathfrak{D}(Q_{f(-A)})$. Analogous assertions hold for the operators B, $f(-B)$ and the quadratic forms generated by them.

Lemma 12.11 shows that $\langle (-A) R_t^A u, u \rangle \leq \langle (-B) R_t^B u, u \rangle$ for all $u \in \mathfrak{H}$. If we apply this to the limit expression above, we get $\langle f(-A)u, u \rangle \leq \langle f(-B)u, u \rangle$ or, more precisely,

$$0 \leq Q_{f(-A)}(u, u) \leq Q_{f(-B)}(u, u) \leq \infty, \quad u \in \mathfrak{H},$$

which means that $f(-A) \leq_{\mathrm{qf}} f(-B)$. $\square$

Corollary 12.13. *Let $f \in \mathcal{CBF}$ and let $(A, \mathfrak{D}(A))$ be a dissipative self-adjoint operator on $\mathfrak{H}$. Then for all $c > 0$ it holds that*

$$\mathfrak{D}\big(f(-cA)\big) = \mathfrak{D}\big(f(-A)\big) \quad \text{and} \quad \mathfrak{D}\big(f(c - A)\big) = \mathfrak{D}\big(f(-A)\big).$$

Proof. Assume first that $c \geq 1$. Observe that for all $\lambda, t > 0$

$$\frac{c\lambda}{t + c\lambda} \leq c \frac{\lambda}{t + \lambda} \quad \text{and} \quad \frac{c + \lambda}{t + (c + \lambda)} \leq \frac{c}{t + c} + \frac{\lambda}{t + \lambda}.$$

Since $f \in \mathcal{CBF}$ is non-decreasing we conclude from this and the Stieltjes representation (6.7) of f that

$$f(\lambda) \le f(c\lambda) \le cf(\lambda) \quad \text{and} \quad f(\lambda) \le f(c + \lambda) \le f(c) + f(\lambda).$$

As both $\lambda \mapsto f(c\lambda)$ and $\lambda \mapsto f(c+\lambda)$ are complete Bernstein functions, the assertion follows from (12.11). The argument for $0 < c < 1$ is similar. □

Corollary 12.14. *Let $f \in \mathcal{CBF}$ and let $(A, \mathfrak{D}(A))$ and $(B, \mathfrak{D}(B))$ be two dissipative self-adjoint operators on $\mathfrak{H}$. If $\mathfrak{D}(B) \subset \mathfrak{D}(A)$, then $\mathfrak{D}(f(-B)) \subset \mathfrak{D}(f(-A))$.*

Proof. Since A and B are closed operators with $\mathfrak{D}(B) \subset \mathfrak{D}(A)$ we know from an inequality due to Hörmander, see [379, Chapter II.6, Theorem 2], that there exists a constant $c > 0$ such that

$$\|Au\| \le c \left(\|Bu\| + \|u\|\right), \quad u \in \mathfrak{D}(B).$$

Using the triangle inequality and dissipativity, (12.5), we find on $\mathfrak{D}(B)$

$$\|Au\| \le c \left(\|(B - \epsilon)u\| + (1 + \epsilon)\|u\|\right) \le c \left(\|(B - \epsilon)u\| + \frac{(1 + \epsilon)}{\epsilon}\|(B - \epsilon)u\|\right).$$

With a suitable constant $c_\epsilon > 0$ we can recast this inequality as

$$(-A)^2 \le_{\text{qf}} c_\epsilon^2 (\epsilon - B)^2.$$

Setting $f = g$ and $\alpha = 1/2$ in Corollary 7.15 (iii), we see that $h(\lambda) := f^2(\sqrt{\lambda})$ is again a complete Bernstein function. By spectral calculus and Theorem 12.12 we get

$$f^2(-A) = f^2\left(\sqrt{(-A)^2}\right) \le_{\text{qf}} f^2\left(\sqrt{c_\epsilon^2 (\epsilon - B)^2}\right) = f^2(c_\epsilon (\epsilon - B)),$$

hence, $\|f(-A)u\| \le \|f(c_\epsilon (\epsilon - B))u\|$. Therefore, the claim follows from Corollary 12.13. □

If we use the complete Bernstein functions $f(\lambda) = \lambda^\alpha$, $0 < \alpha < 1$, we obtain the famous Heinz–Kato-inequality.

Corollary 12.15. *Let $(A, \mathfrak{D}(A))$, $(B, \mathfrak{D}(B))$ be two dissipative self-adjoint operators on $\mathfrak{H}$. Then*

$$\|(-A)^{1/2}u\| \le \|(-B)^{1/2}u\|, \quad u \in \mathfrak{D}\left((-B)^{1/2}\right)$$

implies for all $\alpha \in (0, 1)$

$$\|(-A)^{\alpha/2}u\| \le \|(-B)^{\alpha/2}u\|, \quad u \in \mathfrak{D}\left((-B)^{\alpha/2}\right).$$

In the remaining part of this section we will prove that the family of complete Bernstein functions and the family of positive operator monotone functions coincide. Recall that $f \in P_n$ if for all positive semi-definite symmetric matrices $A, B \in \mathbb{R}^{n \times n}$, $B \leq_{\mathrm{qf}} A$ implies that $f(B) \leq_{\mathrm{qf}} f(A)$. Let $\alpha_1, \ldots, \alpha_n$ be the (non-negative) eigenvalues of A and denote by $D[\alpha_j]$ the $n \times n$ diagonal matrix with diagonal entries $\alpha_1, \ldots, \alpha_n$. By choosing a suitable orthonormal basis we can assume that $A = D[\alpha_j]$; then $f(A) = D[f(\alpha_j)]$. Let $\beta_1, \ldots, \beta_n$ be the eigenvalues of B. There exists an orthogonal $n \times n$ matrix Q such that $B = Q^\top D[\beta_j]Q$ and $f(B) = Q^\top D[f(\beta_j)]Q$. The conditions $\langle Bx, x \rangle \leq \langle Ax, x \rangle$ and $\langle f(B)x, x \rangle \leq \langle f(A)x, x \rangle$, $x \in \mathbb{R}^n$, can be equivalently restated as

$$\sum_{j=1}^{n}(Qx)_j^2 \beta_j \leq \sum_{j=1}^{n} x_j^2 \alpha_j, \quad x \in \mathbb{R}^n,$$

and

$$\sum_{j=1}^{n}(Qx)_j^2 f(\beta_j) \leq \sum_{j=1}^{n} x_j^2 f(\alpha_j), \quad x \in \mathbb{R}^n,$$

respectively.

Define the weighted norm $\|x\|_{(\alpha)} := (\sum_{j=1}^{n} x_j^2 \alpha_j)^{1/2}$ and the corresponding operator norm

$$\|Q\|_{(\alpha, \beta)} := \sup_{x \neq 0} \frac{\|Qx\|_{(\beta)}}{\|x\|_{(\alpha)}}.$$

Then $f \in P_n$ if, and only if, for every orthogonal $n \times n$ matrix Q and all $x \in \mathbb{R}^n$,

$$\|Q\|_{(\alpha, \beta)} \leq 1 \quad \text{implies that} \quad \sum_{j=1}^{n}(Qx)_j^2 f(\beta_j) \leq \sum_{j=1}^{n} x_j^2 f(\alpha_j). \qquad (12.18)$$

We say that a function $f : (0, \infty) \to \mathbb{R}$ is in the set M_n, $n \in \mathbb{N}$, if for all distinct $\lambda_1, \ldots, \lambda_{2n} > 0$ and any choice of $a_1, \ldots, a_{2n} \in \mathbb{R}$ with $\sum_{j=1}^{2n} a_j = 0$

$$\sum_{j=1}^{2n} a_j \frac{\lambda_j t - 1}{\lambda_j + t} \geq 0 \quad \text{for all } t > 0 \text{ implies that} \quad \sum_{j=1}^{2n} a_j f(\lambda_j) \geq 0. \qquad (12.19)$$

Clearly, $M_n \supset M_{n+1}$. Since

$$\sum_{j=1}^{2n} a_j \frac{\lambda_j t - 1}{\lambda_j + t} = \sum_{j=1}^{2n} a_j \left(\left(t + \frac{1}{t} \right) \frac{\lambda_j}{t + \lambda_j} - \frac{1}{t} \right) = \left(t + \frac{1}{t} \right) \sum_{j=1}^{2n} a_j \frac{\lambda_j}{t + \lambda_j},$$

we have $f \in M_n$ if, and only if, for all distinct $\lambda_1, \ldots, \lambda_{2n} > 0$ and any choice of $a_1, \ldots, a_{2n} \in \mathbb{R}$ with $\sum_{j=1}^{2n} a_j = 0$

$$\sum_{j=1}^{2n} a_j \frac{\lambda_j}{t + \lambda_j} \geq 0 \quad \text{for all } t > 0 \text{ implies that} \quad \sum_{j=1}^{2n} a_j f(\lambda_j) \geq 0. \qquad (12.20)$$

We explain now yet another way to characterize when $f \in M_n$. Denote by $\Pi(n)$ the family of polynomials of degree less than or equal to n. For distinct $\lambda_1, \ldots, \lambda_{2n} > 0$ put $\pi(t) := \prod_{j=1}^{2n}(t + \lambda_j)$ and observe that $\pi'(-\lambda_k) = \prod_{j \neq k}(\lambda_j - \lambda_k)$. For any $p \in \Pi(2n - 1)$ set

$$a_k = a_k(p) := \frac{p(-\lambda_k)}{\pi'(-\lambda_k)\lambda_k}, \quad k = 1, \ldots, 2n, \tag{12.21}$$

and let $a = a(p) = (a_k(p))_{1 \leq k \leq 2n} \in \mathbb{R}^{2n}$. Then

$$\frac{p(t)}{\pi(t)} = \sum_{j=1}^{2n} a_j \frac{\lambda_j}{t + \lambda_j} \tag{12.22}$$

holds true and shows that $p \mapsto a(p)$ defines a linear bijection from $\Pi(2n - 1)$ onto $\mathbb{R}^{2n}$. By observing that $p(0) = 0$ is equivalent to $\sum_{j=1}^{2n} a_j(p) = 0$, it follows from (12.20) that $f \in M_n$ if, and only if, for all $p \in \Pi(2n - 1)$ with $p(0) = 0$,

$$p(t) \geq 0 \text{ for all } t > 0 \text{ implies that } \sum_{j=1}^{2n} a_j(p) f(\lambda_j) \geq 0. \tag{12.23}$$

Lemma 12.16. *For every $n \in \mathbb{N}$, $P_{n+1}^+ \subset M_n^+$.*

Proof. Every polynomial $p \in \Pi(2n - 1)$ with $p(0) = 0$ and $p(t) \geq 0$ for all $t > 0$ can be written as $p(t) = tq_1^2(t) + q_2^2(t)$ where $q_1, q_2 \in \Pi(n - 1)$ and $q_2(0) = 0$, cf. Akhiezer [2, p. 77]. Because of the linearity of the map (12.21) it suffices to consider two cases:

(i) $p(t) = tq^2(t), \quad q \in \Pi(n - 1)$;

(ii) $p(t) = q^2(t), \quad q \in \Pi(n - 1), \ q(0) = 0$.

Relabel $\lambda_1, \lambda_2, \ldots, \lambda_{2n} > 0$ as $0 < \beta_1 < \alpha_1 < \beta_2 < \cdots < \beta_n < \alpha_n$. Then $\pi'(-\beta_j) > 0$ and $\pi'(-\alpha_j) < 0$, $j = 1, 2, \ldots, n$.
   (i) Given a polynomial $q \in \Pi(n - 1)$ the partial fractions representation (12.22) of the polynomial $tq^2(t)$ reads

$$\frac{tq^2(t)}{\pi(t)} = -\sum_{j=1}^{n} y_j^2 \frac{\beta_j}{t + \beta_j} + \sum_{j=1}^{n} x_j^2 \frac{\alpha_j}{t + \alpha_j} \tag{12.24}$$

where

$$y_j = \frac{q(-\beta_j)}{(\pi'(-\beta_j))^{1/2}}, \quad x_j = \frac{q(-\alpha_j)}{(-\pi'(-\alpha_j))^{1/2}}, \quad j = 1, \ldots, n. \tag{12.25}$$

Observe that (12.25) defines two linear bijections: $\Phi : \mathbb{R}^n \to \Pi(n-1)$, $\Phi(x) = q$ and $\Psi : \Pi(n-1) \to \mathbb{R}^n$, $\Psi(q) = y$. Let $Q := \Psi \circ \Phi$ be the composition of these two maps. Then Q is a linear bijection from $\mathbb{R}^n$ to $\mathbb{R}^n$. Moreover, for $q = \Phi(x)$ and $y = Qx$, the partial fractions representation (12.24) is valid. By taking $t = 0$ in (12.24) it follows that

$$-\sum_{j=1}^n y_j^2 + \sum_{j=1}^n x_j^2 = 0,$$

implying that Q is orthogonal. By multiplying (12.24) by t and letting $t \to \infty$ we see that $-\sum_{j=1}^n y_j^2 \beta_j + \sum_{j=1}^n x_j^2 \alpha_j \geq 0$, i.e. $\|Q\|_{(\alpha,\beta)} \leq 1$. Let $f \in P_n^+$. By (12.18),

$$-\sum_{j=1}^n y_j^2 f(\beta_j) + \sum_{j=1}^n x_j^2 f(\alpha_j) \geq 0 \tag{12.26}$$

for all $x \in \mathbb{R}^n$ and $y = Qx$, which is precisely (12.23). So far we have only used that $f \in P_n^+$ but not the stronger assumption $f \in P_{n+1}^+$.

(ii) For $q \in \Pi(n-1)$ with $q(0) = 0$ the partial fractions representation (12.22) of the polynomial $q^2(t)$ is

$$\frac{q^2(t)}{\pi(t)} = \sum_{j=1}^n y_j^2 \frac{\beta_j}{t+\beta_j} - \sum_{j=1}^n x_j^2 \frac{\alpha_j}{t+\alpha_j} \tag{12.27}$$

where

$$y_j = \frac{q(-\beta_j)}{(\pi'(-\beta_j)\beta_j)^{1/2}}, \quad x_j = \frac{q(-\alpha_j)}{(-\pi'(-\alpha_j)\alpha_j)^{1/2}}, \quad j = 1,\dots,n. \tag{12.28}$$

In order to verify (12.23) for $p(t) = q^2(t)$ we have to show that

$$-\sum_{j=1}^n x_j^2 f(\alpha_j) + \sum_{j=1}^n y_j^2 f(\beta_j) \geq 0. \tag{12.29}$$

Let $0 < \alpha_0 < \beta_1$, $\beta_{n+1} > \alpha_n$ and $\tilde{\pi}(t) = (t+\alpha_0)\pi(t)(t+\beta_{n+1})$. Then

$$\frac{t}{t+\alpha_0}\frac{q^2(t)}{\pi(t)}\frac{\beta_{n+1}}{t+\beta_{n+1}} = \frac{\beta_{n+1}t q^2(t)}{\tilde{\pi}(t)}$$

$$= -\sum_{j=0}^n \tilde{x}_j^2 \frac{\alpha_j}{t+\alpha_j} + \sum_{j=1}^{n+1} \tilde{y}_j^2 \frac{\beta_j}{t+\beta_j} \tag{12.30}$$

with $\tilde{x}_j$, $j = 0,\dots,n$, and $\tilde{y}_j$, $j = 1,\dots,n+1$, given by the formulae analogous to (12.28). Let $f \in P_{n+1}^+$. From part (i) of the proof we conclude that

$$-\tilde{x}_0^2 f(\alpha_0) - \sum_{j=1}^n \tilde{x}_j^2 f(\alpha_j) + \sum_{j=1}^n \tilde{y}_j^2 f(\beta_j) + \tilde{y}_{n+1}^2 f(\beta_{n+1}) \geq 0. \tag{12.31}$$

Comparing the rational functions in (12.27) and (12.30) we get for $j = 1, \ldots, n$,

$$\lim_{\alpha_0 \to 0} \tilde{x}_j = x_j, \quad \text{and} \quad \lim_{\beta_{n+1} \to \infty} \tilde{y}_j = y_j.$$

Since $q(0) = 0$,

$$\tilde{x}_0^2 = \frac{q^2(-\alpha_0)}{\pi(-\alpha_0)} \frac{\alpha_0}{\alpha_0 - \beta_{n+1}} = O(\alpha_0^2) \quad \text{as} \quad \alpha_0 \to 0$$

and since the degree of q is less than or equal to $n - 1$,

$$\tilde{y}_{n+1}^2 = \frac{\beta_{n+1}}{\beta_{n+1} - \alpha_0} \frac{q^2(-\beta_{n+1})}{\pi(-\beta_{n+1})} = O(\beta_{n+1}^{-2}) \quad \text{as} \quad \beta_{n+1} \to \infty.$$

All $f \in P_{n+1}^+$ are non-decreasing and non-negative, hence $\lim_{\alpha_0 \to 0} \alpha_0^2 f(\alpha_0) = 0$ implying that $\lim_{\alpha_0 \to 0} \tilde{x}_0^2 f(\alpha_0) = 0$. If also $\lim_{\beta_{n+1} \to \infty} f(\beta_{n+1})/\beta_{n+1}^2 = 0$, then we can let $\alpha_0 \to 0$ and $\beta_{n+1} \to \infty$ in (12.31) to get (12.29).

Since $P_{n+1}^+ \subset P_2^+$, we can use formulae (12.26) and (12.25) of part (i) of the proof with $n = 2$, $\beta_1 = 1$, $\alpha_1 = 2$, $\beta_2 = \beta$, $\alpha_2 = 2\beta$ and $q(t) = t + 2\beta$ to arrive at

$$-\frac{2\beta - 1}{\beta - 1} f(1) - \frac{\beta}{(\beta - 1)(\beta - 2)} f(\beta) + \frac{2\beta - 2}{\beta - 2} f(2) \geq 0.$$

Multiplying with $(\beta - 1)(\beta - 2)/\beta^3$ and rearranging yields

$$0 \leq \frac{f(\beta)}{\beta^2} \leq -\frac{(2\beta - 1)(\beta - 2)}{\beta^3} f(1) + \frac{2(\beta - 1)^2}{\beta^3} f(2).$$

Since the right-hand side tends to 0 as $\beta \to \infty$, we find $\lim_{\beta \to \infty} f(\beta)/\beta^2 = 0$. □

It remains to verify that $\bigcap_{n \in \mathbb{N}} M_n^+ \subset \mathcal{CBF}$. This is now mainly a functional analytic argument.

Theorem 12.17. *The families of complete Bernstein and positive operator monotone functions coincide:*

$$\mathcal{CBF} = \bigcap_{n \in \mathbb{N}} P_n^+.$$

In particular, $0 \leq_{qf} (-A) \leq_{qf} (-B)$ implies $0 \leq_{qf} f(-A) \leq_{qf} f(-B)$ if, and only if, $f \in \bigcap_{n \in \mathbb{N}} P_n^+$.

Proof of Theorem 12.17. Combining Theorem 12.12 and Lemma 12.16 shows that

$$\mathcal{CBF} \subset \bigcap_{n \in \mathbb{N}} P_n^+ \subset \bigcap_{n \in \mathbb{N}} M_n^+.$$

It is, therefore, enough to show that the right-hand side is contained in $\mathcal{CBF}$.

Write $\phi_\lambda(t) := (\lambda t - 1)/(\lambda + t)$ and consider the set

$$G := \left\{ g \ : \ g(t) = \sum_{\text{finite}} a_j \, \phi_{\lambda_j}(t), \ a_j \in \mathbb{R}, \ \lambda_j > 0, \ \sum_{\text{finite}} a_j = 0 \right\}.$$

Clearly, $G \subset C[0, \infty]$ and for each $f \in \bigcap_{n \in \mathbb{N}} M_n^+$

$$\Lambda_f : G \to \mathbb{R}, \quad \Lambda_f \left(\sum_{\text{finite}} a_j \, \phi_{\lambda_j} \right) := \sum_{\text{finite}} a_j \, f(\lambda_j)$$

defines a linear functional on G. The defining property (12.19) of the sets M_n just says that Λ_f is positive. Moreover,

$$g_0(t) := 5\big[\phi_2(t) - \phi_1(t)\big] = \frac{5(t^2 + 1)}{t^2 + 3t + 2} \geq \frac{5(t^2 + 1)}{t^2 + 3t^2 + 3 + 2} \geq 1$$

is an element from G. Let us show that Λ_f is bounded. Pick $g \in G$ and define $s := \sup g$. Obviously, $g(t) \leq s^+ \cdot g_0(t)$ where $s^+ := \max\{s, 0\}$. Thus we see that for all $h \in C^+[0, \infty]$

$$\Lambda_f(g) \leq \Lambda_f(s^+ \cdot g_0) = s^+ \Lambda_f(g_0) \leq \sup_{t>0} |g(t) + h(t)| \cdot \Lambda_f(g_0).$$

Since the expression $g \mapsto \sup_{t>0} |g(t) + h(t)| \cdot \Lambda_f(g_0)$ appearing on the right-hand side is a seminorm on $C[0, \infty]$, we can use the Hahn–Banach theorem to find an extension of Λ_f as a continuous linear functional on $C[0, \infty]$ bounded by the same seminorm. Now evaluate Λ_f for $-h$. Then

$$\Lambda_f(-h) \leq \|h - h\|_\infty \Lambda_f(g_0) = 0$$

which shows that Λ_f is a bounded, positive linear functional on $C[0, \infty]$.

We can now appeal to the Riesz representation theorem to get

$$\Lambda_f(h) = \int_{[0,\infty]} h(t)\, \rho(dt), \quad h \in C[0, \infty].$$

for some finite positive measure ρ on $[0, \infty]$. Thus, for $g(t) = \phi_\lambda(t) - \phi_1(t)$,

$$f(\lambda) - f(1) = \Lambda_f(g) = \Lambda_f(\phi_\lambda) - \Lambda_f(\phi_1).$$

This shows

$$f(\lambda) = \int_{[0,\infty]} \frac{\lambda t - 1}{\lambda + t}\, \rho(dt) + f(1) - \Lambda_f(\phi_1)$$

$$= \lambda \rho\{\infty\} + \int_{[0,\infty)} \frac{\lambda t - 1}{\lambda + t}\, \rho(dt) + f(1) - \Lambda_f(\phi_1).$$

Observe that the kernel $\lambda \mapsto (\lambda t - 1)/(\lambda + t)$ is monotonically increasing. Therefore, monotone convergence shows that $f(0+)$ exists. Since it is positive, we see that $\int_{[0,\infty)} t^{-1} \rho(dt) \le f(1) - \Lambda_f(\phi_1) < \infty$; in particular, $\rho\{0\} = 0$. As ρ is finite, this shows that

$$f(\lambda) = \alpha + \beta\lambda + \int_{(0,\infty)} \frac{\lambda t - 1}{\lambda + t}\, \rho(dt)$$

satisfies all conditions of Theorem 6.9, i.e. $f \in \mathcal{CBF}$.   □

Comments 12.18. *Section* 12.1: Standard references for the spectral theorem for normal and self-adjoint operators are the monographs by Yosida [379] and Kato [200]. Our exposition owes a lot to Akhiezer and Glazman [3] and, in particular, to Lax [239, Chapter 32]. Lax [239] is one of the few textbook-style presentations of the Doob and Koopman [101] and Lengyel [240] approach to the spectral theorem. The first proof of the spectral theorem (for bounded operators) is due to Hellinger [155] and Hahn [139]. These papers already contain a version of the Hellinger–Stone formula (12.9). The case of unbounded self-adjoint operators was treated by von Neumann [271] and Stone [343].

We follow Doob and Koopman [101], Lengyel [240] and the beautiful paper by Nevanlinna and Nieminen [275]; see also Akhiezer and Glazman [3] or Lax [239]. This approach works also for general, not necessarily spectrally negative, self-adjoint operators if one uses the version of Theorem 6.2 for $\mathcal{CBF}$-like functions preserving the upper half-plane – these functions are not necessarily positive on the positive real line – see Remark 6.18.

The Examples 12.5–12.7 also have an interpretation in connection with Bochner's subordination which will be discussed in Chapter 13. It turns out that in the Hilbert space setting the generator of the subordinate semigroup, A^f, f is a Bernstein function, can be identified with the operator $-f(-A)$ defined by spectral calculus. An interesting application is the subordination of Dirichlet forms: the subordinate Dirichlet form $(\mathcal{E}^f, \mathfrak{D}(\mathcal{E}^f))$ is generated by $-f(-A)$; using the spectral representation of $f(-A)$, the fact that a Bernstein function is non-decreasing and convex and Jensen's inequality for integrals, it is straightforward to show that

$$\mathcal{E}^f(u, u) \le \|u\|^2\, f\left(\frac{\mathcal{E}(u, u)}{\|u\|^2}\right) \quad \text{for all } u \in \mathfrak{D}(\mathcal{E}).$$

For related results we refer to Ôkura [284] and Section 13.4 below.

Section 12.2: The best reference on quadratic forms is still Kato [200], in particular Sections VI.2–6.

In order to define $B \le A$ using quadratic forms one does actually not need that the quadratic form is bounded from below $Q_B(u, u) \ge -c\langle u, u \rangle$; it is also not necessary to require that $A - B$ is densely defined, see Davies [88, Section 4.2]. Lemma 12.11

is from [88, Theorem 4.17] and [320, Lemma 5.19]. Theorem 12.12 appears for the first time in Heinz [154], our proof is from [320, Satz 5.20]. The Heinz–Kato inequality, Corollary 12.15, goes back to [154]; we give a different proof based on [88, Chapter 4.2] and [320].

The characterization of matrix monotone and operator monotone functions goes back to Löwner [248] who establishes a determinant criterion (in terms of divided differences) for matrix monotonicity. Section 6 of that paper also contains the representation of matrix monotone functions f of finite order by a discrete Stieltjes-like representation of the form $f(\lambda) = a + a_0\lambda + \sum_{j=1}^{n-1} a_j/(a_j + \lambda)$; the limit $n \to \infty$ is discussed and Löwner actually shows that a matrix monotone function (of all orders n) preserves the upper and lower half-planes. Löwner's technique is similar to the interpolation technique of Pick [293] – this paper is the only one quoted in [248] – but he seems to be unaware of the progress made by Nevanlinna on the integral representation of such functions, see Comments 6.22 and 7.24.

Löwner's paper had a huge impact on the literature and Theorem 12.17 is nowadays called *Löwner's theorem*. Several books are devoted to this topic, among them the monographs [100] by Donoghue, [55] and [56] by Bhatia, and the standard treatise by Horn and Johnson [170, 171]. Several quite different proofs of Löwner's theorem are known: Bendat and Sherman [29] use the Hamburger moment problem; Hansen and Pedersen [144] show, using extreme-point methods, that operator monotone functions have a Nevanlinna–Pick-type representation, see also Bhatia [55]. Both Hansen–Pedersen and Bhatia derive the integral representation formula for functions which preserve order relations of operators with spectrum in $[-1, 1]$; the case of operator monotone functions in the sense of Definition 12.10, i.e. for operators with spectrum in $[0, \infty)$, can be recovered by mapping $[-1, 1]$ to $[0, \infty)$ by an affine-linear transformation (which is itself operator monotone). It is possible to determine the extreme points of operator monotone functions (or complete Bernstein functions) directly: if f is operator monotone, then

$$f(\lambda) = \frac{\lambda f(\lambda) - \lambda f(1)}{\lambda - 1} + \frac{\lambda f(1) - f(\lambda)}{\lambda - 1} = g_1(\lambda) + g_2(\lambda), \quad \lambda > 0.$$

Using results from [55] it is possible to show that g_1, g_2 are again operator monotone and that operator monotone functions are two times continuously differentiable. Consider now the set of operator monotone functions with $f'(1) = 1$. This is a basis of the cone of all operator monotone functions. One can show that $|f''(1)| \le 2f'(1) = 2$ and $\alpha := -f''(1)/2 \in (0, 1)$. Since $g_1'(1) = f'(1) + f''(1)/2 = 1 - \alpha$ and $g_2'(1) = -f''(1)/2 = \alpha$, we find that $f(\lambda) = (1-\alpha)g_1(\lambda)/(1-\alpha) + \alpha g_2(\lambda)/\alpha$. If f is extremal, we conclude that $f(\lambda) = g_1(\lambda)/(1-\alpha) = g_2(\lambda)/\alpha$. This shows that the extremal points of the basis are necessarily of the form

$$f(\lambda) = \frac{f(1)\lambda}{\lambda + \frac{1-\alpha}{\alpha}}, \quad \lambda > 0, \ \alpha \in [0, 1].$$

The proof by Korányi [214], see also Sz.-Nagy's remarks [344] and their joint work [215], uses Hilbert space methods and the theory of reproducing kernel spaces. A nice presentation can be found in [100, Chapters X, XI] and [308, Addenda to Chapter 2]. Our proof is a slight simplification of Sparr's elementary approach [334] which itself is inspired by the theory of interpolation spaces; a related account is given by Ameur, Kaijser and Silvestrov [7]. Simon [330, Chapters 6,7 and Theorem 9.10] contains a new treatment of the Bendat–Sherman [29] proof as well as the extreme point argument of Hansen and Pedersen [144].

There is a deep relation between operator monotone functions and operator means. Ando [10] proves that there is a one-to-one relation between the class of operator monotone functions on $[0, \infty)$ with $f(1) = 1$ and the class of operator means. The latter are extremely useful in the theory of electrical networks, see e.g. Anderson and Trapp [8]. Other recent papers on operator means and operator monotone functions include [158] and [182]. Operator inequalities are an active area of research with many contributions by Ando, Furuta [126], Hansen and Pedersen [144], Uchiyama [350, 351, 352, 353, 356] and others. A recent monograph is Zhan [380].

Chapter 13

Subordination and Bochner's functional calculus

In this chapter we consider strongly continuous contraction semigroups of operators on a Banach space. Our focus will be on subordination of such semigroups in the sense of Bochner. We give a proof of Phillips' theorem describing the infinitesimal generator of the subordinate semigroup and discuss the related functional calculus for generators of semigroups and applications to functional inequalities. In the final section of this chapter we discuss the probabilistic counterpart of subordination and use it to obtain eigenvalue estimates for subordinate processes.

13.1 Semigroups and subordination in the sense of Bochner

In this section we will briefly discuss a method to generate new semigroups from a given one. We assume that the reader is familiar with the theory of semigroups on a Banach space $(\mathfrak{B}, \|\cdot\|)$; our standard references are the monographs by Davies [88], Pazy [286] and Yosida [379].

Throughout this section $(\mathfrak{B}, \|\cdot\|)$ denotes a Banach space, $(\mathfrak{B}^*, \|\cdot\|_*)$ its topological dual and $\langle u, \phi \rangle$ stands for the dual pairing between $u \in \mathfrak{B}$ and $\phi \in \mathfrak{B}^*$. Recall that a *strongly continuous* or C_0-*semigroup* of operators on $\mathfrak{B}$ is a family of bounded linear operators $T_t : \mathfrak{B} \to \mathfrak{B}, t \geq 0$, which is strongly continuous (at $t = 0$), i.e.

$$\lim_{t \to 0} \|T_t u - u\| = 0 \quad \text{for all } u \in \mathfrak{B},$$

and has the semigroup property

$$T_t T_s = T_s T_t = T_{t+s}, \quad \text{for all } s, t \geq 0.$$

If $\|T_t u\| \leq \|u\|$ for all $u \in \mathfrak{B}$ and $t \geq 0$, we call $(T_t)_{t \geq 0}$ a C_0-*contraction semigroup*. The (infinitesimal) generator of the semigroup is the operator

$$Au := \text{strong-} \lim_{t \to 0} \frac{T_t u - u}{t},$$

$$\mathfrak{D}(A) := \left\{ u \in \mathfrak{B} \ : \ \lim_{t \to 0} \frac{T_t u - u}{t} \ \text{exists as strong limit} \right\}. \tag{13.1}$$

Since A is the (strong) right derivative of T_t at zero, and since $(T_t)_{t \geq 0}$ is a semigroup, we have

$$T_t u - u = \int_0^t T_s Au \, ds = \int_0^t A T_s u \, ds, \quad u \in \mathfrak{D}(A), \ t > 0. \tag{13.2}$$

This shows, in particular, the following elementary but useful inequality

$$\|T_t u - u\| \leq \min\{t\,\|Au\|, 2\|u\|\}, \quad u \in \mathfrak{D}(A), \ t > 0. \tag{13.3}$$

The generator $(A, \mathfrak{D}(A))$ of a C_0-contraction semigroup is a densely defined linear operator which is *dissipative* and closed. Dissipative means that one of the following equivalent conditions is satisfied

$$\operatorname{Re}\langle Au, \phi\rangle \leq 0 \quad \text{for all } u \in \mathfrak{D}(A), \ \phi \in F(u) \tag{13.4}$$

where $F(u) = \{\phi \in \mathfrak{B}^* : \|\phi\|^2 = \|u\|^2 = \langle u, \phi\rangle\}$,

$$\|\lambda u - Au\| \geq \lambda \|u\| \quad \text{for all } u \in \mathfrak{D}(A), \ \lambda > 0, \tag{13.5}$$

$$\|\zeta u - Au\| \geq \operatorname{Re}\zeta\,\|u\| \quad \text{for all } u \in \mathfrak{D}(A), \ \zeta \in \mathbb{C}, \ \operatorname{Re}\zeta > 0. \tag{13.6}$$

Generators are, in general, unbounded operators. By the Hille–Yosida theorem an operator A generates a C_0-contraction semigroup if, and only if, $\mathfrak{D}(A)$ is dense in $\mathfrak{B}$, A is dissipative and the range $(\lambda - A)(\mathfrak{D}(A)) = \mathfrak{B}$ for some (hence, all) $\lambda > 0$.

The resolvent set ϱ_A of a closed operator A consists of all $\zeta \in \mathbb{C}$ such that the operator $\zeta - A := \zeta\,\mathrm{id} - A$ has a bounded inverse denoted by $R_\zeta := (\zeta - A)^{-1}$; the family $(R_\zeta)_{\zeta \in \varrho_A}$ is the *resolvent* of A. It is well known that ϱ_A is open and that for C_0-contraction semigroups $\{\operatorname{Re}\zeta > 0\} \subset \varrho_A$; in particular, $(0, \infty) \subset \varrho_A$. The set $\sigma_A := \mathbb{C} \setminus \varrho_A$ is called the *spectrum* of A.

The resolvent operators satisfy the *resolvent estimate*

$$\|R_\lambda u\| \leq \frac{1}{\lambda}\|u\| \quad \text{for all } u \in \mathfrak{B}, \ \lambda > 0, \tag{13.7}$$

and the following *resolvent equation* holds for all $\zeta, z \in \varrho_A$

$$R_\zeta - R_z = (z - \zeta)\,R_z R_\zeta. \tag{13.8}$$

This follows exactly as in the Hilbert space setting, cf. page 179, and one sees immediately that $\zeta \mapsto \langle R_\zeta u, \phi\rangle$ is analytic for all $u \in \mathfrak{B}$ and $\phi \in \mathfrak{B}^*$.

There is a one-to-one correspondence between a C_0-contraction semigroup $(T_t)_{t \geq 0}$ and its generator $(A, \mathfrak{D}(A))$ which can formally be expressed by writing $T_t = e^{tA}$. If A is bounded, the exponential e^{tA} may be defined by the exponential series; in the general case one has to use the Yosida approximation, $A_\lambda := \lambda A R_\lambda$, $\lambda > 0$. By the resolvent estimate (13.7)

$$\|A_\lambda u\| = \lambda \|A R_\lambda u\| = \lambda \|\lambda R_\lambda u - u\| \leq 2\lambda \|u\|$$

which shows that A_λ is bounded. Therefore, $\exp(t A_\lambda)$ can be defined as an exponential series and one can show that

$$T_t u = \text{strong-}\lim_{\lambda \to \infty} e^{t A_\lambda} u.$$

In this sense, we may indeed write $T_t = e^{tA}$.

The following method to generate a new C_0-semigroup from a given one is due to S. Bochner. It uses vaguely continuous convolution semigroups introduced in Definition 5.1 and Bernstein functions which appear as Laplace exponents, cf. Theorem 5.2.

Proposition 13.1. *Let* $(T_t)_{t \geq 0}$ *be a C_0-contraction semigroup on the Banach space* $\mathfrak{B}$ *and let* $(\mu_t)_{t \geq 0}$ *be a vaguely continuous convolution semigroup of sub-probability measures on* $[0, \infty)$ *with corresponding Bernstein function* f. *Then the Bochner integral*

$$T_t^f u := \int_{[0,\infty)} T_s u\, \mu_t(ds), \quad t \geq 0, \tag{13.9}$$

defines again a C_0-contraction semigroup on the Banach space $\mathfrak{B}$.

Proof. Since $t \mapsto T_t u$ is strongly continuous, and since

$$\left\| \int_{[0,\infty)} T_s u\, \mu_t(ds) \right\| \leq \int_{[0,\infty)} \|T_s u\|\, \mu_t(ds) \leq \int_{[0,\infty)} \|u\|\, \mu_t(ds) \leq \|u\|,$$

the operators (13.9) are well defined and contractive. In particular, every T_t^f, $t \geq 0$, is a bounded linear operator and we find, using the semigroup property of $(T_t)_{t \geq 0}$, for all $s, t \geq 0$

$$
\begin{aligned}
T_s^f T_t^f u &= \int_{[0,\infty)} T_s^f T_r u\, \mu_t(dr) \\
&= \int_{[0,\infty)} \int_{[0,\infty)} T_\rho T_r u\, \mu_s(d\rho)\, \mu_t(dr) \\
&= \int_{[0,\infty)} \int_{[0,\infty)} T_{\rho+r} u\, \mu_s(d\rho)\, \mu_t(dr) \\
&= \int_{[0,\infty)} T_\sigma u\, \mu_s \star \mu_t(d\sigma).
\end{aligned}
$$

Since $(\mu_t)_{t \geq 0}$ is a convolution semigroup, this proves that $T_s^f T_t^f = T_{s+t}^f$. The strong continuity of $(T_t^f)_{t \geq 0}$ finally follows from

$$\|T_t^f u - u\| \leq \int_{[0,\infty)} \|T_s u - u\|\, \mu_t(ds) + \left(1 - \mu_t[0,\infty)\right) \|u\|$$

which tends to zero as $t \to 0$ since vague-$\lim_{t \to 0} \mu_t = \delta_0$ and $\lim_{t \to 0} \mu_t[0,\infty) = 1$, see the discussion following Definition 5.1. $\qquad\square$

Definition 13.2. Let $(T_t)_{t\geq 0}$ be a C_0-contraction semigroup on the Banach space $\mathfrak{B}$ and let $(\mu_t)_{t\geq 0}$ be a vaguely continuous convolution semigroup of sub-probability measures on $[0,\infty)$ with corresponding Bernstein function f. Then the semigroup $(T_t^f)_{t\geq 0}$ defined by (13.9) is called *subordinate* (in the sense of Bochner) to the semigroup $(T_t)_{t\geq 0}$ with respect to the Bernstein function f.

The infinitesimal generator of $(T_t^f)_{t\geq 0}$ will be denoted by $(A^f, \mathfrak{D}(A^f))$; it is called the *subordinate generator*.

From now on, we will indicate all operators related to the subordinate semigroup $(T_t^f)_{t\geq 0}$ by the superscript f.

Subordination of subordinate semigroups corresponds to the composition of Bernstein functions.

Lemma 13.3. *Let $(T_t)_{t\geq 0}$ be a C_0-contraction semigroup on the Banach space $\mathfrak{B}$ and let $(\mu_t)_{t\geq 0}$ and $(\nu_t)_{t\geq 0}$ be vaguely continuous convolution semigroups of sub-probability measures on $[0,\infty)$ with corresponding Bernstein functions f and g. Then*

$$(T_t^g)^f u = T_t^{f\circ g} u, \quad t\geq 0, \; u\in\mathfrak{B}.$$

Proof. From Theorem 5.27 we know that the sub-probability measures η_t given by the vague integral $\int_{[0,\infty)} \nu_s\, \mu_t(ds)$ form a vaguely continuous convolution semigroup with corresponding Bernstein function $f\circ g$. Therefore,

$$\begin{aligned}
(T_t^g)^f u &= \int_{[0,\infty)}\int_{[0,\infty)} T_r u\, \nu_s(dr)\, \mu_t(ds)\\
&= \int_{[0,\infty)} T_r u \int_{[0,\infty)} \nu_s(dr)\, \mu_t(ds)\\
&= \int_{[0,\infty)} T_r u\, \eta_t(dr) = T_t^{f\circ g} u. \qquad\square
\end{aligned}$$

Remark 13.4. It is instructive to consider Bochner's subordination in the context of the spectral calculus of a dissipative operator $(A, \mathfrak{D}(A))$ on the Hilbert space $\mathfrak{H}$. Denote by $E(d\lambda)$ the spectral resolution of A and by $T_t u = \int_{(-\infty,0]} e^{\lambda t}\, E(d\lambda)u$ the C_0-semigroup generated by A, cf. Example 12.5.

Let $f\in\mathcal{BF}$ and let $(\mu_t)_{t\geq 0}$ be the convolution semigroup associated with it. Then we find

$$\begin{aligned}
T_t^f u &= \int_{[0,\infty)}\int_{(-\infty,0]} e^{\lambda s}\, E(d\lambda)u\, \mu_t(ds)\\
&= \int_{(-\infty,0]}\int_{[0,\infty)} e^{\lambda s}\, \mu_t(ds)\, E(d\lambda)u\\
&= \int_{(-\infty,0]} e^{-tf(-\lambda)}\, E(d\lambda)u = e^{-tf(-A)}u.
\end{aligned}$$

This implies that $(A^f, \mathfrak{D}(A^f))$ coincides with $(-f(-A), \mathfrak{D}(f(-A)))$ and that

$$A^f u = -f(-A)u = -au + bAu + \int_{(0,\infty)} (T_t u - u)\, \mu(dt),$$

as we have seen in Example 12.6.

The formula for A^f from Remark 13.4 remains valid in Banach spaces. This result is due to R. S. Phillips [292, Theorem 4.3].

As usual, we set $(A^0, \mathfrak{D}(A^0)) = (\mathrm{id}, \mathfrak{B})$, $A^k := A \circ A \circ \cdots \circ A$ (k times), and

$$\mathfrak{D}(A^k) := \{u \in \mathfrak{D}(A^{k-1}) : A^{k-1}u \in \mathfrak{D}(A)\}.$$

With this definition $(A^k, \mathfrak{D}(A^k))$ is a closed operator if $(A, \mathfrak{D}(A))$ is a closed operator.

Proposition 13.5. *Let $(T_t)_{t\geq 0}$ be a C_0-contraction semigroup on the Banach space $\mathfrak{B}$ with generator $(A, \mathfrak{D}(A))$ and let f be a Bernstein function. Then for all $k \in \mathbb{N}$ the set $\mathfrak{D}(A^k)$ is an operator core for the subordinate generator $(A^f, \mathfrak{D}(A^f))$.*

Proof. It is well known that a dense subset $\mathfrak{D} \subset \mathfrak{D}(A^f)$ is an operator core for the generator A^f if it is invariant under the semigroup, i.e. if $T_t^f \mathfrak{D} \subset \mathfrak{D}$, see e.g. Davies [88, Theorem 1.9]. Let us show that $T_t^f (\mathfrak{D}(A^k)) \subset \mathfrak{D}(A^k)$ for all $k \in \mathbb{N}$. Note that for all $u \in \mathfrak{D}(A^k)$

$$\text{strong-}\lim_{n\to\infty} \int_{[0,n)} T_s u\, \mu_t(ds) = \int_{[0,\infty)} T_s u\, \mu_t(ds) = T_t^f u, \quad u \in \mathfrak{D}(A^k).$$

Since $(A, \mathfrak{D}(A))$, hence $(A^k, \mathfrak{D}(A^k))$, is a closed operator, we find for all $m, n \in \mathbb{N}$ with $m < n$ and all $u \in \mathfrak{D}(A^k)$

$$\left\| A^k \int_{[m,n)} T_s u\, \mu_t(ds) \right\| = \left\| \int_{[m,n)} T_s A^k u\, \mu_t(ds) \right\|$$

$$\leq \int_{[m,n)} \|T_s A^k u\|\, \mu_t(ds)$$

$$\leq \int_{[m,n)} \|T_s\|\, \mu_t(ds)\, \|A^k u\|,$$

which shows that $(A^k \int_{[0,n)} T_s u\, \mu_t(ds))_{n\in\mathbb{N}}$ is for all $u \in \mathfrak{D}(A^k)$ a Cauchy sequence. The closedness of A^k now shows that $T_t^f u \in \mathfrak{D}(A^k)$ for all $u \in \mathfrak{D}(A^k)$. $\square$

Theorem 13.6 (Phillips). *Let $(T_t)_{t\geq 0}$ be a C_0-contraction semigroup on the Banach space $\mathfrak{B}$ with generator $(A, \mathfrak{D}(A))$ and let f be a Bernstein function with Lévy triplet*

(a, b, μ). *Denote by $(T_t^f)_{t \geq 0}$ and $(A^f, \mathfrak{D}(A^f))$ the subordinate semigroup and its infinitesimal generator. Then $\mathfrak{D}(A)$ is an operator core for $(A^f, \mathfrak{D}(A^f))$ and $A^f|_{\mathfrak{D}(A)}$ is given by*

$$A^f u = -au + bAu + \int_{(0,\infty)} (T_s u - u)\, \mu(ds), \quad u \in \mathfrak{D}(A). \tag{13.10}$$

The integral is understood as a Bochner integral.

Proof. That $\mathfrak{D}(A)$ is an operator core of A^f follows from Proposition 13.5. Using (13.3) we see that the integral term in formula (13.10) converges for all $u \in \mathfrak{D}(A)$,

$$\int_{(0,\infty)} \|T_s u - u\|\, \mu(ds) = \int_{(0,1)} \|T_s u - u\|\, \mu(ds) + \int_{[1,\infty)} \|T_s u - u\|\, \mu(dt)$$

$$\leq \int_{(0,1)} s\, \mu(ds)\, \|Au\| + 2\mu[1,\infty)\, \|u\|,$$

which means that the operator given by (13.10) is defined on $\mathfrak{D}(A)$.

Assume first that $f(0+) = 0$ and that $(T_t)_{t \geq 0}$ satisfies $\|T_t u\| \leq e^{-t\epsilon}\|u\|$ for all $t > 0$ and some $\epsilon > 0$.

We write $f \in \mathcal{BF}$ with $f(0+) = 0$ as

$$f(\lambda) = b\lambda + \int_{(0,\infty)} (1 - e^{-s\lambda})\, \mu(ds),$$

and denote the corresponding vaguely continuous convolution semigroup of probability measures by $(\mu_t)_{t \geq 0}$. By definition, $e^{-tf(\lambda)} = \int_{[0,\infty)} e^{-s\lambda} \mu_t(ds)$, and we see that

$$f_n(\lambda) := \int_{[0,\infty)} (1 - e^{-s\lambda})\, n\mu_{1/n}(ds) = \frac{1 - e^{-\frac{1}{n}f(\lambda)}}{\frac{1}{n}} \xrightarrow{n \to \infty} f(\lambda)$$

defines a sequence of Bernstein functions $f_n \in \mathcal{BF}$ approximating $f \in \mathcal{BF}$. Obviously, the Lévy triplet of f_n is $(0, 0, n\mu_{1/n})$ and we see from Corollary 3.9 that

$$\text{vague-}\lim_{n \to \infty} n\mu_{1/n} = \mu, \tag{13.11}$$

$$0 = \lim_{C \to \infty} \liminf_{n \to \infty} n\mu_{1/n}[C, \infty), \tag{13.12}$$

$$b = \lim_{c \to 0} \liminf_{n \to \infty} \int_{[0,c)} s\, n\mu_{1/n}(ds). \tag{13.13}$$

Throughout the proof we assume, for simplicity, that c and C are always continuity points of the measure μ and that the limits in c and C are taken along sequences of continuity points of μ. The proof of Corollary 3.9 shows that we may then replace $\liminf_n$ by $\lim_n$.

Since $s \mapsto T_s u - u$ is strongly continuous, the function $s \mapsto \langle T_s u - u, \phi \rangle$, $u \in \mathfrak{B}$, $\phi \in \mathfrak{B}^*$, is continuous and we conclude that for all continuity points $0 < c < C < \infty$

$$\lim_{n \to \infty} \int_{(c,C)} \langle T_s u - u, \phi \rangle \, n \mu_{1/n}(ds) = \int_{(c,C)} \langle T_s u - u, \phi \rangle \, \mu(ds).$$

Because of (13.3) we know that

$$|\langle T_s u - u, \phi \rangle| \le \|T_s u - u\| \|\phi\|_* \le \min\{s \|Au\|, \, 2\|u\|\} \|\phi\|_*$$

for all $u \in \mathfrak{D}(A)$ and $s > 0$. The function $1 \wedge s$ is μ-integrable, and by dominated convergence we see that for all $u \in \mathfrak{D}(A)$

$$\lim_{\substack{C \to \infty \\ c \to 0}} \lim_{n \to \infty} \int_{(c,C)} \langle T_s u - u, \phi \rangle \, n \mu_{1/n}(ds) = \int_{(0,\infty)} \langle T_s u - u, \phi \rangle \, \mu(ds). \quad (13.14)$$

For $u \in \mathfrak{D}(A)$ and all $c > 0$ we find

$$\int_{[0,c)} (T_s u - u) \, n \mu_{1/n}(ds) = \int_{[0,c)} \int_0^s T_r A u \, dr \, n \mu_{1/n}(ds)$$

$$= \int_{[0,c)} \int_0^s (T_r A u - A u) \, dr \, n \mu_{1/n}(ds)$$

$$+ \int_{[0,c)} s \, n \mu_{1/n}(ds) \, A u.$$

Because of (13.13), $\int_{[0,1)} s \, n \mu_{1/n}(ds)$, $n \in \mathbb{N}$, is a bounded sequence. Using the strong continuity of the semigroup we find for $c \in (0, 1)$

$$\sup_{n \in \mathbb{N}} \left\| \int_{[0,c)} \int_0^s (T_r A u - A u) \, dr \, n \mu_{1/n}(ds) \right\|$$

$$\le \sup_{r \le c} \|T_r A u - A u\| \sup_{n \in \mathbb{N}} \int_{[0,1)} s \, n \mu_{1/n}(ds) \xrightarrow{c \to 0} 0$$

which implies that

$$\text{strong-} \lim_{c \to 0} \lim_{n \to \infty} \int_{[0,c)} (T_s u - u) \, n \mu_{1/n}(ds) = b A u, \quad u \in \mathfrak{D}(A). \quad (13.15)$$

Finally, we get

$$\int_{[C,\infty)} (T_s u - u) \, n \mu_{1/n}(ds) = \int_{[C,\infty)} T_s u \, n \mu_{1/n}(ds) - n \mu_{1/n}[C, \infty) \, u.$$

Because of (13.12) the sequence $n\mu_{1/n}[1,\infty)$, $n \in \mathbb{N}$, is bounded. Therefore we get for $C > 1$

$$\sup_{n\in\mathbb{N}}\left\|\int_{[C,\infty)} T_s u \, n\mu_{1/n}(ds)\right\| \le \sup_{n\in\mathbb{N}}\int_{[C,\infty)} \|T_s u\| \, n\mu_{1/n}(ds)$$

$$\le \sup_{n\in\mathbb{N}}\int_{[C,\infty)} e^{-s\epsilon} \, n\mu_{1/n}(ds)\|u\|$$

$$\le e^{-C\epsilon} \sup_{n\in\mathbb{N}} n\mu_{1/n}[1,\infty) \xrightarrow{C\to\infty} 0,$$

and we conclude that

$$\text{strong-}\lim_{C\to\infty}\lim_{n\to\infty}\int_{[C,\infty)} (T_s u - u)\, n\mu_{1/n}(ds) = 0, \quad u \in \mathfrak{D}(A). \tag{13.16}$$

If we combine (13.14)–(13.16) we find that for all $u \in \mathfrak{D}(A)$ and $\phi \in \mathfrak{B}^*$

$$\langle A^f u, \phi \rangle = \lim_{n\to\infty} \langle n(T^f_{1/n}u - u), \phi \rangle = \left\langle bAu + \int_{(0,\infty)} (T_s u - u)\,\mu(ds),\ \phi \right\rangle$$

holds, and this implies (13.10).

If $f(0+) = 0$ and if $(T_t)_{t\ge 0}$ is a general C_0-contraction semigroup on $\mathfrak{B}$, then for any $\epsilon > 0$, $T_{\epsilon;t} := e^{-t\epsilon}T_t$ is a C_0-contraction semigroup which satisfies the additional assumption used above. It is not hard to check that the generator of this semigroup is $(A - \epsilon, \mathfrak{D}(A))$. From the first part of the proof we know that

$$(A - \epsilon)^f u = b(A - \epsilon)u + \int_{(0,\infty)} (e^{-\epsilon s}T_s u - u)\,\mu(ds), \quad u \in \mathfrak{D}(A).$$

Note that the expression on the right-hand side still makes sense if $\epsilon = 0$. For every $u \in \mathfrak{D}(A)$ we have

$$\left\| bAu + \int_{(0,\infty)} (T_s u - u)\,\mu(ds) - (A - \epsilon)^f u \right\|$$

$$= \left\| \epsilon bu + \int_{(0,\infty)} (T_s u - e^{-\epsilon s}T_s u)\,\mu(ds) \right\|$$

$$\le \epsilon b\,\|u\| + \int_{(0,\infty)} (1 - e^{-\epsilon s})\,\mu(ds)\,\|u\|$$

and this converges to 0 as $\epsilon \to 0$. This means that

$$\text{strong-}\lim_{\epsilon\to 0}(A - \epsilon)^f u = bAu + \int_{(0,\infty)} (T_s u - u)\,\mu(ds), \quad u \in \mathfrak{D}(A).$$

On the other hand,

$$\left\| \frac{T^f_{1/n}u - u}{\frac{1}{n}} - \frac{T^f_{\epsilon;1/n}u - u}{\frac{1}{n}} \right\| = \left\| \frac{T^f_{1/n}u - T^f_{\epsilon;1/n}u}{\frac{1}{n}} \right\|$$

$$= \left\| \int_{[0,\infty)} (T_s u - e^{-\epsilon s} T_s u)\, n\mu_{1/n}(ds) \right\|$$

$$\leq \int_{[0,\infty)} (1 - e^{-\epsilon s})\, n\mu_{1/n}(ds)\, \|u\|$$

$$= \frac{1 - e^{-\frac{1}{n} f(\epsilon)}}{\frac{1}{n}} \|u\| \xrightarrow{n\to\infty} f(\epsilon)\|u\| \xrightarrow{\epsilon\to0} 0.$$

This shows that $A^f u = $ strong-$\lim_{\epsilon\to0}(A - \epsilon)^f u$ and that A^f is, on $\mathfrak{D}(A)$, given by (13.10).

Finally consider $h \in \mathcal{BF}$ with $h(0+) = a \geq 0$. We write $h = f + a$ where $f \in \mathcal{BF}$ and $f(0+) = 0$. As above we denote by $(\mu_t)_{t\geq0}$ the convolution semigroup corresponding to f; then $\mu^a_t := e^{-ta}\mu_t, t \geq 0$, is the vaguely continuous convolution semigroup associated with h. Indeed, $\mathscr{L}\mu^a_t = e^{-ta}\mathscr{L}\mu_t = e^{-ta}e^{-tf}$. Therefore,

$$T^h_t u = \int_{[0,\infty)} T_s u\, \mu^a_t(ds) = e^{-ta} T^f_t u, \quad u \in \mathfrak{B},\ t \geq 0,$$

and thus $A^h = A^f - a$ id. This completes the proof of the theorem. □

Corollary 13.7. *Let $(A, \mathfrak{D}(A))$ be the generator of a C_0-contraction semigroup on the Banach space $\mathfrak{B}$ and let f be a Bernstein function. Then $\mathfrak{D}(A^f) = \mathfrak{D}(A)$ if, and only if, A is a bounded operator or if $b = \lim_{\lambda\to\infty} f(\lambda)/\lambda > 0$.*

If f is a bounded Bernstein function or if A is a bounded operator, the subordinate generator A^f is bounded, i.e. $\mathfrak{D}(A^f) = \mathfrak{B}$.

Proof. Using (13.3) we can estimate the integral term in Phillips' formula (13.10) in the following way:

$$\left\| \int_{(0,\infty)} (T_s u - u)\, \mu(ds) \right\| \leq \int_{(0,\epsilon)} \|T_s u - u\|\, \mu(ds) + \int_{[\epsilon,\infty)} \|T_s u - u\|\, \mu(dt)$$

$$\leq \int_{(0,\epsilon)} s\, \mu(ds)\, \|Au\| + 2\mu[\epsilon,\infty)\, \|u\|.$$

Therefore we get

$$\|A^f u\| \leq (a + d_\epsilon)\|u\| + (b + c_\epsilon)\|Au\|, \quad u \in \mathfrak{D}(A), \tag{13.17}$$

where $a, b \geq 0$ are from (13.10), $c_\epsilon = \int_{(0,\epsilon)} s\, \mu(ds)$ and $d_\epsilon = 2\mu[\epsilon,\infty)$.

Assume that $\mathfrak{D}(A) = \mathfrak{D}(A^f)$ and that $b = 0$. Since $(A^f, \mathfrak{D}(A))$ and $(A, \mathfrak{D}(A))$ are closed operators we have by a theorem of Hörmander, see [379, Chapter II.6, Theorem 2],

$$\|Au\| \leq c\left(\|A^f u\| + \|u\|\right), \quad u \in \mathfrak{D}(A),$$

with a suitable constant $c > 0$. Choosing ϵ in (13.17) so small that $c\epsilon < 1/(2c)$, we obtain

$$\|Au\| \leq c\left(\|A^f u\| + \|u\|\right) \leq \frac{1}{2}\|Au\| + c\left(d_\epsilon + a\right)\|u\|, \quad u \in \mathfrak{D}(A),$$

which shows that the operator A is bounded.

Conversely, if A is bounded $\mathfrak{D}(A) = \mathfrak{B}$ and by (13.17) A^f is bounded, therefore $\mathfrak{D}(A^f) = \mathfrak{B}$, too. If $b > 0$, we get from (13.10) using the estimate leading to (13.17)

$$\|A^f u\| \geq (b - c_\epsilon)\|Au\| - (a + d_\epsilon)\|u\|, \quad u \in \mathfrak{D}(A).$$

Pick $\epsilon > 0$ such that $b - c_\epsilon > 0$ and observe that then

$$\|Au\| \leq c\|A^f u\| + c'\|u\|, \quad u \in \mathfrak{D}(A),$$

holds for some constants $c, c' > 0$. Since $(A, \mathfrak{D}(A))$ is closed, this immediately implies the closedness of the operator $(A^f, \mathfrak{D}(A))$.

The second part of the assertion follows from (13.17) with $\epsilon = 1$, if A is bounded, and with $\epsilon = 0$ if f is bounded; in the latter case we used $b = 0$ and that μ is a finite measure, cf. Corollary 3.8 (v). $\qquad\square$

Corollary 13.8. *Let $f \in \mathcal{BF}$ and let $(T_t)_{t\geq 0}$ be a C_0-contraction semigroup on the Banach space $\mathfrak{B}$ with generator $(A, \mathfrak{D}(A))$. Then the following estimate holds*

$$\frac{\|A^f u\|}{\|u\|} \leq \frac{2e}{e-1}\, f\left(\frac{1}{2}\frac{\|Au\|}{\|u\|}\right), \quad u \in \mathfrak{D}(A),\, u \neq 0. \qquad (13.18)$$

For the fractional powers $(-A)^\alpha$, $\alpha \in (0,1)$, one has

$$\|(-A)^\alpha u\| \leq 4\|Au\|^\alpha \|u\|^{1-\alpha}, \quad u \in \mathfrak{D}(A). \qquad (13.19)$$

Proof. By Phillips' theorem, Theorem 13.6, we have

$$A^f u = -au + bAu + \int_{(0,\infty)} (T_t u - u)\,\mu(dt), \quad u \in \mathfrak{D}(A).$$

If we combine the elementary convexity estimate

$$\min\{1, ct\} \leq \frac{e}{e-1}\left(1 - e^{-ct}\right), \quad c, t \geq 0,$$

with the operator inequality (13.3),

$$\frac{\|T_t u - u\|}{\|u\|} \leq \min\left\{t\,\frac{\|Au\|}{\|u\|},\, 2\right\}, \quad u \in \mathfrak{D}(A),\, u \neq 0,\, t > 0,$$

we find for $c = \|Au\|/(2\,\|u\|)$ and $u \neq 0$

$$\frac{\|A^f u\|}{\|u\|} \leq a + b \frac{\|Au\|}{\|u\|} + \int_{(0,\infty)} \frac{\|T_t u - u\|}{\|u\|}\, \mu(dt)$$

$$\leq a + b \frac{\|Au\|}{\|u\|} + \int_{(0,\infty)} \min\left\{ t \frac{\|Au\|}{\|u\|}, 2 \right\} \mu(dt)$$

$$\leq a + b \frac{\|Au\|}{\|u\|} + \frac{2e}{e-1} \int_{(0,\infty)} \left(1 - \exp\left[-\frac{t}{2}\frac{\|Au\|}{\|u\|}\right]\right) \mu(dt)$$

$$\leq \frac{2e}{e-1}\, f\left(\frac{1}{2}\frac{\|Au\|}{\|u\|}\right).$$

Note that $2e/(e-1) \leq 4$; if we take $f(\lambda) = \lambda^\alpha$, $\alpha \in (0,1)$, (13.18) becomes (13.19). $\qquad\square$

Corollary 13.9. *Let $f \in \mathcal{BF}$ and let $(T_t^A)_{t\geq 0}$, $(T_t^B)_{t\geq 0}$ be C_0-contraction semigroups on the Banach space $\mathcal{B}$ with generators $(A, \mathfrak{D}(A))$ and $(B, \mathfrak{D}(B))$, respectively. Assume that the operators A, B commute. Then the following estimate holds for $u \in \mathfrak{D}(A) \cap \mathfrak{D}(B)$, $u \neq 0$,*

$$\frac{\|A^f u - B^f u\|}{\|u\|} \leq \frac{2e}{e-1}\, \varphi\left(\frac{1}{2}\frac{\|Au - Bu\|}{\|u\|}\right) \tag{13.20}$$

where $\varphi(\lambda) = f(\lambda) - f(0+)$.

Proof. By Phillips' formula (13.10) we see for $u \in \mathfrak{D}(A) \cap \mathfrak{D}(B)$

$$A^f u - B^f u = b(A - B)u + \int_{(0,\infty)} (T_t^A u - T_t^B u)\, \mu(dt).$$

From the theory of operator semigroups, see e.g. [286, Chapter I, Theorem 2.6, p. 6], we know that

$$T_t^A u - T_t^B u = \int_0^t \frac{d}{ds}(T_{t-s}^B T_s^A)u\, ds = \int_0^t T_{t-s}^B (A - B)T_s^A u\, ds.$$

Since B and A commute, so do the semigroups generated by them, and we find

$$\|T_t^A u - T_t^B u\| \leq \int_0^t \|T_{t-s}^B T_s^A (A - B)u\|\, ds \leq t\|(A - B)u\|$$

on $\mathfrak{D}(A) \cap \mathfrak{D}(B)$. On the other hand, we have always $\|T_t^A u - T_t^B u\| \leq 2\,\|u\|$. From now onwards we can argue exactly as in the proof of Corollary 13.8. $\qquad\square$

Without proof we state the following result on the spectrum of a subordinate generator. It is again from Phillips [292, Theorem 4.4].

Theorem 13.10. *Let $(T_t)_{t\geq 0}$ be a C_0-contraction semigroup on the Banach space $\mathfrak{B}$ with generator $(A, \mathfrak{D}(A))$ and let f be a Bernstein function. Then $\sigma_{A^f} \supset -f(-\sigma_A)$.*

The following improvement of Theorem 13.10 is due to Hirsch [163, Théorème 16] and [162, pp. 195–196]. The *extended spectrum* of an operator A on $\mathfrak{B}$ is the set $\bar{\sigma}_A$ contained in the one-point compactification $\mathbb{C} \cup \{\infty\}$ of $\mathbb{C}$ such that $\bar{\sigma}_A = \sigma_A$ if A is densely defined and bounded and $\bar{\sigma}_A = \sigma_A \cup \{\infty\}$ otherwise. Further, we extend $f \in \mathcal{CBF}$ to $\mathbb{C} \setminus (0, \infty) \cup \{\infty\}$ by setting

$$f(0) := f(0+) \in [0, \infty) \quad \text{and} \quad f(\infty) := \lim_{\lambda \to \infty} f(\lambda) \in (0, \infty].$$

Theorem 13.11. *Let $(T_t)_{t\geq 0}$ be a C_0-contraction semigroup on the Banach space $\mathfrak{B}$ with generator $(A, \mathfrak{D}(A))$ and let f be a complete Bernstein function. Then it holds that $\bar{\sigma}_{A^f} = -f(-\bar{\sigma}_A)$.*

Remark 13.12. If $(T_t)_{t\geq 0}$ gives rise to a transition function of a Markov process $(X_t)_{t\geq 0}$, i.e. if

$$T_t u(x) = P_t u(x) = \mathbb{E}_x u(X_t), \quad u \in \mathcal{B}_b(E),$$

then Bochner's subordination has a probabilistic interpretation which parallels the more specialized situation in Remark 5.28 (ii).

We know from Definition 5.4 and Proposition 5.5 that every $f \in \mathcal{BF}$ defines a convolution semigroup $(\mu_t)_{t\geq 0}$ of sub-probability measures on $[0, \infty)$ or, equivalently, a (killed) subordinator $\widehat{S} = (\widehat{S}_t)_{t\geq 0}$. Without loss of generality we can assume that the processes X and $\widehat{S}$ are defined on the same probability space and that they are independent. It is not difficult to check that the *subordinate process*

$$X_t^f(\omega) := X_{\widehat{S}_t}(\omega) := X_{\widehat{S}_t(\omega)}(\omega)$$

is again a Markov process. Because of independence, the associated operator semigroup is given by

$$P_t^f u(x) = \mathbb{E}_x u(X_{\widehat{S}_t}) = \int_{[0,\infty)} \mathbb{E}_x u(X_s) \, \mathbb{P}_0(\widehat{S}_t \in ds)$$

$$= \int_{[0,\infty)} P_s u(x) \, \mu_t(ds).$$

Since $T_t = P_t$, we have $T_t^f = P_t^f$, and Bochner's subordination can be interpreted as a stochastic time change with respect to an independent subordinator.

An interesting special case is the subordination of Lévy processes.

Example 13.13. A stochastic process $(X_t)_{t\geq0}$ with values in $\mathbb{R}^d$ is called a *Lévy process* if it has independent and stationary increments and if almost all sample paths $t \mapsto X_t(\omega)$ are right-continuous and have left limits. It is well known that the transition function of a Lévy process is characterized by its Fourier transform which is of the form

$$\mathbb{E}_0 e^{i\xi X_t} = e^{-t\psi(\xi)}, \quad \xi \in \mathbb{R}^d.$$

The characteristic exponent $\psi : \mathbb{R}^d \to \mathbb{C}$ is continuous and *negative definite* in the sense of Schoenberg, cf. Chapter 4, in particular (4.8). Every continuous and negative definite function is uniquely determined by its *Lévy–Khintchine formula*, see (4.9) in Theorem 4.15

Denote by $\widehat{S} = (\widehat{S}_t)_{t\geq0}$ and $(\mu_t)_{t\geq0}$ the (killed) subordinator and the convolution semigroup given by the Bernstein function f. Because of independence, we see

$$\mathbb{E} e^{i\xi X_t^f} = \int_{[0,\infty)} \mathbb{E} e^{i\xi X_s} \mu_t(ds) = \int_{[0,\infty)} e^{-s\psi(\xi)} \mu_t(ds) = e^{-tf(\psi(\xi))}.$$

A calculation similar to the one in the proof of Theorem 5.27 shows that $f \circ \psi$ is again given by a Lévy–Khintchine formula. This means that $f \circ \psi$ is a continuous and negative definite function and that a subordinate Lévy process is still a Lévy process.

There is an interesting converse to the last remark of Example 13.13. This is due to Schoenberg [325]; we follow Bochner's presentation [63, p. 99].

Theorem 13.14 (Schoenberg; Bochner). *A function $f : (0,\infty) \to [0,\infty)$ is a Bernstein function if, and only if, for all $d \in \mathbb{N}$ the function $\xi \mapsto f(|\xi|^2)$, $\xi \in \mathbb{R}^d$, is continuous and negative definite.*

Proof. Since $\xi \mapsto |\xi|^2$ is continuous and negative definite in all dimensions, sufficiency follows from Example 13.13.

Conversely, fix $d \in \mathbb{N}$ and assume that $f(|\xi|^2)$ is continuous and negative definite. By Proposition 4.4 this is equivalent to saying that

$$F_t(|\xi|^2) := \exp(-tf(|\xi|^2))$$

is continuous and positive definite for all $t > 0$, and by Bochner's theorem, Theorem 4.14, we know that $F_t(|\xi|^2)$ is for fixed $t > 0$ a Fourier transform. This means that

$$F_t(|\xi|^2) = \int_{\mathbb{R}^d} e^{i\xi \cdot y} v_d^t(dy), \quad t > 0, \tag{13.21}$$

with a measure v_d^t on $\mathbb{R}^d$ which is invariant under rotations and whose total mass $v_d^t(\mathbb{R}^d) = F_t(0) = \exp(-tf(0))$ does not depend on the dimension d.

Write ω_d for the canonical surface measure on the unit sphere S^{d-1} in $\mathbb{R}^d$ and $|\omega_d|$ for the surface volume of S^{d-1}. Set $G_t(\xi) := F_t(|\xi|^2)$; then $G_t(|\xi|\eta) = G_t(\xi)$ for all $\eta \in S^{d-1}$. Therefore, we can radialise the integral in (13.21) to get

$$
\begin{aligned}
F_t(|\xi|^2) = G_t(\xi) &= \frac{1}{|\omega_d|} \int_{|\eta|=1} G_t(|\xi|\eta)\, \omega_d(d\eta) \\
&= \frac{1}{|\omega_d|} \int_{|\eta|=1} \int_{\mathbb{R}^d} e^{i|\xi|\eta \cdot y}\, v_d^t(dy)\, \omega_d(d\eta) \\
&= \int_{\mathbb{R}^d} \frac{1}{|\omega_d|} \int_{|\eta|=1} e^{i|\xi|\eta \cdot y}\, \omega_d(d\eta)\, v_d^t(dy) \\
&= \int_{\mathbb{R}^d} H_{\frac{1}{2}(d-2)}(|\xi| \cdot |y|)\, v_d^t(dy).
\end{aligned}
$$

The function H_κ appearing in the last line is essentially a Bessel function of the first kind, see e.g. Stein and Weiss [337, p. 154],

$$
H_\kappa(x) = c_\kappa J_\kappa(x)\, x^{-\kappa}, \tag{13.22}
$$

where the constant $c_\kappa = \lim_{x\to 0} J_\kappa(x)x^{-\kappa} = 2^\kappa \Gamma(\kappa + 1)$ is chosen in such a way that $H_\kappa(0) = 1$, cf. [134, 8.440]. Therefore,

$$
H_\kappa(x) = \sum_{j=0}^{\infty} (-1)^j \frac{\Gamma(\kappa + 1)}{\Gamma(j + \kappa + 1)} \frac{(x/2)^{2j}}{j!}.
$$

Let $\kappa = \kappa(d) = (d-2)/2$ and let μ_d^t be the image measure of v_d^t under the map $\Phi_d : y \mapsto |y|^2/(4\kappa)$. Then

$$
\int_{\mathbb{R}^d} H_{\frac{1}{2}(d-2)}(|\xi| \cdot |y|)\, v_d^t(dy) = \int_{[0,\infty)} H_\kappa(2\sqrt{\kappa}\, |\xi| \sqrt{r})\, \mu_d^t(dr)
$$

and $\mu_d^t[0, \infty) = v_d^t(\Phi_d^{-1}[0, \infty)) = v_d^t(\mathbb{R}^d) = F_t(0)$ is independent of d. This shows that for fixed $t > 0$ the measures $(\mu_d^t)_{d\in\mathbb{N}}$ are vaguely bounded and by Theorem A.5 vaguely compact. Consequently, there are a subsequence $(d_k)_{k\geq 1}$ and a measure μ^t such that vague-$\lim_{k\to\infty} \mu_{d(k)}^t = \mu^t$. Since all measures $\mu_{d(k)}$ have the same mass, we get from Theorem A.4 that

$$
\text{weak-}\lim_{k\to\infty} \mu_{d_k}^t = \mu^t \quad \text{and} \quad \mu^t[0, \infty) = F_t(0).
$$

Without loss of generality we may assume that the whole sequence converges. Then

$$
\begin{aligned}
&\int_{[0,\infty)} H_\kappa(2\sqrt{\kappa}\, |\xi| \sqrt{r})\, \mu_d^t(dr) \\
&= \int_{[0,\infty)} e^{-|\xi|^2 r}\, \mu_d^t(dr) + \int_{[0,\infty)} \left(H_\kappa(2\sqrt{\kappa}\, |\xi| \sqrt{r}) - e^{-|\xi|^2 r} \right) \mu_d^t(dr).
\end{aligned}
$$

From Lemma 13.15 below we see that the integrand of the second integral converges uniformly to 0 as $d \to \infty$, hence $\kappa \to \infty$, while $\mu_d^t(0, \infty)$ is independent of d. Since μ_d^t converges weakly to μ^t we get

$$\lim_{d \to \infty} \int_{[0,\infty)} H_\kappa(2\sqrt{\kappa}\,|\xi|\sqrt{r})\,\mu_d^t(dr) = \int_{[0,\infty)} e^{-|\xi|^2 r}\,\mu^t(dr)$$

which shows that $F_t(\lambda) = \int_{(0,\infty)} e^{-\lambda r}\,\mu^t(dr)$, i.e. $F_t(\lambda)$ is completely monotone for all $t > 0$. Since by our construction $F_t(\lambda) = e^{-tf(\lambda)}$, we can use Theorem 3.7 to conclude that f is a Bernstein function. $\square$

Lemma 13.15. (i) *There exists a positive constant $C > 0$ such that for all $\kappa > 1$ and all $r > 0$,*

$$\left|H_\kappa(2\sqrt{2\kappa}\,r)\right| \le \frac{C}{r}. \tag{13.23}$$

(ii) $\lim_{\kappa \to \infty} \dfrac{H_\kappa(2\sqrt{\kappa}\,r) - e^{-r^2}}{r^2} = 0$ *locally uniformly in r.*

(iii) $\lim_{\kappa \to \infty} \sup_{r \ge 0} \left|H_\kappa(2\sqrt{\kappa}\,r) - e^{-r^2}\right| = 0$.

Proof. (i) By (13.22) and [134, 8.411.9] the following integral formula is valid

$$H_\kappa(y) = \frac{2\,\Gamma(\kappa+1)}{\Gamma\!\left(\kappa+\frac{1}{2}\right)\Gamma\!\left(\frac{1}{2}\right)} \int_0^1 (1-t^2)^{\kappa-1/2}\cos yt\,dt, \quad y > 0. \tag{13.24}$$

The second mean value theorem for integrals, cf. e.g. [323, Theorem E.22], shows that there exists $\theta \in [0, 1]$ such that

$$\left|\int_0^1 (1-t^2)^{\kappa-1/2}\cos yt\,dt\right| = \left|\int_0^\theta \cos yt\,dt\right| \le \frac{1}{y},$$

implying that

$$|H_\kappa(y)| \le \frac{2\,\Gamma(\kappa+1)}{\Gamma\!\left(\kappa+\frac{1}{2}\right)\Gamma\!\left(\frac{1}{2}\right)}\frac{1}{y}.$$

Hence, for $r > 0$,

$$\left|H_\kappa(2\sqrt{\kappa}\,r)\right| \le \frac{1}{r}\frac{\Gamma(\kappa+1)}{\Gamma\!\left(\kappa+\frac{1}{2}\right)\sqrt{\pi}\,\sqrt{\kappa}}.$$

Since $\lim_{\kappa \to \infty} \dfrac{\Gamma(\kappa+1)}{\Gamma(\kappa+\frac{1}{2})\sqrt{\kappa}} = 1$, see [134, 8.328.2], we get (13.23).

(ii) We have

$$
|H_\kappa(2\sqrt{\kappa}\,r) - e^{-r^2}| = \left| \sum_{j=1}^{\infty} (-1)^j \left(\frac{\Gamma(\kappa+1)\,\kappa^j}{\Gamma(\kappa+1+j)} \frac{r^{2j}}{j!} - \frac{r^{2j}}{j!} \right) \right|
$$

$$
\leq \sum_{j=1}^{\infty} \left(1 - \frac{\Gamma(\kappa+1)\,\kappa^j}{\Gamma(\kappa+1+j)} \right) \frac{r^{2j}}{j!}
$$

$$
\leq r^2 \sum_{j=0}^{\infty} \left(1 - \frac{\Gamma(\kappa+1)\,\kappa^{j+1}}{\Gamma(\kappa+2+j)} \right) \frac{r^{2j}}{(j+1)!}.
$$

Since $1 - \frac{\Gamma(\kappa+1)\kappa^{j+1}}{\Gamma(\kappa+2+j)}$ is positive and converges to zero as $\kappa \to \infty$, the claim follows from the dominated convergence theorem.

(iii) Fix $A > 1$. Because of (i) and the elementary inequality $e^{-r^2} \leq e/r^2$ we get

$$
\sup_{r\geq 0} |H_\kappa(2\sqrt{\kappa}\,r) - e^{-r^2}| \leq \sup_{r\in[0,A]} |H_\kappa(2\sqrt{\kappa}\,r) - e^{-r^2}| + \sup_{r>A} \left| \frac{C}{r} + \frac{e}{r^2} \right|.
$$

By (ii) the first term tends to zero as $\kappa \to \infty$. Letting $A \to \infty$ proves the claim. $\quad\square$

Example 13.16. Let $\psi : \mathbb{R}^d \to \mathbb{C}$ be the characteristic exponent of a Lévy process $(X_t)_{t\geq 0}$ with values in $\mathbb{R}^d$. It is given by the Lévy–Khintchine formula

$$
\psi(\xi) = i\beta \cdot \xi + \frac{1}{2} \xi \cdot Q\xi + \int_{y\neq 0} \left(1 - e^{i\xi\cdot y} + \frac{i\xi \cdot y}{1+|y|^2} \right) v(dy). \qquad (13.25)
$$

Assume that $\psi(\xi)$ depends on the radial part $|\xi|$ only, and that the Lévy measure v is of the form $v(dy) = n(y)\,dy$ with

$$
n(y) = \int_{(0,\infty)} e^{-s|y|^2}\, \rho(ds), \qquad (13.26)
$$

where ρ is a measure on $(0,\infty)$. In particular, ψ is real valued and we have $\beta = 0$, $\frac{1}{2} \xi \cdot Q\xi = b|\xi|^2$ for some $b \geq 0$. Therefore,

$$
\psi(\xi) = b\,|\xi|^2 + \int_{\mathbb{R}^d} \big(1 - \cos(\xi \cdot y)\big) \left(\int_{(0,\infty)} e^{-s|y|^2}\, \rho(ds) \right) dy
$$

$$
= b\,|\xi|^2 + \int_{(0,\infty)} \left(\int_{\mathbb{R}^d} e^{-s|y|^2} \big(1 - \cos(\xi \cdot y)\big)\, dy \right) \rho(ds)
$$

$$
= b\,|\xi|^2 + \mathrm{Re} \int_{(0,\infty)} \left(\int_{\mathbb{R}^d} e^{-s|y|^2} \big(1 - e^{i\xi\cdot y}\big)\, dy \right) \rho(ds)
$$

$$
= b\,|\xi|^2 + \int_{(0,\infty)} \pi^{d/2} s^{-d/2} \left(1 - e^{-\frac{|\xi|^2}{4s}} \right) \rho(ds)
$$

$$
= b\,|\xi|^2 + \int_{(0,\infty)} (4\pi t)^{d/2} \left(1 - e^{-t|\xi|^2} \right) \widehat{\rho}(dt),
$$

where $\widehat{\rho}$ is the image measure of ρ with respect to the mapping $t = (4s)^{-1}$. Let $\mu(dt) = (4\pi t)^{d/2}\widehat{\rho}(dt)$ and set $f(\lambda) = b\lambda + \int_{(0,\infty)}(1 - e^{-\lambda t})\mu(dt)$, $\lambda > 0$. Then f is a Bernstein function and $\psi(\xi) = f(|\xi|^2)$ for all $\xi \neq 0$ and $\psi(0) = f(0+) = 0$.

Conversely, if $\psi(\xi) = f(|\xi|^2)$ with $f \in \mathcal{BF}$, then by retracing the steps we conclude that the Lévy measure ν is of the form $\nu(dy) = n(|y|)dy$ where n is given by (13.26).

The discussion above can be summarized as follows. Let ψ be given by (13.25). Then $\psi(\xi) = f(|\xi|^2)$ for $f \in \mathcal{BF}$ if, and only if, $Q = b\,\text{id}$ and $\nu(dy) = n(|y|)\,dy$ where $n(\sqrt{\lambda})$ is completely monotone. In probabilistic terms this means that X_t is a subordinate Brownian motion, $X_t = B_t^f$, if, and only if, $Q = b\,\text{id}$ and the Lévy measure has a density $n(|y|)$ such that $n(\sqrt{\lambda})$ is completely monotone.

In dimension $d = 1$ we have the following similar statement for complete Bernstein functions: $\psi(\xi) = f(|\xi|^2)$ for $f \in \mathcal{CBF}$ if, and only if, $\beta = 0$, $Q = b\,\text{id}$ and $\nu(dy) = n(|y|)\,dy$ where $n(\lambda)$ is completely monotone. This follows from

$$\int_{\mathbb{R}} e^{-s|y|}\left(1 - \cos(\xi y)\right)dy = \frac{2\xi^2}{(\xi^2 + s^2)\,s},$$

hence,

$$\psi(\xi) = b\,\xi^2 + \int_{(0,\infty)} \frac{\xi^2}{\xi^2 + s^2}\,\frac{2\rho(ds)}{s} = b\,\xi^2 + \int_{(0,\infty)} \frac{\xi^2}{\xi^2 + t}\,\sigma(dt)$$

for some measure σ.

13.2   A functional calculus for generators of semigroups

We continue our study of C_0-contraction semigroups $(T_t)_{t\geq 0}$ on a Banach space $\mathfrak{B}$. In this section we will develop a functional calculus for the generators $(A^f, \mathfrak{D}(A^f))$, $f \in \mathcal{BF}$, of the subordinate semigroup $(T_t^f)_{t\geq 0}$. This calculus is a natural extension of the spectral calculus in Hilbert spaces, and in many cases it is possible to interpret the subordinate generator A^f as the operator $-f(-A)$. In Hilbert spaces and for self-adjoint semigroups this follows immediately from the familiar spectral calculus for self-adjoint operators, see Examples 12.5, 12.6 and 12.7. In Banach spaces, one can use the Dunford–Riesz functional calculus to express A^f as an unbounded Cauchy integral and to identify this operator with $-f(-A)$, see [36, 298] and [320, 321].

The particularly interesting case of fractional powers of dissipative operators is included in this calculus if we take the (complete) Bernstein functions $f_\alpha(\lambda) = \lambda^\alpha$, $\alpha \in (0, 1)$. Phillips' formula (13.10) reads

$$-(-A)^\alpha u = A^{f_\alpha}u = \frac{\alpha}{\Gamma(1-\alpha)}\int_0^\infty (T_t u - u)\,t^{-\alpha-1}\,dt, \quad u \in \mathfrak{D}(A), \quad (13.27)$$

and, if we use the Stieltjes representation (6.7) of $f_\alpha(\lambda)$, we get from Corollary 13.22 below Balakrishnan's famous formula for the fractional power of a dissipative operator

$$-(-A)^\alpha u = A^{f_\alpha} u = \frac{\sin(\alpha\pi)}{\pi} \int_0^\infty AR_t u\, t^{\alpha-1}\, dt, \quad u \in \mathfrak{D}(A). \tag{13.28}$$

The main result for the functional calculus, Theorem 13.23, shows that

$$A^{f_\alpha} A^{f_\beta} = -A^{f_{\alpha+\beta}}, \quad \text{that is} \quad (-A)^\alpha(-A)^\beta = (-A)^{\alpha+\beta},$$

holds on $\mathfrak{D}(A)$ for $\alpha, \beta \geq 0$ such that $\alpha + \beta \leq 1$; Corollary 13.28 extends this to all $\alpha, \beta \geq 0$.

We begin with a few preparations.

Lemma 13.17. *Let f be a Bernstein function, $g_k(\lambda) := k\lambda(k + \lambda)^{-1}$, $k \in \mathbb{N}$, and set $f_k := g_k \circ f$. Then $f_k \in \mathcal{BF}$ and*

$$\frac{f_k}{f} = \frac{k}{k + f} = \mathcal{L}\nu_k$$

for a sub-probability measure ν_k on $[0, \infty)$ such that $\nu_k \xrightarrow{k\to\infty} \delta_0$ weakly.

Proof. Since $g_k, f \in \mathcal{BF}$, the composition $f_k = g_k \circ f$ is in $\mathcal{BF}$. Moreover,

$$\frac{f_k}{f} = \frac{\frac{kf}{k+f}}{f} = \frac{k}{k + f}$$

which is completely monotone since it is the composition of the completely monotone function $\lambda \mapsto k(k + \lambda)^{-1}$ with $f \in \mathcal{BF}$, cf. Theorem 3.7 (ii). Therefore, there exists a measure ν_k on $[0, \infty)$ with $\mathcal{L}\nu_k = f_k/f$. By monotone convergence,

$$\nu_k[0, \infty) = \lim_{\lambda\to 0} \mathcal{L}(\nu_k; \lambda) = \lim_{\lambda\to 0} \frac{f_k(\lambda)}{f(\lambda)} = \lim_{\lambda\to 0} \frac{k}{k + f(\lambda)}$$

$$= \frac{k}{k + f(0+)} \leq 1.$$

Weak convergence follows from $\lim_{k\to\infty} f_k(\lambda)/f(\lambda) = 1$ and Lemma A.9. $\qquad\square$

Denote by $(\mu_t)_{t\geq 0}$ the convolution semigroup of measures on $[0, \infty)$ associated with the Bernstein function f, and write

$$U_k = \int_0^\infty e^{-kt} \mu_t\, dt, \quad k \in \mathbb{N}, \tag{13.29}$$

for the k-potential measure, cf. (5.21).

Lemma 13.18. *Let f_k, f and v_k be as in Lemma 13.17 and let $(T_t)_{t\geq 0}$ be a C_0-contraction semigroup on the Banach space $\mathfrak{B}$ with generator $(A, \mathfrak{D}(A))$ and resolvent $(R_\lambda)_{\lambda>0}$. Then*

$$I_k u := \int_{[0,\infty)} T_s u\, v_k(ds), \quad k \in \mathbb{N},\ u \in \mathfrak{B},$$

defines a family of bounded operators on $\mathfrak{B}$ and

$$I_k = k R_k^f, \quad k \in \mathbb{N},$$

where R_k^f are the subordinate resolvent operators. In particular, we have for all $u \in \mathfrak{B}$ that strong-$\lim_{k\to\infty} I_k u = u$ and $I_k u \in \mathfrak{D}(A^f)$.

Proof. Observe that (5.20) with f replaced by $k + f$ shows

$$\mathscr{L}U_k = \frac{1}{k+f}.$$

Thus, $\mathscr{L}(kU_k) = k(k+f)^{-1} = \mathscr{L}v_k$, and we conclude that $kU_k = v_k$. Since $t \mapsto T_t u$ is strongly continuous, this proves for all $u \in \mathfrak{B}$

$$\begin{aligned}
I_k u = \int_{[0,\infty)} T_s u\, v_k(ds) &= k \int_0^\infty \int_{[0,\infty)} T_s u\, e^{-kt}\, \mu_t(ds)\, dt \\
&= k \int_0^\infty e^{-kt}\, T_t^f u\, dt \\
&= k R_k^f u.
\end{aligned}$$

Using $k \int_0^\infty e^{-kt}\, dt = 1$ we find

$$\|I_k u - u\| \leq \int_0^\infty k e^{-kt} \|T_t^f u - u\|\, dt = \int_0^\infty e^{-s} \|T_{s/k}^f u - u\|\, ds.$$

Because of the strong continuity of the subordinate semigroup we get $I_k u \xrightarrow{k\to\infty} u$ in the strong sense, and a similar calculation proves that $\|I_k\| \leq 1$. That $I_k u \in \mathfrak{D}(A^f)$ follows from the mapping properties of the resolvent R_k^f. $\qquad\square$

Proposition 13.19. *Let f_k, $f \in \mathcal{BF}$ be as in Lemma 13.17, let I_k be the operator from Lemma 13.18 and denote by A^{f_k} and A^f the corresponding subordinate generators and by $(R_\lambda^f)_{\lambda>0}$ the resolvent of A^f. Then A^{f_k} is a bounded operator,*

$$\begin{aligned}
A^{f_k} u = A^f I_k &= k A^f R_k^f u, \quad k \in \mathbb{N},\ u \in \mathfrak{B}, &\text{(13.30)} \\
&= I_k A^f = k R_k^f A^f u, \quad k \in \mathbb{N},\ u \in \mathfrak{D}(A^f), &\text{(13.31)}
\end{aligned}$$

and $u \in \mathfrak{D}(A^f)$ if, and only if, $(A^{f_k}u)_{k\in\mathbb{N}}$ converges strongly; if this is the case, then $A^f u = \lim_{k\to\infty} A^{f_k} u$. Moreover,

$$I_\ell A^{f_k} = A^{f_k} I_\ell \quad \text{and} \quad A^{f_k} I_\ell u = A^{f_\ell} I_k u, \quad k, \ell \in \mathbb{N}, \; u \in \mathfrak{B}. \tag{13.32}$$

Proof. The boundedness of A^{f_k} follows from the boundedness of the function f_k and Corollary 13.7. By Lemma 13.17

$$f_k(\lambda) = \frac{kf(\lambda)}{k + f(\lambda)} = k\big(1 - \mathscr{L}(\nu_k;\lambda)\big) = a_k + \int_{(0,\infty)} (1 - e^{-\lambda s}) \, k\nu_k(ds)$$

with $a_k = k(1 - \nu_k[0,\infty))$. A combination of Phillips' theorem, Theorem 13.6, for f_k and Lemma 13.18 gives for $u \in \mathfrak{D}(A)$

$$A^{f_k} u = -a_k u + \int_{(0,\infty)} (T_s u - u) \, k\nu_k(ds) \tag{13.33}$$
$$= -ku + kI_k u = -ku + k^2 R_k^f u = k A^f R_k^f u.$$

Since the left and right-hand sides define bounded operators, this equality extends to all $u \in \mathfrak{B}$, and (13.30) follows. Using the commutativity of the operators R_k^f and A^f on $\mathfrak{D}(A^f)$, we get (13.31). By Lemma 13.18, $k R_k^f u \xrightarrow{k\to\infty} u$ strongly. If $u \in \mathfrak{D}(A^f)$, (13.31) shows that $\lim_{k\to\infty} A^{f_k} u = A^f u$ strongly. Conversely, if $\lim_{k\to\infty} A^{f_k} u$ converges in the strong sense, (13.30) and the closedness of the operator A^f prove that $u \in \mathfrak{D}(A^f)$.

For (13.32) we use $I_\ell = \ell R_\ell^f$ and (13.30) to get

$$A^{f_k} I_\ell = \ell k A^f R_\ell^f R_k^f = A^{f_\ell} I_k$$

where the second equality follows from the symmetric roles of k and ℓ in the middle term. Since A^f, R_ℓ^f and R_k^f commute, we also get $A^{f_k} I_\ell = I_\ell A^{f_k}$.   □

Corollary 13.20. *Let $(T_t)_{t\geq 0}$ be a C_0-contraction semigroup on the Banach space $\mathfrak{B}$ with resolvent $(R_\lambda)_{\lambda>0}$ and generator $(A, \mathfrak{D}(A))$. For every Bernstein function f it holds that*

$$\mathfrak{D}(A^f) = \left\{ u \in \mathfrak{B} : \lim_{k\to\infty} A^{f_k} u \text{ exists strongly} \right\}$$
$$= \left\{ u \in \mathfrak{B} : \lim_{k\to\infty} \int_{(0,\infty)} (T_t u - u) \, k^2 U_k(dt) \text{ exists strongly} \right\}$$
$$= \left\{ u \in \mathfrak{B} : \lim_{k\to\infty} (k I_k u - ku) \text{ exists strongly} \right\},$$

where $f_k = kf/(k + f)$, $I_k = k R_k^f$, and U_k is the k-potential measure (13.29) associated with the Bernstein function f.

Proof. From Proposition 13.19 we know that $u \in \mathfrak{D}(A^f)$ if, and only if, the limit $\lim_{k\to\infty} A^{f_k} u$ exists strongly. Moreover, the k-potential measure U_k is $k^{-1} \nu_k$ and

$$A^{f_k} u = -k\big(1 - k U_k[0,\infty)\big) u + \int_{(0,\infty)} (T_s u - u)\, k^2 U_k(ds).$$

Since $U_k[0,\infty) = \mathscr{L}(U_k; 0+) = (k + f(0+))^{-1}$, we see

$$k\big(1 - k U_k[0,\infty)\big) = k\left(1 - \frac{k}{k + f(0+)}\right) = \frac{kf(0+)}{k + f(0+)} \xrightarrow{k\to\infty} f(0+).$$

This proves the second characterization of $\mathfrak{D}(A^f)$; the last one follows immediately from (13.33). □

Remark 13.21. A close inspection of the proof of Proposition 13.19 shows that we can replace the approximating operator $I_k = k R_k^f$ by any other bounded operator J_k which is of the form

$$J_k u = \int_{[0,\infty)} T_s u\, \gamma_k(ds), \quad u \in \mathfrak{B},$$

where γ_k is a family of (signed) measures on $[0,\infty)$ which have uniformly bounded variation and satisfy $\mathscr{L}\gamma_k = f_k/f$ where $f_k \in \mathcal{BF}_b$ with $\lim_{k\to\infty} f_k(\lambda) = f(\lambda)$. Any such J_k satisfies strong-$\lim_{k\to\infty} J_k u = u$ for all $u \in \mathfrak{B}$ as well as (13.32), which is all we need for the proofs of Proposition 13.19 and Corollary 13.20. The first assertion follows since $(T_t)_{t\geq 0}$ is strongly continuous and since the γ_k converge weakly to δ_0, the second assertion is immediate from the fact that each T_t, $t \geq 0$, commutes with A, R_λ and the subordinate analogues. An example of such a situation is given in [322] for $f \in \mathcal{CBF}$.

Let us rewrite Corollary 13.20 for a complete Bernstein function f. In this case we have $f(\lambda) = a + b\lambda + \int_{(0,\infty)} \lambda(s + \lambda)^{-1} \sigma(ds)$, with $a, b \geq 0$ and the Stieltjes representation measure σ on $(0,\infty)$ satisfying $\int_{(0,\infty)} (1 + s)^{-1} \sigma(ds) < \infty$, see (6.7); the Lévy measure $\mu(dt)$ has a density $m(t)$ which is given by $m(t) = \mathscr{L}(s \cdot \sigma(ds); t)$.

Corollary 13.22. *Let $(T_t)_{t\geq 0}$ be a C_0-contraction semigroup on the Banach space $\mathfrak{B}$ with resolvent $(R_\lambda)_{\lambda>0}$ and generator $(A, \mathfrak{D}(A))$, and let $f \in \mathcal{CBF}$ with Stieltjes representation given above. Then*

$$\mathfrak{D}(A^f) = \left\{ u \in \mathfrak{B} : \lim_{k\to\infty} \int_{(0,k)} A R_s u\, \sigma(ds)\ \text{exists strongly} \right\}$$

and we have for $u \in \mathfrak{D}(A)$

$$A^f u = -au + bAu + \int_{(0,\infty)} (s R_s u - u)\, \sigma(ds) \tag{13.34}$$

$$= -au + bAu + \int_{(0,\infty)} A R_s u\, \sigma(ds). \tag{13.35}$$

Proof. Applying Phillips' formula (13.10) for $f \in \mathcal{CBF}$ we see for $u \in \mathcal{D}(A)$

$$A^f u = -au + bAu + \int_0^\infty (T_t u - u)\, m(t)\, dt$$

$$= -au + bAu + \int_0^\infty \int_{(0,\infty)} (T_t u - u)e^{-st}\, s\, \sigma(ds)\, dt$$

$$= -au + bAu + \int_{(0,\infty)} \left(\int_0^\infty se^{-st}\, T_t u\, dt - u \right) \sigma(ds)$$

$$= -au + bAu + \int_{(0,\infty)} (sR_s u - u)\, \sigma(ds)$$

and this proves (13.34), and (13.35) follows as $sR_s u - u = AR_s u$. In order to justify the interchange of integrals in the calculation above, we use the estimate (13.3) and observe that the iterated integrals are finite for $u \in \mathcal{D}(A)$:

$$\int_{(0,\infty)} \int_0^\infty \|T_t u - u\|\, e^{-st}\, dt\, s\, \sigma(ds)$$

$$\leq \int_{(0,\infty)} \int_0^\infty \min\{t\|Au\|, 2\|u\|\}\, e^{-st}\, dt\, s\, \sigma(ds)$$

$$\leq 2 \int_{(0,1)} \int_0^\infty s\, e^{-st}\, dt\, \sigma(ds)\|u\| + \int_{[1,\infty)} \int_0^\infty st\, e^{-st}\, dt\, \sigma(ds)\|Au\|$$

$$= 2 \int_{(0,1)} \sigma(ds)\, \|u\| + \int_{[1,\infty)} \frac{1}{s}\sigma(ds)\, \|Au\| < \infty.$$

In order to see the assertion for the domain, we use Remark 13.21 with

$$f_k(\lambda) = \int_{(0,k)} \frac{\lambda}{s+\lambda}\, \sigma(ds).$$

From this it is possible to compute $I(z) = \operatorname{Im}[(f(z) - f_k(z))/f(z)]$ and an elementary but lengthy calculation shows for $z = x + iy \in \mathbb{H}^\uparrow$

$$I(z) = \frac{1}{|f_k(z)|} \int_{t\in(k,\infty)} \int_{s\in(0,k)} \left[\frac{(t-s)y(x^2+y^2)}{((t+x)^2+y^2)((s+x)^2+y^2)} \right] \sigma(ds)\, \sigma(dt).$$

The integrand is positive since $\operatorname{Im} z = y > 0$. Moreover, by monotone convergence we see

$$\frac{f(\lambda) - f_k(\lambda)}{f_k(\lambda)} = \frac{\int_{[k,\infty)} \frac{1}{s+\lambda}\sigma(ds)}{\int_{(0,k)} \frac{1}{s+\lambda}\sigma(ds)} \xrightarrow{\lambda\to 0+} \frac{\int_{[k,\infty)} \frac{1}{s}\sigma(ds)}{\int_{(0,k)} \frac{1}{s}\sigma(ds)} \in [0,\infty).$$

Therefore, Theorem 6.2 (iv) shows that $(f - f_k)/f_k$ is in $\mathcal{CBF}$, and consequently $f/f_k = 1 + (f - f_k)/f_k \in \mathcal{CBF}$, and $f_k/f \in \mathcal{S}$ by Theorem 7.3. This proves that $f_k/f = \mathscr{L}\gamma_k$ for a sub-probability measure γ_k on $[0,\infty)$.

The form of $\mathcal{D}(A^f)$ follows now from Remark 13.21 and Corollary 13.20. □

We can use the knowledge of the structure of the domain $\mathfrak{D}(A^f)$ to construct a functional calculus for semigroup generators.

Theorem 13.23. *Let* $(A, \mathfrak{D}(A))$ *be the generator of a* C_0-*semigroup* $(T_t)_{t \geq 0}$ *on the Banach space* $\mathfrak{B}$, *and let* $f, g \in \mathcal{BF}$. *Then we have*

(i) $A^{cf} = c A^f$ *for all* $c > 0$;

(ii) $A^{f+g} = \overline{A^f + A^g}$;

(iii) $A^{f \circ g} = (A^g)^f$;

(iv) $A^{c+\mathrm{id}+f} = -c\,\mathrm{id} + A + A^f$ *for all* $c \geq 0$;

(v) *if* $fg \in \mathcal{BF}$, *then* $A^{fg} = -A^f A^g = -A^g A^f$.

The equalities in (i)–(v) *are identities in the sense of closed operators, including their domains which are the usual domains for sums, compositions etc. of closed operators.*

Before we proceed with the proof of Theorem 13.23 let us mention the following useful corollaries. Both of them are immediate consequences of Theorem 13.23 (v) in conjunction with Proposition 7.13 and Definition 11.1, respectively.

Corollary 13.24. *Let* $(A, \mathfrak{D}(A))$ *be as in Theorem* 13.23 *and assume that* f *and* g *are in* $\mathcal{CBF}$. *Then* $f^\alpha g^{1-\alpha} \in \mathcal{CBF}$ *for all* $\alpha \in (0,1)$ *and*

$$A^{f^\alpha g^{1-\alpha}} = -A^{f^\alpha} A^{g^{1-\alpha}} = -A^{g^{1-\alpha}} A^{f^\alpha}.$$

Corollary 13.25. *Let* $(A, \mathfrak{D}(A))$ *be as in Theorem* 13.23 *and assume that* $f, f^\star$ *are conjugate functions from* $\mathcal{SBF}$, *cf. Definition* 11.1, *i.e.* $f^\star(\lambda) = \lambda/f(\lambda)$. *Then*

$$A = -A^f A^{f^\star} = -A^{f^\star} A^f.$$

Proof of Theorem 13.23. For $f, g \in \mathcal{BF}$ we know that $cf, f + g, f \circ g \in \mathcal{BF}$. Since $\mathfrak{D}(A) \subset \mathfrak{D}(A^f) \cap \mathfrak{D}(A^g)$, all of the above operators are densely defined. By the linearity of the definition of $(A^f, \mathfrak{D}(A^f))$, see Corollary 13.20, (i) and (ii) are immediate, (iii) is a consequence of the transitivity of the subordination: $T_t^{f \circ g} = (T_t^g)^f$, see Lemma 13.3.

For (iv) we note that A^f is A-bounded in the sense that $\|A^f u\| \leq c \|Au\| + c' \|u\|$ for $c, c' > 0$ and all $u \in \mathfrak{D}(A)$, cf. (13.17). Thus, $(c + A + A^f, \mathfrak{D}(A))$ is a closed operator, and the assertion follows from Phillips' formula (13.10).

For (v) we assume that $fg \in \mathcal{BF}$. Write h for any of the Bernstein functions f, g or fg and set, as in Lemma 13.17,

$$h_k := \frac{kh}{k+h} \quad \text{and} \quad \frac{h_k}{h} = \mathscr{L}\nu_{h,k}, \quad k \in \mathbb{N}, \tag{13.36}$$

for a suitable sub-probability measure $\nu_{h,k}$ on $[0, \infty)$. Note that

$$f_k \, g_\ell \, \mathscr{L} \nu_{fg,m} = (fg)_m \, \mathscr{L} \nu_{f,k} \, \mathscr{L} \nu_{g,\ell} \tag{13.37}$$

holds for all $k, \ell, m \in \mathbb{N}$. By (13.36),

$$h_k = k \left(1 - \frac{k}{k+h} \right) = \mathscr{L} \big(k \delta_0 - k \nu_{h,k} \big)$$

allows to rewrite the identity (13.37) in terms of convolutions of (signed) measures

$$k\ell (\nu_{f,k} - \delta_0) \star (\nu_{g,\ell} - \delta_0) \star \nu_{fg,m} = -m(\nu_{fg,m} - \delta_0) \star \nu_{f,k} \star \nu_{g,\ell}. \tag{13.38}$$

We want to show that

$$m A^{f_k} A^{g_\ell} R_m^{fg} u = -k\ell A^{(fg)_m} R_k^f R_\ell^g u. \tag{13.39}$$

Applying Phillips' formula (13.10) to A^{h_k} shows

$$A^{h_k} u = -k \big(1 - \nu_{h,k}[0,\infty) \big) u + \int_{(0,\infty)} (T_s u - u) \, k \nu_{h,k}(ds)$$

$$= \int_{[0,\infty)} T_s u \, k \nu_{h,k}(ds) - ku$$

$$= k \int_{[0,\infty)} T_s u \, (\nu_{h,k} - \delta_0)(ds),$$

and from Lemma 13.18 we know that

$$k R_k^h u = \int_{[0,\infty)} T_s u \, \nu_{h,k}(ds).$$

Since the operators A^{h_k} and R_k^h, where h stands for f, g or fg, are bounded we can freely interchange the order of integration in the calculation below.

$$m A^{f_k} A^{g_\ell} R_m^{fg} u$$

$$= k\ell \int_{[0,\infty)} T_r u \, (\nu_{f,k} - \delta_0)(dr) \int_{[0,\infty)} T_s u \, (\nu_{g,\ell} - \delta_0)(ds) \int_{[0,\infty)} T_t u \, \nu_{fg,m}(dt)$$

$$= k\ell \int_{[0,\infty)} \int_{[0,\infty)} \int_{[0,\infty)} T_{r+s+t} u \, (\nu_{f,k} - \delta_0)(dr) \, (\nu_{g,\ell} - \delta_0)(ds) \, \nu_{fg,m}(dt)$$

$$= k\ell \int_{[0,\infty)} T_s u \, (\nu_{f,k} - \delta_0) \star (\nu_{g,\ell} - \delta_0) \star \nu_{fg,m}(ds).$$

A very similar calculation leads to

$$-k\ell A^{(fg)_m} R_k^f R_\ell^g u = -m \int_{[0,\infty)} T_s u \, (\nu_{fg,m} - \delta_0) \star \nu_{f,k} \star \nu_{g,\ell}(ds)$$

and the convolution identity (13.38) shows that (13.39) holds. Since (13.39) is an equality between bounded operators, it extends from $\mathfrak{D}(A)$ to all $u \in \mathcal{B}$.

Using (13.39) we can now prove (v). Recall from operator theory that $A^f A^g$ is a closed operator on $\mathfrak{D}(A^f A^g) = \{u \in \mathcal{B} : u \in \mathfrak{D}(A^g) \text{ and } A^g u \in \mathfrak{D}(A^f)\}$. If $u \in \mathfrak{D}(A^f A^g)$, we find by letting $\ell \to \infty$, $k \to \infty$ and then $m \to \infty$ in (13.39) that $u \in \mathfrak{D}(A^{fg})$, see Corollary 13.20. This shows that $-A^{fg}$ extends $A^f A^g$.

Conversely, if $f \in \mathfrak{D}(A^{fg})$, see Corollary 13.20, we first let $m \to \infty$ in (13.39) and then $\ell \to \infty$ and $k \to \infty$. This proves that $u \in \mathfrak{D}(A^f A^g)$ and that $A^f A^g$ extends $-A^{fg}$. Therefore $A^f A^g = -A^{fg}$ and, since f and g play symmetric roles, also $A^g A^f = -A^{fg}$.   □

Remark 13.26. The proof of Theorem 13.23 (v) shows that the functional calculus is also an operational calculus. In fact, (13.39) is derived from the identity (13.37) between functions from $\mathcal{BF}_b$ and $\mathcal{CM}_b$ or from the corresponding identity for (signed) measures (13.38). In the proof of Theorem 13.23 we see that it is sufficient to verify the operator identity at the level of $\mathcal{CM}_b$ and $\mathcal{BF}_b$, i.e. for the trivial semigroup $(e^{-\lambda t})_{t \geq 0}$ rather than the operator semigroup $(T_t)_{t \geq 0}$.

The requirement that $fg \in \mathcal{BF}$ for $f, g \in \mathcal{BF}$ in Theorem 13.23 (v) is quite restrictive. We can overcome this difficulty if f, g are complete Bernstein functions by using the log-convexity of $\mathcal{CBF}$, see Proposition 7.13, i.e.

$$f^\alpha g^{1-\alpha} \in \mathcal{CBF} \quad \text{for all } f, g \in \mathcal{CBF}, \ \alpha \in (0, 1). \tag{13.40}$$

This, together with Theorem 13.23 enables us to write down a consistent formula for the operator $(fg)(-A)$ and all $f, g \in \mathcal{CBF}$ – even if A^{fg} has no longer any meaning since it is not the generator of a subordinate semigroup if $fg \notin \mathcal{BF}$.

Definition 13.27. For $f_1, f_2, \cdots, f_n \in \mathcal{CBF}, n \in \mathbb{N}$, we set

$$(f_1 \cdot f_2 \cdots f_n)(-A) = \left[(-A^{f_1^{1/n}})(-A^{f_2^{1/n}}) \cdots (-A^{f_n^{1/n}}) \right]^n \tag{13.41}$$

on the natural domain for compositions of closed operators.

Let $\mathcal{A} = \{f_1 \cdot f_2 \cdots f_n : n \in \mathbb{N}, \ f_j \in \mathcal{CBF}, \ 0 \leq j \leq n\}$ denote the set of all finite products of complete Bernstein functions. We want to show that the above definition extends Theorem 13.23 (v) to the set $\mathcal{A}$. First, however, we have to check that (13.41) is independent of the representation of $F \in \mathcal{A}$. Assume that $F = f_1 f_2 \cdots f_n$ and $F = g_1 g_2 \cdots g_m$, $m \geq n$, $f_j, g_j \in \mathcal{CBF}$, are two representations of F. Extending (13.40) to n-fold products we see by Theorem 13.23 (v)

$$\left[(-A^{f_1^{1/n}})(-A^{f_2^{1/n}}) \cdots (-A^{f_n^{1/n}}) \right]^n = (-1)^{(n^2)} [A^{F^{1/n}}]^n = [-A^{F^{1/n}}]^n,$$

where we used that $(-1)^n = (-1)^{(n^2)}$. A similar formula holds for the other representation of F. In particular, $F^{1/n}, F^{1/m} \in \mathcal{CBF}$ and it remains to show that $[-A^{F^{1/n}}]^n = [-A^{F^{1/m}}]^m$. This, however, follows from

$$\left[[-A^{F^{1/n}}]^n\right]^{1/m} = [-A^{F^{1/n}}]^{n/m} = -A^{F^{1/m}},$$

where the first equality is well known for (arbitrary) powers of closed operators, while the second one comes from Theorem 13.23 (iii). Thus, (13.41) is a well defined closed operator for all $F \in \mathcal{A}$ and the following corollary shows that Definition 13.27 extends Theorem 13.23 (v).

Corollary 13.28. *Let $f, g \in \mathcal{CBF}$ and let $(T_t)_{t \geq 0}$ be a C_0-contraction semigroup on the Banach space $\mathcal{B}$ with generator $(A, \mathfrak{D}(A))$. For $H = fg \in \mathcal{A}$ it holds that*

$$H(-A) = A^f A^g = A^g A^f. \tag{13.42}$$

More generally, if $H = FG$ with $F, G \in \mathcal{A}$, then

$$H(-A) = F(-A)G(-A) = G(-A)F(-A). \tag{13.43}$$

The above equalities are identities between closed operators.

Proof. Put $B = A^{\sqrt{fg}}$. By definition, $H(-A) = B^2$, and by Theorem 13.23 (v) $B = A^{\sqrt{fg}} = -A^{\sqrt{f}} A^{\sqrt{g}} = -A^{\sqrt{g}} A^{\sqrt{f}}$. Since

$$\mathfrak{D}(B^2) = \left\{ u \in \mathcal{B} \, : \, u \in \mathfrak{D}(A^{\sqrt{f}} \sqrt{g}) \quad \text{and} \quad A^{\sqrt{f}} \sqrt{g} u \in \mathfrak{D}(A^{\sqrt{f}} \sqrt{g}) \right\}$$
$$= \left\{ u \in \mathcal{B} \, : \, u \in \mathfrak{D}(A^{\sqrt{g}}), \quad A^{\sqrt{g}} u \in \mathfrak{D}(A^{\sqrt{f}}), \right.$$
$$\left. A^{\sqrt{f}} A^{\sqrt{g}} u \in \mathfrak{D}(A^{\sqrt{g}}) \quad \text{and} \quad A^{\sqrt{g}} A^{\sqrt{f}} A^{\sqrt{g}} u \in \mathfrak{D}(A^{\sqrt{f}}) \right\},$$

we get

$$B^2 = (A^{\sqrt{f}} A^{\sqrt{g}})(A^{\sqrt{f}} A^{\sqrt{g}}) = (A^{\sqrt{f}} A^{\sqrt{f}})(A^{\sqrt{g}} A^{\sqrt{g}}) = A^f A^g = A^g A^f.$$

This proves (13.42). Iterating this identity we get (13.43) first for $F \in \mathcal{CBF}$ and $G = g_1 g_2 \cdots g_m$ with $g_j \in \mathcal{CBF}$, and then for $F, G \in \mathcal{A}$.   □

We continue with a study of the domain $\mathfrak{D}(A^f)$ of the subordinate generator. For this it is important to understand the asymptotic behaviour of $\|T_t u - u\|$ for $t \to 0$ and $u \in \mathfrak{D}(A^f)$.

Let $f \in \mathcal{BF}$ and write U for the potential measure, i.e. $\mathscr{L}U = 1/f$. Then

$$\beta_s := f(s)(U - \delta_{1/s} \star U), \quad s > 0, \tag{13.44}$$

is a family of (signed) measures and

$$\mathscr{L}(\beta_s; \lambda) = \frac{f(s)}{f(\lambda)} (1 - e^{-\lambda/s}).$$

The next lemma is stated for special Bernstein functions; recall that $\mathcal{CBF} \subset \mathcal{SBF}$.

Lemma 13.29. *Let $(\beta_s)_{s>0}$ be as in (13.44) and assume that $f \in \mathcal{SBF}$. Then the total variation of the measures β_s is uniformly bounded by $2(e + 1)$.*

Proof. We know from Theorem 11.3 that $U(dt) = c\,\delta_0(dt) + u(t)\,dt$ with a non-increasing density $u(t)$, $t > 0$, such that $\int_0^1 u(t)\,dt < \infty$. Writing β_s^{ac} for the absolutely continuous part of β_s we get

$$\|\beta_s\|_{TV} \le f(s)\|c\delta_0 - c\delta_{1/s}\|_{TV} + \|\beta_s^{ac}\|_{TV} \le 2cf(s) + \|\beta_s^{ac}\|_{TV}$$

and

$$\|\beta_s^{ac}\|_{TV} = \int_0^\infty \left| \frac{d\beta_s(t)}{dt} \right| dt$$

$$= f(s) \left[\int_0^{1/s} u(t)\,dt + \int_{1/s}^\infty \left(u(t - 1/s) - u(t) \right) dt \right]$$

$$= 2f(s) \int_0^{1/s} u(t)\,dt$$

$$\le 2f(s)\,e \int_0^{1/s} e^{-st}\,u(t)\,dt$$

$$\le 2\,ef(s)\,\frac{1}{f(s)} = 2e.$$

As $f(\lambda) = 1/\mathscr{L}(c\delta_0(dt) + u(t)\,dt; \lambda)$, we have $\sup_{\lambda>0} f(\lambda) = \lim_{\lambda\to\infty} f(\lambda) = c^{-1}$, i.e. f is bounded if, and only if, $c > 0$. This shows that the total variation is bounded by $2 + 2e$ or by $2e$, depending on f being bounded or unbounded.   □

Lemma 13.30. *Let $f \in \mathcal{SBF}$ with potential measure U, set $f_k = kf/(k+f)$ and let $(\beta_s)_{s>0}$ be as in (13.44). For every C_0-contraction semigroup $(T_t)_{t\ge0}$ on the Banach space $\mathfrak{B}$ with generator $(A, \mathfrak{D}(A))$ the operators*

$$\int_{[0,\infty)} T_t u\, \beta_s(dt), \quad u \in \mathfrak{B},$$

are uniformly bounded, and the following identity holds

$$f(s)\,(T_{1/s} - \mathrm{id}) \int_{[0,\infty)} T_t u\, v_k(dt) = A^{f_k} \int_{[0,\infty)} T_t u\, \beta_s(dt), \qquad (13.45)$$

where $f_k/f = \mathscr{L}v_k$ as in Lemma 13.17.

Proof. The uniform boundedness follows directly from Lemma 13.29, while (13.45) is a consequence of Remark 13.26 and the identity

$$f(s)\,(e^{-\lambda/s}-1)\mathscr{L}(\nu_k;\lambda) = f(s)\,(e^{-\lambda/s}-1)\frac{f_k(\lambda)}{f(\lambda)}$$

$$= f_k(\lambda)\,f(s)\,\mathscr{L}(\delta_{1/s}\star U - U;\lambda)$$

$$= -f_k(\lambda)\,\mathscr{L}(\beta_s;\lambda). \qquad \square$$

We will now combine Lemma 13.30 with the technique developed in Proposition 13.19 and the proof of Theorem 13.23 (v) to derive the next result.

Theorem 13.31. *Let* $f \in \mathcal{SBF}$, *let* $(\beta_s)_{s>0}$ *be as above and let* $(T_t)_{t\geq 0}$ *be a* C_0-*contraction semigroup on the Banach space* $\mathfrak{B}$ *with generator* $(A, \mathfrak{D}(A))$. *Then*

$$f(s)\,(T_{1/s}u - u) = A^f \int_{[0,\infty)} T_t u\, \beta_s(dt) \qquad s > 0,\; u \in \mathfrak{B}, \tag{13.46}$$

$$= \int_{[0,\infty)} T_t A^f u\, \beta_s(dt) \qquad s > 0,\; u \in \mathfrak{D}(A^f). \tag{13.47}$$

Proof. Recall that $\lim_{k\to\infty} A^{f_k} v$ exists if, and only if, $v \in \mathfrak{D}(A^f)$. Therefore, the limit $k \to \infty$ in (13.45) shows that $v = \int_{[0,\infty)} T_t u\, \beta_s(dt) \in \mathfrak{D}(A^f)$ and that (13.46) holds. This implies (13.47) for all $u \in \mathfrak{D}(A^f)$, since A^f is a closed operator. $\quad\square$

Corollary 13.32. *Let* $f \in \mathcal{SBF}$ *and let* $(T_t)_{t\geq 0}$ *be a* C_0-*contraction semigroup on the Banach space* $\mathfrak{B}$ *with generator* $(A, \mathfrak{D}(A))$. *Then we have*

$$\|T_t u - u\| \leq \frac{2(e+1)}{f(t^{-1})}\, \|A^f u\|, \quad u \in \mathfrak{D}(A^f). \tag{13.48}$$

Proof. If $u \in \mathfrak{D}(A^f)$, (13.47) gives for $t = 1/s$,

$$\|T_t u - u\| \leq \frac{1}{f(t^{-1})}\|\beta_{1/t}\|_{TV}\,\|A^f u\|,$$

and the assertion follows because $\sup_{s>0} \|\beta_s\|_{TV} \leq 2(e+1)$, see Lemma 13.29. $\quad\square$

The following corollary contains the converse of Corollary 13.7.

Corollary 13.33. *Let* $f \in \mathcal{SBF}$ *and let* $(T_t)_{t\geq 0}$ *be a* C_0-*contraction semigroup on the Banach space* $\mathfrak{B}$ *with generator* $(A, \mathfrak{D}(A))$. *The subordinate generator* A^f *is bounded if, and only if, A is bounded or f is bounded.*

Proof. The sufficiency has already been established (under less restrictive assumptions) in Corollary 13.7. In order to see the necessity, let A^f be a bounded operator, i.e. $\|A^f u\| \le c \|u\|$ for some $c > 0$ and all $u \in \mathfrak{B}$. Suppose that $\lim_{\lambda \to \infty} f(\lambda) = \infty$. By Corollary 13.32,

$$\frac{\|T_\epsilon u - u\|}{\|u\|} \le \frac{2(e+1)}{f(\epsilon^{-1})} \frac{\|A^f u\|}{\|u\|} \le \frac{2(e+1)c}{f(\epsilon^{-1})}$$

for all $u \in \mathfrak{B}$, $u \ne 0$. Thus,

$$\lim_{\epsilon \to 0} \sup_{0 \ne u \in \mathfrak{B}} \frac{\|T_\epsilon u - u\|}{\|u\|} \le \lim_{\epsilon \to 0} \frac{2(e+1)c}{f(\epsilon^{-1})} = \lim_{\lambda \to \infty} \frac{2(e+1)c}{f(\lambda)} = 0.$$

which shows that $(T_t)_{t \ge 0}$ is continuous in the uniform operator topology. This means that the generator A of $(T_t)_{t \ge 0}$ is bounded, cf. [286, p. 2, Theorem 1.2].   □

Our next corollary is a generalization of the well known relation for fractional powers $\mathfrak{D}((-A)^\beta) \subset \mathfrak{D}((-A)^\alpha)$ if $0 < \alpha \le \beta \le 1$.

Corollary 13.34. *Let $(T_t)_{t \ge 0}$ be a C_0-contraction semigroup on the Banach space $\mathfrak{B}$ with generator $(A, \mathfrak{D}(A))$ and let $f \in \mathcal{SBF}$, $g \in \mathcal{BF}$ with Lévy triplets (a, b, μ^f) and $(\alpha, 0, \mu^g)$. If*

$$\int_{(0,\epsilon)} \frac{1}{f(t^{-1})} \mu^g(dt) < \infty \quad \text{for some } \epsilon > 0, \tag{13.49}$$

then $\mathfrak{D}(A^f) \subset \mathfrak{D}(A^g)$.

Proof. If $u \in \mathfrak{D}(A) \subset \mathfrak{D}(A^f)$, we can use Corollary 13.32 and get for sufficiently small $\epsilon > 0$

$$\|A^g u\| = \left\| -\alpha u + \int_{(0,\infty)} (T_t u - u) \, \mu^g(dt) \right\|$$

$$\le \int_{(0,\epsilon)} \|T_t u - u\| \, \mu^g(dt) + \left(\alpha + 2 \int_{[\epsilon,\infty)} \mu^g(dt) \right) \|u\|$$

$$\le 2(e+1) \int_{(0,\epsilon)} \frac{1}{f(t^{-1})} \mu^g(dt) \|A^f u\| + \left(\alpha + 2\mu^g[\epsilon,\infty) \right) \|u\|.$$

Because of our assumption, the integral above converges, and since $\mathfrak{D}(A)$ is an operator core for A^f and A^g, the inequality extends to all $u \in \mathfrak{D}(A^f)$. This shows that $u \in \mathfrak{D}(A^g)$.   □

Corollary 13.35. *Let $(T_t)_{t\geq 0}$ be a C_0-contraction semigroup on the Banach space $\mathcal{B}$ with generator $(A, \mathcal{D}(A))$ and let $f \in \mathcal{SBF}$, $g \in \mathcal{BF}$ with Lévy triplets (a, b, μ^f) and $(\alpha, 0, \mu^g)$. If*

$$\int_1^\infty \frac{f'(\lambda)g(\lambda)}{f^2(\lambda)}\, d\lambda < \infty \quad and \quad \frac{g(\lambda)}{f(\lambda)} = O(1) \quad as\ \lambda \to \infty, \tag{13.50}$$

then $\mathcal{D}(A^f) \subset \mathcal{D}(A^g)$.

Proof. In the notation of Corollary 13.34 it is easy to see that

$$\mu^g[\delta, \epsilon) \leq \frac{1}{(1-e^{-1})}\int_{[\delta,\infty)}(1 - e^{-s/\delta})\,\mu^g(ds) \leq \frac{e}{e-1}\,g(\delta^{-1})$$

holds for all $0 < \delta < \epsilon$. Note that

$$\lim_{\delta \to 0}\int_{[\delta,\epsilon)} \frac{1}{f(t^{-1})}\,\mu^g(dt) = \int_{(0,\epsilon)} \frac{1}{f(t^{-1})}\,\mu^g(dt),$$

whenever the integral on the right-hand side exists. Integrating by parts we get

$$\begin{aligned}
\int_{[\delta,\epsilon)} \frac{1}{f(t^{-1})}\,\mu^g(dt) &= \frac{\mu^g[\delta,\epsilon)}{f(\delta^{-1})} + \int_{[\delta,\epsilon)} \frac{t^{-2}f'(t^{-1})}{f^2(t^{-1})}\,\mu^g[t,\epsilon)\,dt \\
&\leq \frac{e}{e-1}\left(\frac{g(\delta^{-1})}{f(\delta^{-1})} + \int_{[\delta,\epsilon)} \frac{t^{-2}f'(t^{-1})g(t^{-1})}{f^2(t^{-1})}\,dt\right)
\end{aligned}$$

for $\epsilon > 0$ and $\delta \in (0, \epsilon)$. This shows that (13.50) implies (13.49) and the assertion follows from Corollary 13.34. $\qquad\square$

Corollary 13.35 can be further simplified. Since Bernstein functions are non-decreasing and concave, we have

$$\frac{f'(\lambda)}{f(\lambda)} \leq \frac{1}{\lambda}.$$

Therefore the integrand appearing in (13.50) can be estimated by

$$\frac{f'(\lambda)g(\lambda)}{f^2(\lambda)} \leq \frac{g(\lambda)}{\lambda} \cdot \frac{1}{f(\lambda)}.$$

The expression on the right-hand side is the product of two completely monotone functions, see Corollary 3.8 (iv) and Theorem 3.7 (ii) for the first and the second factor, respectively. In particular, $g(\lambda)/(\lambda f(\lambda))$ is non-increasing.

Using a standard Abelian argument shows that $\int_1^\infty g(\lambda)/(\lambda f(\lambda))\, d\lambda < \infty$ always entails that, for some constant $C > 0$,

$$\frac{g(\lambda)}{\lambda f(\lambda)} \leq \frac{C}{\lambda} \quad \text{for all}\ \lambda > 1.$$

Thus, $g(\lambda)/f(\lambda) = O(1)$ as $\lambda \to \infty$. If (α, β, μ^g) is the Lévy triplet of g, then by Corollary 3.8 (viii), there exists $c > 0$ such that

$$\frac{g(\lambda)}{\lambda f(\lambda)} \geq \frac{\alpha + \beta\lambda}{c\,\lambda^2} \geq \frac{\beta}{c\,\lambda}, \qquad \lambda > 1.$$

From this we conclude that $\beta = 0$, and we have proved the following result.

Corollary 13.36. *Let $f \in \mathcal{SBF}$, $g \in \mathcal{BF}$ and let $(T_t)_{t \geq 0}$ be a C_0-semigroup on the Banach space $\mathcal{B}$ with generator $(A, \mathfrak{D}(A))$. If*

$$\int_1^\infty \frac{g(\lambda)}{\lambda f(\lambda)}\, d\lambda < \infty, \tag{13.51}$$

then $\mathfrak{D}(A^f) \subset \mathfrak{D}(A^g)$.

In particular, if for some $\kappa > 0$ the ratio $g(\lambda)/f(\lambda) = O(\lambda^{-\kappa})$ as $\lambda \to \infty$, the condition (13.51) holds.

The following result shows that the functional calculus is stable under pointwise limits of Bernstein functions.

Theorem 13.37. *Let $(T_t)_{t \geq 0}$ be a C_0-contraction semigroup on the Banach space $\mathcal{B}$ with generator $(A, \mathfrak{D}(A))$. Assume that $(f^n)_{n \in \mathbb{N}}$ is a sequence of Bernstein functions such that the limit $f(\lambda) = \lim_{n \to \infty} f^n(\lambda)$ exists. Then*

$$\text{strong-}\lim_{n \to \infty} A^{f^n} u = A^f u, \quad \text{for all } u \in \mathfrak{D}(A^2).$$

Proof. Let $f(0+) = a > 0$ and write $f_k = kf/(k + f)$ and $f_k^n = kf^n/(k + f^n)$. Each $k/(k + f^n)$ is a completely monotone function which is bounded by 1; therefore, it is the Laplace transform of a sub-probability measure v_k^n and we see

$$\frac{f_k^n}{f_k} = \frac{kf^n}{k + f^n}\,\frac{k + f}{kf} = \left(\frac{k}{f} + 1\right)\left(1 - \frac{k}{k + f^n}\right)$$

$$= \left(\frac{k}{f} + 1\right) - \mathscr{L}v_k^n \cdot \left(\frac{k}{f} + 1\right).$$

Since $1/f = \mathscr{L}U \in \mathcal{CM}$, cf. (5.20), we see that $f_k^n/f_k = \mathscr{L}\gamma_k^n$ with the signed measure $\gamma_k^n = (kU + \delta_0) - v_k^n \star (kU + \delta_0)$. The total variations of the signed measures $(\gamma_k^n)_{n \in \mathbb{N}}$ satisfy

$$\sup_{n \in \mathbb{N}} \|\gamma_k^n\|_{TV} \leq 2\left(\frac{k}{f(0+)} + 1\right) < \infty \quad \text{for every } k \in \mathbb{N}.$$

Now Remark 13.21 applies to the sequence $(f_k^n)_{n \in \mathbb{N}}$ and shows that

$$\text{strong-}\lim_{n \to \infty} A^{f_k^n} u = A^{f_k} u \quad \text{for all } k \in \mathbb{N},\ u \in \mathfrak{D}(A).$$

Let (a^n, b^n, μ^n) denote the Lévy triplet of f^n. Recall from the proof of Phillips' theorem, Theorem 13.6, the estimate

$$\|A^{f^n} u\| \le \left(b^n + \int_{(0,1)} s\, \mu^n(ds) \right) \|Au\| + \left(a^n + \mu^n[1, \infty) \right) \|u\|.$$

Combining this with the elementary convexity inequality $\min\{1, s\} \le \frac{e}{e-1}(1 - e^{-s})$, $s > 0$, we get

$$\|A^{f^n} u\| \le \frac{2e\, f^n(1)}{e - 1}\, (\|Au\| + \|u\|) \le c\, (\|Au\| + \|u\|), \quad n \in \mathbb{N},\ u \in \mathfrak{D}(A),$$

where c is independent of $n \in \mathbb{N}$. Because of Proposition 13.19 we have

$$A^{f^n} u - A^{f^n_k} u = A^{f^n} u - k R^{f^n}_k A^{f^n} u = -R^{f^n}_k A^{f^n} A^{f^n} u$$

which implies that for all $u \in \mathfrak{D}(A^2)$

$$\|A^{f^n} u - A^{f^n_k} u\| \le \frac{1}{k} \|A^{f^n} A^{f^n} u\| \le \frac{C}{k}\, (\|A^2 u\| + \|Au\| + \|u\|).$$

The same argument shows that the inequality is valid for $\|A^f u - A^{f_k} u\|$ with the same constant C. Therefore,

$$\|A^{f^n} u - A^f u\| \le \|A^{f^n} u - A^{f^n_k} u\| + \|A^{f^n_k} u - A^{f_k} u\| + \|A^{f_k} u - A^f u\|$$

$$\le \frac{2C}{k}\, (\|A^2 u\| + \|Au\| + \|u\|) + \|A^{f^n_k} u - A^{f_k} u\|.$$

Letting $n \to \infty$ and then $k \to \infty$ proves that $\lim_{n\to\infty} A^{f^n} u = A^f u$ on $\mathfrak{D}(A^2)$.

If $f(0+) = 0$, we replace f and f^n in the above calculations by $f + \epsilon$ and $f^n + \epsilon$, respectively. Because of Theorem 13.23

$$-\epsilon u + A^{f^n} u = A^{\epsilon + f^n} u \xrightarrow{n \to \infty} A^{\epsilon + f} u = -\epsilon u + A^f u, \quad u \in \mathfrak{D}(A^2),$$

and the general case follows. $\qquad\qquad\qquad\qquad\qquad\qquad\qquad\qquad\qquad\qquad\quad \square$

Remark 13.38. It is straightforward to extend the functional calculus to Stieltjes functions.

Let $(A, \mathfrak{D}(A))$ be the generator of a C_0-contraction semigroup $(T_t)_{t \ge 0}$ on the Banach space $\mathfrak{B}$ and denote by $(R_\lambda)_{\lambda > 0}$ the resolvent. We define the *zero-resolvent* and the *potential operator* by

$$R_0 u := \lim_{\lambda \to 0} R_\lambda u, \quad \text{and} \quad V u := \lim_{n \to \infty} \int_0^n T_t u\, dt,$$

with domains $\mathfrak{D}(R_0)$ and $\mathfrak{D}(V)$, respectively, comprising all $u \in \mathfrak{B}$ where the limits above exist in the strong sense. The following facts are well known, see e.g. [39, Chapter II.11]. The zero-resolvent $(R_0, \mathfrak{D}(R_0))$ is an extension of the potential operator $(V, \mathfrak{D}(V))$ and R_0 is densely defined if, and only if, $\lim_{\lambda \to 0} \lambda R_\lambda u = 0$ for all $u \in \mathfrak{B}$. Moreover, the range $\mathfrak{R}(R_0) \subset \mathfrak{D}(A)$, and on $\mathfrak{D}(R_0)$ we have $A R_0 u = -u$. Since

$$\overline{\mathfrak{R}(A)} = \left\{ u \in \mathfrak{B} : \lim_{\lambda \to 0} \lambda R_\lambda u = 0 \right\},$$

R_0 is densely defined if, and only if, the range $\mathfrak{R}(A)$ is dense in $\mathfrak{B}$. If R_0 is densely defined we have $R_0 = -A^{-1}$.

Assume, therefore, that $\mathfrak{D}(R_0)$ is dense in $\mathfrak{B}$.

For every $\phi \in \mathcal{S}$ given by $\phi(\lambda) = a\lambda^{-1} + b + \int_{(0,\infty)} (\lambda + t)^{-1} \sigma(dt)$ we define

$$\phi(-A)u := a R_0 u + bu + \int_{(0,\infty)} R_t u \, \sigma(dt), \quad u \in \mathfrak{R}(A).$$

Since for $u = Av$, $v \in \mathfrak{D}(A)$, it holds that $R_0 u = R_0 A v = -v$ and $R_t u = R_t A v$, we see that

$$\phi(-A)u = A^f v, \quad \text{for all } u \in \mathfrak{R}(A), \ u = Av,$$

where f is the complete Bernstein function given by

$$f(\lambda) = a + b\lambda + \int_{(0,\infty)} \frac{\lambda}{t + \lambda} \sigma(dt).$$

Denote by $f^\star$ the conjugate of the complete Bernstein function f, i.e. the unique function $f^\star \in \mathcal{CBF}$ such that $f(\lambda) f^\star(\lambda) = \lambda$. Then we see for all $u \in \mathfrak{R}(A)$

$$u = Av = -A^{f^\star} A^f v = -A^{f^\star} \phi(-A)u$$

which means that $-\phi(-A)$ is the right-inverse of $A^{f^\star}$ on $\mathfrak{R}(A)$.

Now let $v = A^{f^\star} w$ for $w \in \mathfrak{D}(A A^{f^\star})$. Then $u = Av = A A^{f^\star} w = A^{f^\star} Aw$ and

$$\phi(-A) A^{f^\star} Aw = A^f A^{f^\star} w = -Aw$$

which shows that $-\phi(-A)$ is the left-inverse of $A^{f^\star}$ on $\mathfrak{R}(A) \cap \mathfrak{D}(A^{f^\star})$. This proves the following result:

Let $(A, \mathfrak{D}(A))$ be the generator of some C_0-contraction semigroup on the Banach space $\mathfrak{B}$ such that the range $\mathfrak{R}(A)$ is dense in $\mathfrak{B}$ and let $f, f^\star \in \mathcal{CBF}$ be conjugate, i.e. $f(\lambda) f^\star(\lambda) = \lambda$. Then $\phi(\lambda) := f(\lambda)/\lambda$ is a Stieltjes function and

$$-(A^{f^\star})^{-1} u = \phi(-A)u \quad \text{for all } u \in \mathfrak{R}(A) \cap \mathfrak{D}(A^{f^\star}).$$

13.3   Subordination and functional inequalities

In this section we show that certain functional inequalities are preserved under subordination. Functional inequalities are important tools to prove regularity results for semigroups. The following deep result, taken from [362], illustrates this nicely. Let m be a Radon measure on a locally compact separable metric space E such that $\operatorname{supp} m = E$. Assume that $(T_t)_{t\geq0}$ is a symmetric sub-Markovian C_0-contraction semigroup on $L^2(E,m)$ with generator $(A, \mathfrak{D}(A))$. For $p > 2$ and $n := 2p/(p-2)$ the following inequalities are equivalent:

$$\|u\|_{L^p}^2 \leq c \, \langle Au, u\rangle_{L^2} \quad \text{for all } u \in \mathfrak{D}(A); \tag{13.52}$$

$$\|u\|_{L^2}^2 \left(\frac{\|u\|_{L^2}^2}{\|u\|_{L^1}^2} \right)^{2/n} \leq c' \langle Au, u\rangle_{L^2} \quad \text{for all } u \in \mathfrak{D}(A) \cap L^1; \tag{13.53}$$

$$\|T_t\|_{L^1 \to L^\infty} \leq c'' \, t^{-n/2} \quad \text{for all } t > 0. \tag{13.54}$$

The inequality (13.52) is often called a *Sobolev-type inequality*, (13.53) a *Nash-type inequality* and (13.54) is the *ultracontractivity* of the semigroup $(T_t)_{t\geq0}$. If the semigroup is ultracontractive, then there exists a family of kernels $p_t(x, y)$, $t \geq 0$, such that $T_t u(x) = \int u(y) p_t(x, y)\, dy$ and we can interpret (13.54) as an on-diagonal estimate: $\operatorname{ess\,sup}_{x,y} p_t(x, y) \leq c'' \, t^{-n/2}$.

Functional inequalities can also be used to study spectral properties of A and $(T_t)_{t\geq0}$. For a state-of-the art survey we refer to the monograph [364].

Throughout this section $(\mathfrak{B}, \|\cdot\|)$ denotes a Banach space, $(\mathfrak{B}^*, \|\cdot\|_*)$ its topological dual and $\langle u, \phi\rangle$ stands for the dual pairing between $u \in \mathfrak{B}$ and $\phi \in \mathfrak{B}^*$. By u^* we denote a normalized tangent functional (or duality map) of $u \in \mathfrak{B}$, i.e. any element $u^* \in \mathfrak{B}^*$ such that $\|u\|^2 = \|u^*\|_*^2 = \langle u, u^*\rangle$. By the Hahn–Banach theorem, each $u \in \mathfrak{B}$ has at least one normalized tangent functional. The map $u \mapsto u^*$ is single-valued if, and only if, $\mathfrak{B}$ is smooth; this is the same as to say that $u \mapsto \|u\|$ is Gateaux-differentiable, see e.g. [218, §26]. A sufficient, and in the reflexive case also necessary, condition is that $\mathfrak{B}^*$ is strictly convex. Note that the duality mapping $u \mapsto u^*$ is homogeneous.

As in the previous sections, $(T_t)_{t\geq0}$ is a C_0-contraction semigroup on the Banach space $(\mathfrak{B}, \|\cdot\|)$, the generator is denoted by $(A, \mathfrak{D}(A))$ and Φ is a (not necessarily linear) functional $\Phi : \mathfrak{B} \to [0, \infty]$ such that

$$\Phi(u) = 0 \implies u = 0, \quad \Phi(cu) = |c|\Phi(u), \quad \text{and} \quad \Phi(T_t u) \leq \Phi(u) \tag{13.55}$$

hold for all $c \in \mathbb{R}$, $t \geq 0$ and $u \in \mathfrak{B}$.

Functional inequalities are often considered in the Hilbert space $L^2(E,m)$. In this case the normalized tangent functional u^* is unique and can be identified with u, and $\Phi(u)$ is usually taken to be the norm in $L^1(E,m)$ or $L^\infty(E,m)$.

For the main result, Theorem 13.41 below, we need a few preparations. We begin with a differential inequality which is an abstract version of the well known Gronwall–Bellman–Bihari inequality.

Recall that for an increasing function $G : (0, \infty) \to \mathbb{R}$ the *generalized (right continuous) inverse* $G^{-1} : \mathbb{R} \to [0, +\infty]$ is defined as

$$G^{-1}(y) := \inf \{t > 0 : G(t) > y\}, \quad \inf \emptyset := \infty.$$

If G is bijective, e.g. strictly increasing and onto, then G^{-1} is just the usual inverse.

Lemma 13.39. *Let $h : [0, \infty) \to [0, \infty)$ be a differentiable function. Suppose that there exists an increasing function $\varphi : [0, \infty) \to [0, \infty)$ such that $\varphi(t) > 0$ for $t > 0$, $\int_{0+} 1/\varphi(t)\, dt = \infty$ and*

$$h'(t) \le -\varphi(h(t)) \quad \text{for all } t \ge 0. \tag{13.56}$$

Then, we have
$$h(t) \le G^{-1}(G(h(0)) - t) \quad \text{for all } t \ge 0, \tag{13.57}$$

where G^{-1} is the (generalized right continuous) inverse of

$$G(t) = \begin{cases} \displaystyle\int_1^t \frac{dr}{\varphi(r)}, & \text{if } t \ge 1, \\[2mm] \displaystyle-\int_t^1 \frac{dr}{\varphi(r)}, & \text{if } t \le 1. \end{cases}$$

Proof. Since $h'(t) \le -\varphi(h(t)) \le 0$, the set $I = \{t : h(t) > 0\}$ is a (bounded or unbounded) interval; for $t \in I$, the function $h(t)$ is strictly decreasing. With the convention $\int_a^b = -\int_b^a$, we see for all $t \in I$ that

$$G(h(t)) = \int_1^{h(t)} \frac{1}{\varphi(r)}\, dr = G(h(0)) + \int_{h(0)}^{h(t)} \frac{1}{\varphi(r)}\, dr$$

$$= G(h(0)) + \int_0^t \frac{h'(r)\, dr}{\varphi(h(r))}$$

$$\le G(h(0)) - t.$$

If $t \notin I$, $G(h(t)) = G(0) = -\infty$, and the above inequality is trivial. The claim follows from the definition of the generalized inverse G^{-1}. $\qquad\square$

An abstract Nash inequality can be characterized in terms of estimates for the underlying semigroup. This is the essence of the next proposition.

Proposition 13.40. *Let* $(T_t)_{t\geq 0}$ *be a* C_0-*contraction semigroup on the Banach space* $\mathfrak{B}$. *Denote by* $(A, \mathfrak{D}(A))$ *its generator and let* $\Phi : \mathfrak{B} \to [0, \infty]$ *be a (not necessarily linear) functional satisfying (13.55). Then the following abstract Nash inequality*

$$\|u\|^2 \, B\left(\|u\|^2\right) \leq \text{Re}\,\langle -Au, u^*\rangle, \quad u \in \mathfrak{D}(A), \ \Phi(u) = 1, \tag{13.58}$$

with some increasing function $B : (0, \infty) \to (0, \infty)$ *holds if, and only if,*

$$\|T_t u\|^2 \leq G^{-1}\left(G(\|u\|^2) - t\right) \quad \text{for all } t \geq 0, \ u \in \mathfrak{D}(A), \ \Phi(u) = 1, \tag{13.59}$$

where

$$G(t) = \begin{cases} \displaystyle\int_1^t \frac{dr}{2rB(r)}, & \text{if } t \geq 1, \\[3mm] \displaystyle-\int_t^1 \frac{dr}{2rB(r)}, & \text{if } t \leq 1. \end{cases} \tag{13.60}$$

Proof. Assume that (13.58) holds. Then,

$$\|u\|^2 \, B\left(\frac{\|u\|^2}{\Phi^2(u)}\right) \leq \text{Re}\,\langle -Au, u^*\rangle, \quad u \in \mathfrak{D}(A).$$

For all $u \in \mathfrak{D}(A)$ with $\Phi(u) = 1$ we have

$$\begin{aligned}
\frac{d}{dt}\|T_t u\|^2 &= 2\|T_t u\| \frac{d}{dt}\|T_t u\| \\
&= 2\|T_t u\| \lim_{h\to 0+} \frac{1}{h}\left(\frac{\langle T_t u, (T_t u)^*\rangle}{\|T_t u\|} - \frac{\langle T_{t-h}u, (T_{t-h}u)^*\rangle}{\|T_{t-h}u\|}\right) \\
&\leq 2 \lim_{h\to 0+} \frac{1}{h}\Big(\langle T_t u, (T_t u)^*\rangle - \text{Re}\,\langle T_{t-h}u, (T_t u)^*\rangle\Big) \\
&= 2 \lim_{h\to 0+} \text{Re}\,\left\langle \frac{T_t u - T_{t-h}u}{h}, (T_t u)^*\right\rangle \\
&= 2\,\text{Re}\,\langle AT_t u, (T_t u)^*\rangle.
\end{aligned}$$

The inequality in the above calculation follows from the definition of a normalized tangent functional,

$$\text{Re}\,\langle T_{t-h}u, (T_t u)^*\rangle \leq \|T_{t-h}u\| \cdot \|(T_t u)^*\|_* = \frac{\langle T_{t-h}u, (T_{t-h}u)^*\rangle}{\|T_{t-h}u\|}\|T_t u\|.$$

Since the function B is increasing and $\Phi(T_t u) \leq \Phi(u) = 1$, we have

$$\frac{d}{dt}\|T_t u\|^2 \leq -2\,\|T_t u\|^2 \, B\left(\|T_t u\|^2\right).$$

This, together with Lemma 13.39, proves (13.59).

Conversely, assume that (13.59) holds. Then, for all $u \in \mathfrak{D}(A)$ with $\Phi(u) = 1$,

$$
\begin{aligned}
2 \operatorname{Re} \langle -Au, u^* \rangle &= \lim_{t \to 0+} \left(\frac{\|u\| + \|T_t u\|}{\|u\|} \operatorname{Re} \langle -Au, u^* \rangle \right) \\
&= \lim_{t \to 0+} \left(\frac{\|u\| + \|T_t u\|}{\|u\|} \operatorname{Re} \left\langle \frac{1}{t}(u - T_t u), u^* \right\rangle \right) \\
&= \lim_{t \to 0+} \left(\frac{\|u\| + \|T_t u\|}{\|u\|} \frac{\langle u, u^* \rangle - \operatorname{Re} \langle T_t u, u^* \rangle}{t} \right) \\
&\geq \lim_{t \to 0+} \left(\frac{\|u\| + \|T_t u\|}{\|u\|} \frac{\|u\|^2 - \|T_t u\| \|u\|}{t} \right) \\
&= \lim_{t \to 0+} \frac{\|u\|^2 - \|T_t u\|^2}{t} \\
&\geq \lim_{t \to 0+} \frac{\|u\|^2 - G^{-1} \left(G(\|u\|^2) - t \right)}{t} \\
&= -\frac{d}{dt} G^{-1} \left(G(\|u\|^2) - t \right) \Big|_{t=0} \\
&= 2 \left[G^{-1} \left(G(\|u\|^2) - t \right) \cdot B \left(G^{-1}(G(\|u\|^2) - t) \right) \right] \Big|_{t=0} \\
&= 2 \|u\|^2 B(\|u\|^2),
\end{aligned}
$$

which is just the abstract Nash inequality (13.58).   □

Theorem 13.41. *Let $(T_t)_{t \geq 0}$ be a C_0-contraction semigroup on the Banach space $\mathfrak{B}$. Denote by $(A, \mathfrak{D}(A))$ its generator and let $\Phi : \mathfrak{B} \to [0, \infty]$ be a (not necessarily linear) functional satisfying (13.55). Assume that the following abstract Nash inequality holds*

$$
\|u\|^2 B \left(\|u\|^2 \right) \leq \operatorname{Re} \langle -Au, u^* \rangle, \quad u \in \mathfrak{D}(A), \ \Phi(u) = 1, \tag{13.61}
$$

where $B : (0, \infty) \to (0, \infty)$ is an increasing function. Then, for any Bernstein function f, the generator A^f of the subordinate semigroup satisfies

$$
\frac{\|u\|^2}{4} f \left(2B \left(\frac{\|u\|^2}{2} \right) \right) \leq \operatorname{Re} \langle -A^f u, u^* \rangle, \quad u \in \mathfrak{D}(A^f), \ \Phi(u) = 1. \tag{13.62}
$$

Proof. For the Bernstein function $a + b\lambda$ it is obvious that (13.61) implies (13.62). Therefore, it is enough to prove the theorem for Bernstein functions of the form $f(\lambda) = \int_{(0,\infty)} (1 - e^{-t\lambda}) \, \mu(dt)$. Since $\mathfrak{D}(A)$ is an operator core for $(A^f, \mathfrak{D}(A^f))$, cf. Theorem 13.6, we only need to consider (13.62) for $u \in \mathfrak{D}(A)$.

We know from (13.61) and Proposition 13.40 that for all $t \geq 0$, $u \in \mathfrak{D}(A)$ with $\Phi(u) = 1$ and for G given by (13.60)

$$
\|T_t u\|^2 \leq G^{-1} \left(G(\|u\|^2) - t \right).
$$

From the definition of a normalized tangent functional u^* we see

$$\frac{\mathrm{Re}\,\langle T_t u, u^*\rangle}{\|u\|^2} \leq \frac{|\langle T_t u, u^*\rangle|}{\|u\|^2} \leq \frac{\|T_t u\| \cdot \|u^*\|_*}{\|u\|^2} \leq \frac{\sqrt{G^{-1}(G(\|u\|^2) - t)}}{\|u\|}.$$

Phillips' formula (13.10) for A^f yields for any $u \in \mathfrak{D}(A)$ with $\Phi(u) = 1$,

$$\begin{aligned}
\mathrm{Re}\,\langle -A^f u, u^*\rangle &= \int_{(0,\infty)} \mathrm{Re}\,\langle u - T_s u, u^*\rangle\,\mu(ds) \\
&= \|u\|^2 \int_{(0,\infty)} \left(1 - \frac{\mathrm{Re}\,\langle T_s u, u^*\rangle}{\|u\|^2}\right) \mu(ds) \\
&\geq \|u\|^2 \int_{(0,\infty)} \left(1 - \frac{\sqrt{G^{-1}(G(\|u\|^2) - s)}}{\|u\|}\right) \mu(ds) \\
&= \|u\|^2 \int_{(0,\infty)} \left(\frac{1 - \dfrac{G^{-1}(G(\|u\|^2) - s)}{\|u\|^2}}{1 + \dfrac{\sqrt{G^{-1}(G(\|u\|^2) - s)}}{\|u\|}}\right) \mu(ds) \\
&\geq \frac{\|u\|^2}{2} \int_{(0,\infty)} \left(1 - \frac{G^{-1}(G(\|u\|^2) - s)}{\|u\|^2}\right) \mu(ds),
\end{aligned}$$

since G^{-1} is increasing. Thus,

$$\mathrm{Re}\,\langle -A^f u, u^*\rangle \geq g(\|u\|^2),$$

where

$$g(r) = \frac{r}{2} \int_{(0,\infty)} \left(1 - \frac{G^{-1}(G(r) - s)}{r}\right) \mu(ds).$$

For all $r > 0$ we have

$$\begin{aligned}
g(r) &= \frac{1}{2} \int_{(0,\infty)} \left(r - G^{-1}(G(r) - s)\right) \mu(ds) \\
&= \frac{1}{2} \int_{(0,\infty)} \left(\int_{G(r)-s}^{G(r)} dG^{-1}(t)\right) \mu(ds) \\
&= \frac{1}{2} \int_{(0,\infty)} \int_{\mathbb{R}} \mathbb{1}_{(-\infty,G(r))}(t) \mathbb{1}_{(G(r)-t,\infty)}(s)\, dG^{-1}(t)\,\mu(ds) \\
&= \frac{1}{2} \int_{(-\infty,G(r))} \mu(G(r) - t, \infty)\, dG^{-1}(t) \\
&= \frac{1}{2} \int_0^r \mu(G(r) - G(t), \infty)\, dt.
\end{aligned}$$

In the last equality we used that B is increasing, $G(t) > -\infty$ for all $t > 0$ and $G(0) = -\infty$; this follows from

$$G(0) = -\int_0^1 \frac{dr}{2rB(r)} \leq \frac{-1}{2B(1)} \int_0^1 \frac{dr}{r} = -\infty.$$

Using again the monotonicity of B, we find from the mean value theorem

$$\frac{G(r) - G(t)}{r - t} \leq \frac{1}{2tB(t)} \quad \text{for all} \quad 0 < t < r. \tag{13.63}$$

Therefore,

$$\begin{aligned}
g(r) &\geq \frac{1}{2} \int_0^r \mu\left(\frac{1}{2tB(t)}(r-t), \infty\right) dt \\
&\geq \frac{1}{2} \int_{r/2}^r \mu\left(\frac{1}{2tB(t)}(r-t), \infty\right) dt \\
&\geq \frac{1}{2} \int_{r/2}^r \mu\left(\frac{1}{rB(r/2)}(r-t), \infty\right) dt \\
&= \frac{1}{2} \int_0^{r/2} \mu\left(\frac{v}{rB(r/2)}, \infty\right) dv \\
&= \frac{1}{2} rB(r/2) \int_0^{1/2B(r/2)} \mu(s, \infty)\, ds.
\end{aligned}$$

Now we can use Lemma 3.4 to deduce that

$$g(r) \geq \frac{r}{4} f\left(2B\left(\frac{r}{2}\right)\right) \quad \text{for all} \quad r > 0,$$

which implies (13.62).   □

Remark 13.42. If $(T_t)_{t \geq 0}$ is a self-adjoint C_0-contraction semigroup on a Hilbert space, then we can replace the left-hand side of (13.62) by

$$\frac{\|u\|^2}{2} f\left(B\left(\frac{\|u\|^2}{2}\right)\right).$$

Since the Bernstein function f is subadditive, the estimate $\frac{1}{2}f(2\lambda) \leq f(\lambda)$, $\lambda > 0$, shows that this improves (13.62).

We can use Theorem 13.41 to show that super-Poincaré and weak Poincaré inequalities are also preserved under subordination.

Corollary 13.43. *Let $(A, D(A))$ be the generator of a C_0-contraction semigroup on the Banach space $\mathfrak{B}$, let $\Phi : \mathfrak{B} \to [0, \infty]$ be a (not necessarily linear) functional satisfying (13.55), and let f be a Bernstein function. Assume that A satisfies the following super-Poincaré inequality*

$$\|u\|^2 \le r \operatorname{Re} \langle -Au, u^* \rangle + \beta(r)\, \Phi^2(u), \quad r > 0,\ u \in \mathfrak{D}(A), \tag{13.64}$$

where $\beta : (0, \infty) \to (0, \infty)$ is a decreasing function such that $\lim_{r\to\infty} \beta(r) = 0$ and $\beta(0) := \lim_{r\to 0} \beta(r) = \infty$. Then the generator A^f of the subordinate semigroup also satisfies a super-Poincaré inequality

$$\|u\|^2 \le r \operatorname{Re} \langle -A^f u, u^* \rangle + \beta_f(r)\, \Phi^2(u), \quad r > 0,\ u \in \mathfrak{D}(A^f), \tag{13.65}$$

where $\beta_f(r) = 4\beta(1/f^{-1}(4/r))$ and $f^{-1}(y) = \infty$ if $y \notin \mathfrak{R}(f)$.

Proof. Set $B(x) := \sup_{s>0} \frac{1}{s}\left(1 - \frac{\beta(s)}{x}\right)$. Then we can rewrite (13.64) in the following form

$$\|u\|^2 B(\|u\|^2) \le \operatorname{Re} \langle -Au, u^* \rangle, \quad u \in \mathfrak{D}(A),\ \Phi(u) = 1.$$

Clearly, $B(x)$ is an increasing function on $(0, \infty)$. Since $\beta^{-1} : (0, \infty) \to (0, \infty)$, we infer from

$$\frac{1}{2\,\beta^{-1}\left(\frac{x}{2}\right)} = \frac{1 - \frac{\beta\left(\beta^{-1}\left(\frac{x}{2}\right)\right)}{x}}{\beta^{-1}\left(\frac{x}{2}\right)} \le B(x) = \sup_{s \ge \beta^{-1}\left(\frac{x}{2}\right)} \frac{1 - \frac{\beta(s)}{x}}{s} \le \frac{1}{\beta^{-1}(x)} \tag{13.66}$$

that $B : (0, \infty) \to (0, \infty)$. Theorem 13.41 shows that

$$C(\|u\|^2) \le \operatorname{Re} \langle -A^f u, u^* \rangle, \quad u \in \mathfrak{D}(A),\ \Phi(u) = 1,$$

where

$$C(x) = \frac{x}{4} f\left(2B\left(\frac{x}{2}\right)\right) = \frac{x}{4} \sup_{s>0} f\left(\frac{2}{s}\left(1 - \frac{2\beta(s)}{x}\right)\right).$$

For $r > 0$ we define $\widetilde{\beta}(r) := \sup_{s>0}\{C^{-1}(s) - rs\}$. Then

$$\|u\|^2 \le r \operatorname{Re} \langle -A^f u, u^* \rangle + \widetilde{\beta}(r)\, \Phi^2(u), \quad r > 0,\ u \in \mathfrak{D}(A).$$

Let us now estimate $\widetilde{\beta}(r)$. By (13.66),

$$C(x) \ge \frac{x}{4} f\left(\frac{1}{\beta^{-1}(x/4)}\right) := C_0(x), \quad \text{hence,} \quad C^{-1}(y) \le C_0^{-1}(y).$$

By its very definition, $C_0 : (0, \infty) \to (0, \infty)$ is a strictly increasing function such that $\lim_{x\to 0} C_0(x) = 0$ and $\lim_{x\to\infty} C_0(x) = \infty$. Therefore,

$$C_0^{-1}(y) = 4y \left[f\left(\frac{1}{\beta^{-1}(C_0^{-1}(y)/4)}\right) \right]^{-1}. \tag{13.67}$$

On the other hand,

$$\widetilde{\beta}(r) \leq \sup_{y>0} \left\{ C_0^{-1}(y) - ry \right\} = \sup_{y>0,\, C_0^{-1}(y) \geq ry} C_0^{-1}(y),$$

and from (13.67) we see that $C_0^{-1}(y) \geq ry$ leads to

$$\frac{1}{f^{-1}(4/r)} \leq \beta^{-1}\left(C_0^{-1}(y)/4\right).$$

Since β is decreasing, we can rewrite this as

$$C_0^{-1}(y) \leq 4\beta\left(\frac{1}{f^{-1}(4/r)}\right),$$

and so

$$\widetilde{\beta}(r) \leq \sup_{y>0,\, C_0^{-1}(y) \leq 4\beta(1/f^{-1}(4/r))} C_0^{-1}(y) \leq 4\beta\left(\frac{1}{f^{-1}(4/r)}\right).$$

This finishes the proof.   □

Corollary 13.44. *Let $(A, D(A))$ be the generator of a C_0-contraction semigroup on the Banach space $\mathfrak{B}$, let $\Phi : \mathfrak{B} \to [0, \infty]$ be a (not necessarily linear) functional satisfying (13.55), and let f be a Bernstein function. Assume that A satisfies the following weak Poincaré inequality*

$$\|u\|^2 \leq \alpha(s)\mathrm{Re}\,\langle -Au, u^* \rangle + s\,\Phi^2(u), \quad s > 0, \ u \in \mathfrak{D}(A), \tag{13.68}$$

where $\alpha : (0, \infty) \to (0, \infty)$ is a decreasing function such that $\lim_{s\to\infty} \alpha(s) = 0$ and $\alpha(0) := \lim_{s\to 0} \alpha(s) = \infty$. Then the generator A^f of the subordinate semigroup also satisfies a weak Poincaré inequality

$$\|u\|^2 \leq \alpha_f(s)\,\mathrm{Re}\,\langle -A^f u, u^* \rangle + s\,\Phi^2(u), \quad s > 0, \ u \in \mathfrak{D}(A), \tag{13.69}$$

where $\alpha_f(s) = 4/f(\frac{1}{\alpha(s/4)})$.

Proof. Since the proof is similar to the proof of Corollary 13.43, we only provide a formal argument for the form of $\alpha_f(s)$. Note that $s = \alpha^{-1}(r) = \beta(r)$ transforms the weak Poincaré inequality (13.68) into the super-Poincaré inequality (13.64) and vice versa. Thus, we can use Corollary 13.43 to infer that the weak Poincaré inequality remains valid under subordination with $\alpha_f(s) = \beta_f^{-1}(s)$. Since

$$\beta_f(r) = 4\beta\left(\frac{1}{f^{-1}(4/r)}\right)$$

we may insert $r = 4/f(\frac{1}{\alpha(s/4)})$ and get $\beta_f(4/f(\frac{1}{\alpha(s/4)})) = s$. This shows that $\alpha_f(s) = \beta_f^{-1}(s) = 4/f(\frac{1}{\alpha(s/4)})$.   □

We close this section with a few applications which illustrate how contractivity and decay properties of a semigroup are influenced by subordination. Let E be a locally compact separable metric space and m a Radon measure on E with full support. A symmetric sub-Markovian semigroup $(T_t)_{t\geq 0}$ on $L^2(E,m)$ is said to be

- *hypercontractive*, if $\|T_t\|_{L^2\to L^4} < \infty$ for some $t > 0$;

- *supercontractive*, if $\|T_t\|_{L^2\to L^4} < \infty$ for all $t > 0$;

- *ultracontractive*, if $\|T_t\|_{L^2\to L^\infty} < \infty$ for all $t > 0$.

Corollary 13.45. *Let f be a Bernstein function and $(T_t)_{t\geq 0}$ be an ultracontractive symmetric sub-Markovian semigroup on $L^2(X,m)$. If there are constants $c > 0$ and $p > 1$ such that*

$$\|T_t\|_{L^2\to L^\infty} \leq \exp\left(c\, t^{-1/(p-1)}\right) \quad \text{for all} \quad t > 0,$$

then the following assertions hold:

(i) *If $\int_1^\infty \frac{d\lambda}{f(\lambda^p)} < \infty$, then $(T_t^f)_{t\geq 0}$ is ultracontractive.*

(ii) *If $\lim_{\lambda\to\infty} \frac{f^{-1}(\lambda)}{\lambda^p} = 0$, then $(T_t^f)_{t\geq 0}$ is supercontractive.*

(iii) *If $\lim_{\lambda\to\infty} \frac{f^{-1}(\lambda)}{\lambda^p} \in (0,\infty)$, then $(T_t^f)_{t\geq 0}$ is hypercontractive.*

Proof. Denote by A and A^f the generators of the semigroups $(T_t)_{t\geq 0}$ and $(T_t^f)_{t\geq 0}$, respectively. By [364, Proposition 3.3.16, p. 157], we know that the following super-Poincaré inequality holds,

$$\|u\|_{L^2}^2 \leq r\langle -Au, u\rangle_{L^2} + \beta(r)\|u\|_{L^1}^2, \quad r > 0,\ u \in \mathfrak{D}(A)\cap L^1(E,m),$$

where

$$\beta(r) = c_1\left[\exp\left(c_2\, r^{-1/p}\right) - 1\right]$$

for some $c_1, c_2 > 0$. By Corollary 13.43,

$$\|u\|_{L^2}^2 \leq r\langle -A^f u, u\rangle + \beta_f(r)\|u\|_{L^1}^2, \quad r > 0,\ u \in \mathfrak{D}(A),$$

where

$$\beta_f(r) = 4c_1\left\{\exp\left[c_2\left(f^{-1}(4/r)\right)^{1/p}\right] - 1\right\}.$$

Therefore, the assertions follow from [364, Theorem 3.3.14, p. 156; Theorem 3.3.13, p. 155]. □

Corollary 13.46. *Let* $(T_t)_{t \geq 0}$ *be a symmetric sub-Markovian semigroup on* $L^2(X, m)$ *and* f *a Bernstein function. Assume that there is a constant* $p > 0$ *such that*

$$\|T_t u\|_{L^2}^2 \leq \frac{\Phi^2(u)}{t^p} \quad \text{for all } t > 0, \ u \in L^2(E, m),$$

where $\Phi : L^2(E, m) \to [0, \infty]$ *is a (not necessarily linear) functional satisfying* (13.55). *If*

$$\eta(t) := \int_t^\infty \frac{d\lambda}{\lambda f(\lambda)} < \infty \quad \text{for all } t > 0,$$

then there are constants $c_1, c_2 > 0$ *such that*

$$\|T_t^f u\|_{L^2}^2 \leq c_1 \left[\eta^{-1}(c_2 t) \right]^p \Phi^2(u).$$

Proof. Denote by A and A^f the generators of $(T_t)_{t \geq 0}$ and $(T_t^f)_{t \geq 0}$, respectively. From [364, Corollary 4.1.8 (1), p. 189; Corollary 4.1.5 (2), p. 186] we know that the following weak Poincaré inequality holds

$$\|u\|_{L^2}^2 \leq \alpha(r) \langle -Au, u \rangle_{L^2} + r \, \Phi^2(u), \quad r > 0, \ u \in \mathfrak{D}(A),$$

where $\alpha(r) = c_3 \, r^{-1/\delta}$ for some constant $c_3 > 0$. Corollary 13.44 shows that

$$\|u\|_{L^2}^2 \leq \alpha_f(r) \langle -A^f u, u \rangle_{L^2} + r \, \Phi^2(u), \quad r > 0, \ u \in \mathfrak{D}(A^f),$$

where $\alpha_f(r) = 4 / f \left(c_4 r^{1/p} \right)$ for some constant $c_4 > 0$. Therefore, the assertion follows from [364, Theorem 4.1.7, p. 188]. $\qquad\square$

13.4   Eigenvalue estimates for subordinate processes

For $\alpha \in (0, 2]$, the generator of a killed symmetric α-stable process in an open subset D of $\mathbb{R}^d$ is the Dirichlet fractional Laplacian $-(-\Delta)^{\alpha/2}|_D$. The spectrum of $-(-\Delta)^{\alpha/2}|_D$ is very important both in theory and in applications. Although a lot is known in the case $\alpha = 2$, until recently very little was known for $\alpha < 2$. In this section we discuss the spectrum of the generator of a general killed subordinate process. The main tools of this section are subordinate killed processes and the theory of Dirichlet forms.

Let E be a locally compact separable metric space, $\mathscr{B} = \mathscr{B}(E)$ the Borel σ-algebra on E, and m a positive Radon measure on $(E, \mathscr{B})$ with full support $\operatorname{supp} m = E$. Let $X = (X_t, \mathbb{P}_x)_{t \geq 0, x \in E}$ be an m-symmetric Hunt process on E. Adjoined to the state space E is a cemetery point $\partial \notin E$; the process X retires to ∂ at its lifetime $\zeta := \inf\{t \geq 0 : X_t = \partial\}$. Denote by $P_t(x, dy)$ the transition function of X. The transition semigroup $(P_t)_{t \geq 0}$ and the resolvent $(G_\beta)_{\beta > 0}$ of X are defined by

$$P_t g(x) := \mathbb{E}_x\big[g(X_t)\big] = \mathbb{E}_x\big[g(X_t)\mathbb{1}_{\{t < \zeta\}}\big],$$

$$G_\beta g(x) := \int_0^\infty e^{-\beta t} P_t g(x)\, dt = \mathbb{E}_x\left[\int_0^\infty e^{-\beta t} g(X_t)\, dt\right]$$

$$= \mathbb{E}_x\left[\int_0^\zeta e^{-\beta t} g(X_t)\, dt\right].$$

Here g is a non-negative Borel measurable function on E and we are using the convention that any function g defined on E is automatically extended to $E_\partial = E \cup \{\partial\}$ by setting $g(\partial) = 0$. The L^2-generator of $(P_t)_{t \geq 0}$ will be denoted by A and its domain by $\mathfrak{D}(A)$, see Appendix A.2 for details. Let $(\mathcal{E}, \mathfrak{D}(\mathcal{E}))$ be the Dirichlet form associated with X, i.e.

$$\begin{cases} \mathcal{E}(u, v) = \big\langle (-A)^{1/2}u, \, (-A)^{1/2}v \big\rangle_{L^2}, \\ \mathfrak{D}(\mathcal{E}) = \mathfrak{D}((-A)^{1/2}). \end{cases} \tag{13.70}$$

In general, $(\mathfrak{D}(\mathcal{E}), \mathcal{E})$ is not a Hilbert space. Therefore, one considers the forms

$$\mathcal{E}_\beta(u, v) := \mathcal{E}(u, v) + \beta \, \langle u, v \rangle_{L^2}, \quad u, v \in \mathfrak{D}(\mathcal{E}), \ \beta > 0,$$

which make $(\mathfrak{D}(\mathcal{E}), \mathcal{E}_\beta)$ into a Hilbert space. For suitable constants $c_\beta, C_\beta > 0$ we have $c_\beta \mathcal{E}_1(u, u) \leq \mathcal{E}_\beta(u, u) \leq C_\beta \, \mathcal{E}_1(u, u)$; this means that the spaces $(\mathfrak{D}(\mathcal{E}), \mathcal{E}_\beta)$ coincide.

Let $(T_t)_{t \geq 0}$ and $(R_\beta)_{\beta > 0}$ be the semigroup and resolvent associated with the operator A or, equivalently, the Dirichlet form $(\mathcal{E}, \mathfrak{D}(\mathcal{E}))$. It follows from [123, Theorem 4.2.3] that $(T_t)_{t \geq 0}$ can be identified with $(P_t)_{t \geq 0}$ and that $(R_\beta)_{\beta > 0}$ can be identified with $(G_\beta)_{\beta > 0}$. So in this section, we will use $(T_t)_{t \geq 0}$, $(R_\beta)_{\beta > 0}$ and $(P_t)_{t \geq 0}$, $(G_\beta)_{\beta > 0}$ interchangeably.

For any open subset D of E, let $\tau_D = \inf\{t > 0 : X_t \notin D\}$ be the first exit time of X from D. The process X^D obtained by killing the process X upon exiting from D is defined by

$$X_t^D := \begin{cases} X_t, & t < \tau_D, \\ \partial, & t \geq \tau_D. \end{cases}$$

The killed process X^D is again an m-symmetric Hunt process on D, see [123] or Appendix A.2. The associated Dirichlet form is $(\mathcal{E}^D, \mathfrak{D}(\mathcal{E}^D))$ where $\mathcal{E}^D = \mathcal{E}$ and

$$\mathfrak{D}(\mathcal{E}^D) = \{u \in \mathfrak{D}(\mathcal{E}) : u = 0 \ \mathcal{E}\text{-q.e. on } E \setminus D\}. \tag{13.71}$$

Here $\mathcal{E}$-q.e. stands for *quasi everywhere* with respect to the capacity associated with the Dirichlet form $(\mathcal{E}, \mathfrak{D}(\mathcal{E}))$. Let $(P_t^D)_{t \geq 0}$ and $(G_\beta^D)_{\beta > 0}$ denote the semigroup and resolvent of X^D, and let A^D be the L^2-generator of X^D.

The following result will play an important role in this section. Note that the semigroup version of this result is not true, see [91].

Lemma 13.47. *For any $\beta > 0$ and every $g \in L^2(E, m)$ with $g = 0$ m-a.e. on $E \setminus D$, we have*

$$\langle G_\beta^D g, g \rangle_{L^2} \leq \langle G_\beta g, g \rangle_{L^2}. \qquad (13.72)$$

Proof. For every $\beta > 0$ and every $g \in L^2(E, m)$ with $g = 0$ m-a.e. on $E \setminus D$, the strong Markov property of X shows

$$G_\beta g(x) = G_\beta^D g(x) + H_\beta^D G_\beta g(x) \quad \text{for all } x \in D.$$

This is the $\mathcal{E}_\beta$-orthogonal decomposition of $G_\beta g$ into $G_\beta^D g \in \mathfrak{D}(\mathcal{E}^D)$ and its orthogonal complement, see (A.20) in Appendix A.2. Here

$$H_\beta^D g(x) := \mathbb{E}_x \big[e^{-\beta \tau_D} g(X_{\tau_D}) \big]$$

is the β-hitting operator of X. Hence

$$\begin{aligned}
\langle G_\beta g - G_\beta^D g, g \rangle_{L^2} &= \langle H_\beta^D G_\beta g, g \rangle_{L^2} \\
&= \frac{1}{\beta} \, \mathcal{E}_\beta (H_\beta^D G_\beta g, G_\beta g) \\
&= \frac{1}{\beta} \, \mathcal{E}_\beta (H_\beta^D G_\beta g, H_\beta^D G_\beta g) \geq 0.
\end{aligned}$$

This establishes (13.72). $\qquad\qquad\qquad\qquad\qquad\qquad\qquad\qquad\qquad\qquad\qquad\qquad\square$

Suppose that $S = (S_t)_{t \geq 0}$ is a subordinator with Laplace exponent f given by

$$f(\lambda) = b\lambda + \int_{(0,\infty)} (1 - e^{-\lambda t}) \mu(dt), \quad \lambda > 0,$$

with $b > 0$ or $\mu(0, \infty) = \infty$. Let $(\mu_t)_{t \geq 0}$ be the convolution semigroup corresponding to S. Throughout this section we will assume that X and S are independent. The process $X^f = (X_t^f)_{t \geq 0}$ defined by $X_t^f := X(S_t)$ is called the subordinate process of X with respect to S; this is again an m-symmetric Hunt process on E. The subordinate transition function $P_t^f(x, dy)$ of X^f is given by

$$P_t^f(x, B) = \int_0^\infty P_s(x, B) \mu_t(ds) \quad \text{for } t > 0, \ x \in E \ \text{and} \ B \in \mathcal{B}(E). \quad (13.73)$$

Let $(\mathcal{E}^f, \mathfrak{D}(\mathcal{E}^f))$ be the Dirichlet form corresponding to X^f. Then $\mathfrak{D}(\mathcal{E}) \subset \mathfrak{D}(\mathcal{E}^f)$. When $b > 0$,

$$\mathfrak{D}(\mathcal{E}) = \mathfrak{D}(\mathcal{E}^f) \qquad (13.74)$$

and for $u \in \mathfrak{D}(\mathcal{E})$,

$$
\begin{aligned}
\mathcal{E}^f(u,u) &= b\mathcal{E}(u,u) + \int_{(0,\infty)} \langle u - P_s u, u \rangle_{L^2} \, \mu(ds) \\
&= b\mathcal{E}(u,u) + \int_{E \times E} \big(u(x) - u(y)\big)^2 J^f(dx,dy) \\
&\quad + \int_E u^2(x) \kappa^f(dx),
\end{aligned} \tag{13.75}
$$

where

$$
J^f(dx,dy) := \frac{1}{2} \int_{(0,\infty)} P_s(x,dy) \, \mu(ds) \, m(dx)
$$

and

$$
\kappa^f(dx) := \int_{(0,\infty)} \big(1 - P_s(x,E)\big) \mu(ds) \, m(dx).
$$

When $b = 0$,

$$
\mathfrak{D}(\mathcal{E}^f) = \left\{ u \in L^2(E,m) : \int_{(0,\infty)} \langle u - P_s u, u \rangle_{L^2} \, \mu(ds) < \infty \right\} \tag{13.76}
$$

and for any $u \in \mathfrak{D}(\mathcal{E}^f)$,

$$
\begin{aligned}
\mathcal{E}^f(u,u) &= \int_{(0,\infty)} \langle u - P_s u, u \rangle_{L^2} \, \mu(ds) \\
&= \int_{E \times E} \big(u(x) - u(y)\big)^2 J^f(dx,dy) + \int_E u^2(x) \kappa^f(dx),
\end{aligned} \tag{13.77}
$$

with J^f and κ^f from above. For the above facts regarding $(\mathcal{E}^f, \mathfrak{D}(\mathcal{E}^f))$ we refer to [284]. Let A^f be the L^2-infinitesimal generator of X^f. From Theorem 13.6, see also Example 12.6, it follows that

$$
\mathfrak{D}(A) \subset \mathfrak{D}(A^f), \tag{13.78}
$$

$$
A^f u = b A u + \int_{(0,\infty)} (P_s u - u) \, \mu(ds) \quad \text{for } u \in \mathfrak{D}(A). \tag{13.79}
$$

By Proposition 13.5 we know that $\mathfrak{D}(A)$ is a core for the generator A^f; in particular, $\mathfrak{D}(A)$ is dense in $(\mathfrak{D}(\mathcal{E}^f), \mathcal{E}^f)$.

For every open subset D of E let $X^{f,D}$ denote the process obtained by killing X^f upon exiting from D. We will use $(P_t^{f,D})_{t \geq 0}$ to denote the transition semigroup of $X^{f,D}$ and $A^{f,D}$ to denote the generator of $X^{f,D}$. The Dirichlet form of $X^{f,D}$ is $(\mathcal{E}^{f,D}, \mathfrak{D}(\mathcal{E}^{f,D}))$ where $\mathcal{E}^{f,D} = \mathcal{E}^f$ and

$$
\mathfrak{D}(\mathcal{E}^{f,D}) = \{ u \in \mathfrak{D}(\mathcal{E}^f) : u = 0 \ \mathcal{E}^f\text{-q.e. on } E \setminus D \}.
$$

For $u \in \mathfrak{D}(\mathcal{E}^{f,D})$ we can rewrite $\mathcal{E}^f(u,u)$ as

$$
\mathcal{E}^f(u,u) = b\mathcal{E}(u,u) + \int_{D \times D} \left(u(x) - u(y)\right)^2 J^f(dx, dy)
$$
$$
+ \int_D u(x)^2 \kappa^{f,D}(x) m(dx),
$$

(13.80)

where

$$
\kappa^{f,D}(x) = \int_{(0,\infty)} P_s(x, E \setminus D) \mu(ds) + \kappa^f(x) \quad \text{for } x \in D.
$$

Let $X^{D,f}$ denote the process obtained by subordinating X^D with the subordinator S. The semigroup of $X^{D,f}$ will be denoted by $(P_t^{D,f})_{t \geq 0}$ and its generator by $A^{D,f}$. The Dirichlet form of $X^{D,f}$ is $(\mathcal{E}^{D,f}, \mathfrak{D}(\mathcal{E}^{D,f}))$, where $\mathcal{E}^{D,f}$ is defined in the same way through X^D as $\mathcal{E}^f$ is defined through X. For any $u \in \mathfrak{D}(\mathcal{E}^{D,f})$,

$$
\mathcal{E}^{D,f}(u,u) = b\mathcal{E}(u,u) + \int_{D \times D} \left(u(x) - u(y)\right)^2 J^{D,f}(dx, dy)
$$
$$
+ \int_D u^2(x) \kappa^{D,f}(x) m(dx),
$$

(13.81)

where

$$
J^{D,f}(dx, dy) = \frac{1}{2} \int_{(0,\infty)} P_s^D(x, dy) \mu(ds) m(dx) \quad \text{for } x, y \in D,
$$

and

$$
\kappa^{D,f}(x) = \int_{(0,\infty)} \left(1 - P_s^D 1(x)\right) \mu(ds) \quad \text{for } x \in D.
$$

Theorem 13.48. *Suppose that the Laplace exponent f of S is a complete Bernstein function. Then for any open subset D of E, we have $\mathfrak{D}(\mathcal{E}^{D,f}) \subset \mathfrak{D}(\mathcal{E}^{f,D})$ and*

$$
\mathcal{E}^f(u,u) \leq \mathcal{E}^{D,f}(u,u) \quad \text{for every } u \in \mathfrak{D}(\mathcal{E}^{D,f}).
$$

Proof. Since f is a complete Bernstein function, it has the following representation

$$
f(\lambda) = b\lambda + \int_0^\infty (1 - e^{-\lambda t}) m(t) \, dt,
$$

where $m(t) = \int_{(0,\infty)} e^{-ts} s \, \sigma(ds)$ and σ is the Stieltjes measure of f, see Remark 6.10. From Corollary 13.22 we know that A^f has the following representation in terms of the L^2-generator A and the resolvent $(G_\beta)_{\beta > 0}$ of X:

$$
A^f u = bAu + \int_{(0,\infty)} (\beta G_\beta u - u) \sigma(d\beta) \quad \text{for } u \in \mathfrak{D}(A).
$$

(13.82)

For $u \in \mathfrak{D}(A)$, it follows from (13.82) that

$$\mathcal{E}^f(u,u) = \langle -A^f u, u \rangle_{L^2} = b\mathcal{E}(u,u) + \int_{(0,\infty)} \beta \langle u - \beta G_\beta u, u \rangle_{L^2} \, \sigma(d\beta).$$

Since $\mathfrak{D}(A)$ is $\mathcal{E}_1$-dense in $\mathfrak{D}(\mathcal{E})$,

$$\mathcal{E}^f(u,u) = b\mathcal{E}(u,u) + \int_{(0,\infty)} \beta \langle u - \beta G_\beta u, u \rangle_{L^2} \, \sigma(d\beta) \quad \text{for } u \in \mathfrak{D}(\mathcal{E}).$$

Similarly, we have

$$\mathcal{E}^{D,f}(u,u) = b\mathcal{E}^D(u,u) + \int_{(0,\infty)} \beta \langle u - \beta G_\beta^D u, u \rangle_{L^2} \, \sigma(d\beta) \quad \text{for } u \in \mathfrak{D}(\mathcal{E}^D).$$

As $\mathfrak{D}(\mathcal{E}^D) \subset \mathfrak{D}(\mathcal{E})$, we deduce from these formulae and (13.72) that for $u \in \mathfrak{D}(\mathcal{E}^D)$,

$$\mathcal{E}^{D,f}(u,u) - \mathcal{E}^f(u,u) = \beta \int_{(0,\infty)} \beta \langle G_\beta u - G_\beta^D u, u \rangle_{L^2} \, \sigma(d\beta) \geq 0.$$

Since $\mathfrak{D}(A^D) \subset \mathfrak{D}(\mathcal{E}^D)$ is $\mathcal{E}_1^f$-dense in $\mathfrak{D}(\mathcal{E}^{D,f})$ – see the comments after (13.79) –, we conclude that $\mathfrak{D}(\mathcal{E}^{D,f}) \subset \mathfrak{D}(\mathcal{E}^f)$ and, therefore, that $\mathfrak{D}(\mathcal{E}^{D,f}) \subset \mathfrak{D}(\mathcal{E}^{f,D})$. Moreover $\mathcal{E}^f(u,u) \leq \mathcal{E}^{D,f}(u,u)$ holds for every $u \in \mathfrak{D}(\mathcal{E}^{D,f})$. $\qquad\square$

The semigroup $(P_t)_{t\geq 0}$ is *ultracontractive* if each P_t, $t > 0$, is a bounded operator from $L^2(E,m)$ to $L^\infty(E,m)$. Under this condition, for all $t > 0$ the transition operator P_t has an integral kernel $p(t,x,y)$ with respect to the measure m, which satisfies

$$0 \leq p(t,x,y) \leq c_t \text{ a.e. on } E \times E \qquad (13.83)$$

with $t \mapsto c_t$ being a non-increasing function on $(0,\infty)$, see e.g. [89, (2.1.1) and Lemma 2.1.2]. Conversely, if for every $t > 0$ the transition operator P_t has a bounded density function, i.e. if (13.83) holds for some $c_t > 0$, then for a.e. $x \in E$ and every $f \in L^2(E,m)$

$$|P_t f(x)|^2 \leq \int_E p(t,x,y) \, f^2(y) \, m(dy) \leq c_t \| f \|_{L^2(E,m)}^2.$$

So $(P_t)_{t\geq 0}$ is ultracontractive. Therefore the semigroup $(P_t)_{t\geq 0}$ is ultracontractive if, and only if, P_t has a bounded integral kernel for every $t > 0$.

In the remainder of this section we will assume that the semigroup $(P_t)_{t\geq 0}$ of the process X is ultracontractive. Assume further that D is an open subset of E with $m(D) < \infty$. It is well known that under these assumptions the semigroup $(P_t^D)_{t\geq 0}$ of X^D has a density with respect to m. We will denote this density by $p^D(t,x,y)$.

Under the assumption (13.83), it is easy to see that the subordinate process X^f has a transition density $p^f(t, x, y)$ given by

$$p^f(t, x, y) = \int_0^\infty p(s, x, y)\, \mu_t(ds).$$

Under the assumptions stated in the paragraph above, P_t^D is a Hilbert–Schmidt operator for $t > 0$. Thus for any $t > 0$, both P_t^D and A^D have discrete spectra. We will use $(e^{-\lambda_n t})_{n \in \mathbb{N}}$ and $(-\lambda_n)_{n \in \mathbb{N}}$ to denote the eigenvalues of P_t^D and A^D respectively, arranged in decreasing order and repeated according to multiplicity.

There are examples, see [83, Section 3], where for every $t > 0$, $P_t^{f,D}$ is not a Hilbert–Schmidt operator. But, for every $t > 0$, $P_t^{f,D}$ is still a compact operator.

Theorem 13.49. *For every $t > 0$, $P_t^{f,D}$ is a compact operator in $L^2(D, m)$ and therefore it has discrete spectrum.*

Proof. For every $s > 0$ we define the operator Q_s on $L^2(D, m)$ by

$$Q_s g(x) := \int_D p(s, x, y)\, g(y)\, m(dy), \quad g \in L^2(D, m).$$

For $g \in L^2(D, m)$ with $\|g\|_{L^2(D,m)} \le 1$ and any Borel subset A of D, we have

$$\int_A (Q_s g(x))^2\, m(dx) \le \int_A Q_s(g^2)(x)\, m(dx)$$

$$= \int_E g^2(x)\, P_s \mathbb{1}_A(x)\, m(dx)$$

$$\le \|P_s \mathbb{1}_A\|_\infty \int_E g^2(x)\, m(dx)$$

$$\le \|P_s \mathbb{1}_A\|_\infty$$

$$\le \min\{1, c_s m(A)\}. \tag{13.84}$$

Now we fix $t > 0$. For $g \in L^2(D, m)$ with $\|g\|_{L^2(D,m)} \le 1$ and all Borel subsets A of D, we have

$$\int_A \left(\int_0^\infty Q_s g(x) \mu_t(ds)\right)^2 m(dx) \le \int_A \int_0^\infty |Q_s g(x)|^2\, \mu_t(ds)\, m(dx)$$

$$= \int_0^\infty \int_A |Q_s g(x)|^2\, m(dx)\, \mu_t(ds).$$

For every $\epsilon > 0$ we find some $s_0 > 0$ such that $\mu_t[0, s_0] < \frac{1}{2}\epsilon$. Then by (13.84),

$$\int_A \left(\int_0^\infty Q_s g(x)\, \mu_t(ds)\right)^2 dx \le \frac{\epsilon}{2} + c_{s_0} m(A).$$

This shows that for Borel sets $A \subset E$ with $m(A) < \epsilon / 2c_{s_0}$, we have

$$\int_A \left(\int_0^\infty Q_s g(x) \mu_t(ds) \right)^2 m(dx) \le \epsilon$$

which entails that the family of functions

$$\left\{ \left(\int_0^\infty Q_s g(\cdot) \mu_t(ds) \right)^2 : \|g\|_{L^2(D,m)} \le 1 \right\}$$

is uniformly integrable.

Let $p^{f,D}(t, x, y)$ be the density function of $P_t^{f,D}$. Clearly,

$$p^{f,D}(t, x, y) \le p^f(t, x, y), \quad (t, x, y) \in (0, \infty) \times D \times D.$$

Thus for all $t > 0$ and all Borel functions g on D, we have

$$|P_t^{f,D} g(x)| \le \int_D p^f(t, x, y) |g|(y) m(dy) \le \int_0^\infty Q_s |g|(x) \mu_t(ds).$$

Consequently, the family of functions

$$\{(P_t^{f,D} g)^2 : \|g\|_{L^2(D,m)} \le 1\}$$

is uniformly integrable on $(D, \mathscr{B}(D), m)$. By using this one can show that, for each $t > 0$, $P_t^{f,D}$ is a compact operator in $L^2(D, m)$ and hence has discrete spectrum.

Indeed: assume that $P_t^{f,D}$ is not compact. Then there exist some $\epsilon > 0$ and a sequence $(g_n)_{n \in \mathbb{N}}$ in $\mathcal{A} := \{P_{t/2}^{f,D} g : \|g\|_{L^2(D,m)} \le 1\}$ such that

$$\|P_{t/2}^{f,D} g_n - P_{t/2}^{f,D} g_m\|_{L^2(D,m)} \ge \epsilon, \quad n \ne m.$$

Since we know that $\{g^2 : g \in \mathcal{A}\}$ is uniformly integrable, we have that

$$\lim_{K \to \infty} \sup_{g \in \mathcal{A}} \int_D g^2(x) \mathbb{1}_{\{|g| > K\}}(x) m(dx) = 0,$$

and consequently, by the boundedness of $P_{t/2}^{f,D}$ in $L^2(D, m)$,

$$\lim_{K \to \infty} \sup_{g \in \mathcal{A}} \int_D \left(P_{t/2}^{f,D}(g \mathbb{1}_{\{|g| > K\}})(x) \right)^2 m(dx) = 0.$$

Therefore there exists some $K > 0$ such that

$$\|P_{t/2}^{f,D} g_{n,K} - P_{t/2}^{f,D} g_{k,K}\|_{L^2(D,m)} \ge \frac{\epsilon}{2}, \quad n \ne k, \tag{13.85}$$

where $g_{n,K} = g_n \mathbb{1}_{\{|g_n| \leq K\}}$. Let $\mathscr{C}$ be the σ-algebra generated by $(g_{n,K})_{n\in\mathbb{N}}$. Since $L^1(D,\mathscr{C},m)$ is separable, the set $(g_{n,K})_{n\in\mathbb{N}} \subset L^\infty(D,\mathscr{C},m)$ is weakly compact and metrizable with respect to the weak topology $\sigma(L^\infty(D,\mathscr{C},m), L^1(D,\mathscr{C},m))$, cf. [70, IV.36, Lemma 1]. Therefore there exist $g \in L^\infty(D,\mathscr{C},m)$ and a subsequence $(g_{n_j,K})_{j\in\mathbb{N}}$ such that for $h \in L^1(D,\mathscr{C},m)$ we have

$$\lim_{j\to\infty} \int_D g_{n_j,K}(x)\,h(x)\,m(dx) = \int_D g(x)\,h(x)\,m(dx).$$

Without loss of generality, we will assume in the rest of this proof that $m(D) = 1$, that is, m is a probability measure on D. For any $h \in L^1(D,m)$ we will use $m(h\,|\,\mathscr{C})$ to denote the conditional expectation of h with respect to $\mathscr{C}$. Noting that for any $h \in L^1(D,m)$ and $k \in L^\infty(D,\mathscr{C},m)$ we have

$$\int_D k(x)\,h(x)\,m(dx) = \int_D k(x)\,m(h\,|\,\mathscr{C})(x)\,m(dx),$$

we obtain that

$$\lim_{j\to\infty} \int_D g_{n_j,K}(x)\,h(x)\,m(dx) = \int_D g(x)\,h(x)\,m(dx) \quad \text{for all } h \in L^1(D,m).$$

On the other hand, for every $x \in D$, we have

$$P_{t/2}^{f,D} g(x) - P_{t/2}^{f,D} g_{n_j,K}(x) = \int_D p^{f,D}(t/2,x,y)\,\big(g(y) - g_{n_j,K}(y)\big)\,m(dy).$$

The fact that $p^{f,D}(t/2,x,\cdot)$ is integrable and the weak convergence of the sequence $(g_{n_j,K})_{j\in\mathbb{N}}$ imply $P_{t/2}^{f,D} g_{n_j,K}(x) \xrightarrow{j\to\infty} P_{t/2}^{f,D} g(x)$ for all $x \in D$. Since the family $((P_{t/2}^{f,D} g_{n_j,K})^2)_{j\in\mathbb{N}}$ is uniformly integrable, we get $P_{t/2}^{f,D} g_{n_j,K} \xrightarrow{j\to\infty} P_{t/2}^{f,D} g$ in $L^2(D,m)$. This contradicts (13.85) and the proof is complete. $\qquad\square$

In the remainder of this section, we will use $(-\tilde{\lambda}_n)_{n\in\mathbb{N}}$ and $(e^{-\tilde{\lambda}_n t})_{n\in\mathbb{N}}$ to denote the eigenvalues of $A^{f,D}$ and $P_t^{f,D}$, respectively, arranged in decreasing order and repeated according to multiplicity.

Theorem 13.50. *If the Laplace exponent f of the subordinator S is a complete Bernstein function, then*

$$\tilde{\lambda}_n \leq f(\lambda_n) \quad \text{for every } n \in \mathbb{N}.$$

Proof. Since A^D has discrete spectrum $(\lambda_n)_{n\in\mathbb{N}}$, the subordinate generator $A^{D,f}$ has discrete spectrum $(f(\lambda_n))_{n\in\mathbb{N}}$ with the same eigenfunctions as A^D. By the min-max principle for eigenvalues, see [301, Section XIII.1], and by Theorem 13.48 we get for every $n \in \mathbb{N}$

$$\tilde{\lambda}_n = \inf_{\substack{L:\, \text{subspace of } L^2(D,m) \\ \text{with } \dim L = n}} \sup \left\{ \mathcal{E}^f(u,u) \,:\, u \in L \text{ and } \langle u, u \rangle_{L^2(D,m)} = 1 \right\}$$

$$\leq \inf_{\substack{L:\, \text{subspace of } L^2(D,m) \\ \text{with } \dim L = n}} \sup \left\{ \mathcal{E}^{D,f}(u,u) \,:\, u \in L \text{ and } \langle u, u \rangle_{L^2(D,m)} = 1 \right\}$$

$$= f(\lambda_n).$$

Here we are using the convention that for a Dirichlet form $(\mathcal{E}, \mathfrak{D}(\mathcal{E}))$, $\mathcal{E}(u,u) = \infty$ when $u \in L^2(D,m) \setminus \mathfrak{D}(\mathcal{E})$. This proves the theorem. $\qquad\square$

Proposition 13.51. *Suppose that S is a subordinator with Laplace exponent f. If there exists some $C \in (0,1)$ such that*

$$\mathbb{P}_x(X_t \in D) \leq C, \quad \text{for every } t > 0 \text{ and every } x \in E \setminus D, \tag{13.86}$$

then

$$(1-C)\,\kappa^{D,f}(x) \leq \kappa^{f,D}(x) \leq \kappa^{D,f}(x) \quad \text{for } x \in D. \tag{13.87}$$

Proof. Recall that μ is the Lévy measure of S. For any $x \in D$ we have

$$\kappa^{f,D}(x) = \int_{(0,\infty)} \int_{E \setminus D} P_s(x, dy)\, \mu(ds) + \int_{(0,\infty)} \left(1 - P_s(x, E)\right) \mu(ds)$$

$$= \int_{(0,\infty)} \left(1 - P_s \mathbb{1}_D(x)\right) \mu(ds)$$

and

$$\kappa^{D,f}(x) = \int_{(0,\infty)} \left(1 - P_s^D \mathbb{1}_E(x)\right) \mu(ds).$$

Thus, the second inequality of (13.87) follows because $p^D(s, x, \cdot) \leq p(s, x, \cdot)$ for all $(s,x) \in (0,\infty) \times D$.

In order to prove the first inequality, let τ_D be the first exit time of D for the process X and let $u(x,s) = \mathbb{P}_x(X_s \in D)$. According to the assumption (13.86), $u(x,t) \leq C$ for every $x \in E \setminus D$ and $t > 0$. By the strong Markov property of X we have for all $x \in D$ and $t > 0$

$$\mathbb{P}_x(\tau_D \leq t, X_t \in D) = \mathbb{E}_x\left[\mathbb{1}_{\{\tau_D \leq t\}} u(X_{\tau_D}, \tau_D)\right] \leq C\, \mathbb{P}_x(\tau_D \leq t).$$

Therefore for all $s > 0$ and $x \in D$,

$$1 - P_s \mathbb{1}_D(x) = 1 - P_s^D 1(x) - \mathbb{P}_x(\tau_D \leq t, X_t \in D)$$

$$\geq 1 - P_s^D 1(x) - C\, \mathbb{P}_x(\tau_D \leq t)$$

$$= (1-C)\left(1 - P_s^D 1(x)\right).$$

This proves the first inequality of (13.87). $\qquad\square$

A symmetric α-stable process in $\mathbb{R}^d$, $0 < \alpha < 2$, is a Lévy process $X = (X_t)_{\geq 0}$ taking values in $\mathbb{R}^d$ and with the characteristic exponent $\psi(\xi) = |\xi|^\alpha$, cf. Example 13.13. It is well known that the semigroup of this process is ultracontractive.

Proposition 13.52. *Let X be a symmetric α-stable process in $\mathbb{R}^d$ and let D be a bounded convex open and connected subset of $\mathbb{R}^d$. Then*

$$\frac{1}{2} \kappa^{D,f}(x) \leq \kappa^{f,D}(x) \leq \kappa^{D,f}(x).$$

Proof. Because of the spherical symmetry of X we find for every bounded convex open connected subset $D \subset \mathbb{R}^d$

$$\mathbb{P}_x(X_t \in D) < \frac{1}{2}, \quad \text{for every } t > 0 \text{ and } x \in E \setminus D.$$

The assertion now follows from Proposition 13.51. $\square$

It follows from Theorem 13.48 that $\mathfrak{D}(\mathcal{E}^{D,f}) \subset \mathfrak{D}(\mathcal{E}^{f,D})$. From the expressions for the subordinate and killed subordinate jump measures J^f and $J^{D,f}$ one easily concludes that

$$J^{D,f}(\cdot,\cdot) \leq J^f(\cdot,\cdot). \tag{13.88}$$

Theorem 13.53. *Suppose that the Laplace exponent f of S is a complete Bernstein function. If X is a symmetric α-stable process in $\mathbb{R}^d$ and D is a bounded convex open and connected subset of $\mathbb{R}^d$, then $\mathfrak{D}(\mathcal{E}^{D,f}) = \mathfrak{D}(\mathcal{E}^{f,D})$ and*

$$\frac{1}{2} \mathcal{E}^{D,f}(u,u) \leq \mathcal{E}^f(u,u) \leq \mathcal{E}^{D,f}(u,u), \quad u \in \mathfrak{D}(\mathcal{E}^{f,D}). \tag{13.89}$$

Proof. We know from Theorem 13.48 that $\mathfrak{D}(\mathcal{E}^{D,f}) \subset \mathfrak{D}(\mathcal{E}^{f,D})$. On the other hand, by (13.74)–(13.77), (13.80)–(13.81) and Proposition 13.52, $\mathfrak{D}(\mathcal{E}^{f,D}) \subset \mathfrak{D}(\mathcal{E}^{D,f})$. Thus $\mathfrak{D}(\mathcal{E}^{D,f}) = \mathfrak{D}(\mathcal{E}^{f,D})$. The inequality (13.89) follows immediately by combining Proposition 13.52 with (13.88) and Theorem 13.48. $\square$

Theorem 13.53 together with the min-max principle for eigenvalues, see the proof of Theorem 13.50, immediately gives the following result for the eigenvalues $f(\lambda_n)$ of A^f and $\tilde{\lambda}_n$ of $A^{f,D}$.

Theorem 13.54. *Suppose that the Laplace exponent f of S is a complete Bernstein function. If X is a symmetric α-stable process in $\mathbb{R}^d$ and D is a bounded convex open and connected subset of $\mathbb{R}^d$, then*

$$\frac{1}{2} f(\lambda_n) \leq \tilde{\lambda}_n \leq f(\lambda_n) \quad \text{for every } n \in \mathbb{N}.$$

Comments 13.55. *Sections* 13.1 *and* 13.2: Subordination was introduced by Bochner in the short note [62] in connection with diffusion equations and semigroups. Using Hilbert space spectral calculus, Bochner points out that the generator of a subordinate semigroup is a function of the original generator. The name *subordination* seems to originate in Bochner's monograph [63, Chapters 4.3, 4.4] where subordination of stochastic processes, their transition semigroups and generators is developed. A rigorous functional analytic and stochastic account is in the paper by Nelson [270] where, for the first time, the interpretation of subordination as stochastic time-change can be found; this was, independently, also pointed out by Woll [375] and Blumenthal [59]. Nelson is also the first to make the link between Bochner's subordination and the rapidly developing theory of operator semigroups pioneered by Hille and Phillips. Phillips' paper [292] contains the modern formulation of subordination of semigroups on abstract Banach spaces as well as the famous result on the form of the subordinate generator. An operational calculus for bounded generators of semigroups, based on the Dunford–Riesz functional calculus, is described in Hille and Phillips [160, Chapter XV]. For C_0-semigroups and unbounded generators this calculus is further developed in Balakrishnan [16] and [17]. The origins of subordination in potential theory can be traced back to the papers by Feller [117] and Bochner [62] where Riesz potentials of fractional order are discussed, and to Faraut [111] who studies fractional powers of Hunt kernels and a symbolic calculus induced by subordination and Bernstein functions, see also [112]. This line of research is continued by Hirsch who develops in a series of papers, [162, 163, 164, 165], a Bochner-type functional calculus for Hunt kernels and abstract potentials; his main motivation is the question which functions operate on Hunt kernels which was raised in the note by M. Itô [179] and generalized in [161]. In functional analysis, the study of fractional powers and functions of infinitesimal generators was continued by Nollau [280, 281, 282], Westphal [367], Pustyl'nik [298] and later by Berg, Boyadzihev and deLaubenfels [36] and Schilling [320, 321, 322]. Up-to-date accounts of subordination, mainly for Lévy processes and convolution semigroups, can be found in Sato [314, Chapter 6] and Bertoin [49]; fractional powers and related functional calculi are discussed in Martínez-Carracedo and Sanz-Alix [260].

The material on C_0-semigroups is mostly standard and can be found in many books on semigroup theory and functional analysis. Our sources are Davies [88], Ethier and Kurtz [108], Pazy [286] and Yosida [379]. An excellent exposition of dissipativity and fractional powers is in Tanabe [345, Chapter 2]. Remark 13.4 is modeled after Bochner [62], the proof of Proposition 13.5 is taken from [320], see also [183, Vol. 1]. Phillips' theorem appears for the first time in [292] where it is proved in the context of Banach algebras. Our proof of this theorem seems to be new. The results for the domains of subordinate generators are from [322]; Corollaries 13.8 and 13.9 are generalizations of the moment inequality in Pustyl'nik [298, Theorem 6] and Berg–Boyadzhiev–deLaubenfels [36, Prop. 5.14]; our method of proof is an improvement of [113, Prop. 1.4.9]. Observe that a combination of Corollary 7.15 or

Proposition 7.16 and Theorem 13.23 (v) with the estimate (13.19) leads to interesting new interpolation-type estimates. Take, for example $f, g \in \mathcal{CBF}$ and $\alpha \in (0, 1)$. Then $f^\alpha g^{1-\alpha} \in \mathcal{CBF}$ and we have for all $u \in \mathfrak{D}(A)$

$$\|A^{f^\alpha g^{1-\alpha}} u\| = \|A^{f^\alpha} A^{g^{1-\alpha}} u\| \leq 4 \|A^f u\|^\alpha \|A^g u\|^{1-\alpha}, \qquad (13.90)$$

or, with $f^\alpha(\lambda^\beta) g^{1-\alpha}(\lambda^{1-\beta})$ and $\alpha, \beta \in [0, 1]$, we have for all $u \in \mathfrak{D}(A)$

$$\left\| f\big((-A)^\beta\big)^\alpha g\big((-A)^{1-\beta}\big)^{1-\alpha} u \right\| \leq 4 \left\| f\big((-A)^\beta\big) u \right\|^\alpha \left\| g\big((-A)^{1-\beta}\big) u \right\|^{1-\alpha}. \qquad (13.91)$$

Theorem 13.10 on the spectrum of the subordinate generator is from [292], the $\mathcal{CBF}$-version is from Hirsch [163]. A predecessor of this is Balakrishnan [16, 17].

The proof of Theorem 13.14 follows Schoenberg's original argument [325] who considers completely monotone functions; the statement of the theorem is due to Bochner, but his proof in [63, pp. 99–100] is, unfortunately, incorrect. Lemma 13.15 is taken from [325]. The Bochner–Schoenberg theorem still holds on Abelian groups if we cast it in the following form: determine all functions f which operate on the negative (resp. positive) definite functions, i.e. $f \circ \psi$ is negative (resp. positive) definite for all negative (resp. positive) definite functions ψ. Note that $\psi(\xi) = |\xi|^2$ is indeed negative definite in the sense of Schoenberg, cf. Chapter 4. This is due to Herz [157] for positive definite functions and Harzallah [147, 148, 149] for negative definite functions; further generalizations are in Kahane [197]. Example 13.16 is due to Bochner [63, Theorem 4.3.3, p. 94], see also [185, Lemma 2.1]; note that the integrability condition in [185] is automatically satisfied. The second part of Example 13.16 appears in [233, Proposition 2.14] who hints to Rogers [306].

The section on the functional calculus is a simplified and generalized version of [320, 322], see also [183, Vol. 1, Chapter 4.3]. The presentation owes a lot to F. Hirsch who pointed out that the operator I_k appearing in Lemma 13.18 coincides with kR_k^f. The idea to reduce the functional calculus to an operational calculus on the level of the trivial semigroup $t \mapsto e^{-ta}$, cf. Remark 13.26, is from Westphal [367] where this is worked out for fractional powers. That paper also contains the fractional power analogues of Theorem 13.31 and Corollary 13.32. This approach is heavily influenced by (abstract) approximation theory, see e.g. Butzer and Berens [73], in particular Chapter 2.3. Corollary 13.34 is from [322], and Corollary 13.36, with a different proof and for complete Bernstein functions, is due to Pustyl'nik [298, Theorem 8]. The remarkable pointwise stability assertion for the functional calculus, Theorem 13.37, is an improvement of a similar result in [322]. The extension of the functional calculus sketched in Remark 13.38 is in line with the calculus developed in [298], and links our results with Hirsch's theory on abstract potential operators: note that $-\phi(-A)$ is essentially the inverse of the generator of a C_0-contraction semigroup, hence an abstract potential operator.

Section 13.3: Functional inequalities have become indispensable in the analysis of Markovian semigroups, especially in estimates for semigroups and the corresponding heat kernels. Among the earliest contributions in this area are the papers by Federbush [114], Gross [138], and [122] by Fukushima who exploits ideas by Stampaccia [336] and uses the Sobolev-type inequality (13.52) to derive L^p, $p > 2$, estimates for the resolvent of a symmetric Markovian semigroup on L^2. In the hands of Varopoulos, Saloff-Coste and Coulhon – cf. [362] and the references given there – and Carlen, Kusuoka and Stroock – see, e.g. [74] – functional inequalities became a most powerful tool. For the history we refer to the surveys in Jacob [183, Vol. 2, Chapter 3.6] and Varopoulos et al. [362, p. 25]. The monographs [362], [313] and [311] are very good introductions to the field, while [364] is currently the most comprehensive treatise on functional inequalities and their applications.

Our presentation follows the paper [324] where all results are proved in a Hilbert space context. This paper extends earlier results by Bendikov and Maheux [30] who consider only fractional powers and the Bernstein functions $f(\lambda) = \lambda^\alpha, 0 < \alpha < 2$. Lemma 13.39 is a generalization of Bihari's inequality [57, Section 3], see also Walter [363, Chapter I, 1.VI and 6.IX] and Beckenbach-Bellman [27, Chapter 4, §4 and §5]. The idea of the proof comes from [80, Appendix A, Lemma A.1, A.3 pp. 193-4]; note, however, that the right hand side of the inequality (13.56) may be negative. This is the main difference compared to the earlier papers. For symmetric semigroups on a Hilbert space, Theorem 13.41 has a converse: if (13.62) holds for some $f \in \mathcal{BF}$, then (13.61) is also true, see [324, Proposition 11]. Similar converse statements hold for Corollaries 13.43 and 13.44. Therefore, it is possible to extend Corollary 13.45: $(T_t^f)_{t \geq 0}$ is *not* hypercontractive if $\lim_{\lambda \to \infty} f^{-1}(\lambda)/\lambda^p = \infty$.

Section 13.4: The recent study on the spectrum of the Dirichlet fractional Laplacian was initiated by Bañuelos and Kulczycki in [19]. Suppose that D is a bounded open subset of $\mathbb{R}^d$, let $(-\tilde{\lambda}_n^{(\alpha)})_{n \in \mathbb{N}}$ be the eigenvalues of $-(-\Delta)^{\alpha/2}|_D$, arranged in decreasing order and repeated according to multiplicity. Using a mixed Steklov problem for the Laplacian in $\mathbb{R}^{d+1}$ it was shown in [19] that for bounded Lipschitz domains $D \subset \mathbb{R}^d$

$$\tilde{\lambda}_n^{(1)} \leq \sqrt{\tilde{\lambda}_n^{(2)}} \quad \text{for all } n \in \mathbb{N}.$$

By considering a higher order Steklov problem, DeBlassie shows in [91] that for bounded Lipschitz domains $D \subset \mathbb{R}^d$ and for any rational $\alpha \in (0, 2)$

$$\tilde{\lambda}_n^{(\alpha)} \leq (\tilde{\lambda}_n^{(2)})^{\alpha/2} \quad \text{for all } n \in \mathbb{N}.$$

The upper bound above is shown for general subordinate Markov processes in [81], where a lower bound is also established. The exposition of this section is a combination of [81] and [83]. The last part of the proof of Theorem 13.49 is adapted from the proofs of [133, Lemma 3.1, Theorem 3.2].

For a general subordinator S whose Laplace exponent f is not necessarily a complete Bernstein function we have a weaker version of Theorem 13.48: that is, we always have $\mathfrak{D}(\mathcal{E}^{D,f}) \subset \mathfrak{D}(\mathcal{E}^{f,D})$ and

$$\mathcal{E}^f(u,u) \leq 4\mathcal{E}^{D,f}(u,u) \quad \text{for all } u \in \mathfrak{D}(\mathcal{E}^{D,f}). \tag{13.92}$$

This is due to the fact that $P_s^D u \leq P_s u$ for all non-negative $u \in \mathfrak{D}(\mathcal{E}^{D,f})$, which allows to conclude from (13.73)–(13.77) that $u \in \mathfrak{D}(\mathcal{E}^f)$ and $\mathcal{E}^f(u,u) \leq \mathcal{E}^{D,f}(u,u)$. For a general $u \in \mathfrak{D}(\mathcal{E}^{D,f})$, u^+ and u^- are both in $\mathfrak{D}(\mathcal{E}^{D,f})$, and so we have that both u^+ and u^- are in $\mathfrak{D}(\mathcal{E}^f)$ and that

$$\mathcal{E}^f(u^+,u^+) \leq \mathcal{E}^{D,f}(u^+,u^+), \quad \mathcal{E}^f(u^-,u^-) \leq \mathcal{E}^{D,f}(u^-,u^-).$$

Thus $u = u^+ - u^- \in \mathfrak{D}(\mathcal{E}^f)$. The contraction property of Dirichlet forms yields

$$\mathcal{E}^{D,f}(u^+,u^+) \leq \mathcal{E}^{D,f}(u,u), \quad \mathcal{E}^{D,f}(u^-,u^-) \leq \mathcal{E}^{D,f}(u,u).$$

Therefore we have $\mathfrak{D}(\mathcal{E}^{D,f}) \subset \mathfrak{D}(\mathcal{E}^{f,D})$ as well as (13.92). Using (13.92), we can see that the conclusions of Theorems 13.53 and 13.54 hold for a general subordinator S, with the exception that the upper bounds for $\mathcal{E}^f(u,u)$ in (13.89) and for $\tilde{\lambda}_n$ have an additional multiplicative factor 4. Using the following formula for the first eigenvalue

$$\tilde{\lambda}_1 = \inf\left\{\mathcal{E}^f(|u|,|u|) : u \in \mathfrak{D}(\mathcal{E}^{f,D}) \text{ with } \|u\|_{L^2(D,m)} = 1\right\}$$

and the analogous formula for the first eigenvalue $f(\lambda_1)$ for $(\mathcal{E}^{D,f}, \mathfrak{D}(\mathcal{E}^{D,f}))$, we get that $\tilde{\lambda}_1 \leq f(\lambda_1)$ for a general subordinator.

One can also prove that the eigenvalues of the generator of a killed subordinate Markov process depend continuously on the Laplace exponent of the subordinator. For this and related results, we refer to [82].

Chapter 14

Potential theory of subordinate killed Brownian motion

Recall that a Bernstein function f is a special Bernstein function if its conjugate function $f^\star(\lambda) := \lambda / f(\lambda)$ is again a Bernstein function. Thus, the identity function factorizes into the product of two Bernstein functions. In this chapter we will use this factorization and show, as a consequence of Theorem 11.9, that the potential kernel of a strong Markov process Y is the composition of the potential kernels of two subordinate processes. As before, we call the subordinators associated with special Bernstein functions *special subordinators*. If the underlying process Y is a killed Brownian motion with an intrinsically ultracontractive semigroup, the factorization of the potential kernel implies a representation of excessive functions of Y as potentials of one of the subordinate processes.

Let E be a locally compact separable metric space with the Borel σ-algebra $\mathscr{B}(E)$. By $Y = (Y_t, \mathbb{P}_x)_{t \geq 0, x \in E}$ we denote a strong Markov process with state space E and semigroup $(P_t)_{t \geq 0}$. We will assume that the process Y is *transient*, i.e. there exists a positive measurable function $g : E \to (0, \infty)$ such that $x \mapsto \int_0^\infty P_t g(x)\, dt$ is not identically infinite. The *potential operator* of the process Y is defined by

$$Gg(x) = \int_0^\infty P_t g(x)\, dt, \tag{14.1}$$

where $x \in E$ and $g : E \to [0, \infty)$ is a measurable function. The function Gg is called the *potential of g*. Let $f \in \mathcal{SBF}$ with the representation (11.1) and let $f^\star$ be its conjugate function given by (11.2). Further, let $S = (S_t)_{t \geq 0}$ and $T = (T_t)_{t \geq 0}$ be the special subordinators with Laplace exponents $f, f^\star \in \mathcal{SBF}$ and assume that S, T and Y are independent. The potential measures of S and T will be denoted by U and V respectively. We define two subordinate processes $Y^f = (Y_t^f)_{t \geq 0}$ and $Y^{f^\star} = (Y_t^{f^\star})_{t \geq 0}$ by

$$Y_t^f = Y(S_t) \quad \text{and} \quad Y_t^{f^\star} = Y(T_t), \quad t \geq 0.$$

Then Y^f and $Y^{f^\star}$ are again transient strong Markov processes on $(E, \mathscr{B}(E))$, see [69].

Lemma 14.1. *The potential operators G^f and $G^{f^\star}$ of Y^f and $Y^{f^\star}$ are given by*

$$G^f g(x) = \int_{[0,\infty)} P_t g(x)\, U(dt), \qquad (14.2)$$

$$G^{f^\star} g(x) = \int_{[0,\infty)} P_t g(x)\, V(dt). \qquad (14.3)$$

Proof. We will only show (14.2), the proof of (14.3) is similar. Let $(\mu_t)_{t\geq 0}$ be the convolution semigroup on $[0,\infty)$ corresponding to f. The transition operators $(P_t^f)_{t\geq 0}$ of Y^f are given by

$$P_t^f(x, A) = \mathbb{P}_x(Y_t^f \in A) = \mathbb{P}_x(Y(S_t) \in A)$$

$$= \int_{[0,\infty)} \mathbb{P}_x(Y_s \in A)\, \mu_t(ds)$$

$$= \int_{[0,\infty)} P_s(x, A)\, \mu_t(ds),$$

where $x \in E$ and $A \in \mathscr{B}(E)$. For every non-negative Borel function $g : E \to [0,\infty)$ it follows that

$$G^f g(x) = \int_0^\infty P_t^f g(x)\, dt = \int_0^\infty \int_{[0,\infty)} P_s g(x)\, \mu_t(ds)\, dt$$

$$= \int_{[0,\infty)} P_s g(x)\, U(ds). \qquad \square$$

Throughout this chapter we will assume that $b > 0$ or $\mu(0,\infty) = \infty$. This implies that $U(dt) = u(t)\, dt$ and $V(dt) = b\delta_0(dt) + v(t)\, dt$ where $u, v : (0,\infty) \to (0,\infty)$ are non-increasing functions. Hence, the potential operators G^f and $G^{f^\star}$ can be written as

$$G^f g(x) = \int_0^\infty P_t g(x)\, u(t)\, dt, \qquad (14.4)$$

$$G^{f^\star} g(x) = bg(x) + \int_0^\infty P_t g(x)\, v(t)\, dt. \qquad (14.5)$$

Moreover, by Theorem 11.9, we have

$$bu(t) + \int_0^t u(s)v(t - s)\, ds = bu(t) + \int_0^t v(s)u(t - s)\, ds = 1, \quad t > 0.$$

If the transition kernels $(P_t(x, dy))_{t\geq 0}$ have densities $p(t, x, y)$ with respect to a reference measure $m(dy)$ on $(E, \mathscr{B}(E))$, then the potential operator G will also have a density $G(x, y)$ given by

$$G(x, y) = \int_0^\infty p(t, x, y)\, dt.$$

In this case one can define the *potential of a measure* γ on $(E, \mathscr{B}(E))$ by

$$G\gamma(x) = \int_E G(x, y)\, \gamma(dy) = \int_0^\infty P_t \gamma(x)\, dt.$$

Note that the potential of a function g can be regarded as the potential of the measure $g(x)\, m(dx)$. Further, the potential operator G^f will have a density $G^f(x, y)$ given by the formula

$$G^f(x, y) = \int_0^\infty p(t, x, y) u(t)\, dt.$$

The factorization in the next proposition is similar in spirit to Corollary 13.25.

Proposition 14.2. (i) *For any non-negative Borel function g on E we have*

$$G^f G^{f\star} g(x) = G^{f\star} G^f g(x) = Gg(x), \quad x \in E.$$

(ii) *If the transition kernels $(P_t(x, dy))_{t\geq 0}$ admit densities, then for any Borel measure γ on $(E, \mathscr{B}(E))$ we have*

$$G^{f\star} G^f \gamma(x) = G\gamma(x), \quad x \in E.$$

Proof. (i) We are only going to show that $G^f G^{f\star} g(x) = Gg(x)$ for all $x \in E$. For the proof of $G^{f\star} G^f g(x) = Gg(x)$ we refer to part (ii). Let g be any non-negative Borel function on E. Using (14.4), (14.5), Fubini's theorem and Theorem 11.9 we get

$$
\begin{aligned}
G^f G^{f\star} g(x) &= \int_0^\infty P_t G^{f\star} g(x) u(t)\, dt \\
&= \int_0^\infty P_t \left(bg(x) + \int_0^\infty P_s g(x) v(s)\, ds \right) u(t)\, dt \\
&= bG^f g(x) + \int_0^\infty P_t \left(\int_0^\infty P_s g(x) v(s) ds \right) u(t)\, dt \\
&= bG^f g(x) + \int_0^\infty \int_0^\infty P_{t+s} g(x)\, v(s) u(t)\, ds\, dt \\
&= bG^f g(x) + \int_0^\infty \left(\int_t^\infty P_r g(x) v(r - t) dr \right) u(t)\, dt \\
&= bG^f g(x) + \int_0^\infty \left(\int_0^r u(t) v(r - t) dt \right) P_r g(x)\, dr \\
&= \int_0^\infty \left(bu(r) + \int_0^r u(t) v(r - t) dt \right) P_r g(x)\, dr \\
&= \int_0^\infty P_r g(x)\, dr \\
&= Gg(x).
\end{aligned}
$$

(ii) Similarly as above,

$$
\begin{aligned}
G^{f^{\star}} G^{f} \gamma(x) &= bG^{f}\gamma(x) + \int_{0}^{\infty} P_{t} G^{f}\gamma(x)\, v(t)\, dt \\
&= bG^{f}\gamma(x) + \int_{0}^{\infty} P_{t}\left(\int_{0}^{\infty} P_{s}\gamma(x)\, u(s)\, ds \right) v(t)\, dt \\
&= bG^{f}\gamma(x) + \int_{0}^{\infty}\int_{0}^{\infty} P_{t+s}\gamma(x)\, u(s)v(t)\, ds\, dt \\
&= bG^{f}\gamma(x) + \int_{0}^{\infty}\left(\int_{t}^{\infty} P_{r}\gamma(x)\, u(r-t)\, dr \right) v(t)\, dt \\
&= bG^{f}\gamma(x) + \int_{0}^{\infty}\left(\int_{0}^{r} u(r-t)v(t)\, dt \right) P_{r}\gamma(x)\, dr \\
&= \int_{0}^{\infty}\left(b + \int_{0}^{r} u(r-t)v(t)\, dt \right) P_{r}\gamma(x)\, dr \\
&= \int_{0}^{\infty} P_{r}\gamma(x)\, dr \\
&= G\gamma(x). \qquad\qquad\qquad\qquad\qquad\qquad\qquad \square
\end{aligned}
$$

In the rest of this chapter, we will always assume that the underlying Markov process Y is a Brownian motion killed upon exiting a bounded open connected subset $D \subset \mathbb{R}^{d}$. To be more precise, let $X = (X_{t}, \mathbb{P}_{x})_{t\geq 0, x\in\mathbb{R}^{d}}$ be a Brownian motion in $\mathbb{R}^{d}$ with transition density

$$
p(t, x, y) = (4\pi t)^{-d/2} \exp\left(-\frac{|x-y|^{2}}{4t} \right), \quad t > 0,\ x, y \in \mathbb{R}^{d},
$$

and let $\tau_{D} = \inf\{t > 0 : X_{t} \notin D\}$ be the first exit time of X from D. The process $X^{D} = (X_{t}^{D}, \mathbb{P}_{x})_{t\geq 0, x\in D}$ defined by

$$
X_{t}^{D} = \begin{cases} X_{t}, & t < \tau_{D}, \\ \partial, & t \geq \tau_{D}, \end{cases}
$$

where ∂ is the cemetery point, is called the *Brownian motion killed upon exiting D*. The process X^{D} is a Hunt process, and since D is bounded, it is transient. It will play the role of the process Y from the beginning of this chapter. The semigroup of X^{D} will be denoted by $(P_{t}^{D})_{t\geq 0}$, and its transition density with respect to Lebesgue measure by $p^{D}(t, x, y)$, $t > 0$, $x, y \in D$. The potential operator of X^{D} is given by

$$
G^{D} g(x) = \int_{0}^{\infty} P_{t}^{D} g(x)\, dt
$$

and has a density $G^{D}(x, y) = \int_{0}^{\infty} p^{D}(t, x, y)\, dt$, $x, y \in D$.

Since the transition density $p^D(t, x, y)$ is strictly positive, the eigenfunction φ_1 of the operator $-\Delta|_D$ corresponding to the smallest eigenvalue λ_1 can be chosen to be strictly positive, see for instance [89]. We assume that φ_1 is normalized so that $\int_D \varphi_1^2(x)\,dx = 1$.

From now on we will assume that $(P_t^D)_{t\geq 0}$ is *intrinsically ultracontractive*, that is, for each $t > 0$ there exists a constant c_t such that

$$p^D(t, x, y) \leq c_t \varphi_1(x)\varphi_1(y), \quad x, y \in D.$$

Intrinsic ultracontractivity of $(P_t^D)_{t\geq 0}$ is a very weak geometric condition on D. It is well known, cf. [90], that for a bounded Lipschitz domain D the semigroup $(P_t^D)_{t\geq 0}$ is intrinsically ultracontractive. For every intrinsically ultracontractive $(P_t^D)_{t\geq 0}$ there is some $\tilde{c}_t > 0$ such that

$$p^D(t, x, y) \geq \tilde{c}_t \varphi_1(x)\varphi_1(y), \quad x, y \in D.$$

Let $S = (S_t)_{t\geq 0}$ and $T = (T_t)_{t\geq 0}$ be two special subordinators with conjugate Laplace exponents $f, f^\star \in \mathcal{SBF}$ given by (11.1) and (11.2), respectively. We retain the assumption that $b > 0$ or $\mu(0, \infty) = \infty$, and keep the notation for the potential measures $U(dt) = u(t)\,dt$ and $V(dt) = b\delta_0(dt) + v(t)\,dt$. Suppose that X, S and T are independent and define the subordinate processes $X^{D,f} = (X_t^{D,f})_{t\geq 0}$ and $X^{D,f^\star} = (X_t^{D,f^\star})_{t\geq 0}$ by

$$X_t^{D,f} = X^D(S_t) \quad \text{and} \quad X_t^{D,f^\star} = X^D(T_t), \quad t \geq 0.$$

Then $X^{D,f}$ and $X^{D,f^\star}$ are strong Markov processes on D. We call $X^{D,f}$ and $X^{D,f^\star}$ subordinate killed Brownian motions. Their semigroups and potential operators will be denoted by $(P_t^{D,f})_{t\geq 0}$, $G^{D,f}$ and $(P_t^{D,f^\star})_{t\geq 0}$, $G^{D,f^\star}$, respectively. Clearly,

$$G^{D,f} g(x) = \int_{[0,\infty)} P_t^D g(x)\, U(dt) = \int_0^\infty P_t^D g(x)\, u(t)\, dt,$$

$$G^{D,f^\star} g(x) = \int_{[0,\infty)} P_t^D g(x)\, V(dt) = bg(x) + \int_0^\infty P_t^D g(x)\, v(t)\, dt,$$

where $x \in D$ and g is a non-negative Borel measurable function on D. By Proposition 14.2 we have $G^{D,f} G^{D,f^\star} g = G^{D,f^\star} G^{D,f} g = G^D g$ for every Borel function $g : D \to [0, \infty)$, and $G^{D,f^\star} G^{D,f} \gamma = G^D \gamma$ for every measure γ on $(D, \mathcal{B}(D))$.

Recall that a Borel function $s : D \to [0, \infty]$ is *excessive* for $X^{D,f}$ (or $(P_t^{D,f})_{t\geq 0}$), if $P_t^{D,f} s \leq s$ for all $t \geq 0$ and $s = \lim_{t \to 0} P_t^{D,f} s$. We will denote the family of all excessive function for $X^{D,f}$ by $\mathcal{S}(X^{D,f})$. The notation $\mathcal{S}(X^D)$ and $\mathcal{S}(X^{D,f^\star})$ is now self-explanatory.

A Borel function $h : D \to [0, \infty]$ is *harmonic* for $X^{D,f}$ if h is not identically infinite in D and if for every relatively compact open subset $O \subset \overline{O} \subset D$,

$$h(x) = \mathbb{E}_x\big[h\big(X^{D,f}(\tau_O)\big)\big] \quad \text{for all } x \in O,$$

where $\tau_O = \inf \{t > 0 : X_t^{D,f} \notin O\}$ is the first exit time of $X^{D,f}$ from O. We denote the family of all non-negative harmonic functions for $X^{D,f}$ by $\mathbb{H}^+(X^{D,f})$. Similarly, $\mathbb{H}^+(X^D)$ denotes the family of all non-negative harmonic functions for X^D. These are precisely those non-negative functions in D which satisfy $\Delta h = 0$ in D. Recall that if $h \in \mathbb{H}^+(X^D)$ then both h and $P_t^D h$ are continuous functions. It is easy to show that $\mathbb{H}^+(\cdot) \subset \mathbb{S}(\cdot)$, see the proof of Lemma 2.1 in [331].

Let $(G_\lambda^{D,f})_{\lambda>0}$ be the resolvent of the semigroup $(P_t^{D,f})_{t\geq 0}$. Then $G_\lambda^{D,f}$ is given by a kernel which is absolutely continuous with respect to Lebesgue measure. Moreover, one can easily show that for a bounded Borel function g vanishing outside a compact subset of D, the functions $x \mapsto G_\lambda^{D,f} g(x)$, $\lambda > 0$, and $x \mapsto G^{D,f} g(x)$ are continuous. This implies, see e.g. [60, p. 266], that all functions in $\mathbb{S}(X^{D,f})$ are lower semicontinuous.

Proposition 14.3. *If $g \in \mathbb{S}(X^{D,f})$, then $G^{D,f^\star} g \in \mathbb{S}(X^D)$.*

Proof. First observe that if $g \in \mathbb{S}(X^{D,f})$, then g is the increasing limit of potentials $G^{D,f} g_n$ for suitable non-negative Borel functions g_n, cf. [87, p. 86]. Hence it follows from Proposition 14.2 that

$$G^{D,f^\star} g = \lim_{n\to\infty} G^{D,f^\star} G^{D,f} g_n = \lim_{n\to\infty} G^D g_n.$$

Therefore, $G^{D,f^\star} g$ is either in $\mathbb{S}(X^D)$ or identically infinite since it is an increasing limit of excessive functions of X^D. We prove now that $G^{D,f^\star} g$ is not identically infinite. In fact, since $g \in \mathbb{S}(X^{D,f})$, there exists some $x_0 \in D$ such that for every $t > 0$,

$$\infty > g(x_0) \geq P_t^{D,f} g(x_0) = \int_0^\infty P_s^D g(x_0) \, \mu_t(ds),$$

where μ_t is the distribution of S_t. Thus there is some $s > 0$ such that $P_s^D g(x_0)$ is finite. Hence

$$\infty > P_s^D g(x_0) = \int_D p^D(s, x_0, y) g(y) \, dy \geq \tilde{c}_s \varphi_1(x_0) \int_D \varphi_1(y) g(y) \, dy,$$

so we have $\int_D \varphi_1(y) g(y)\, dy < \infty$. Consequently

$$
\begin{aligned}
\int_D G^{D,f^\star} g(x)\, \varphi_1(x)\, dx &= \int_D g(x)\, G^{D,f^\star} \varphi_1(x)\, dx \\
&= \int_D g(x) \left(b\varphi_1(x) + \int_0^\infty P_t^D \varphi_1(x)\, v(t) dt \right) dx \\
&= \int_D g(x) \left(b\varphi_1(x) + \int_0^\infty e^{-\lambda_1 t} \varphi_1(x)\, v(t)\, dt \right) dx \\
&= \int_D \varphi_1(x) g(x)\, dx \left(b + \int_0^\infty e^{-\lambda_1 t} v(t)\, dt \right) < \infty.
\end{aligned}
$$

Therefore $G^{D,f^\star} g$ is not identically infinite in D.   □

Remark 14.4. Note that the proposition above is valid with $X^{D,f}$ and $X^{D,f^\star}$ interchanged: if $g \in \$(X^{D,f^\star})$, then $G^{D,f} g \in \$(X^D)$.

Proposition 14.2 (ii) shows that if $s = G^D \gamma$ is the potential of a measure, then $s = G^{D,f^\star} g$ where $g = G^{D,f} \gamma$ is in $\$(X^{D,f})$. The function g can be written in the following way:

$$
\begin{aligned}
g(x) &= \int_0^\infty P_r^D \gamma(x)\, u(r)\, dr \\
&= \int_0^\infty P_r^D \gamma(x) \left(u(\infty) + \int_{(r,\infty)} (-du(t)) \right) dr \\
&= \int_0^\infty P_r^D \gamma(x) u(\infty)\, dr + \int_0^\infty P_r^D \gamma(x) \left(\int_{(r,\infty)} (-du(t)) \right) dr \\
&= u(\infty) s(x) + \int_{(0,\infty)} \left(\int_0^t P_r^D \gamma(x)\, dr \right) (-du(t)) \\
&= u(\infty) s(x) + \int_{(0,\infty)} \left(P_t^D s(x) - s(x) \right) du(t). \tag{14.6}
\end{aligned}
$$

In Proposition 14.8 below we will show that every $s \in \$(X^D)$ can be represented as a potential $G^{D,f^\star} g$, where g, given by (14.6), is in $\$(X^{D,f})$. In order to accomplish this, we will first show that g is continuous when $s \in \mathbb{H}^+(X^D)$. Before we do this, we prove two preliminary results about non-negative harmonic functions of X^D.

For any $x_0 \in D$, choose $r > 0$ such that $B(x_0, 2r) \subset D$. Put $B = B(x_0, 2r)$.

Lemma 14.5. If $h \in \mathbb{H}^+(X^D)$, then there is a constant $c > 0$ such that

$$
h(x) - P_t^D h(x) \le ct, \quad x \in \overline{B}, \ t > 0.
$$

Proof. For any $x \in B$, $h(X_{t \wedge \tau_B})$ is a $\mathbb{P}_x$-martingale. Therefore,

$$
\begin{aligned}
0 \leq h(x) - P_t^D h(x) &= \mathbb{E}_x\big[h(X_{t \wedge \tau_B})\big] - \mathbb{E}_x\big[h(X_t)\mathbb{1}_{\{t < \tau_D\}}\big] \\
&= \mathbb{E}_x\big[h(X_t)\mathbb{1}_{\{t < \tau_B\}}\big] + \mathbb{E}_x\big[h(X_{\tau_B})\mathbb{1}_{\{\tau_B \leq t\}}\big] \\
&\quad - \mathbb{E}_x\big[h(X_t)\mathbb{1}_{\{t < \tau_B\}}\big] - \mathbb{E}_x\big[h(X_t)\mathbb{1}_{\{\tau_B \leq t < \tau_D\}}\big] \\
&= \mathbb{E}_x\big[h(X_{\tau_B})\mathbb{1}_{\{\tau_B \leq t\}}\big] - \mathbb{E}_x\big[h(X_t)\mathbb{1}_{\{\tau_B \leq t < \tau_D\}}\big] \\
&\leq \mathbb{E}_x\big[h(X_{\tau_B})\mathbb{1}_{\{\tau_B \leq t\}}\big] \\
&\leq M\,\mathbb{P}_x(\tau_B \leq t),
\end{aligned}
$$

where M is a constant such that $h(y) \leq M$ for all $y \in \overline{B}$. It is a standard fact, cf. [24, Lemma I.(8.1)], that there exists a constant $c > 0$ such that for every $x \in \overline{B}$ it holds that $\mathbb{P}_x(\tau_B \leq t) \leq ct$, for all $t > 0$. Therefore, $0 \leq h(x) - P_t^D h(x) \leq Mct$, for all $x \in \overline{B}$ and all $t > 0$.   □

Lemma 14.6. *For every $h \in \mathbb{H}^+(X^D)$, we define $k_n(x) := n\big(h(x) - P_{1/n}^D h(x)\big)$ and $s_n := G^D k_n(x)$. Then s_n increases to h as $n \to \infty$, and there exists $c > 0$ such that*

$$
s_n(y) - P_t^D s_n(y) \leq c\,t, \quad y \in \overline{B},\ n \in \mathbb{N}.
$$

Proof. The assertion that s_n increases to h as $n \to \infty$ follows from [23, Proposition II.(6.2)]. We only prove the second assertion. For any $t > 0$, $n \in \mathbb{N}$ and $y \in \overline{B}$, we have

$$
\begin{aligned}
s_n(y) - P_t^D s_n(y) &= \int_0^t P_r^D k_n(y)\, dr \\
&= n \int_0^t P_r^D (h - P_{1/n}^D h)(y)\, dr \\
&= n \int_0^t P_r^D h(y)\, dr - n \int_{\frac{1}{n}}^{t+\frac{1}{n}} P_r^D h(y)\, dr.
\end{aligned}
$$

If $t < 1/n$, then it follows from Lemma 14.5 that

$$
\begin{aligned}
s_n(y) - P_t^D s_n(y) &= n \int_{\frac{1}{n}}^{t+\frac{1}{n}} \big(h(y) - P_r^D h(y)\big)\, dr - n \int_0^t \big(h(y) - P_r^D h(y)\big)\, dr \\
&\leq n \int_{\frac{1}{n}}^{t+\frac{1}{n}} \big(h(y) - P_r^D h(y)\big)\, dr \\
&\leq c_1 n \int_{\frac{1}{n}}^{t+\frac{1}{n}} r\, dr \leq c_2 t.
\end{aligned}
$$

If $t \geq 1/n$, then again by Lemma 14.5 we have

$$s_n(y) - P_t^D s_n(y) = n \int_0^{\frac{1}{n}} P_r^D h(y) \, dr - n \int_t^{t+\frac{1}{n}} P_r^D h(y) \, dr$$

$$= n \int_t^{t+\frac{1}{n}} \left(h(y) - P_r^D h(y) \right) dr - n \int_0^{\frac{1}{n}} \left(h(y) - P_r^D h(y) \right) dr$$

$$\leq n \int_t^{t+\frac{1}{n}} \left(h(y) - P_r^D h(y) \right) dr$$

$$\leq h(y) - P_{t+\frac{1}{n}}^D h(y) \leq c_3 \left(t + \frac{1}{n} \right) \leq c_4 t. \qquad \square$$

Lemma 14.7. *If $h \in \mathbb{H}^+(X^D)$, the function g defined by*

$$g(x) = u(\infty)h(x) + \int_{(0,\infty)} \left(P_t^D h(x) - h(x) \right) du(t), \quad x \in D, \qquad (14.7)$$

is continuous.

Proof. For $x_0 \in D$ choose $r > 0$ such that $B(x_0, 2r) \subset D$, and let $B := B(x_0, r)$. It follows from the continuity and harmonicity of h that there exists a constant $c_1 > 0$ such that $0 \leq h(x) - P_t^D h(x) \leq c_1$ on B. Using Lemma 14.5 we know that there exists some $c_2 > 0$ such that

$$0 \leq h(x) - P_t^D h(x) \leq c_2 t, \quad x \in B.$$

Since $\int_{(0,\infty)} (1 \wedge t)(-du(t)) < \infty$, we can apply the dominated convergence theorem to get the continuity of g. $\qquad \square$

Proposition 14.8. *If $s \in \mathbb{S}(X^D)$, then*

$$s(x) = G^{D,f^\star} g(x), \quad x \in D,$$

where $g \in \mathbb{S}(X^{D,f})$ and is given by the formula

$$g(x) = u(\infty)s(x) + \int_0^\infty \left(P_t^D s(x) - s(x) \right) du(t). \qquad (14.8)$$

Proof. We know that the result is true when s is the potential of a measure. Let $s \in \mathbb{S}(X^D)$ be arbitrary. By the Riesz decomposition theorem, $s = G^D \gamma + h$, where γ is a measure on D and h is in $\mathbb{H}^+(X^D)$, see e.g. [60, Chapter 6]. By linearity, it is enough to prove the result for $s \in \mathbb{H}^+(X^D)$.

In the rest of the proof we assume therefore that $s \in \mathbb{H}^+(X^D)$. Define the function g by formula (14.8). We have to prove that g is an excessive function for $X^{D,f}$ and $s = G^{D,f^\star} g$. By Lemma 14.7, we know that g is continuous.

Define $k_n := (s(x) - P_{1/n}^D s(x))$ and $s_n := G^D k_n(x)$. It follows from Lemma 14.6 that s_n increases to s. Thus $P_t^D s_n \uparrow P_t^D s$ and $s_n - P_t^D s_n \xrightarrow{n \to \infty} s - P_t^D s$. If

$$g_n(x) = u(\infty)s_n(x) + \int_{(0,\infty)} \left(s_n(x) - P_t^D s_n(x)\right) \left(-du(t)\right),$$

then we know that $s_n = G^{D,f^\star} g_n$ and g_n is excessive for $X^{D,f}$. It follows from Lemma 14.6 that, for every $x \in D$, $0 \le s_n(x) - P_t^D s_n(x) \le s(x)$. Again by Lemma 14.6 there exists for every $x \in D$ some $c > 0$ such that $0 \le s_n(x) - P_t^D s_n(x) \le ct$. Since $\int_{(0,\infty)} (1 \wedge t)(-du(t)) < \infty$, we can apply the dominated convergence theorem to get

$$g(x) = u(\infty)s(x) + \int_{(0,\infty)} \left(s(x) - P_t^D s(x)\right) \left(-du(t)\right)$$

$$= \lim_{n \to \infty} u(\infty)s_n(x) + \int_{(0,\infty)} \lim_{n \to \infty} \left(s_n(x) - P_t^D s_n(x)\right) \left(-du(t)\right)$$

$$= \lim_{n \to \infty} g_n(x).$$

By Fatou's lemma and the fact that g_n is $G^{D,f}$-excessive, we get for any $x \in D$ and $\lambda > 0$

$$\lambda G_\lambda^{D,f} g(x) = \lambda G_\lambda^{D,f} \left(\lim_{n \to \infty} g_n\right)(x) \le \liminf_{n \to \infty} \lambda G_\lambda^{D,f} g_n(x)$$

$$\le \liminf_{n \to \infty} g_n(x) = g(x);$$

this means that g is supermedian. Since it is well known that a supermedian function which is lower semicontinuous is excessive, cf. [87, p. 81], this proves that g is excessive for $X^{D,f}$. By Proposition 14.3 we then have that $G^{D,f^\star} g$ is excessive for X^D.

Set $G_1^D h(x) = \int_0^\infty e^{-t} P_t^D h(x)\, dt$ for any non-negative measurable function h and define $s^1 := s - G_1^D s$. Using an argument similar to that of the proof of Proposition 14.3, we can show that $G^D s$ is not identically infinite. By the resolvent equation $G^D s^1 = G^D s - G^D G_1^D s = G_1^D s$, or equivalently,

$$s(x) = s^1(x) + G_1^D s(x) = s^1(x) + G^D s^1(x), \quad x \in D.$$

Formula (14.6) for the potential $G^D s^1$, Fubini's theorem and the fact that $G^{D,f^\star}$ and G_1^D commute, show

$$G_1^D s = G^D s^1 = G^{D,f^\star} \left(u(\infty)G^D s^1 + \int_{(0,\infty)} (P_t^D G^D s^1 - G^D s^1)\, du(t)\right)$$

$$= G^{D,f^\star} \left(u(\infty)G_1^D s + \int_{(0,\infty)} (P_t^D G_1^D s - G_1^D s)\, du(t)\right)$$

$$= G_1^D G^{D,f^\star} \left(u(\infty)s + \int_{(0,\infty)} (P_t^D s - s)\, du(t)\right).$$

By the uniqueness principle, see e.g. [87, p. 140], it follows that

$$s = G^{D,f^\star}\left(u(\infty)s + \int_0^\infty (P_t^D s - s)\,du(t)\right) = G^{D,f^\star}g \quad \text{a.e. in } D.$$

Since both s and $G^{D,f^\star}g$ are excessive for X^D, they are equal everywhere.   □

Propositions 14.2 and 14.8 can be combined in the following theorem containing additional information on harmonic functions.

Theorem 14.9. *For $s \in \$(X^D)$, there is a function $g \in \$(X^{D,f})$ with $s = G^{D,f^\star}g$. The function g is given by the formula (14.6). Furthermore, if $s \in \mathbb{H}^+(X^D)$, then $g \in \mathbb{H}^+(X^{D,f})$.*
Conversely, if $g \in \$(X^{D,f})$, then the function s defined by $s = G^{D,f^\star}g$ is in $\$(X^D)$. If, moreover, $g \in \mathbb{H}^+(X^{D,f})$, then $s \in \mathbb{H}^+(X^D)$.
Finally, every non-negative harmonic function for $X^{D,f}$ is continuous.

Proof. It remains to show the statements about harmonic functions. First note that every $g \in \$(X^{D,f})$ admits the Riesz decomposition $g = G^{D,f}\gamma + h$ where γ is a Borel measure on D and $h \in \mathbb{H}^+(X^{D,f})$, see [60, Chapter 6]. We have already mentioned that functions in $\$(X^D)$ admit such a decomposition. Since $\$(X^D)$ and $\$(X^{D,f})$ are in one-to-one correspondence, and since potentials of measures of X^D and $X^{D,f}$ are in one-to-one correspondence, the same must hold for $\mathbb{H}^+(X^D)$ and $\mathbb{H}^+(X^{D,f})$.

The continuity of non-negative harmonic functions for $X^{D,f}$ follows from Lemma 14.7 and Proposition 14.8.   □

Comments 14.10. The study of the potential theory of subordinate killed Brownian motions was initiated in [128] for stable subordinators and carried out satisfactorily in this case in [129]. The general case of special subordinators was dealt with in [332]. Our exposition follows that of [64, Section 5.5.2], but we provide a more elementary and direct proof of Proposition 14.8 here. Using Theorem 14.9, one can generalize some deep and important potential theoretic results for killed Brownian motions, like the Martin boundary identification and the boundary Harnack principle, to subordinate killed Brownian motions. For these results we refer to [129], [64, Section 5.5] and [332] where the underlying process X is a killed rotationally invariant α-stable process.

Intrinsic ultracontractivity was introduced by Davies and Simon in [90]. It is a very weak regularity condition on D. For example, see [90], the killed semigroup $(P_t^D)_{t\geq 0}$ is intrinsically ultracontractive whenever D is a bounded Lipschitz domain. For weaker conditions on D guaranteeing that $(P_t^D)_{t\geq 0}$ is intrinsically ultracontractive, we refer to [18] and [25].

Chapter 15

Applications to generalized diffusions

Complete Bernstein functions and the corresponding Bondesson class of sub-probability measures play a significant role in various parts of probability theory. In this chapter we will explain their appearance in the theory of generalized one-dimensional diffusions (also called gap diffusions or quasi-diffusions). We have two main objectives: firstly, we want to characterize the Laplace exponents of the inverse local times at zero of generalized diffusions. It turns out that these are precisely the complete Bernstein functions. Secondly, we want to study the first passage distributions for generalized diffusions and to explain their connection with the Bondesson class and its subclasses. The crucial tool in these investigations will be Kreĭn's theory of strings which we will describe in some detail.

In this chapter we will use Dirichlet forms and Hunt processes. The basic definitions and results are collected in Appendix A.2.

15.1 Inverse local time at zero

We start by introducing the family $\mathfrak{M}_+$ of the so-called strings.

Definition 15.1. A non-decreasing and right-continuous function $m : \mathbb{R} \to [0, \infty]$ is a *string*, if

(a) $m(x) = m(0-) = 0$ for all $x < 0$,

(b) $m(x_0) < \infty$ for some $x_0 \geq 0$,

(c) $m(x) > 0$ for all $x > 0$.

The family of all strings will be denoted by $\mathfrak{M}_+$.

Slightly abusing notation, we will use the same letter m to denote the measure on $\mathbb{R}$ induced by the non-decreasing function $m \in \mathfrak{M}_+$. Clearly, $m(-\infty, 0) = 0$. Let $r := \sup\{x : m(x) < \infty\}$ be the *length* of the string, and let E_m denote the support of the measure m restricted to $[0, r)$. Note that the point r of a possibly infinite jump is excluded from the support while $0 \in E_m$. Finally, let $l := \sup E_m \leq r$.

Next we introduce generalized diffusions. For this let $B^+ = (B_t^+, \mathbb{P}_x)_{t \geq 0, 0 \leq x < r}$ be a reflected Brownian motion on $[0, \infty)$ which is killed upon hitting the point r; if $r = \infty$, then B^+ is simply a reflected Brownian motion. After being killed, the process goes to the cemetery ∂ which we adjoin to the state space $[0, r)$ as an isolated point. Note that B^+ is a Hunt process which is symmetric with respect to Lebesgue measure dx. In particular, there exists an associated Dirichlet form on $L^2([0,r),dx)$. Let $L = (L(t,x))_{t\geq 0,\, 0\leq x<r}$ be the jointly continuous local time of B^+ normalized such that for all Borel functions $g : [0,\infty) \to [0,\infty)$ the following occupation time formula holds:

$$\int_0^t g(B_s^+)\,ds = \int_{[0,r)} g(x)L(t,x)\,dx, \quad t \geq 0.$$

For $m \in \mathfrak{M}_+$ the corresponding measure m on $[0,r)$ is a smooth measure in the sense of Definition A.15; note that no non-empty set is of capacity zero for B^+. We define a positive continuous additive functional $C = (C_t)_{t\geq 0}$ of the process B^+ by

$$C_t := \int_{[0,r)} L(t,x)\,m(dx). \tag{15.1}$$

Clearly, the Revuz measure of C is equal to m. Let $\tau = (\tau_t)_{t\geq 0}$ be the right-continuous inverse of C: $\tau_t = \inf\{s > 0 : C_s > t\}$. Define a process $X = (X_t)_{t\geq 0}$ via the time-change of B^+ by the process τ, $X_t := B^+(\tau_t)$. By [123, Theorem 6.2.1] X is an m-symmetric Hunt process on E_m which is associated with a regular Dirichlet form on $L^2(E_m,m)$. Note that the lifetime ζ of X is equal to C_∞.

Definition 15.2. The process X is called a *generalized diffusion* on $[0,r)$ (or, more precisely, on E_m).

Clearly, $X_t = \partial$ for $t \geq \zeta$. The class of such processes includes regular diffusions on $[0,r)$ and $[0,l]$ (in the sense of Itô–McKean), as well as birth-and-death processes. We will use the standard convention that any function g on the state space is automatically extended to ∂ by setting $g(\partial) = 0$. The generalized diffusion X admits the local time process $L^X = (L^X(t,x))_{t\geq 0, x\in E_m}$ which can be realized as a time-change of the local time of B^+. More precisely, in the same way as in [307, Theorem V.49.1], it follows that $L^X(t,x) = L(\tau_t,x)$ and

$$\int_0^s g(X_t)\,dt = \int_{[0,r)} g(x)L^X(s,x)\,m(dx), \tag{15.2}$$

for every bounded measurable function g, and every $s \geq 0$. In particular, it follows from the joint continuity of L that $x \mapsto L^X(s,x)$ is continuous.

For simplicity, let us denote the local time at zero for X, $L^X(t,0)$, by $\ell^X(t)$, and let $(\ell^X)^{-1}(t) := \inf\{s > 0 : \ell^X(s) > t\}$ be the inverse local time at zero. The inverse local time is a (possibly killed) subordinator, see, e.g. [47, p. 114]. Therefore we may assume that the Laplace exponent f of $(\ell^X)^{-1}$ has the representation

$$f(\lambda) = a + b\lambda + \int_{(0,\infty)} (1 - e^{-\lambda t})\,\mu(dt).$$

An interesting problem is to characterize the family of Bernstein functions that can arise as Laplace exponents of inverse local times at zero of generalized diffusions. This question was raised in [178, p. 217] and solved, independently, in [209] and [217]. The answer is given in the following theorem.

Theorem 15.3. *Let X be a generalized diffusion corresponding to $m \in \mathfrak{M}_+$ and let $(\ell^X)^{-1}$ be its inverse local time at zero. Then the Laplace exponent f of $(\ell^X)^{-1}$ belongs to $\mathcal{CBF}$. Conversely, given any function $f \in \mathcal{CBF}$, there exists a generalized diffusion such that f is the Laplace exponent of its inverse local time at zero.*

The proof of this theorem relies heavily on Kreĭn's theory of strings. A full exposition of this theory would lead us far away from the topics of this book. Therefore, we will only outline the main ideas, and give more details for the parts of the proof where the methods are close to our subject.

The first ingredient of the proof is probabilistic: for $\lambda > 0$ the λ-potential operator of X is defined by

$$G_\lambda g(x) := \mathbb{E}_x \int_0^\infty e^{-\lambda t} g(X_t)\, dt, \tag{15.3}$$

where $x \in E_m$ and g is a non-negative Borel function on $[0, r)$. By Fubini's theorem and (15.2), the λ-potential can be rewritten as

$$\begin{aligned}
G_\lambda g(x) &= \mathbb{E}_x \int_0^\infty \lambda e^{-\lambda s} \left(\int_0^s g(X_t)\, dt \right) ds \\
&= \mathbb{E}_x \int_0^\infty \lambda e^{-\lambda s} \left(\int_{[0,r)} g(y) L^X(s, y)\, m(dy) \right) ds \\
&= \int_{[0,r)} \mathbb{E}_x \left(\int_0^\infty \lambda e^{-\lambda s} L^X(s, y)\, ds \right) g(y)\, m(dy) \\
&= \int_{[0,r)} \mathbb{E}_x \left(\int_{[0,\infty)} e^{-\lambda s} L^X(ds, y) \right) g(y)\, m(dy) \\
&= \int_{[0,r)} G_\lambda(x, y) g(y)\, m(dy). \tag{15.4}
\end{aligned}$$

Here $G_\lambda(x, y)$ is the kernel of the λ-potential operator G_λ defined by

$$G_\lambda(x, y) := \mathbb{E}_x \left(\int_{[0,\infty)} e^{-\lambda s} L^X(ds, y) \right), \quad x, y \in E_m. \tag{15.5}$$

Note that for each $x \in E_m$, the function $y \mapsto G_\lambda(x, y)$ is continuous on E_m. From now on, each function u defined on E_m is *linearly* extended to $[0, r) \setminus E_m$; note that this does not uniquely determine the extension on (l, r). This applies also to $G_\lambda(x, \cdot)$.

Proposition 15.4. *For every $\lambda > 0$ it holds that*

$$G_\lambda(0,0) = \frac{1}{f(\lambda)}.$$

Proof. From (15.5) we have

$$G_\lambda(0,0) = \mathbb{E}_0 \int_{[0,\infty)} e^{-\lambda t}\,\ell^X(dt) = \mathbb{E}_0 \int_0^\infty e^{-\lambda\,(\ell^X)^{-1}(t)}\,dt$$

$$= \int_0^\infty \mathbb{E}_0(e^{-\lambda\,(\ell^X)^{-1}(t)})\,dt = \int_0^\infty e^{-tf(\lambda)}\,dt = \frac{1}{f(\lambda)}. \qquad \square$$

By Proposition 15.4 and Theorem 7.3, $f \in \mathcal{CBF}$ if and only if $\lambda \mapsto G_\lambda(0,0) \in \mathcal{S}$. In order to show that $\lambda \mapsto G_\lambda(0,0) \in \mathcal{S}$, we are going to identify $G_\lambda(0,0)$ with the so-called characteristic function of the string and show that the latter is a Stieltjes function. To do so we need the analytic counterpart of generalized diffusions, namely the second-order differential operator $A = \frac{d}{dm}\frac{d}{dx}$ which is formally defined in the following way. The domain $\mathfrak{D}^0(A)$ of A consists of functions $u : \mathbb{R} \to \mathbb{C}$ such that

$$u(x) = \alpha + \beta x + \int_{[0,x]} (x-y)g(y)\,m(dy)$$

$$= \alpha + \beta x + \int_0^x \int_{[0,y]} g(w)\,m(dw)\,dy, \tag{15.6}$$

where $\alpha, \beta \in \mathbb{C}$ and $g : \mathbb{R} \to \mathbb{C}$. If $x < 0$ we interpret $[0,x]$ as the empty set $\emptyset$, therefore $u(x) = \alpha + \beta x$ when $x < 0$. Then A is defined by $Au := g$ for $u \in \mathfrak{D}^0(A)$. Every function u in $\mathfrak{D}^0(A)$ is absolutely continuous, linear outside of E_m, and has right u'_+ and left derivatives u'_-. More precisely,

$$u'_+(x) = \lim_{y\downarrow x} \frac{u(y)-u(x)}{y-x} = \beta + \int_{[0,x]} g(y)\,m(dy), \tag{15.7}$$

$$u'_-(x) = \lim_{y\uparrow x} \frac{u(y)-u(x)}{y-x} = \beta + \int_{[0,x)} g(y)\,m(dy). \tag{15.8}$$

In particular, $\beta = u'_-(0)$. Moreover, for $0 \le x \le l$,

$$u'_+(x) - u'_-(x) = m\{x\}Au(x). \tag{15.9}$$

For $z \in \mathbb{C}$, let $\phi = \phi(x,z)$ and $\psi = \psi(x,z)$ satisfy the following integral equations for $x \in [0,r)$:

$$\phi(x,z) = 1 + z \int_{[0,x]} (x-y)\,\phi(y,z)\,m(dy), \tag{15.10}$$

$$\psi(x,z) = x + z \int_{[0,x]} (x-y)\,\psi(y,z)\,m(dy). \tag{15.11}$$

It can be shown that both equations above have unique solutions. We sketch the construction of ϕ. Let $\phi_0(x) \equiv 1$ and define for $n \in \mathbb{N}$

$$\phi_n(x) := \int_{[0,x]} (x - y)\,\phi_{n-1}(y)\,m(dy). \tag{15.12}$$

Expressing ϕ_n as an n-fold integral, one can prove the following estimate, see [196, p. 32] or [105, p. 162] for details:

$$\phi_n(x) \leq \frac{(x\,m(x))^n}{n^n\,n!} \leq \frac{(2x\,m(x))^n}{(2n)!}. \tag{15.13}$$

Then

$$\phi(x, z) := \sum_{n=0}^{\infty} \phi_n(x) z^n \tag{15.14}$$

converges and satisfies (15.10).

Note that the equations (15.10) and (15.11) can be rewritten as $A\phi(\cdot, z) = z\phi(\cdot, z)$, $A\psi(\cdot, z) = z\psi(\cdot, z)$, with the boundary conditions $\phi(0, z) = 1$, $\phi'_-(0, z) = 0$, $\psi(0, z) = 0$ and $\psi'_-(0, z) = 1$. Here, and in the remaining part of this chapter, the derivatives are with respect to the first variable.

In the remainder of this chapter we will use the following notational convention when working with the functions ϕ and ψ: if the second argument z is a real number, we will write λ instead of z. The functions ϕ and ψ have the following properties:

(a) for each fixed $x \in [0, r)$, $z \mapsto \phi(x, z)$ and $z \mapsto \psi(x, z)$ are real entire functions, i.e. they are analytic functions on $\mathbb{C}$ which are real if $z \in \mathbb{R}$;

(b) for all $x \in (0, r)$ and $\lambda > 0$ we have $\phi(x, \lambda) > 0$, $\phi'_+(x, \lambda) > 0$, $\psi(x, \lambda) > 0$ and $\psi'_+(x, \lambda) > 0$;

(c) for every $z \in \mathbb{C}$ with $\operatorname{Im} z \neq 0$ we have $\phi(x, z) \neq 0$, $\phi'_+(x, z) \neq 0$, $\psi(x, z) \neq 0$ and $\psi'_+(x, z) \neq 0$.

Property (a) follows from (15.14) and the analogous equation for ψ, and (b) follows from (15.12) and the analogous equations for ψ; (c) can be deduced from the following Lagrange identity: if $u, v \in \mathfrak{D}^0(A)$, then for all $x \in [0, r)$,

$$\int_{[0,x]} \big(v(y)Au(y) - u(y)Av(y)\big)\,m(dy)$$
$$= \big(u'_+(x)v(x) - u(x)v'_+(x)\big) - \big(u'_-(0)v(0) - u(0)v'_-(0)\big), \tag{15.15}$$

cf. [196, p. 24]. It follows from this that for any $x \in [0, r)$ and $z \in \mathbb{C}$,

$$\phi(x, z)\psi'_+(x, z) - \phi'_+(x, z)\psi(x, z)$$
$$= \phi(x, z)\psi'_-(x, z) - \phi'_-(x, z)\psi(x, z) = 1. \tag{15.16}$$

Proposition 15.5. *For $x \in [0, r)$ and $z \in \mathbb{C} \setminus (-\infty, 0]$ define*

$$h(x, z) := \frac{\psi(x, z)}{\phi(x, z)}.$$

Then $h(x, \cdot)$ restricted to $(0, \infty)$ belongs to $\mathcal{S}$. Moreover, for $0 \le x < r$,

$$h(x, z) = \int_0^x \frac{1}{\phi(y, z)^2} \, dy. \tag{15.17}$$

Proof. If $x = 0$, then $h(0, \cdot) \equiv 0$, so there is nothing to prove. Thus we assume that $x > 0$. Since the denominator does not vanish, $h(x, z)$ is well defined for all $z \in \mathbb{C} \setminus (-\infty, 0]$ and it is analytic on $\mathbb{C} \setminus (-\infty, 0]$. We will show that $h(x, \cdot)$ switches $\mathbb{H}^\uparrow$ and $\mathbb{H}^\downarrow$, which proves the claim because of Corollary 7.4. For $0 \le y < r$ define

$$\chi_x(y, z) := -\psi(y, z) + h(x, z)\phi(y, z), \tag{15.18}$$

and note that $A\chi_x(\cdot, z) = z\chi_x(\cdot, z)$ on $[0, x]$. A straightforward calculation yields that

$$\chi_x(x, z) = 0, \qquad (\chi_x)'_+(x, z) = -\frac{1}{\phi(x, z)}, \tag{15.19}$$

$$\chi_x(0, z) = h(x, z), \qquad (\chi_x)'_-(0, z) = -1.$$

Let $\operatorname{Im} z \ne 0$. By the Lagrange identity (15.15) applied to $\chi_x(\cdot, z)$ and $\chi_x(\cdot, \bar{z})$, and by (15.19), we get

$$(z - \bar{z}) \int_{[0,x]} \chi_x(y, z)\chi_x(y, \bar{z}) \, m(dy)$$
$$= \big((\chi_x)'_+(x, z)\chi_x(x, \bar{z}) - (\chi_x)'_+(x, \bar{z})\chi_x(x, z)\big)$$
$$\quad - \big((\chi_x)'_-(0, z)\chi_x(0, \bar{z}) - (\chi_x)'_-(0, \bar{z})\chi(0, z)\big)$$
$$= -\big(h(x, z) - h(x, \bar{z})\big) = -\big(h(x, z) - \overline{h(x, z)}\big).$$

Hence

$$\operatorname{Im} z \int_{[0,x]} |\chi_x(y, z)|^2 \, m(dy) = -\operatorname{Im}\big(h(x, z)\big).$$

Since $\int_{[0,x]} |\chi_x(y, z)|^2 \, m(dy) > 0$, we have that $\operatorname{Im} z \cdot \operatorname{Im} h(x, z) \le 0$.

The last statement of the proposition is a consequence of

$$h(x, z) = \frac{\psi(x, z)}{\phi(x, z)} = \int_0^x \left(\frac{\psi(y, z)}{\phi(y, z)}\right)' dy = \int_0^x \frac{1}{\phi(y, z)^2} \, dy, \tag{15.20}$$

where the last equality follows from (15.16). $\qquad\qquad\qquad\qquad\qquad\qquad \square$

Let $\lambda > 0$ and $0 < x \leq y < r$. Then

$$
\phi(y, \lambda) \geq 1 + \lambda \phi_1(y) = 1 + \lambda \int_0^y m(w) \, dw
$$

$$
\geq \lambda \int_{x/2}^y m(w) \, dw \geq \lambda \frac{y}{2} m\left(\frac{x}{2}\right). \tag{15.21}
$$

Therefore,

$$
\int_x^r \frac{dy}{\phi(y, \lambda)^2} \leq \frac{4}{\lambda^2 m(x/2)^2} \int_x^r \frac{dy}{y^2} < \infty.
$$

Since $\phi(\cdot, \lambda)$ is bounded away from zero on $[0, x]$, it follows that

$$
\int_0^r \frac{dy}{\phi(y, \lambda)^2} < \infty, \tag{15.22}
$$

and it can be easily shown that

$$
\lim_{\lambda \to \infty} \int_0^r \frac{dy}{\phi(y, \lambda)^2} = 0. \tag{15.23}
$$

Definition 15.6. The function $h : (0, \infty) \to [0, \infty)$ defined by

$$
h(\lambda) := \lim_{x \to r} h(x, \lambda) = \lim_{x \to r} \frac{\psi(x, \lambda)}{\phi(x, \lambda)} = \int_0^r \frac{dy}{\phi(y, \lambda)^2} \tag{15.24}
$$

is called the *characteristic function* of the string $m \in \mathfrak{M}_+$.

It follows from Proposition 15.5 and (15.22) that h is well defined. The next theorem is one of the fundamental results of Kreĭn's theory of strings.

Theorem 15.7. (i) *Let $m \in \mathfrak{M}_+$ be a string. Then its characteristic function h is a Stieltjes function such that $h(+\infty) = 0$, and hence it has the representation*

$$
h(\lambda) = \int_{[0,\infty)} \frac{1}{\lambda + t} \sigma(dt),
$$

where σ is a measure on $[0, \infty)$ such that $\int_{(0,\infty)} (1 + t)^{-1} \sigma(dt) < \infty$.

(ii) *Conversely, for any $h \in \mathcal{S}$ such that $h(+\infty) = 0$ there exists a unique string $m \in \mathfrak{M}_+$ such that h is the characteristic function of m.*

Proof. (i) It was shown in Proposition 15.5 that $h(x, \cdot) \in \mathcal{S}$ for every $x \in [0, r)$. By Definition 15.6, h is a pointwise limit of Stieltjes functions which is, by Theorem 2.2 (iii), again a Stieltjes function. Formula (15.23) implies that $h(+\infty) = 0$, whence the form of the representation.

(ii) The proof of the converse is beyond the scope of this book and we refer the interested reader to [105] and [196]. □

Remark 15.8. If we drop the requirement that $m(x) > 0$ for every $x > 0$ in the definition of $m \in \mathfrak{M}_+$, then $c := \inf E_m$ may be strictly positive. The discussions leading to Theorem 15.7 remain almost literally the same, except that the limit in (15.23) turns out to be equal to c. Theorem 15.7 is then slightly extended to encompass all Stieltjes functions.

Proposition 15.9. *Let* $m \in \mathfrak{M}_+$ *be a string and define for* $0 \le x < r$ *and* $\lambda > 0$ *a function* $\chi(x, \lambda) := -\psi(x, \lambda) + h(\lambda)\phi(x, \lambda)$. *Then*

$$\chi(x, \lambda) = \phi(x, \lambda) \int_x^r \frac{dy}{\phi(y, \lambda)^2}, \tag{15.25}$$

$A\chi(\cdot, \lambda) = \lambda\chi(\cdot, \lambda)$ *and* $\chi'_-(0, \lambda) = -1$.

Proof. By (15.24) and (15.20)

$$\begin{aligned}
\chi(x, \lambda) &= -\psi(x, \lambda) + \phi(x, \lambda) \int_0^r \frac{dy}{\phi(y, \lambda)^2} \\
&= -\phi(x, \lambda) \int_0^x \frac{dy}{\phi(y, \lambda)^2} + \phi(x, \lambda) \int_0^r \frac{dy}{\phi(y, \lambda)^2} \\
&= \phi(x, \lambda) \int_x^r \frac{dy}{\phi(y, \lambda)^2}.
\end{aligned}$$

That $A\chi(\cdot, \lambda) = \lambda\chi(\cdot, \lambda)$ follows immediately from the same property of $\phi(\cdot, \lambda)$ and $\psi(\cdot, \lambda)$, while $\chi'_-(0, \lambda) = -1$ is a straightforward computation. □

For $\lambda > 0$ let

$$R_\lambda(x, y) := \begin{cases} \phi(x, \lambda)\chi(y, \lambda), & 0 \le x \le y < r, \\ \chi(x, \lambda)\phi(y, \lambda), & 0 \le y \le x < r, \end{cases} \tag{15.26}$$

and note that by Proposition 15.9

$$R_\lambda(0, 0) = \phi(0, \lambda)\chi(0, \lambda) = h(\lambda). \tag{15.27}$$

Thus, $\lambda \mapsto R_\lambda(0, 0) \in \mathcal{S}$, see also Remark 12.8.

For $g \in L^2((-\infty, r), m)$ define the λ-resolvent operator

$$R_\lambda g(x) := \int_{[0,r)} R_\lambda(x, y)g(y)\, m(dy). \tag{15.28}$$

It is easy to prove that R_λ is a bounded linear operator from $L^2((-\infty, r), m)$ to itself, see [105, p. 168]. In what follows we will need the L^2-domain of the operator A that appears in [105, p. 151].

Definition 15.10. Let $L^2 := L^2((-\infty, r), m)$.

(i) If $l + m(l-) = \infty$ and $\int_{[0,r)} x^2 m(dx) = \infty$, let

$$\mathfrak{D}(A) := \mathfrak{D}^0(A) \cap \{u \in L^2 : Au \in L^2, u'_-(0) = 0\}.$$

(ii) If $l + m(l-) = \infty$ and $\int_{[0,r)} x^2 m(dx) < \infty$, let

$$\mathfrak{D}(A) := \mathfrak{D}^0(A) \cap \{u \in L^2 : Au \in L^2, u'_-(0) = u'_+(l-) = 0\}.$$

(iii) If $l + m(l-) < \infty$, let

$$\mathfrak{D}(A) := \mathfrak{D}^0(A) \cap \{u \in L^2 : Au \in L^2, u'_-(0) = (r-l)u'_+(l) + u(l) = 0\}.$$

Note, that in case (ii) of Definition 15.10, the assumption $\int_{[0,r)} x^2 m(dx) < \infty$ implies $m(l-) < \infty$, hence $r = l = \infty$. Therefore, $u'_+(l-) = 0$ means that $\lim_{x\to\infty} u'_+(x) = 0$. In case (iii) it is more convenient to write the assumption in terms of the function u on $[0, l]$. Note that by (15.9), $u'_+(l) = u'_-(l) + m\{l\}Au(l)$. Thus, if $r < \infty$, the condition $(r-l)u'_+(l) + u(l) = 0$ can be restated as

$$-m\{l\}Au(l) = u'_-(l) + \frac{1}{r-l}u(l). \tag{15.29}$$

If $r = \infty$, we may interpret the condition $(r-l)u'_+(l) + u(l) = 0$ as $u'_+(l) = 0$, which is equivalent to

$$-m\{l\}Au(l) = u'_-(l). \tag{15.30}$$

Lemma 15.11. *Let $g \in L^2((-\infty, r), m)$. Then $u = R_\lambda g$ is a solution to the equation $\lambda u - Au = g$. Moreover, $R_\lambda g \in \mathfrak{D}(A)$, and it is the unique solution of $\lambda u - Au = g$ belonging to $\mathfrak{D}(A)$.*

A proof of this lemma can be found in [105, Theorem on p. 167] and [196].

The next step is the identification of the probabilistic concept of the λ-potential with the analytic concept of the λ-resolvent.

Let $C_b(E_m)$ denote the space of bounded continuous functions on $E_m \subset [0, r)$. For $x \in E_m$, and $g \in C_b(E_m)$, let

$$G_\lambda g(x) = \mathbb{E}_x \int_0^\infty e^{-\lambda t} g(X_t) \, dt.$$

Since $G_\lambda 1 \le 1/\lambda$, $G_\lambda g$ is well defined for all $g \in C_b(E_m)$. For $x \in E_m$ we define $T_x^{B^+} = \inf\{t > 0 : B_t^+ = x\}$ and $T_x^X = \inf\{t > 0 : X_t = x\}$. Then $T_x^X = C_{T_x^{B^+}}$.

Lemma 15.12. *For all $x \in E_m$ we have*

$$\lim_{E_m \ni y \to x} T_y^X = T_x^X, \quad \mathbb{P}_x - a.s. \tag{15.31}$$

Proof. Note that by the monotone convergence theorem, $C_{t_n} \to C_t$ as $t_n \to t$. This proves that (15.31) is true as $y \to x$ from the left. Suppose now that $y \to x$ from the right. Then

$$T_y^X = C(T_y^{B^+}) = \int_{[0,l]} L(T_y^{B^+}, w) \, m(dw) = \int_{[0,y]} L(T_y^{B^+}, w) \, m(dw)$$

$\mathbb{P}_x$-a.s. where the last equality uses $L(T_y^{B^+}, w) = 0$ $\mathbb{P}_x$-a.s. for all $w > y$. Since m is finite on compact intervals, the claim follows by the dominated convergence theorem and the fact that $T_y^{B^+} \to T_x^{B^+}$ $\mathbb{P}_x$-a.s. as $y \to x$. $\square$

Lemma 15.13. $G_\lambda : C_b(E_m) \to C_b(E_m)$ *and* $G_\lambda : B_b(E_m) \to C_b(E_m)$.

Proof. Let $g \in C_b(E_m)$ and $x \in E_m$. Assume that x is not isolated from the right within E_m. We will prove the right-continuity of $G_\lambda g$ at x. Pick $y > x$; by the strong Markov property we find for all $x, y \in E_m$ that

$$G_\lambda g(x) = \mathbb{E}_x \left(\int_0^{T_y^X} e^{-\lambda t} g(X_t) \, dt \right) + \mathbb{E}_x (e^{-\lambda T_y^X}) G_\lambda g(y). \tag{15.32}$$

Letting $y \to x$ we can use Lemma 15.12 to see that the first term on the right-hand side of (15.32) converges to zero, while $\mathbb{E}_x(e^{-\lambda T_y^X})$ converges to 1; therefore, we have $G_\lambda g(y) \to G_\lambda g(x)$. In the same way one checks the left-continuity at points in E_m which are not isolated from the left.

The above argument uses only the fact that $g \in B_b(E_m)$, i.e. the second assertion of the lemma follows at once. $\square$

Because of the resolvent equation, the range $\mathfrak{D} := G_\lambda(C_b(E_m))$ of G_λ does not depend on $\lambda > 0$. Moreover,

$$\lim_{\lambda \to \infty} \lambda G_\lambda g(x) = \mathbb{E}_x \left(\lim_{\lambda \to \infty} \lambda \int_0^\infty e^{-\lambda t} g(X_t) \, dt \right)$$

$$= \mathbb{E}_x \left(\lim_{\lambda \to \infty} \int_0^\infty e^{-s} g(X_{s/\lambda}) \, ds \right)$$

$$= \mathbb{E}_x \left(\int_0^\infty e^{-s} g(X_0) \, ds \right)$$

$$= g(x),$$

implying that G_λ is one-to-one, hence invertible. Define

$$\tilde{A} := \lambda I - G_\lambda^{-1} : \mathfrak{D} \to C_b(E_m)$$

and note that, again by the resolvent equation, $\tilde{A}$ does not depend on $\lambda > 0$.

For $0 < y < l$ and $x, w \in [0, y)$ let

$$G_y^{B^+}(x, w) = (y - x) \wedge (y - w) \tag{15.33}$$

be the Green function of B^+ killed upon hitting y.

Lemma 15.14. *Let* $g : [0, y] \to \mathbb{R}$ *be a bounded Borel function. Then for every* $x \in [0, y)$

$$\mathbb{E}_x \int_0^{T_y^X} g(X_t)\, dt = \int_{[0,y]} G_y^{B^+}(x, w) g(w)\, m(dw).$$

Proof. By a change of variables and Fubini's theorem

$$\mathbb{E}_x \int_0^{T_y^X} g(X_t)\, dt = \mathbb{E}_x \int_0^{C(T_y^{B^+})} g(B_{\tau(t)}^+)\, dt$$

$$= \mathbb{E}_x \int_{[0, T_y^{B^+}]} g(B_t^+)\, C(dt)$$

$$= \mathbb{E}_x \int_{[0, T_y^{B^+}]} g(B_t^+) \int_{[0,r)} L(dt, w)\, m(dw)$$

$$= \int_{[0,y]} \left(\mathbb{E}_x \int_{[0, T_y^{B^+}]} g(B_t^+)\, L(dt, w) \right) m(dw)$$

$$= \int_{[0,y]} \left(g(w)\, \mathbb{E}_x\big[L(T_y^{B^+}, w) \big] \right) m(dw)$$

$$= \int_{[0,y]} g(w)\, G_y^{B^+}(x, w)\, m(dw).$$

In the penultimate line we used that $L(dt, w)$ is supported on $\{B_t^+ = w\}$, and in the last line we used the well-known formula relating the Green function and the local time for (reflected) Brownian motions.   □

Let $x \in E_m$ be positive and apply Dynkin's formula for the stopping time T_x^X, see [178, pp. 98–99]. For $u \in \mathfrak{D}$

$$\mathbb{E}_0\big(u(X(T_x^X)) \big) - u(0) = \mathbb{E}_0 \left(\int_0^{T_x^X} \tilde{A}u(X_t)\, dt \right).$$

By Lemma 15.14 and the obvious equality $X(T_x^X) = x$ we get

$$u(x) = u(0) + \int_{[0,x]} G_x^{B^+}(0, y)\, \tilde{A}u(y)\, m(dy)$$

$$= u(0) + \int_{[0,x]} (x - y)\, \tilde{A}u(y)\, m(dy).$$

This formula clearly shows that $u \in \mathfrak{D}^0(A)$, $Au = \tilde{A}u$ and $u'_-(0) = 0$.

Take $g \in C_b(E_m) \cap L^2((-\infty, r), m)$ and $\lambda > 0$. Then $G_\lambda g \in \mathfrak{D} \subset \mathfrak{D}^0(A)$, and

$$AG_\lambda g = \tilde{A}G_\lambda g = \lambda G_\lambda g - g,$$

i.e. $\lambda G_\lambda g - AG_\lambda g = g$. Moreover, $(G_\lambda g)'_-(0) = 0$.

Proposition 15.15. *For every $g \in C_b(E_m) \cap L^2((-\infty, r), m)$ we have $G_\lambda g = R_\lambda g$.*

Proof. Let $g \in C_b(E_m) \cap L^2((-\infty, r), m)$. In the paragraph preceding Lemma 15.14 it was shown that $G_\lambda g$ solves the equation $\lambda u - Au = g$. If $G_\lambda g \in \mathfrak{D}(A)$, the assertion follows by the uniqueness of the solution, cf. Lemma 15.11. Observe that $(G_\lambda g)'_-(0) = 0$, $G_\lambda g \in L^2((-\infty, r), m)$ and $AG_\lambda g = \lambda G_\lambda g - g \in L^2((-\infty, r), m)$.

Suppose that we are in case (i) of Definition 15.10, that is $l + m(l-) = \infty$ and $\int_{[0,r)} x^2\, m(dx) = \infty$. Then there is nothing else to show, i.e. $G_\lambda g \in \mathfrak{D}(A)$.

Suppose now that we are in case (ii) of Definition 15.10, that is $l + m(l-) = \infty$ and $\int_{[0,r)} x^2\, m(dx) < \infty$. By the remark following Definition 15.10, it remains to show that $\lim_{x\to\infty} u'_+(x) = 0$. Suppose that $g \geq 0$ and g vanishes on $[y, \infty)$ for some $y \in E_m$. By the strong Markov property we see for all $x \in E_m$ that

$$G_\lambda g(x) = \mathbb{E}_x\left(\int_0^{T_y^X} e^{-\lambda t} g(X_t)\, dt\right) + \mathbb{E}_x(e^{-\lambda T_y^X})G_\lambda g(y), \qquad (15.34)$$

and if $x > y$ it follows that

$$G_\lambda g(x) = \mathbb{E}_x(e^{-\lambda T_y^X})G_\lambda g(y). \qquad (15.35)$$

Thus $G_\lambda g(\cdot)$ is proportional to $\mathbb{E}_\cdot(e^{-\lambda T_y^X})$ on $[y, \infty) \cap E_m$. Let $y, x, \upsilon \in E_m$ such that $y < x < \upsilon$. Again by the strong Markov property,

$$\mathbb{E}_\upsilon(e^{-\lambda T_y^X}) = \mathbb{E}_\upsilon(e^{-\lambda T_x^X})\, \mathbb{E}_x(e^{-\lambda T_y^X}),$$

showing that $x \mapsto \mathbb{E}_x(e^{-\lambda T_y^X})$ is decreasing. Furthermore, let $w \in E_m$ be such that $y < w < x$. Then

$$
\begin{aligned}
\mathbb{E}_x(e^{-\lambda T_y^X}) &= \mathbb{E}_x(e^{-\lambda T_y^X} \mathbb{1}_{\{T_w^X < T_v^X\}}) + \mathbb{E}_x(e^{-\lambda T_y^X} \mathbb{1}_{\{T_v^X < T_w^X\}}) \\
&= \mathbb{E}_x\big(e^{-\lambda T_w^X}\, \mathbb{E}_w(e^{-\lambda T_y^X})\mathbb{1}_{\{T_w^X < T_v^X\}}\big) \\
&\quad + \mathbb{E}_x\big(e^{-\lambda T_v^X}\, \mathbb{E}_v(e^{-\lambda T_y^X})\mathbb{1}_{\{T_v^X < T_w^X\}}\big) \\
&\le \mathbb{P}_x(T_w^X < T_v^X)\mathbb{E}_w(e^{-\lambda T_y^X}) + \mathbb{P}_x(T_v^X < T_w^X)\mathbb{E}_v(e^{-\lambda T_y^X}).
\end{aligned}
$$

Since

$$
\mathbb{P}_x(T_v^X < T_w^X) = \mathbb{P}_x(T_v^{B^+} < T_w^{B^+}) = \frac{v - x}{v - w},
$$

it follows that $x \mapsto \mathbb{E}_x(e^{-\lambda T_y^X})$ is convex in $[y, \infty) \cap E_m$ (in the sense that, if we interpolate it linearly in $[y, \infty) \setminus E_m$, it would be a convex function on $[y, \infty)$). Thus we know that $x \mapsto G_\lambda g(x)$ is a non-negative decreasing and convex function on $[y, \infty) \cap E_m$. This implies that $\lim_{x \to \infty}(G_\lambda g)'_+(x) = 0$. The case of g taking both signs follows by using $g = g^+ - g^-$. Thus, $G_\lambda g = R_\lambda g$ for all $g \in C_b(E_m) \cap L^2((-\infty, r), m)$ having compact support. To finish the proof we use the fact that $C_b(E_m) \cap L^2((-\infty, r), m) \cap C_c(-\infty, r)$ is dense in $L^2((-\infty, r), m)$ and the continuity of G_λ and R_λ.

Finally, assume that we are in case (iii) of Definition 15.10. We restrict ourselves to the case $l < r < \infty$; the other two cases, $l = r < \infty$ and $l < r = \infty$, are similar. Assume that $g \ge 0$ and that g vanishes on $[y, l]$ for some $y \in E_m$, $y < l$. For simplicity we write $u = G_\lambda g$ and we remark that $\lambda u = Au$ on $[y, l]$. By the same argument as in case (ii) it follows that $x \mapsto G_\lambda g(x)$ is decreasing on $[y, l]$. Moreover, since $\mathbb{P}_l(T_y^X < \infty) > 0$, we get that

$$
u(l) = \mathbb{E}_l(e^{-\lambda T_y^X})u(y) > 0.
$$

Since for $x > y$

$$
u'_-(x) = \int_{[0,x)} Au(w)\, m(dw) = \int_{[0,y)} Au(w)\, m(dw) + \lambda \int_{[y,x)} u(w)\, m(dw),
$$

we see that u'_- is non-decreasing on $[y, l]$. Further,

$$
\begin{aligned}
u'_-(l) - u'_-(y) &= \int_{[y,l)} Au(w)\, m(dw) \\
&= \lambda \int_{[y,l)} u(w)\, m(dw) \\
&\le \lambda u(y)\, m[y, l) < \infty,
\end{aligned}
$$

implying that $u'_-(l) < \infty$. By Dynkin's formula we get

$$Au(l) = \lim_{x \to l} \frac{\mathbb{P}_l(T_x^X < \infty)u(x) - u(l)}{\mathbb{E}_l(T_x^X)}. \tag{15.36}$$

This can be rewritten in the following form

$$\frac{Au(l)\,\mathbb{E}_l(T_x^X)}{\mathbb{P}_l(T_x^X < \infty)(x - l)} \tag{15.37}$$
$$= \frac{u(x) - u(l)}{x - l} + \frac{1 - \mathbb{P}_l(T_x^X < \infty)^{-1}}{x - l}\,u(l) + \frac{\mathbb{E}_l(T_x^X)}{x - l}\,o(1),$$

as $x \to l$. Note that

$$\mathbb{P}_x(T_y^X < \infty) = \mathbb{P}_x(T_x^{B^+} < T_r^{B^+}) = \frac{r - l}{r - x}.$$

Therefore,

$$\lim_{x \to l} \frac{1 - \mathbb{P}_l(T_x^X < \infty)^{-1}}{x - l} = \lim_{x \to l} \frac{1 - \frac{r-x}{r-l}}{x - l} = \frac{1}{r - l}.$$

Let $T_r^{B^+}$ denote the lifetime of B^+ (which is the same as the first hitting time to r of reflected (non-killed) Brownian motion), and recall that

$$\mathbb{E}_l L(T_x^{B^+} \wedge T_r^{B^+}, w) = G_{(x,r)}^B(l, w) = \frac{(l - x)(r - w)}{r - x},$$

where $G_{(x,r)}^B$ denotes the Green function of Brownian motion killed upon exiting the interval (x, r). Therefore,

$$\begin{aligned}
\mathbb{E}_l(T_x^X) = \mathbb{E}_l(C_{T_x^{B^+}}) &= \mathbb{E}_l \int_{[0,r)} L(T_x^{B^+}, w)\, m(dw) \\
&= \mathbb{E}_l \int_{[x,l]} L(T_x^{B^+} \wedge T_r^{B^+}, w)\, m(dw) \\
&= \int_{[x,l]} \mathbb{E}_l\big(L(T_x^{B^+} \wedge T_r^{B^+}, w)\big)\, m(dw) \\
&= \int_{[x,l]} \frac{(l - x)(r - w)}{r - x}\, m(dw),
\end{aligned}$$

and we conclude that

$$\lim_{x \to l} \frac{\mathbb{E}_l(T_x^X)}{x - l} = -\lim_{x \to l} \int_{[x,l]} \frac{r - w}{r - x}\, m(dw) = -m\{l\}.$$

If we let $x \to l$ in (15.37) we get

$$-m\{l\} \, Au(l) = u'_-(l) + \frac{1}{r-l} \, u(l),$$

which is (15.29). From this point onwards we can argue as in case (ii). This finishes the proof.   □

Proof of Theorem 15.3. By (15.4) and (15.28), the statement of Proposition 15.15 can be written as

$$\int_{[0,r)} G_\lambda(x, y) g(y) \, m(dy) = \int_{[0,r)} R_\lambda(x, y) g(y) \, m(dy)$$

for all (non-negative) continuous functions $g \in L^2((-\infty, r), m)$. For $x \in E_m$, both $G_\lambda(x, \cdot)$ and $R_\lambda(x, \cdot)$ are continuous on $[0, r)$. Hence it follows that

$$G_\lambda(x, y) = R_\lambda(x, y) \quad \text{for all } (x, y) \in E_m \times [0, r). \tag{15.38}$$

In particular,

$$G_\lambda(0, 0) = R_\lambda(0, 0) = h(\lambda),$$

which shows that $\lambda \mapsto G_\lambda(0, 0) \in \mathcal{S}$. This completes the proof of the direct part of Theorem 15.3.

In order to prove the converse, let $f \in \mathcal{CBF}$ and define $h : (0, \infty) \to (0, \infty)$ by $h(\lambda) = 1/f(\lambda)$. Then $h \in \mathcal{S}$ by Theorem 7.3, and Theorem 15.7 shows that there exists a unique string $m \in \mathfrak{M}_+$ having h as its characteristic function. Let $X = (X_t)_{t \geq 0}$ be a generalized diffusion corresponding to m. It is clear that the Laplace exponent of the inverse local time at zero of X is equal to f.   □

Note that Theorem 15.3 and Theorem 15.7 give a one-to-one correspondence between $f \in \mathcal{CBF}$, $m \in \mathfrak{M}_+$ and $h \in \mathcal{S}$ with $h(+\infty) = 0$. Unfortunately, the proofs do not give any insight into the explicit relationship between the Lévy triplet (a, b, μ) of f, the string m, and the Stieltjes measure σ of h. In the remainder of this section, we are going to establish some connections among them. Throughout the rest of this section we adopt the convention that $\frac{1}{0} = +\infty$ and $\frac{1}{+\infty} = 0$.

We start with a probabilistic explanation of the connections between the Lévy triplet (a, b, μ) and the string m. Let us denote the local time at zero $L(t, 0)$ for B^+ by $\ell(t)$, and let $\ell^{-1}(t) := \inf\{s > 0 : \ell(s) > t\}$ be the inverse local time of B^+ at zero. The inverse local time of X at zero, $(\ell^X)^{-1}$, has the following representation:

$$(\ell^X)^{-1}(t) = C_{\ell^{-1}(t)} = \int_{[0,\infty)} L(\ell^{-1}(t), x) \, m(dx). \tag{15.39}$$

Note that $\inf\{t > 0 : (\ell^X)^{-1}(t) = \infty\} = \ell^X(\zeta) = \ell(T_r^{B^+})$ where ζ is the lifetime of X and $T_r^{B^+}$ the lifetime of B^+. Recall also the well-known fact that $\ell(T_r^{B^+})$ is an exponential random variable with parameter r^{-1}, cf. [304, Proposition VI 4.6]. Hence, the lifetime of $(\ell^X)^{-1}$ is exponentially distributed with parameter r^{-1}, that is $a = r^{-1}$. Now we get from (15.39) and $L(\ell^{-1}(t), 0) = t$ that

$$(\ell^X)^{-1}(t) = m\{0\} t + \int_{(0,\infty)} L(\ell^{-1}(t), x) m(dx).$$

The second term is a pure jump subordinator, hence the drift of $(\ell^X)^{-1}$ is $m\{0\}$ and $b = m\{0\}$. Finally, we use the fact that $(L(\ell^{-1}(t), x))_{t \geq 0}$ is for every $x \geq 0$ a subordinator with the Lévy density $x^{-2} e^{-\lambda/x}$, see [178, Problem 4, p. 73]. This implies that $\mathbb{E}_0[L(\ell^{-1}(1), x)] = 1$ for every $x \geq 0$. Hence, by (15.39)

$$\mathbb{E}_0[(\ell^X)^{-1}(1)] = \int_{[0,\infty)} \mathbb{E}_0[L(\ell^{-1}(1), x)] m(dx) = m[0, \infty).$$

The connection between the Lévy triplet (a, b, μ) and the measure σ is straightforward. Indeed,

$$1 = f(\lambda) h(\lambda) = \left(a + b\lambda + \int_{(0,\infty)} (1 - e^{-\lambda t}) \mu(dt) \right) \left(\int_{[0,\infty)} \frac{1}{\lambda + t} \sigma(dt) \right),$$

and letting $\lambda \to 0$ proves that

$$a = \left(\int_{[0,\infty)} \frac{\sigma(dt)}{t} \right)^{-1} \in [0, \infty).$$

Similarly,

$$1 = \frac{f(\lambda)}{\lambda} \lambda h(\lambda) = \left(\frac{a}{\lambda} + b + \frac{1}{\lambda} \int_{(0,\infty)} (1 - e^{-\lambda t}) \mu(dt) \right) \left(\int_{[0,\infty)} \frac{\lambda}{\lambda + t} \sigma(dt) \right),$$

and letting $\lambda \to \infty$ gives $b = \sigma[0, \infty)^{-1} \in [0, \infty)$; letting $\lambda \to 0$ we get

$$\mathbb{E}_0[(\ell^X)^{-1}(1)] = b + \int_{(0,\infty)} t \, \mu(dt) = \frac{1}{\sigma\{0\}} \in (0, \infty].$$

This proves already our next result.

Proposition 15.16. *The following relations hold:*

$$m[0, \infty) = \frac{1}{\sigma\{0\}} = \mathbb{E}_0[(\ell^X)^{-1}(1)],$$

$$m\{0\} = \frac{1}{\sigma[0, \infty)} = b,$$

$$r = \left(\int_{[0,\infty)} \frac{\sigma(dt)}{t} \right) = \frac{1}{a}.$$

If the characteristic function h of the string is known, one can work out the Lévy triplet (a, b, μ) of f: let

$$h^\star(\lambda) := \frac{1}{\lambda h(\lambda)}$$

be the Stieltjes function which is conjugate to h in the sense that $f^\star := 1/h^\star$ is conjugate to f. Write

$$h^\star(\lambda) = \frac{c^\star}{\lambda} + d^\star + \int_{(0,\infty)} \frac{1}{\lambda + t} \sigma^\star(dt)$$

so that

$$f(\lambda) = \lambda h^\star(\lambda) = c^\star + d^\star \lambda + \int_{(0,\infty)} \frac{\lambda}{\lambda + t} \sigma^\star(dt)$$

$$= c^\star + d^\star \lambda + \int_0^\infty (1 - e^{-\lambda t}) n^\star(t)\, dt,$$

where

$$n^\star(t) = \int_{(0,\infty)} s e^{-st} \sigma^\star(ds)$$

is the completely monotone density of μ. It can be shown, see [224] or [105], that $h^\star$ is the characteristic function of the so-called *dual string* $m^\star$ which is, by definition, the right-continuous inverse of the string m: $m^\star(x) := \inf\{y > 0 : m(y) > x\}$.

Finding explicitly the string $m \in \mathfrak{M}_+$ such that a given $f \in \mathcal{CBF}$ is the Laplace exponent of the inverse local time at zero of a generalized diffusion X corresponding to m is known as the *Kreĭn representation problem*. Equivalently, the problem can be stated as follows: given $g \in \mathcal{CM}$, the density of the Lévy measure μ of f, $\mu(d\lambda) = g(\lambda)\, d\lambda$, find the operator $A = \frac{d}{dm} \frac{d}{dx}$ which generates the generalized diffusion X. Surprisingly, there are very few examples of (classes of) completely monotone functions for which this problem has been solved. In Table 15.1 we record those classes with the corresponding operator A. Table 15.1 was established by C. Donati-Martin and M. Yor in [97] and [98], except for the first row which is from [267].

Let $Y = (Y_t)_{t \geq 0}$ be a reflected diffusion on $[0, \infty)$ with the speed measure m and the scale function s normalized such that $s(0) = 0$. The infinitesimal generator of Y is $\frac{d}{dm} \frac{d}{ds}$. The process $s(Y) = (s(Y_t))_{t \geq 0}$ is also a reflected diffusion on $[0, \infty)$ having the same local time at 0 as Y. Therefore, the Laplace exponents of the inverse local times coincide. Let $\widetilde{m} = m \circ s^{-1}$. The generalized diffusion X corresponding to the string $\widetilde{m}$ and the process $s(Y)$ have the same infinitesimal generator $A = \frac{d}{d\widetilde{m}} \frac{d}{dx}$. This shows that functions in the left column of Table 15.1 are densities of Lévy measures of the inverse local time at zero of generalized diffusions.

Table 15.1. Explicit Kreĭn representations.

$g(\lambda)$	$A = \dfrac{d}{dm}\dfrac{d}{ds}$
$\dfrac{c}{\lambda^{\alpha+1}}, \quad 0 < \alpha < 1$	$A_{-\alpha} := \dfrac{1}{2}\dfrac{d^2}{dx^2} + \dfrac{\delta-1}{2x}\dfrac{d}{dx},$ $\delta = 2(1-\alpha)$
$\dfrac{c}{\lambda^{\alpha+1}}e^{-\rho\lambda}, \quad 0 < \alpha < 1, \ \rho > 0$	$A_{-\alpha} + \sqrt{2\rho}\,\dfrac{\widehat{K}'_\alpha(\sqrt{2\rho x})}{\widehat{K}_\alpha(\sqrt{2\rho x})}\dfrac{d}{dx},$ $\widehat{K}_\alpha(x) := x^\alpha K_\alpha(x), x > 0$
$\dfrac{c}{\lambda}e^{-\rho\lambda}, \quad \rho > 0$	$\dfrac{1}{2}\dfrac{d^2}{dx^2} + \left(\dfrac{1}{2x} + \sqrt{2\rho}\,\dfrac{\widehat{K}'_0(\sqrt{2\rho x})}{\widehat{K}_0(\sqrt{2\rho x})}\right)\dfrac{d}{dx}$
$c\left(\dfrac{\rho}{\sinh(\rho\lambda)}\right)^{\alpha+1}e^{(1-\alpha)\rho\lambda},$ $0 < \alpha < 1, \ \rho > 0$	$A_{-\alpha} - \rho x\dfrac{d}{dx}$
$c\left(\dfrac{\rho}{\sinh(\rho\lambda)}\right)^{\alpha+1}, \quad 0 < \alpha < 1$	$A_{-\alpha} + \rho x\,\dfrac{\widehat{K}'_{\alpha/2}(\rho x^2/2)}{\widehat{K}_{\alpha/2}(\rho x^2/2)}\dfrac{d}{dx}$

15.2 First passage times

We are now going to study the distributions of the first passage times of the generalized diffusion X. For $y \in E_m$ define

$$T_y^X := \inf\{t > 0 \ : \ X_t = y\}.$$

Formula (15.5) states that

$$G_\lambda(x, y) = \mathbb{E}_x\left(\int_{[0,\infty)} e^{-\lambda s}\, L^X(ds, y)\right), \quad x, y \in E_m,$$

and since the local time $L^X(s, y)$ is equal to zero for all times $s < T_y^X$, it follows from the strong Markov property that

$$G_\lambda(x, y) = \mathbb{E}_x\left(\int_{[T_y^X,\infty)} e^{-\lambda s}\, L^X(ds, y)\right) = \mathbb{E}_x(e^{-\lambda T_y^X})\, G_\lambda(y, y).$$

Because of (15.26) and (15.38) we get

$$\mathbb{E}_x(e^{-\lambda T_y^X}) = \frac{G_\lambda(x,y)}{G_\lambda(y,y)} = \begin{cases} \frac{\phi(x,\lambda)}{\phi(y,\lambda)}, & 0 \le x \le y < r, \\ \frac{\chi(x,\lambda)}{\chi(y,\lambda)}, & 0 \le y \le x < r. \end{cases} \tag{15.40}$$

Letting $x \to 0+$ we see

$$\mathbb{P}_x(T_y^X < \infty) = \begin{cases} 1, & 0 \le x \le y < r, \\ \frac{r-x}{r-y}, & 0 \le y < x < r < \infty, \\ 1, & 0 \le y < x < r = \infty. \end{cases}$$

We could also derive this formula from $\mathbb{P}_x(T_y^{B^+} < T_r^{B^+}) = (y-x)/(y-r)$. Let π_{xy} be the distribution of T_y^X under $\mathbb{P}_x$. Then $\mathbb{E}_x[e^{-\lambda T_y^X}] = \mathscr{L}(\pi_{xy}; \lambda)$. If $0 \le x < y$, the first row of (15.40) gives that

$$\mathscr{L}(\pi_{xy}; \lambda) = \frac{\phi(x,\lambda)}{\phi(y,\lambda)}. \tag{15.41}$$

In order to identify the class of distributions having this Laplace transform, we must study the zeroes of $\phi(x, \cdot)$ in greater detail.

Assume that $x > 0$ and recall that for $z \in \mathbb{C}$ with $\operatorname{Im} z \neq 0$ we have $\phi(x,z) \neq 0$ and $\phi'_+(x,z) \neq 0$. Hence, the only possible zeroes of $\phi(x, \cdot)$ are on the negative real axis. Since $\phi(x, \cdot)$ is analytic, the zeroes are discrete. Let $(z_{j,x})_{j \ge 1}$ be the family of all zeroes of $\phi(x, \cdot)$ arranged in decreasing order. Let $\lambda < 0$ and apply the Lagrange identity (15.15) to $\phi(x,\lambda)$ and $\phi(x, \lambda + h)$, $h \in \mathbb{R}$ sufficiently small, to get that

$$h \int_{[0,x]} \phi(y, \lambda + h)\phi(y,\lambda)\, m(dy) = \phi'_+(x, \lambda + h)\phi(x,\lambda) - \phi(x, \lambda + h)\phi'_+(x,\lambda).$$

Dividing by h and letting $h \to 0$ yields

$$\int_{[0,x]} \phi(y,\lambda)^2\, m(dy) = \phi(x,\lambda)\frac{\partial}{\partial\lambda}\phi'_+(x,\lambda) - \phi'_+(x,\lambda)\frac{\partial}{\partial\lambda}\phi(x,\lambda).$$

This means that all zeroes of $\phi(x, \cdot)$ are simple. In particular, for $\lambda = z_{j,x}$ the above equation becomes

$$\int_{[0,x]} \phi(y, z_{j,x})^2\, m(dy) = -\phi'_+(x, z_{j,x})\frac{\partial}{\partial\lambda}\phi(x, z_{j,x}), \tag{15.42}$$

and we conclude that $\phi'_+(x, z_{j,x})$ and $\frac{\partial}{\partial\lambda}\phi(x, z_{j,x})$ have opposite signs.

Lemma 15.17. *For each $x \in [0, r)$,*

$$\phi(x, z) = \prod_{j} \left(1 - \frac{z}{z_{j,x}} \right). \tag{15.43}$$

If m has at least n points of increase in $(0, x)$, then $\phi(x, \cdot)$ grows at least as fast as a polynomial of degree n.

Proof. By (15.14) and (15.13)

$$|\phi(x, z)| \le \sum_{n=0}^{\infty} \frac{(2x\, m(x))^n}{(2n)!} |z|^n = \sum_{n=0}^{\infty} \frac{(\sqrt{2x m(x)|z|})^{2n}}{(2n)!}$$

$$\le \exp\left(\sqrt{2x\, m(x)} \sqrt{|z|} \right).$$

This shows that $\phi(x, \cdot)$ is an entire function of order at most $1/2$, and the factorization (15.43) follows from Hadamard's factorization theorem, cf. [259, Vol. 2, p. 289].

We show now the second statement of the lemma. First recall that both $\phi(\cdot, \lambda)$ and $\phi'_+(\cdot, \lambda)$ are increasing for $\lambda > 0$. Let $0 < u < v < w$. Since

$$\phi'_+(v, \lambda) = \lambda \int_{[0,v]} \phi(y, \lambda)\, m(dy),$$

we conclude that for all $\lambda > 0$

$$\phi'_+(v, \lambda) \ge \lambda \int_{(u,v]} \phi(y, \lambda)\, m(dy) \ge \lambda \int_{(u,v]} \phi(u, \lambda)\, m(dy)$$

$$= \lambda \phi(u, \lambda) m(u, v].$$

This gives

$$\phi(w, \lambda) = \phi(v, \lambda) + \int_{v}^{w} \phi'_+(y, \lambda)\, dy$$

$$\ge \int_{v}^{w} \phi'_+(v, \lambda)\, dy \tag{15.44}$$

$$= (w - v)\phi'_+(v, \lambda)$$

$$\ge \lambda (w - v) m(u, v]\, \phi(u, \lambda).$$

We show now that $\phi(x, \cdot)$ grows at least as fast as a polynomial of degree n. Indeed, choose $2n + 1$ points $0 < x_0 < x_1 < \cdots < x_{2n} = x$ so that every interval (x_{2k-2}, x_{2k-1}) contains a point of increase of m. Then $m(x_{2k-2}, x_{2k-1}] > 0$ for all $k = 1, 2, \ldots, n$. Using repeatedly the inequality (15.44) with $x_{2k-2} < x_{2k-1} < x_{2k}$, we find

$$\phi(x,\lambda) \geq \lambda^n \prod_{k=1}^{n} (x_{2k} - x_{2k-1}) \prod_{k=1}^{n} m(x_{2k-2}, x_{2k-1}] \phi(x_0, \lambda)$$

$$\geq \lambda^n \prod_{k=1}^{n} (x_{2k} - x_{2k-1}) \prod_{k=1}^{n} m(x_{2k-2}, x_{2k-1}],$$

where we used that $\phi(\cdot, \lambda) \geq 1$ for every $\lambda > 0$. This concludes the proof. $\qquad\square$

Remark 15.18. One can adapt the argument used in in Lemma 15.17 to show that

$$\frac{\psi(x,z)}{x} = \prod_j \left(1 - \frac{z}{w_{j,x}}\right), \tag{15.45}$$

where $(w_{j,x})_{j\geq 1}$ is the family of all zeroes of $\psi(x, \cdot)$ arranged in decreasing order. The zeroes are also simple. Since $\psi'_+(x,z)\phi(x,z) - \psi(x,z)\phi'_+(x,z) = 1$, it follows that $\phi(x, \cdot)$ and $\psi(x, \cdot)$ have no common zeroes.

As a consequence of the convergence of the infinite product (15.43), the series $\sum_j 1/|z_{j,x}| < \infty$ converges. Moreover, if $0 < y < x$, then $\phi(x, \cdot)$ may have more zeroes than $\phi(y, \cdot)$. If there are infinitely many points of increase of m in $(0, x)$, then $\phi(x, \cdot)$ will have an infinite, but countable, number of zeroes.

Let us now discuss the relative position of the zeroes of $\phi(x, \cdot)$ and $\phi(y, \cdot)$ for $0 < y < x$. For simplicity assume that both $(z_{j,x})_{j\in\mathbb{N}}$ and $(z_{j,y})_{j\in\mathbb{N}}$ are infinite sequences. The argument when these sequences are of different length is similar, the difference being that some of the zeroes do not exist.

Lemma 15.19. *Assume that* $0 < y < x$. *Then* $z_{j,y} < z_{j,x} < 0$ *for all* $j \geq 1$.

Proof. We begin with a few remarks that will be useful in the proof. First, since the zeroes of $\phi(x, \cdot)$ are simple, the function $\lambda \mapsto \phi(x, \lambda)$ changes sign every time it passes through a zero. Secondly, the function $(x, \lambda) \mapsto \phi(x, \lambda)$ is jointly continuous. Thirdly, $\phi(0, \cdot) \equiv 1$, and also $\phi(\cdot, 0) \equiv 1$.

Now fix $x > 0$. We claim that $\phi(y, \lambda) > 0$ for all $(y, \lambda) \in A_{1,x} := [0, x] \times [z_{1,x}, 0]$ such that $(y, \lambda) \neq (x, z_{1,x})$. This will prove that $z_{1,y} < z_{1,x}$ for all $0 < y < x$. Indeed, as $z_{1,x}$ is the largest zero of $\phi(x, \cdot)$, we have $\phi(x, \lambda) > 0$ for all $\lambda \in (z_{1,x}, 0]$. Suppose that there is some $(v, \lambda) \in A_{1,x}$, $(v, \lambda) \neq (x, z_{1,x})$, such that $\phi(v, \lambda) = 0$. Because of the continuity of ϕ the minimum

$$v_1 = \min \left\{ v \in [0, x] \ : \ \exists \lambda \in [z_{1,x}, 0] \text{ such that } \phi(v, \lambda) = 0 \text{ and } (v, \lambda) \in A_{1,x} \right\}$$

is attained. Since $\phi(0, \cdot) \equiv 1$ and $\phi(x, \lambda) > 0$ for $\lambda \in (z_{1,x}, 0]$, we have $0 < v_1 < x$. Choose $\lambda_1 \in [z_{1,x}, 0)$ such that $\phi(v_1, \lambda_1) = 0$. Assume first that $\lambda_1 > z_{1,x}$. Using the fact that the zeroes of $\phi(v_1, \cdot)$ are discrete and that $\phi(v_1, \cdot)$ changes sign every time it passes through a zero, we have, for sufficiently small $\epsilon > 0$, either $\phi(v_1, \lambda) < 0$ for $\lambda \in (\lambda_1 - \epsilon, \lambda_1)$, or $\phi(v_1, \lambda) < 0$ for $\lambda \in (\lambda_1, \lambda_1 + \epsilon)$.

Assume that $\phi(v_1, \lambda) < 0$, $\lambda \in (\lambda_1 - \epsilon, \lambda_1)$. Since $\{\phi < 0\}$ is open in $A_{1,x}$, there would exist a point (v, λ) with $v < v_1$ such that $\phi(v, \lambda) < 0$. This, however, contradicts the minimality of v_1 which entails that $\phi(v_1, \lambda) > 0$ for $\lambda \in (\lambda_1 - \epsilon, \lambda_1)$.

The same argument yields $\phi(v_1, \lambda) > 0$ for $\lambda \in (\lambda_1, \lambda_1 + \epsilon)$ and sufficiently small $\epsilon > 0$. We also get $\lambda_1 \leq z_{1,x}$. Hence, $\phi(v, \lambda) \neq 0$ for all $(v, \lambda) \in [0, x] \times (z_{1,x}, 0]$.

We now rule out the possibility that $\lambda_1 = z_{1,x}$. Indeed, since $\frac{\partial}{\partial \lambda}\phi(x, z_{1,x}) > 0$ and $\frac{\partial}{\partial \lambda}\phi(v_1, z_{1,x}) > 0$, we get from (15.42) that $\phi'_+(x, z_{1,x}) < 0$ and $\phi'_+(v_1, z_{1,x}) < 0$. This implies that there is a $v \in (v_1, x)$ with $\phi(v, z_{1,x}) = 0$ and $\phi'_+(v, z_{1,x}) > 0$. On the other hand, we know already that $z_{1,x}$ is the largest zero of $\phi(v, \cdot)$, i.e. $z_{1,v} = z_{1,x}$. By (15.42), $\phi'_+(v, z_{1,x}) = \phi'_+(v, z_{1,v}) = -\frac{\partial}{\partial z}\phi(v, z_{1,v}) < 0$, which is impossible. This concludes the proof that $\phi(v, \lambda) \neq 0$ for every $(v, \lambda) \in A_{1,x}$, $(v, \lambda) \neq (x, z_{1,x})$. Thus, we have $z_{1,y} < z_{1,x}$ for $0 < y < x$.

We continue the proof by showing that $z_{2,y} < z_{2,x}$. Define

$$A_{2,x} := \{(v, \lambda) : 0 \leq v \leq x, z_{2,x} \leq \lambda \leq z_{1,v}\}$$

and note that $\phi < 0$ immediately below the curve $\{(v, z_{1,v}) : 0 \leq v \leq x\}$. As above, by continuity there exists

$$v_2 = \min\{v : \exists \lambda \in [0, x) \text{ such that } \phi(v, \lambda) = 0 \text{ and } (v, \lambda) \in A_{2,x}\}.$$

Choose λ_2 such that $\phi(v_2, \lambda_2) = 0$ and assume that $\lambda_2 > z_{2,x}$. We argue in the same way as in the first part of the proof. Since the zeroes of $\phi(v_2, \cdot)$ are discrete and since $\phi(v_2, \cdot)$ changes sign every time it passes through a zero, we have, for sufficiently small $\epsilon > 0$, either $\phi(v_2, \lambda) > 0$ for $\lambda \in (\lambda_2 - \epsilon, \lambda_2)$, or $\phi(v_2, \lambda) > 0$ for $\lambda \in (\lambda_2, \lambda_2 + \epsilon)$. Suppose that $\phi(v_2, \lambda) > 0$, $\lambda \in (\lambda_2 - \epsilon, \lambda_2)$. Since $\{\phi > 0\}$ is open in $A_{2,x}$, there exists a point (v, λ) with $v < v_2$ such that $\phi(v, \lambda) > 0$. But this contradicts the minimality of v_2. Therefore, $\phi(v_2, \lambda) < 0$ for $\lambda \in (\lambda_2 - \epsilon, \lambda_2)$. The same argument yields that $\phi(v_2, \lambda) < 0$ for $\lambda \in (\lambda_2, \lambda_2 + \epsilon)$ and small enough $\epsilon > 0$. This proves that it cannot happen that $\lambda_2 > z_{2,x}$; consequently, $\phi(v, \lambda) \neq 0$ for all $(v, \lambda) \in A_{2,x}$, with $\lambda < z_{2,x}$. To rule out the possibility that $\lambda_2 = z_{2,x}$ we argue as in the first part of the proof. Thus, $\phi(v, \lambda) \neq 0$ and, in fact, $\phi(v, \lambda) < 0$ for all $(v, \lambda) \in A_{2,x}$, $(v, \lambda) \neq (x, z_{2,x})$. This proves that we have $z_{2,y} < z_{2,x}$ for $0 < y < x$.

The inequalities $z_{j,y} < z_{j,x}$, $j \geq 3$, can be seen in the same way.   □

Recall that π_{xy} is the distribution of the first passage time T_y^X at level y of the process X started at x.

Theorem 15.20. *If* $0 \leq x < y < r$, *then* $\pi_{xy} \in$ *BO and* $\pi_{0y} \in$ *CE* $\subset$ *GGC.*

Proof. We begin with the case $x = 0$, that is π_{0y}. By (15.41) and Lemma 15.17, we have

$$\mathscr{L}(\pi_{0y};\lambda) = \frac{1}{\phi(y,\lambda)} = \prod_j \left(1 + \frac{\lambda}{|z_{j,y}|}\right)^{-1}.$$

Since the factors on the right-hand side are Laplace transforms of exponential distributions, Definition 9.15 shows that $\pi_{0y} \in$ CE.

Now suppose that $x \neq 0$. Since $x < y$, Lemma 15.19 shows that $|z_{j,x}| > |z_{j,y}|$. A consequence of Lemma 15.17 is that $\sum_j 1/|z_{j,y}| < \infty$. Hence,

$$\mathscr{L}(\pi_{xy};\lambda) = \frac{\phi(x,\lambda)}{\phi(y,\lambda)} = \prod_j \left(1 + \frac{\lambda}{|z_{j,x}|}\right)\left(1 + \frac{\lambda}{|z_{j,y}|}\right)^{-1}$$

is well defined and strictly positive. Corollary 9.17 tells us that $\pi_{xy} \in$ BO.   □

In the remaining part of this chapter we will give a more precise description of the first passage distributions of generalized diffusions. All generalized diffusions which we have studied so far live on $E_m \subset [0, r)$ and are, in fact, generalized *reflected* diffusions. We will now consider generalized diffusions on open intervals of the type $(-r_1, r_2), 0 < r_1, r_2 \leq \infty$. The definition of these processes follows closely the one of generalized reflected diffusions.

Let $m : \mathbb{R} \to [-\infty, \infty]$ be a non-decreasing, right-continuous function with $m(0-) = 0$. The family of all such functions will be denoted by $\mathfrak{M}$. We use the same letter m to denote the measure on $\mathbb{R}$ corresponding to $m \in \mathfrak{M}$. We associate with $m \in \mathfrak{M}$ two functions $m_1, m_2 : [0, \infty) \to [0, \infty]$ defined by

$$m_1(x) = \begin{cases} -m((-x)-), & x \geq 0, \\ 0, & x < 0, \end{cases} \quad \text{and} \quad m_2(x) = \begin{cases} m(x), & x \geq 0, \\ 0, & x < 0. \end{cases}$$

Both m_1 and m_2 are non-decreasing and right-continuous, $m_1(0) = -m(0-) = 0$ and $m_2(0) \geq 0$. For $k = 1, 2$, $r_k = \sup\{x : m_k(x) < \infty\} \leq \infty$, let E_{m_k} be the support of m_k restricted to $[0, r_k)$, and note that $0 \in E_{m_k}$ need not hold. Set $l_k := \sup E_{m_k} \leq \infty$. In view of Remark 15.8, we may regard m_k as a string of length r_k.

Let $B = (B_t, \mathbb{P}_x)_{t \geq 0, -r_1 < x < r_2}$ be a standard one-dimensional Brownian motion killed upon hitting $\{-r_1, r_2\}$; if $r_1 = r_2 = \infty$, then B is just a Brownian motion. Let $L := (L(t, x))_{t \geq 0, -r_1 < x < r_2}$ be its jointly continuous local time and define

$$C_t = \int_{(-r_1, r_2)} L(t, x) m(dx).$$

By $\tau = (\tau_t)_{t \geq 0}$ we denote the right-continuous inverse of C. Define a process $(X_t)_{t \geq 0}$ via the time-change of B by the process τ, $X_t := B_{\tau(t)}$. In the same way as in the beginning of this chapter we conclude that X is a strong Markov process with state space $E_m = \text{supp}(m_{|(-r_1, r_2)})$ which is called a *generalized diffusion* on $(-r_1, r_2)$ or, more precisely, on E_m. We write ζ to denote the lifetime of X.

For $z \in \mathbb{C}$ let $\phi_k(x, z)$ and $\psi_k(x, z)$, $k = 1, 2$, be the solutions of the equations

$$\phi_k(x, z) = 1 + z \int_{[0,x]} (x - y)\phi_k(y, z)\, m_k(dy), \quad 0 \le x < r_k,$$

$$\psi_k(x, z) = x + z \int_{[0,x]} (x - y)\psi_k(y, z)\, m_k(dy), \quad 0 \le x < r_k.$$

Define $\phi(\cdot, z)$ and $\psi(\cdot, z)$ on $(-r_1, r_2)$ by

$$\phi(x, z) = \begin{cases} \phi_1(-x, z), & x < 0, \\ \phi_2(x, z), & x \ge 0, \end{cases} \quad \text{and} \quad \psi(x, z) = \begin{cases} -\psi_1(-x, z), & x < 0, \\ \psi_2(x, z), & x \ge 0. \end{cases}$$

Then it is straightforward to check that

$$\phi(x, z) = \begin{cases} 1 - z \int_{[0,x]} (x - y)\phi(y, z)\, m(dy), & -r_1 < x < 0, \\ 1 + z \int_{[0,x]} (x - y)\phi(y, z)\, m(dy), & 0 \le x < r_2, \end{cases} \tag{15.46}$$

$$\psi(x, z) = \begin{cases} x - z \int_{[0,x]} (x - y)\psi(y, z)\, m(dy), & -r_1 < x < 0, \\ x + z \int_{[0,x]} (x - y)\psi(y, z)\, m(dy), & 0 \le x < r_2. \end{cases} \tag{15.47}$$

For $\lambda > 0$ let

$$h_1(\lambda) = \lim_{x \to r_1} \frac{\psi_1(x, \lambda)}{\phi_1(x, \lambda)} = \lim_{x \to r_1} \frac{-\psi(-x, \lambda)}{\psi(-x, \lambda)} = -\lim_{x \to -r_1} \frac{\psi(x, \lambda)}{\phi(x, \lambda)}$$

be the characteristic function of the string m_1, and define $h_2(\lambda)$ in an analogous way for the string m_2. Set

$$h(\lambda) := \left(h_1(\lambda)^{-1} + h_2(\lambda)^{-1} \right)^{-1}.$$

For a bounded Borel function $g : (-r_1, r_2) \to \mathbb{R}$ and $x \in E_m$, let

$$G_\lambda g(x) = \mathbb{E}_x \int_0^\infty g(X_t)\, dt, \quad \lambda > 0.$$

As in the case of generalized reflected diffusions, one can show that the operator G_λ admits a kernel $G_\lambda(x, y)$ given by the formula

$$G_\lambda(x, y) = h(\lambda)\left(\phi(x, \lambda) + h_1(\lambda)^{-1}\psi(x, \lambda)\right)\left(\phi(y, \lambda) - h_2(\lambda)^{-1}\psi(y, \lambda)\right),$$

for all $x, y \in E_m$ with $x \leq y$, see [217, p. 247]. For every $y \in E_m$ we define $T_y^X = \inf\{t > 0 : X_t = y\}$. In the same way as before one derives that for all $x, y \in E_m$

$$G_\lambda(x, y) = \mathbb{E}_x[e^{-\lambda T_y^X}] G_\lambda(y, y). \tag{15.48}$$

If $r_2 < \infty$ and $r_2 \in \overline{E}_m$, define $\tilde{E}_m = E_m \cup \{r_2\}$; otherwise, set $\tilde{E}_m = E_m$. In case $r_2 \in \tilde{E}_m$, set $T_{r_2}^X := \lim_{y \to r_2} T_\eta^X$. Later in this chapter we will need the following lemma.

Lemma 15.21.   (i) *If $r_2 \in \tilde{E}_m$, then $\int_0^{r_2} m(x)\, dx < \infty$ if, and only if, $\phi(r_2-, \lambda)$ is finite.*

(ii) *If the representing measure of h_2 is supported by a sequence of positive numbers $(a_j)_{j\geq 1}$ such that $\sum_{j\geq 1} 1/a_j < \infty$, and if $r_2 < \infty$, then either $m(r_2-) < \infty$, or $m(r_2-) = \infty$ and $\int_0^{r_2} m(x)\, dx < \infty$.*

Proof. (i) One direction follows from (15.21), while the other can be found in [105, p. 163].

(ii) For this part we refer the reader to the paper [194].   □

Proposition 15.22.   (i) *For $\lambda > 0$ and $y \in (-r_1, r_2)$ let*

$$h_2(y, \lambda) = \frac{\psi(y, \lambda)}{\phi(y, \lambda)} \quad and \quad h(y, \lambda) := \frac{1}{\frac{1}{h_1(\lambda)} + \frac{1}{h_2(y,\lambda)}}.$$

Then

$$\mathbb{E}_0[e^{-\lambda T_y^X}] = \frac{h(y, \lambda)}{\psi(y, \lambda)} \quad for\ every\ \lambda > 0 \tag{15.49}$$

and

$$\mathbb{P}_0(T_y^X < \infty) = \begin{cases} 1, & r_1 = \infty, \\ \frac{r_1}{r_1+y}, & r_1 < \infty. \end{cases}$$

(ii) *Suppose that $r_2 \in \tilde{E}_m$. Then $\mathbb{P}_0(T_{r_2}^X < \infty) > 0$ if, and only if, $\int_0^{r_2} m(x)\, dx$ is finite. In this case*

$$\mathbb{E}_0[e^{-\lambda T_{r_2}^X}] = \frac{h(\lambda)}{\psi(r_2, \lambda)} \quad for\ every\ \lambda > 0$$

where $\psi(r_2, \lambda) := \psi(r_2-, \lambda) < \infty$.

Proof. (i) Recall that $\phi(0,\lambda) = 1$ and $\psi(0,\lambda) = 0$. It follows from (15.48) that

$$\mathbb{E}_0[e^{-\lambda T_y^X}] = \frac{G_\lambda(0,y)}{G_\lambda(y,y)} = \frac{1}{\phi(y,\lambda) + h(\lambda)^{-1}\psi(y,\lambda)} \qquad (15.50)$$

$$= \frac{1}{\psi(y,\lambda)(h_2(y,\lambda)^{-1} + h_1(\lambda)^{-1})}$$

$$= \frac{h(y,\lambda)}{\psi(y,\lambda)}.$$

For the second statement note that $\mathbb{P}_0(T_y^X < \infty) = \lim_{\lambda \to 0} \mathbb{E}_0[e^{-\lambda T_y^X}]$. It follows from the last formula of Proposition 15.16 that $\lim_{\lambda \to 0} h_1(\lambda) = h_1(0) = r_1$, while $\lim_{\lambda \to 0} h_2(y,\lambda) = \psi(y,0)/\phi(y,0) = y$. Hence

$$\lim_{\lambda \to 0} \frac{h(y,\lambda)}{\psi(y,\lambda)} = \begin{cases} 1, & r_1 = \infty, \\ \frac{r_1}{r_1 + y}, & r_1 < \infty. \end{cases}$$

(ii) Under the assumption $r_2 \in \tilde{E}_m$ Lemma 15.21 proves that $\int_0^{r_2} m(x)\,dx < \infty$ if, and only if, $\phi(r_2-,\lambda) < \infty$. Suppose that $\int_0^{r_2} m(x)\,dx = \infty$ and let $y \to r_2$ in (15.50). We get $\mathbb{E}_0[e^{-\lambda T_{r_2}^X}] = 0$, hence $\mathbb{P}_0(T_{r_2}^X < \infty) = 0$. Assume now that $\int_0^{r_2} m(x)\,dx < \infty$. Since $h(\lambda) = \lim_{x \to r_2} \phi(x,\lambda)/\psi(x,\lambda)$ and $\phi(r_2-,\lambda) < \infty$ by Lemma 15.21 (i), we have that $\psi(r_2-,\lambda) < \infty$. Let $y \to r_2$ in (15.49). It follows that

$$\mathbb{E}_0[e^{-\lambda T_{r_2}^X}] = \frac{h(\lambda)}{\psi(r_2,\lambda)},$$

and since the right-hand side is strictly positive, $\mathbb{P}(T_{r_2}^X < \infty) > 0$.   □

Remark 15.23. Since both h_1 and $h_2(y,\cdot)$ are in $\mathcal{S}$, it follows that $h(y,\cdot) \in \mathcal{S}$ as well.

For $x \in E_m$ and $y \in \tilde{E}_m$ such that $x < y$ let π_{xy} be the $\mathbb{P}_x$-distribution of T_y^X. Define

$$\mathsf{H} := \{\pi_{xy} : x \in E_m,\ y \in \tilde{E}_m,\ x < y,\ m \in \mathfrak{M}\}$$

to be the family of all such first passage distributions. Translating the function m, if necessary, it is enough to consider $x = 0$.

Theorem 15.24. *A sub-probability measure π belongs to H if, and only if, there exist $\pi_1 \in \mathsf{CE}$ and $\pi_2 \in \mathsf{ME}$ with $\pi_2\{0\} = 0$ such that $\pi = \pi_1 \star \pi_2$, the representing sequence $(b_j)_{j \geq 1}$ of π_1 is either empty or strictly increasing, and the representing measure of $\lambda \mapsto (\lambda \mathcal{L}(\pi_2; \lambda))^{-1}$ has a point mass at every b_j.*

Proof. Suppose that $\pi = \pi_{0y} \in H$. By (15.17) we have that

$$\frac{y}{\psi(y,z)} = \prod_j \left(1 - \frac{z}{w_{j,y}}\right)^{-1} = \prod_j \left(1 + \frac{z}{b_j}\right)^{-1}, \qquad (15.51)$$

where $b_j := |w_{j,y}|$ and $(w_{j,y})_{j\geq 1}$ are the zeroes of $\psi(y,\cdot)$. Since the zeroes $(w_{j,y})_{j\geq 1}$ are simple and decreasing, the sequence $(b_j)_{j\geq 1}$ is increasing. Note that the right-hand side of (15.51) is the Laplace transform of a probability measure from CE, cf. Corollary 9.17, so we may define $\pi_1 \in$ CE by $\mathscr{L}(\pi_1;\lambda) := y/\psi(y,\lambda)$, $\lambda \geq 0$. Moreover, $\lambda \mapsto h(y,\lambda)/y \in \mathcal{S}$ and $h(y,0)/y \leq 1$ by Proposition 15.22. It follows by Definition 9.4 that there exists $\pi_2 \in$ ME such that $\mathscr{L}(\pi_2;\lambda) = h(y,\lambda)/y$. From Proposition 15.22 we have that

$$\mathscr{L}(\pi;\lambda) = \frac{h(y,\lambda)}{\psi(y,\lambda)} = \frac{y}{\psi(y,\lambda)} \frac{h(y,\lambda)}{y} = \mathscr{L}(\pi_1;\lambda)\mathscr{L}(\pi_2;\lambda),$$

that is, $\pi = \pi_1 \star \pi_2$.

Since $h_2(y,\cdot) \in \mathcal{S}$, the function $\lambda \mapsto (\lambda h_2(y,\lambda))^{-1} = \phi(y,\lambda)/(\lambda\psi(y,\lambda))$ also belongs to $\mathcal{S}$. By Remark 15.18, $\phi(y,\cdot)$ and $\psi(y,\cdot)$ have no common zeroes, i.e. the set of poles of $\lambda \mapsto \phi(y,\lambda)/(\lambda\psi(y,\lambda))$ is precisely $\{0\} \cup \{w_{j,y} : j \geq 1\}$. Hence, the set $\{0\} \cup \{b_j : j \geq 1\}$ is the support of the representing measure of $\lambda \mapsto (\lambda h_2(y,\lambda))^{-1}$. Since

$$\frac{1}{\lambda h(y,\lambda)} = \frac{1}{\lambda h_1(\lambda)} + \frac{1}{\lambda h_2(y,\lambda)},$$

the representing measure of $\lambda \mapsto (\lambda h(y,\lambda))^{-1}$ has a point mass at each point in the set $\{0\} \cup \{b_j : j \geq 1\}$. But

$$\frac{1}{\lambda\mathscr{L}(\pi_2;\lambda)} = \frac{y}{\lambda h(y,\lambda)},$$

showing that the representing measure of $\lambda \mapsto (\lambda\mathscr{L}(\pi_2;\lambda))^{-1}$ has a point mass at every b_n. It remains to check that $\pi_2\{0\} = 0$. Note that $\pi_2\{0\} = \lim_{\lambda\to\infty}\mathscr{L}(\pi_2;\lambda) = \lim_{\lambda\to\infty}h(y,\lambda)/y$. By the definition of $h(y,\lambda)$ it suffices to show that at least one of the limits $\lim_{\lambda\to\infty}h_1(\lambda)$ and $\lim_{\lambda\to\infty}h_2(y,\lambda)$ is equal to zero. But this follows from (15.23) and the fact that 0 is a point of increase for at least one of the measures m_1 and m_2, cf. Remark 15.8.

Sufficiency: By assumption we have that

$$\mathscr{L}(\pi_1;\lambda) = \begin{cases} 1, & \text{if } \{b_j : j \geq 1\} = \emptyset, \\ \prod_j \left(1 + \frac{\lambda}{b_j}\right)^{-1}, & \text{otherwise.} \end{cases}$$

Since $\pi_2 \in \text{ME}$ it follows that $\mathscr{L}(\pi_2; \cdot) \in \mathcal{S}$ as well as $\lambda \to (\lambda \mathscr{L}(\pi_2, \lambda))^{-1} \in \mathcal{S}$. Therefore, there exist $c \geq 0$ and a measure $\tilde{\sigma}$ satisfying $\int_{[0,\infty)} (1+t)^{-1} \tilde{\sigma}(dt) < \infty$ such that

$$\frac{1}{\lambda \mathscr{L}(\pi_2; \lambda)} = c + \int_{[0,\infty)} \frac{1}{\lambda + t} \tilde{\sigma}(dt)$$

and that $\tilde{\sigma}$ has a point mass at each point in $\{0\} \cup \{b_j \; : \; j \geq 1\}$. Since

$$\mathscr{L}(\pi_2; \lambda) = \left(c\lambda + \int_{[0,\infty)} \frac{\lambda}{\lambda + t} \tilde{\sigma}(dt) \right)^{-1},$$

we conclude that $\tilde{\sigma}\{0\} = \mathscr{L}(\pi_2; 0)^{-1} \geq 1$. As $\pi_2\{0\} = 0$ we can let $\lambda \to \infty$ to find $c > 0$ or $\tilde{\sigma}[0, \infty) = \infty$. Write $\tilde{\sigma}$ as $\tilde{\sigma}_1 + \tilde{\sigma}_2$ where $\text{supp}(\tilde{\sigma}_2) = \{0\} \cup \{b_j \; : \; j \geq 1\}$ and $\tilde{\sigma}_2\{0\} = \tilde{\sigma}\{0\} \geq 1$. Thus $\tilde{\sigma}_1\{0\} = 0$.

By (7.3) we see that there exist measures σ_1 and σ_2 on $[0, \infty)$, and real numbers $c_1 \geq 0$, $c_2 \geq 0$, such that

$$c_1 + \int_{[0,\infty)} \frac{1}{\lambda + t} \sigma_1(dt) = \left(\int_{(0,\infty)} \frac{\lambda}{\lambda + t} \tilde{\sigma}_1(dt) \right)^{-1}, \tag{15.52}$$

$$c_2 + \int_{[0,\infty)} \frac{1}{\lambda + t} \sigma_2(dt) = \left(c\lambda + \int_{[0,\infty)} \frac{\lambda}{\lambda + t} \tilde{\sigma}_2(dt) \right)^{-1}. \tag{15.53}$$

Denote the functions defined in (15.52) and (15.53) by g_1 and g_2 respectively. Then

$$\mathscr{L}(\pi_2; \lambda) = \left(g_1(\lambda)^{-1} + g_2(\lambda)^{-1} \right)^{-1}. \tag{15.54}$$

We record the following facts that follow from (15.52) and (15.53) by letting $\lambda \to \infty$:

(a) $c_1 = 1/\tilde{\sigma}_1(0, \infty)$;

(b) if $c = 0$, then $c_2 = 1/\tilde{\sigma}_2[0, \infty)$, while if $c > 0$, then $c_2 = 0$;

(c) $c_1 c_2 = 0$; indeed, if $c = 0$, then $\tilde{\sigma}[0, \infty) = \infty$, so that either $\tilde{\sigma}_1[0, \infty) = \infty$ or $\tilde{\sigma}_2[0, \infty) = \infty$.

Further, $g_1(0) = \tilde{\sigma}_1\{0\} = \infty$ and $g_2(0) = \tilde{\sigma}_2\{0\}^{-1} \in (0, 1]$. Note that we can write

$$\frac{1}{g_2(\lambda)} = c\lambda + \gamma_0 + \sum_{j \geq 1} \gamma_j \frac{\lambda}{\lambda + b_j},$$

where γ_j is the mass of $\tilde{\sigma}_2$ at b_j, $j \geq 0$. This shows that $1/g_2$ has real poles at $-b_j$, $j \geq 1$, hence also real zeroes $-a_j$, $j \geq 1$, which intertwine with the poles: $0 < a_1 < b_1 < a_2 < b_2 < \cdots$. Therefore,

$$g_2(\lambda) = \frac{\prod_j \left(1 + \frac{\lambda}{b_j} \right)}{\prod_j \left(1 + \frac{\lambda}{a_j} \right)}. \tag{15.55}$$

This shows that the representing measure of g_2 is supported in $(a_j)_{j\geq 1}$ and $\sum_{j\geq 1} 1/a_j < \infty$. Let m_1 and m_2 be strings having g_1 and g_2 as their characteristic functions. Since $g_2(0) \in (0,1]$, it follows that the length r_2 of the string m_2 satisfies $r_2 = g_2(0) \in (0,1]$. Moreover, by Lemma 15.21 (ii) we conclude that either $m(r_2-) < \infty$, or $m(r_2-) = \infty$ and $\int_0^{r_2} m(x)\,dx < \infty$. In the second case it holds that $r_2 \in \tilde{E}_m$. In the first case one can extend m_2 without changing its characteristic function g_2 such that the length of the extended m_2 is strictly larger than r_2 – e.g. extend m_2 to be the constant $m(r_2-)$ on an interval to the right of r_2. We use the same letter for the extended string.

Next

$$\sigma_1[0,\infty) = \lim_{\lambda\to\infty} \lambda g_1(\lambda) = \frac{1}{\lim_{\lambda\to\infty} \int_{(0,\infty)} \frac{1}{\lambda+t}\tilde{\sigma}_1(dt)} = \infty,$$

hence by the second line in Proposition 15.16 we get that $m_1\{0\} = 0$.

Define m by

$$m(x) = \begin{cases} -m_1((-x)-), & x < 0, \\ m_2(x), & x \geq 0. \end{cases}$$

Then $m \in \mathfrak{M}$. Let $\phi(\cdot,\lambda)$ and $\psi(\cdot,\lambda)$ be the solutions of equations (15.46) and (15.47) with this m. Then clearly g_1 and g_2 can be expressed in terms of ϕ and ψ as

$$g_1(\lambda) = -\lim_{x\to -r_1} \frac{\psi(x,\lambda)}{\phi(x,\lambda)},$$

$$g_2(\lambda) = \lim_{x\to r_2} \frac{\psi(x,\lambda)}{\phi(x,\lambda)} = \frac{\psi(r_2,\lambda)}{\phi(r_2,\lambda)}.$$

Comparing the last formula with (15.55) we see that the family of all the zeroes of $\psi(r_2,\cdot)$ is equal to $\{-b_j : j \geq 1\}$. Hence,

$$\psi(r_2,\lambda) = \prod_j \left(1 + \frac{\lambda}{b_j}\right),$$

implying that

$$\mathscr{L}(\pi_1;\lambda) = \frac{1}{\psi(r_2,\lambda)}.$$

Together with (15.54) this gives that

$$\mathscr{L}(\pi;\lambda) = \mathscr{L}(\pi_1;\lambda)\mathscr{L}(\pi_2;\lambda) = \frac{g(\lambda)}{\psi(r_2,\lambda)}$$

where $g(\lambda) := (g_1(\lambda)^{-1} + g_2(\lambda)^{-1})^{-1}$. Since $c_1 = 0$ or $c_2 = 0$, we see that 0 is in the support of m. Hence, it follows from Proposition 15.22 that π is the hitting time distribution of r_2 starting at 0 for the generalized diffusion corresponding to m. □

Comments 15.25. Kreĭn's theory of strings was developed by M. G. Kreĭn in the period from 1951 to 1954 in a series of papers [220, 221, 222, 223, 224]. An excellent exposition of these results is given in [196] which also contains historical and bibliographical remarks on the subject. An equally good source for the theory of strings are Chapters 5 and 6 of [105]. Our presentation is a combination of those two references. A complete proof of Theorem 15.7 can be found in either of these references. Definition 15.10 is from [105] as well as Lemma 15.11 which says that $R_\lambda : L^2((-\infty, r), m) \to \mathfrak{D}(A)$ is one-to-one and onto. The image of $L^2((-\infty, r), m)$ under R_λ is dense in $L^2((-\infty, r), m)$ and both R_λ and A are self-adjoint.

Applications of Kreĭn's string theory to Markov processes are studied in the 1970's. It seems that Langer, Partzsch and Schütze are the first to use this connection in [238] where it is shown that the operator A (acting on an appropriate domain) is the infinitesimal generator of a C_0-semigroup. Path properties of the corresponding Feller process are studied by Groh in [137]. S. Watanabe, [365] gives the first construction of the process as a time-change of (reflected) Brownian motion and introduces the name *quasi-diffusion*. This name is adopted by Küchler in [228, 229] and [230] who looks at the properties of transition densities of the process. The name *generalized diffusion* is probably introduced by N. Ikeda, S. Kotani and S. Watanabe in [172], and is used again by Tomisaki, [349]. Related works are [198, 216] and [199]. The term *gap diffusion* is used by Knight in [209].

The question of characterizing Lévy measures of inverse local times of diffusions was posed in [178, p. 217]. In its original form, which requires that m has full support, the problem is very difficult and still unsolved. In the context of generalized diffusions it was independently solved by Knight [209] and Kotani and Watanabe [217]. Bertoin [46] revisited the problem and gave another proof. The proof we give follows essentially the approach from [209] and [217]. One of the key steps in the proof is the identification of the probabilistically defined Green function $G_\lambda(x, y)$ with the analytically defined $R_\lambda(x, y) = \phi(x, \lambda)\chi(y, \lambda)$. This seems to be part of the folklore; as we were unable to give a precise reference, we decided to include a proof of Proposition 15.15. This proof follows [178, pp. 126–7].

The relations between m and σ, given in Proposition 15.16, appear in [224] and can be derived using the spectral theory of strings, see [105, p. 192]. Note that a consequence of the last formula is that $r = \lim_{\lambda \to 0} h(\lambda) = h(0+)$.

The problem of the characterization of the hitting time distributions in the case of nonsingular diffusion on an interval $(-r_1, r_2)$ was studied by Kent in [204], see also [203] for complete proofs, and [206]. He showed that $\pi_{xy} \in$ BO if the boundary $-r_1$ is not natural. The case presented in Theorem 15.20 covers an instantaneously reflecting, non-killing left boundary. The complete characterization of hitting time distributions for generalized diffusions, Theorem 15.24, was obtained by Yamazato in [377]. A slight drawback of this result is that the condition on π_2 is given in terms of the representing measure of $(\lambda \mathscr{L}(\pi_2; \lambda))^{-1}$. In [378] Yamazato expressed the condition on π_2 in terms of the function η_2 appearing in the representation (9.4).

Chapter 16

Examples of complete Bernstein functions

Below we provide a list of examples of complete Bernstein functions. We restrict ourselves to complete Bernstein functions of the form

$$f(\lambda) = \int_{(0,\infty)} (1 - e^{-\lambda t}) m(t) \, dt = \int_{(0,\infty)} \frac{\lambda}{\lambda + t} \sigma(dt).$$

Whenever possible, we provide expressions for the representing density $m(t)$ and measure $\sigma(dt)$, respectively; for simplicity, we do not distinguish between a density and the measure given in terms of the density. We also state whether f is a Thorin–Bernstein function, i.e. whether f is of the form

$$f(\lambda) = \int_{(0,\infty)} \log\left(1 + \frac{\lambda}{t}\right) \tau(dt).$$

By 'unknown if in $\mathcal{TBF}$' we indicate that we were not able to determine whether $f \in \mathcal{TBF}$ or not.

The third column of the tables contains references and remarks how to derive the corresponding assertions. Some more lengthy comments can be found in Sections 16.12.1–16.12.3 below.

We record here again the relations between the Lévy density m and the Stieltjes measure σ and the Thorin measure τ, cf. Remark 6.10 (iii) and Remark 8.3 (ii):

$$m(t) = \int_{(0,\infty)} e^{-ts} s \, \sigma(ds);$$

$$m(t) = \int_{(0,\infty)} e^{-ts} \tau(0, s) \, ds;$$

$$t \cdot m(t) = \int_{(0,\infty)} e^{-ts} \tau(ds);$$

$$t \cdot \sigma(dt) = \tau(0, t) \, dt.$$

Only a few publications contain extensive lists of (complete) Bernstein functions. We used the papers by Jacob and Schilling [185] (29 entries) and Berg [35] (11 entries) as a basis for the tables in this chapter.

16.1 Special functions used in the tables

Our standard references for this section are Abramowitz and Stegun [1], Erdélyi et al. [106, 107] and Gradshteyn and Ryzhik [134].

Trigonometric functions

$$\sec(x) = \frac{1}{\cos(x)}, \qquad \csc(x) = \frac{1}{\sin(x)}$$

Hyperbolic functions

$$\sinh(x) = \frac{e^x - e^{-x}}{2}, \qquad \cosh(x) = \frac{e^x + e^{-x}}{2}, \qquad \tanh(x) = \frac{\sinh(x)}{\cosh(x)}$$

Gamma and related functions

Gamma function

$$\Gamma(x) = \int_0^\infty e^{-t} t^{x-1} \, dt$$

Digamma function

$$\Psi_0(x) = \Psi(x) = \frac{d}{dx} \log \Gamma(x) = \frac{\Gamma'(x)}{\Gamma(x)}$$

Trigamma function

$$\Psi_1(x) = \frac{d^2}{dx^2} \log \Gamma(x) = \sum_{m=0}^\infty \frac{1}{(x+m)^2}$$

Tetragamma function

$$\Psi_2(x) = \frac{d^3}{dx^3} \log \Gamma(x) = -\sum_{m=0}^\infty \frac{1}{(x+m)^3}$$

Incomplete Gamma functions

$$\Gamma(\alpha, x) = \int_x^\infty t^{\alpha-1} e^{-t} \, dt = x^{(\alpha-1)/2} e^{-x/2} W_{(\alpha-1)/2, \alpha/2}(x)$$

$$\gamma(\alpha, x) = \int_0^x t^{\alpha-1} e^{-t} \, dt = \alpha^{-1} x^\alpha \, {}_1F_1(\alpha, \alpha+1, -x)$$

Beta and related functions

Beta function

$$B(x, y) = \int_0^1 t^{x-1} (1-t)^{y-1} \, dt = \frac{\Gamma(x)\,\Gamma(y)}{\Gamma(x+y)}$$

Incomplete Beta functions

$$B_x(\mu, v) = \int_0^x t^{\mu-1}(1-t)^{v-1}\, dt$$

$$I_x(\mu, v) = \frac{B_x(\mu, v)}{B_1(\mu, v)}$$

Exponential integrals

$$\mathrm{Ei}(x) = -\int_{-x}^{\infty} \frac{e^{-t}}{t}\, dt \quad \text{(Cauchy principal value integral if } x > 0)$$

$$\mathrm{E}_1(x) = \int_{x}^{\infty} \frac{e^{-t}}{t}\, dt = -\mathrm{Ei}(-x) = \Gamma(0, x), \quad x > 0$$

$$\mathrm{E}_m(x) = \int_{1}^{\infty} \frac{e^{-xt}}{t^m}\, dt, \quad m \geq 2,\ x > 0$$

Sine and cosine integrals

$$\mathrm{si}(x) = -\int_{x}^{\infty} \frac{\sin t}{t}\, dt, \quad x > 0$$

$$\mathrm{ci}(x) = -\int_{x}^{\infty} \frac{\cos t}{t}\, dt, \quad x > 0$$

Error functions

$$\mathrm{Erf}(x) = \frac{2}{\sqrt{\pi}} \int_0^x e^{-t^2}\, dt$$

$$\mathrm{Erfc}(x) = 1 - \mathrm{Erf}(x)$$

Bessel and modified Bessel functions

Bessel functions

$$J_v(x) = \sum_{n=0}^{\infty} (-1)^n \frac{(x/2)^{v+2n}}{n!\,\Gamma(v+n+1)}$$

$$Y_v(x) = \frac{J_v \cos(v\pi) - J_{-v}(x)}{\sin(v\pi)}$$

Modified Bessel functions

$$I_v(x) = \sum_{n=0}^{\infty} \frac{(x/2)^{v+2n}}{n!\,\Gamma(v+n+1)}$$

$$K_v(x) = \frac{\pi}{2} \frac{I_{-v}(x) - I_v(x)}{\sin(v\pi)}$$

Bessel function zeroes

$$j_{v,n}, \quad n = 1, 2, \ldots \quad \text{zeroes of } J_v$$

Mittag-Leffler function

$$E_{\alpha,\beta}(x) = \sum_{n=0}^{\infty} \frac{x^n}{\Gamma(n\alpha + \beta)}$$

Hypergeometric functions

Confluent hypergeometric functions

$$_1F_1(a, b, x) = \sum_{k=0}^{\infty} \frac{(a)_k}{(b)_k} \frac{x^k}{k!}$$

Hypergeometric functions

$$_2F_1(a, b, c, x) = \sum_{k=0}^{\infty} \frac{(a)_k (b)_k}{(c)_k} \frac{x^k}{k!}$$

Whittaker functions

Whittaker functions of the first kind

$$M_{\kappa,\mu}(x) = x^{\mu+\frac{1}{2}} e^{-x/2} \, _1F_1\left(\mu - \kappa + \frac{1}{2}, \, 2\mu + 1, \, x\right)$$

Whittaker functions of the second kind

$$W_{\kappa,\mu}(x) = \frac{\Gamma(-2\mu)}{\Gamma(\frac{1}{2} - \kappa - \mu)} M_{\kappa,\mu}(x) + \frac{\Gamma(2\mu)}{\Gamma(\frac{1}{2} - \kappa + \mu)} M_{\kappa,-\mu}(x)$$

modified Whittaker functions

$$\Phi_{\alpha,\beta}(x) = x^{-\kappa-1} e^{x/2} W_{\kappa,\mu}(x),$$

$$\text{where } \alpha = \mu - \kappa, \ \beta = \mu - \kappa, \ \kappa = \frac{-(\alpha+\beta)}{2}, \ \mu = \frac{\alpha-\beta}{2}$$

Parabolic cylinder functions

$$D_\nu(x) = 2^{\nu/2+1/2} x^{-1/2} W_{\nu/2+1/4, \, 1/4}(x^2/2)$$

Tricomi functions

$$\Psi(a,b,x) = \frac{\Gamma(1-b)}{\Gamma(a-b+1)} \, _1F_1(a,b,x)$$

$$+ \frac{\Gamma(b-1)}{\Gamma(a)} x^{1-b} \, _1F_1(a-b+1, \, 2-b, \, x)$$

$$= \frac{\pi}{\sin(b\pi)} \left(\frac{_1F_1(a,b,x)}{\Gamma(a-b+1)\Gamma(b)} - x^{1-b} \frac{_1F_1(a-b+1, \, 2-b, \, x)}{\Gamma(a)\Gamma(2-b)} \right)$$

Lommel's functions of two variables

$$U_\nu(y, x) = \sum_{n=0}^{\infty} (-1)^n \left(\frac{y}{x}\right)^{\nu+2n} J_{\nu+2n}(x)$$

$$V_\nu(y, x) = \cos\left(\frac{y}{2} + \frac{x^2}{2y} + \frac{\nu\pi}{2}\right) + U_{2-\nu}(y, x)$$

Partial sums of the exponential function

$$e_n(x) = \sum_{k=0}^{n} \frac{x^k}{k!}, \quad n = 0, 1, 2, \ldots$$

Hermite polynomials

$$H_n(x) = (-1)^n e^{x^2} \frac{d^n}{dx^n} e^{-x^2}, \quad n = 0, 1, 2, \ldots$$

Generalized Laguerre polynomials

$$L_n^{(\gamma)}(x) = \sum_{j=0}^{n} (-1)^j \binom{n+\gamma}{n-j} \frac{x^j}{j!}, \quad n = 0, 1, 2, \ldots$$

Lambert function

The Lambert W function is defined as the unique solution of the equation

$$W(x) e^{W(x)} = x$$

such that $W : [-e^{-1}, \infty) \to [-1, \infty)$, $W(0) = 0$ and, $W|_{(0,\infty)} > 0$.

16.2   Algebraic functions

No	Function $f(\lambda)$	Comment
1	$\lambda^\alpha,\quad 0 < \alpha < 1$	
2	$(\lambda + 1)^\alpha - 1,\quad 0 < \alpha < 1$	
3	$1 - (1 + \lambda)^{\alpha-1},\quad 0 < \alpha < 1$	
4	$\dfrac{\lambda}{\lambda + a},\quad a > 0$	
5	$\dfrac{\lambda}{\sqrt{\lambda + a}},\quad a > 0$	Theorem 8.2 (v)

Lévy (L), Stieltjes (S) and Thorin (T) representation measures

L: $\dfrac{\alpha}{\Gamma(1-\alpha)}\, t^{-1-\alpha}$

S: $\dfrac{\sin(\alpha\pi)}{\pi}\, t^{\alpha-1}$

T: $\dfrac{\alpha\sin(\alpha\pi)}{\pi}\, t^{\alpha-1}$

L: $\dfrac{\alpha}{\Gamma(1-\alpha)}\, e^{-t} t^{-1-\alpha}$

S: $\dfrac{\sin(\alpha\pi)}{\pi} \dfrac{(t-1)^{\alpha}}{t}\, \mathbb{1}_{(1,\infty)}(t)$

T: $\dfrac{\alpha\sin(\alpha\pi)}{\pi}\, (t-1)^{\alpha-1}\, \mathbb{1}_{(1,\infty)}(t)$

L: $\dfrac{1}{\Gamma(1-\alpha)}\, e^{-t} t^{-\alpha}$

S: $\dfrac{\sin(\alpha\pi)}{\pi} \dfrac{(t-1)^{\alpha-1}}{t}\, \mathbb{1}_{(1,\infty)}(t)$

T: $f \notin \mathcal{TBF}$

L: $a e^{-at}$

S: $\delta_a(dt)$

T: $f \notin \mathcal{TBF}$

L: $\dfrac{e^{-at}(2at+1)}{2\sqrt{\pi}\, t^{3/2}}$

S: $\dfrac{1}{\pi\sqrt{t-a}}\, \mathbb{1}_{(a,\infty)}(t)$

T: $f \notin \mathcal{TBF}$

No	Function $f(\lambda)$	Comment
6	$\dfrac{\lambda}{(\lambda + a)^\alpha}, \quad a > 0,\ 0 < \alpha < 1$	Theorem 8.2 (v)
7	$\sqrt{\dfrac{\lambda}{\lambda + a}}, \quad a > 0$	(1.18) of [50], Theorem 1 of [262], Theorem 8.2 (v). See §16.12.1
8	$\dfrac{\lambda^\alpha}{(1 + \lambda)^\alpha}, \quad 0 < \alpha < 1$	(1.49) of [50], Theorem 8.2 (v). See §16.12.1
9	$\dfrac{\lambda^\alpha - \lambda\,(1 + \lambda)^{\alpha-1}}{(1 + \lambda)^\alpha - \lambda^\alpha}, \quad 0 < \alpha < 1$	(1.18) of [50], Theorem 8.2 (v). See §16.12.1
10	$\dfrac{(1 + \lambda)^\alpha}{(1 + \lambda)^\alpha - 1} - \dfrac{1 + \alpha\lambda\,(1 + \lambda)^{\alpha-1}}{(1 + \lambda)^\alpha - 1},$ $0 < \alpha < 1$	Theorem 1.7.3 of [50] and Theorem 3.4* of [187], Theorem 8.2 (v). See §16.12.1

Lévy (L), Stieltjes (S) and Thorin (T) representation measures

L: $\dfrac{\sin(\alpha\pi)\,\Gamma(1-\alpha)}{\pi}\,e^{-at}t^{\alpha-2}(at+1-\alpha)$

S: $\dfrac{\sin(\alpha\pi)}{\pi}\,(t-a)^{-\alpha}\,\mathbb{1}_{(a,\infty)}(t)$

T: $f \notin \mathcal{TBF}$

L: $\dfrac{a}{2}\,e^{-at/2}\left(I_0\left(\dfrac{at}{2}\right) - I_1\left(\dfrac{at}{2}\right)\right)$

S: $\dfrac{1}{\pi\sqrt{at-t^2}}\,\mathbb{1}_{(0,a)}(t)$

T: $f \notin \mathcal{TBF}$

L: $\alpha\pi\,\csc(\alpha\pi)\,{}_1F_1(1+\alpha,\,2,\,-t)$

S: $\dfrac{\sin(\alpha\pi)}{\pi}\,t^{\alpha-1}(1-t)^{-\alpha}\,\mathbb{1}_{(0,1)}(t)$

T: $f \notin \mathcal{TBF}$

L: closed expression unknown

S: $\dfrac{\sin(\alpha\pi)}{\pi}\,\dfrac{t^{\alpha-1}(1-t)^{\alpha-1}}{(1-t)^{2\alpha} - 2(1-t)^{\alpha}t^{\alpha}\cos(\alpha\pi) + t^{2\alpha}}\,\mathbb{1}_{(0,1)}(t)$

T: $f \notin \mathcal{TBF}$

L: $\dfrac{\alpha\sin(\alpha\pi)}{\pi}\,\dfrac{e^{-t}(t-1)^{\alpha}}{(t-1)^{2\alpha} - 2(t-1)^{\alpha}\cos(\alpha\pi) + 1}$

S: $\dfrac{\alpha\sin(\alpha\pi)}{\pi}\,\dfrac{(t-1)^{\alpha-1}}{(t-1)^{2\alpha} - 2(t-1)^{\alpha}\cos(\alpha\pi) + 1}\,\mathbb{1}_{(1,\infty)}(t)$

T: $f \notin \mathcal{TBF}$

No	Function $f(\lambda)$	Comment
11	$\dfrac{\lambda^\beta - 1}{\lambda^\alpha - 1} - 1, \quad 0 < \alpha < \beta < 1$	Theorem 2 of [143]
12	$\dfrac{\lambda^\alpha}{(1+\lambda)^\alpha - \lambda^\alpha} - \dfrac{(1+\lambda)^{\beta-1}\lambda^{\alpha-\beta+1}}{(1+\lambda)^\alpha - \lambda^\alpha},$ $0 < \alpha, \beta < 1$	(5.2), (5.3) of [50], Theorem 8.2 (v). See §16.12.1
13	$\dfrac{\alpha - 1}{\alpha} \dfrac{\lambda^\alpha - 1}{\lambda^{\alpha-1} - 1}, \quad -1 \le \alpha \le 2, \alpha \ne 0, 1$	Theorem A of [126], Corollary 6.3
14	$\dfrac{\pi}{\sin(\alpha\pi)} \dfrac{\lambda(c^\alpha - \lambda^\alpha)}{(c - \lambda)}, \quad c > 0, -1 < \alpha < 1$	4.3(7) and 14.2(6) of [107], Theorem 8.2 (v)
15	$\dfrac{\lambda}{c^2 + \lambda^2}\left(\dfrac{c^{\alpha-1}\lambda}{2\cos(\alpha\pi/2)} + \dfrac{c^\alpha}{2\sin(\alpha\pi/2)} - \dfrac{\lambda^\alpha}{\sin(\alpha\pi)}\right),$ $c > 0, -1 < \alpha < 2, \alpha \ne 1$	4.3(9) and 14.2(7) of [107], Theorem 8.2 (v)

Lévy (L), Stieltjes (S) and Thorin (T) representation measures

---

L: closed expression unknown

S: $$\dfrac{t^{\alpha+\beta}\sin((\beta-\alpha)\pi)-t^{\beta}\sin(\beta\pi)+t^{\alpha}\sin(\alpha\pi)}{\pi t\,(t^{2\alpha}-2t^{\alpha}\cos(\alpha\pi)+1)}$$

T: unknown if in $\mathcal{TBF}$

L: closed expression unknown

S: $$\dfrac{\dfrac{(1-t)^{\alpha}}{t^{1-\alpha}}\sin(\pi\alpha)+\dfrac{(1-t)^{\beta-1}}{t^{\beta-2\alpha}}\sin(\pi\beta)+\dfrac{(1-t)^{\alpha+\beta-1}}{t^{\beta-\alpha}}\sin(\pi(\alpha-\beta))}{\pi\left[(1-t)^{2\alpha}-2(1-t)^{\alpha}\,t^{\alpha}\cos(\pi\alpha)+t^{2\alpha}\right]}\mathbb{1}_{(0,1)}(t)$$

T: $f\notin\mathcal{TBF}$

L: closed expression unknown

S: $$\dfrac{1-\alpha}{\alpha\pi}\,\dfrac{t^{\alpha-2}(1+t)\sin(\alpha\pi)}{t^{2(\alpha-1)}+2t^{\alpha-1}\cos(\alpha\pi)+1}$$

T: unknown if in $\mathcal{TBF}$

L: $\Gamma(\alpha+2)\,c^{\alpha+1}e^{ct}\,\Gamma(-\alpha-1,ct)$

S: $$\dfrac{t^{\alpha}}{t+c}$$

T: $\dfrac{t^{\alpha}(c+ct+\alpha t)}{(t+c)^2}$, $0<\alpha<1;\,f\notin\mathcal{TBF},\,-1<\alpha<0$

L: $c^{\alpha}\csc(\alpha\pi)\,V_{\alpha+2}(2ct,0)$

S: $$\dfrac{1}{\pi}\dfrac{t^{\alpha}}{c^2+t^2}$$

T: $\dfrac{1}{\pi}\dfrac{(\alpha+1)\,c^2 t^{\alpha}+(\alpha-1)t^{\alpha+2}}{(c^2+t^2)^2}$, $1<\alpha<2;$ $f\notin\mathcal{TBF},\,-1<\alpha<1$

No	Function $f(\lambda)$	Comment
16	$(\lambda^{-\alpha_1} + \lambda^{-\alpha_2} + \cdots + \lambda^{-\alpha_n})^{-1}$, $$0 \le \alpha_1, \ldots, \alpha_n \le 1$$	Corollary 6.3
17	$\left(\dfrac{1}{\lambda^{\alpha_1} + \cdots + \lambda^{\alpha_m}} + \dfrac{1}{\lambda^{\beta_1} + \cdots + \lambda^{\beta_n}} \right)^{-1}$, $$0 \le \alpha_1, \ldots, \alpha_m \le 1,\ 0 \le \beta_1, \ldots, \beta_n \le 1$$	Corollary 6.3

Lévy (L), Stieltjes (S) and Thorin (T) representation measures

L: closed expression unknown

S:

$$\frac{\frac{1}{t\pi}\left[t^{\sum_{j=1}^{n}\alpha_j}\sin(\sum_{j=1}^{n}\alpha_j\pi)\right]\left[\sum_{j=1}^{n}t^{\sum_{k\neq j}\alpha_k}\cos(\sum_{k\neq j}\alpha_k\pi)\right]}{\left[\sum_{j=1}^{n}t^{\sum_{k\neq j}\alpha_k}\cos(\sum_{k\neq j}\alpha_k\pi)\right]^2+\left[\sum_{j=1}^{n}t^{\sum_{k\neq j}\alpha_k}\sin(\sum_{k\neq j}\alpha_k\pi)\right]^2}$$

$$-\frac{\frac{1}{t\pi}\left[t^{\sum_{j=1}^{n}\alpha_j}\cos(\sum_{j=1}^{n}\alpha_j\pi)\right]\left[\sum_{j=1}^{n}t^{\sum_{k\neq j}\alpha_k}\sin(\sum_{k\neq j}\alpha_k\pi)\right]}{\left[\sum_{j=1}^{n}t^{\sum_{k\neq j}\alpha_k}\cos(\sum_{k\neq j}\alpha_k\pi)\right]^2+\left[\sum_{j=1}^{n}t^{\sum_{k\neq j}\alpha_k}\sin(\sum_{k\neq j}\alpha_k\pi)\right]^2}$$

T: depends on the choice of the parameters $\alpha_1,\dots,\alpha_n$

L: closed expression unknown

S:

$$\frac{\frac{1}{t\pi}\left[\sum_{j,k}t^{\alpha_j+\beta_k}\sin((\alpha_j+\beta_k)\pi)\right]\left[\sum_{j=1}^{m}t^{\alpha_j}\cos(\alpha_j\pi)+\sum_{k=1}^{n}t^{\beta_k}\cos(\beta_k\pi)\right]}{\left[\sum_{j=1}^{m}t^{\alpha_j}\cos(\alpha_j\pi)+\sum_{k=1}^{n}t^{\beta_k}\cos(\beta_k\pi)\right]^2+\left[\sum_{j=1}^{m}t^{\alpha_j}\sin(\alpha_j\pi)+\sum_{k=1}^{n}t^{\beta_k}\sin(\beta_k\pi)\right]^2}$$

$$-\frac{\frac{1}{t\pi}\left[\sum_{j,k}t^{\alpha_j+\beta_k}\cos((\alpha_j+\beta_k)\pi)\right]\left[\sum_{j=1}^{m}t^{\alpha_j}\sin(\alpha_j\pi)+\sum_{k=1}^{n}t^{\beta_k}\sin(\beta_k\pi)\right]}{\left[\sum_{j=1}^{m}t^{\alpha_j}\cos(\alpha_j\pi)+\sum_{k=1}^{n}t^{\beta_k}\cos(\beta_k\pi)\right]^2+\left[\sum_{j=1}^{m}t^{\alpha_j}\sin(\alpha_j\pi)+\sum_{k=1}^{n}t^{\beta_k}\sin(\beta_k\pi)\right]^2}$$

T: depends on the choice of the parameters $\alpha_1,\dots,\alpha_m,\beta_1,\dots,\beta_n$

16.3 Exponential functions

No	Function $f(\lambda)$	Comment
18	$\sqrt{\lambda}\,(1 - e^{-2a\sqrt{\lambda}}),\quad a > 0$	14.2(43) of [107], 2.25 and 7.78 in [283], Theorem 8.2 (v)
19	$\sqrt{\lambda}\,(1 + e^{-2a\sqrt{\lambda}}),\quad a > 0$	14.60(3) of [107], 2.25 and 7.78 in[283], Theorem 8.2 (v)
20	$\dfrac{\lambda\,(1 - e^{-2\sqrt{\lambda+a}})}{\sqrt{\lambda + a}},\quad a > 0$	Appendix 1.17 of [68], Theorem 8.2 (v). See §16.12.2
21	$\lambda\,(1 + \lambda)^{1/\lambda} - \lambda - \dfrac{\lambda}{\lambda + 1}$	[5], p. 457, Theorem 8.2 (v)
22	$e\lambda - \lambda\left(1 + \dfrac{1}{\lambda}\right)^{\lambda} - \dfrac{\lambda}{\lambda + 1}$	Theorem 3 of [5], Theorem 8.2 (v)

| **Lévy (L), Stieltjes (S) and Thorin (T) representation measures** |

L: $\dfrac{1}{2\sqrt{\pi}t^{5/2}}\left((2a^2-t)e^{-a^2/t}+t\right)$

S: $\dfrac{2}{\pi\sqrt{t}}\sin^2(a\sqrt{t})$

T: $f\notin \mathfrak{TBF}$

L: $\dfrac{1}{2\sqrt{\pi}t^{5/2}}\left((t-2a^2)e^{-a^2/t}+t\right)$

S: $\dfrac{2}{\pi\sqrt{t}}\cos^2(a\sqrt{t})$

T: $f\notin \mathfrak{TBF}$

L: $\dfrac{1}{2\sqrt{\pi}\,t^{5/2}}\left(e^{-1/t-at}\left(2+t\left(e^{1/t}-1\right)(1+2at)\right)\right)$

S: $\dfrac{2}{\pi}\dfrac{\sin^2(\sqrt{t-a})}{\sqrt{t-a}}\,\mathbb{1}_{(a,\infty)}(t)$

T: $f\notin \mathfrak{TBF}$

L: closed expression unknown

S: $\dfrac{1}{\pi}\dfrac{\sin(\pi/t)}{(t-1)^{1/t}}\,\mathbb{1}_{(1,\infty)}(t)$

T: $f\notin \mathfrak{TBF}$

L: closed expression unknown

S: $\dfrac{1}{\pi}\left(\dfrac{t}{1-t}\right)^{t}\sin(\pi t)\,\mathbb{1}_{(0,1)}(t)$

T: $f\notin \mathfrak{TBF}$

No	Function $f(\lambda)$	Comment
23	$\lambda\left(1+\dfrac{1}{\lambda}\right)^{\lambda+1} - e\lambda - 1$	Theorem 3 of [5], Theorem 8.2 (v)
24	$\lambda(\lambda+1)\left(e-\left(1+\dfrac{1}{\lambda}\right)^{\lambda}\right) - \dfrac{e\lambda}{2}$	Lemma 1 of [5], Theorem 8.2 (v)
25	$\lambda(\lambda+1)\left(\left(1+\dfrac{1}{\lambda}\right)^{\lambda+1} - e\right) - \dfrac{e\lambda}{2} - 1$	Remark before Theorem 3 of [5], Theorem 8.2 (v)

16.4   Logarithmic functions

No	Function $f(\lambda)$	Comment
26	$\log\left(1+\dfrac{\lambda}{a}\right), \quad a>0$	

Lévy (L), Stieltjes (S) and Thorin (T) representation measures

---

L: closed expression unknown

S: $\dfrac{1}{\pi} \left(\dfrac{t}{1-t} \right)^{t-1} \sin(\pi t) \, \mathbb{1}_{(0,1)}(t)$

T: $f \notin \mathcal{TBF}$

L: closed expression unknown

S: $\dfrac{1}{\pi} t^t (1-t)^{1-t} \sin(\pi t) \, \mathbb{1}_{(0,1)}(t)$

T: $f \notin \mathcal{TBF}$

L: closed expression unknown

S: $\dfrac{1}{\pi} t^{t-1} (1-t)^{-t} \sin(\pi t) \, \mathbb{1}_{(0,1)}(t)$

T: $f \notin \mathcal{TBF}$

Lévy (L), Stieltjes (S) and Thorin (T) representation measures

---

L: $e^{-at} \dfrac{1}{t}$

S: $\dfrac{1}{t} \mathbb{1}_{(a,\infty)}(t)$

T: $\delta_a(dt)$

No	Function $f(\lambda)$	Comment
27	$\lambda \log \left(1 + \dfrac{a}{\lambda}\right), \quad a > 0$	Theorem 8.2 (v)
28	$\log \dfrac{b(\lambda + a)}{a(\lambda + b)}, \quad 0 < a < b$	Theorem 8.2 (v)
29	$\dfrac{1}{a} - \dfrac{1}{\lambda} \log \left(1 + \dfrac{\lambda}{a}\right), \quad a > 0$	Theorem 1.7.3 of [50], Theorem 3.8* of [187], Theorem 8.2 (v). See §16.12.1
30	$\dfrac{a\lambda}{a + \lambda} \log \left(1 + \dfrac{\lambda}{a}\right) - \dfrac{1}{a}, \quad a > 0$	Theorem 3.8* of [187], Remark 8.3 (ii), (iii). See §16.12.1
31	$(\lambda + a) \log(\lambda + a) - \lambda \log \lambda - a \log a,$ $\quad a > 0$	Theorem 3.8 of [187]. See §16.12.1

Lévy (L), Stieltjes (S) and Thorin (T) representation measures

L: $\dfrac{1 - e^{-at}(1 + at)}{t^2}$

S: $\mathbb{1}_{(0,a)}(t)$

T: $f \notin \mathcal{TBF}$

L: $(e^{-at} - e^{-bt})\dfrac{1}{t}$

S: $\dfrac{1}{t}\,\mathbb{1}_{(a,b)}(t)$

T: $f \notin \mathcal{TBF}$

L: $-\operatorname{Ei}(-at)$

S: $\dfrac{1}{t^2}\,\mathbb{1}_{(a,\infty)}(t)$

T: $f \notin \mathcal{TBF}$

L: $\dfrac{e^{-at}}{at} + \operatorname{Ei}(-at)$

S: $\dfrac{t - a}{at^2}\,\mathbb{1}_{(a,\infty)}(t)$

T: $\dfrac{1}{t^2}\,\mathbb{1}_{(a,\infty)}(t)$

L: $\dfrac{1 - e^{-at}}{t^2}$

S: $\dfrac{t \wedge a}{t}$

T: $\mathbb{1}_{(0,a)}(t)$

No	Function $f(\lambda)$	Comment
32	$(\lambda + b)\log(\lambda + b) - b\log b$ $- (\lambda + a)\log(\lambda + a) + a\log a,$ $0 < a < b$	
33	$\dfrac{\lambda}{a - \lambda}\log\left(\dfrac{a}{\lambda}\right), \quad a > 0$	14.2(2) in [107], 4.2(11) in [107]
34	$\dfrac{\lambda}{a^2 + \lambda^2}\left(\dfrac{\pi\lambda}{2a} - \log\left(\dfrac{\lambda}{a}\right)\right),\; a > 0$	14.2(3) in [107], 4.2(14) in [107]
35	$\dfrac{\lambda}{a^2 + \lambda^2}\left(\dfrac{\pi a}{2} + \lambda\log\left(\dfrac{\lambda}{a}\right)\right),\; a > 0$	14.2(4) in [107], 4.1(6) in [107]
36	$\dfrac{\lambda}{a + \lambda}\left(\pi^2 + \log^2\left(\dfrac{\lambda}{a}\right)\right), \quad a > 0$	14.2(26) in [107]

Lévy (L), Stieltjes (S) and Thorin (T) representation measures

L: $\dfrac{e^{-at} - e^{-bt}}{t^2}$

S: $\dfrac{(t-a)^+ \wedge (b-a)}{t}$

T: $\mathbb{1}_{(a,b)}(t)$

L: $ae^{at}\,\mathrm{Ei}(-at) + t^{-1}$

S: $\dfrac{1}{a+t}$

T: $\dfrac{a}{(a+t)^2}$

L: $\cos(at)\,\mathrm{ci}(at) - \sin(at)\,\mathrm{si}(at)$

S: $\dfrac{1}{t^2 + a^2}$

T: $f \notin \mathfrak{TBF}$

L: $-\dfrac{d}{dt}\big(\cos(at)\,\mathrm{ci}(at) - \sin(at)\,\mathrm{si}(at)\big)$

S: $\dfrac{t}{t^2 + a^2}$

T: $\dfrac{2a^2 t}{(t^2 + a^2)^2}$

L: closed expression unknown

S: $\dfrac{1}{2}\dfrac{1}{t-a}\log\left(\dfrac{t}{a}\right)$

T: $\dfrac{1}{2}\dfrac{t - a - a\log(t/a)}{(t-a)^2}$

No	Function $f(\lambda)$	Comment
37	$\dfrac{\log(\sqrt{a\lambda} + \sqrt{b})}{\sqrt{\lambda}}, \quad a > 0,\ b \geq 1$	14.2(27) in [107]
38	$\dfrac{1}{2}(a^2 + a\sqrt{a^2 + 2\lambda})$ $+ \lambda \log\left(\sqrt{1 + \dfrac{a^2}{2\lambda}} + \dfrac{a}{\sqrt{2\lambda}}\right) - a^2,$ $a > 0$	Theorem 2.2 of [210]
39	$\log\left(2(1 - a)\right)$ $- \log\left(1 + \lambda - \sqrt{(1 + \lambda)^2 - 4a(1 - a)}\right),$ $\dfrac{1}{2} < a < 1$	Example 3.2.3 of [67]
40	$\dfrac{1}{(1 + \lambda)\log(1 + 1/\lambda)}$	(1.30) of [50], Theorem 8.2 (v). See §16.12.1

Lévy (L), Stieltjes (S) and Thorin (T) representation measures

---

L: closed expression unknown

S: $\dfrac{\log(at+b)}{2\pi\sqrt{t}}$

T: $\dfrac{a\sqrt{t}}{2\pi(at+b)} + \dfrac{\log(at+b)}{4\pi\sqrt{t}}$

L: $\dfrac{1}{2t^2}\,\mathrm{Erf}(a\sqrt{t/2})$

S: $\begin{cases} \frac{1}{2}, & t \le a^2/2 \\ \frac{1}{\pi}\left(\frac{\pi}{2} - \cos^{-1}(a(2t)^{-1/2}) + a(2t)^{-1/2}\sqrt{1-a^2/2t}\right), & t > a^2/2 \end{cases}$

T: $\begin{cases} \frac{1}{2}, & t \le a^2/2 \\ \frac{1}{\pi}\sin^{-1}\left(\frac{a}{\sqrt{2t}}\right), & t > a^2/2 \end{cases}$

L: $\dfrac{1}{t}B(1/2,1/2)\left(4\sqrt{a(1-a)}\right)^{-1/2}t^{-1/2}e^{-t}M_{0,0}\!\left(4t\sqrt{a(1-a)}\right)$

S: closed expression unknown

T: $\dfrac{1}{\pi}\,\dfrac{\mathbb{1}_{\left(1-2\sqrt{a(1-a)},\,1+2\sqrt{a(1-a)}\right)}(t)}{\sqrt{t-1+2\sqrt{a(1-a)}}\,\sqrt{1+2\sqrt{a(1-a)}-t}}$

L: closed expression unknown

S: $\dfrac{1}{t(1-t)}\,\dfrac{1}{\pi^2 + \log^2(-1+1/t)}\,\mathbb{1}_{(0,1)}(t)$

T: $f \notin \mathfrak{TBF}$

No	Function $f(\lambda)$	Comment
41	$\dfrac{\lambda - 1}{\log \lambda}$	Theorem A of [126], Corollary 6.3
42	$\dfrac{\lambda \log \lambda - \lambda + 1}{\log^2 \lambda}$	Theorem A of [125]
43	$\dfrac{\lambda - 1 - \log \lambda}{\log^2 \lambda}$	[126]
44	$\dfrac{\lambda \log^2 \lambda}{\lambda - 1 - \log \lambda}$	[126]
45	$\lambda(1 + \lambda) \log \left(1 + \dfrac{1}{\lambda} \right) - \lambda$	Theorem 2.1 of [126], Corollary 6.3, Theorem 8.2 (v)
46	$\lambda \left(1 - \lambda \log \left(1 + \dfrac{1}{\lambda} \right) \right)$	
47	$\dfrac{\lambda(\lambda + 1)}{(\lambda + 2) \log(\lambda + 2)}$	Theorem 2.4 of [126]

Lévy (L), Stieltjes (S) and Thorin (T) representation measures

---

L: closed expression unknown

S: $\dfrac{t+1}{t(\log^2 t + \pi^2)}$

T: $\dfrac{\pi^2 + \log^2 t - \frac{2}{t}\log t - 2\log t}{(\pi^2 + \log^2 t)^2}$

L, T: closed expression unknown

S: $\dfrac{\pi^2 - 2(1 + 1/t)\log t + \log^2 t}{(\pi^2 + \log^2 t)^2}$

L, S, T: closed expression unknown

L, S: closed expression unknown   —   **T:**   unknown if in $\mathcal{TBF}$

L: $\dfrac{e^{-t}(2+t) + t - 2}{t^3}$

S: $(1-t)\,\mathbb{1}_{(0,1)}(t)$

T: $f \notin \mathcal{TBF}$

L: $\dfrac{2 - e^{-t}(2 + 2t + t^2)}{t^3}$

S: $t\,\mathbb{1}_{(0,1)}(t)$

T: $f \notin \mathcal{TBF}$

L, S, T: closed expression unknown

No	Function $f(\lambda)$	Comment
48	$\dfrac{\lambda(\lambda + 2) \log(\lambda + 2)}{(\lambda + 1)^2}$	Theorem 2.3 of [126]
49	$\dfrac{\lambda}{(\lambda + a)(\lambda + a + 1) \log \frac{\lambda+a+1}{\lambda+a}}$, $a > 0$	Theorem 1.7.3 of [50], Theorem 3.6 of [187], Theorem 8.2 (v). See §16.12.1
50	$1 - \dfrac{(a + 1)\lambda}{(1 + (a + 1)\lambda) \log(1 + (a + 1)\lambda)}$ $+ \dfrac{a\lambda}{(1 + a\lambda) \log(1 + a\lambda)}$, $a > 0$	Theorem 1.7.3 of [50], Theorem 3.6* of [187], Theorem 8.2 (v). See §16.12.1
51	$\log \log \left(1 + \dfrac{1}{a}\right) - \log \log \left(1 + \dfrac{1}{\lambda + a}\right)$, $a > 0$	Theorem 3.6 of [187], Theorem 1.2 of [50]. See §16.12.1

Lévy (L), Stieltjes (S) and Thorin (T) representation measures

L, S: closed expression unknown   —   **T:**   unknown if in $\mathcal{TBF}$

L: closed expression unknown

S: $\dfrac{1}{(t-a)(1+a-t)} \dfrac{1}{\pi^2 + (\log \frac{1+a-t}{t-a})^2} \mathbb{1}_{(a,1+a)}(t)$

T: $f \notin \mathcal{TBF}$

L: closed expression unknown

S: $\dfrac{1}{(1-at)((1+a)t - 1)} \dfrac{1}{\pi^2 + (\log \frac{(1+a)t-1}{1-at})^2} \mathbb{1}_{(1/(1+a),\, 1/a)}(t)$

T: $f \notin \mathcal{TBF}$

L: closed expression unknown

S: $\begin{cases} 0, & 0 < t < a \\[2mm] \dfrac{1}{t}\left(\dfrac{1}{2} - \dfrac{\arctan\left(\frac{1}{\pi} \log \frac{a-t+1}{t-a}\right)}{\pi} \right), & a \le t \le a+1 \\[3mm] \dfrac{1}{t}, & t > a+1 \end{cases}$

T: $\dfrac{1}{(t-a)(1+a-t)} \dfrac{1}{\pi^2 + \left(\log \frac{1+a-t}{t-a}\right)^2} \mathbb{1}_{(a,1+a)}(t)$

No	Function $f(\lambda)$	Comment
52	$\log \lambda - \log\log \dfrac{1 + (1+a)\lambda}{1 + a\lambda}$, $a > 0$	Theorem 3.6* of [187], Theorem 1.2 of [50]. See §16.12.1
53	$\dfrac{\left(1 - \lambda\, e_n\left(\log\frac{1}{\lambda}\right)\right)}{\log^{n+1}\left(\frac{1}{\lambda}\right)}, \quad n \in \mathbb{N}$	[124, 125, 126]
54	$\dfrac{\lambda - e_n(\log \lambda)}{\log^{n+1}\lambda}, \quad n \in \mathbb{N}$	[124, 125, 126]
55	$\dfrac{\lambda \log^{n+1}\lambda}{\lambda - e_n(\log \lambda)}, \quad n \in \mathbb{N}$	[124, 125, 126]
56	$\dfrac{\lambda \log^{n+1}\left(\frac{1}{\lambda}\right)}{1 - \lambda e_n\left(\log\frac{1}{\lambda}\right)}, \quad n \in \mathbb{N}$	[124, 125, 126]
57	$\log\left(1 + 2\lambda + 2\sqrt{\lambda(1+\lambda)}\right)$	Theorem 3.1 of [187]. See §16.12.1

Lévy (L), Stieltjes (S) and Thorin (T) representation measures

---

L: closed expression unknown

S:
$$\begin{cases} 0, & 0 < t < 1/(1+a) \\ \dfrac{1}{t}\left(\dfrac{1}{2} + \dfrac{\arctan\left(\frac{1}{\pi}\log\frac{(1+a)t-1}{1-at}\right)}{\pi}\right), & 1/(1+a) \le t \le 1/a \\ \dfrac{1}{t}, & t > 1/a \end{cases}$$

T: $\dfrac{1}{(1-at)((1+a)t-1)}\ \dfrac{1}{\pi^2 + \left(\log\frac{(1+a)t-1}{1-at}\right)^2}\ \mathbb{1}_{(1/(1+a),1/a)}(t)$

L, S, T: closed expression unknown

L, S: closed expression unknown   —   **T:**   unknown if in $\mathcal{TBF}$

L, S: closed expression unknown   —   **T:**   unknown if in $\mathcal{TBF}$

L, S: closed expression unknown   —   **T:**   unknown if in $\mathcal{TBF}$

L: $t^{-1}e^{-t/2}I_0\left(\dfrac{t}{2}\right)$

S:
$$\begin{cases} \dfrac{2}{\pi t}\arcsin(\sqrt{t}), & 0 < t < 1 \\ \dfrac{1}{t}, & t \ge 1 \end{cases}$$

T: $\dfrac{1}{\pi}\ \dfrac{1}{\sqrt{t(1-t)}}\ \mathbb{1}_{(0,1)}(t)$

No	Function $f(\lambda)$	Comment
58	$2\log\dfrac{1+\sqrt{1+\lambda}}{2}$	Theorem 3.1* of [187]. See §16.12.1
59	$\log(1+\lambda^{\alpha}), \quad 0 < \alpha < 1$	Theorem 2.2 of [188], Example 3.2.4 of [67], Theorem 8.2 (v)
60	$\log\dfrac{1}{(1+\lambda)^{\alpha}-\lambda^{\alpha}}, \quad 0 < \alpha < 1$	(3.7) in Section 3.1.a of [187]. See §16.12.1
61	$\log\dfrac{\alpha\lambda}{(1+\lambda)^{\alpha}-1}, \quad 0 < \alpha < 1$	Theorem 3.4* of [187]. See §16.12.1
62	$\dfrac{\lambda}{c+\lambda}\left(\dfrac{\lambda^{\alpha}}{\sin(\alpha\pi)} - c^{\alpha}\cot(\alpha\pi) + \dfrac{c^{\alpha}}{\pi}\log\left(\dfrac{c}{\lambda}\right)\right),$ $c > 0,\ 0 < \alpha < 1$	14.2(8) in [107]

Lévy (L), Stieltjes (S) and Thorin (T) representation measures

---

L: closed expression unknown

S: $\dfrac{1}{t}\left(1-\dfrac{2}{\pi}\arcsin\left(\dfrac{1}{\sqrt{t}}\right)\right)\mathbb{1}_{(1,\infty)}(t)$

T: $\dfrac{1}{\pi}\dfrac{1}{t\sqrt{t-1}}\mathbb{1}_{(1,\infty)}(t)$

L: $\dfrac{\alpha}{2t}E_{\alpha/2,1}(-t^{\alpha/2})$

S: $\dfrac{1}{\pi t}\arccos\dfrac{1+t^{\alpha}\cos(\alpha\pi)}{(t^{2\alpha}+2t^{\alpha}\cos(\alpha\pi)+1)^{1/2}}$

T: $\dfrac{1}{\pi}\dfrac{\alpha t^{\alpha-1}\sin(\alpha\pi)}{1+t^{2\alpha}+2t^{\alpha}\cos(\alpha\pi)}$

L, S: closed expression unknown

T: $\dfrac{\alpha\sin(\alpha\pi)}{\pi}\dfrac{t^{\alpha-1}(1-t)^{\alpha-1}}{(1-t)^{2\alpha}-2(1-t)^{\alpha}t^{\alpha}\cos(\pi\alpha)+t^{2\alpha}}\mathbb{1}_{(0,1)}(t)$

L, S: closed expression unknown

T: $\dfrac{\alpha\sin(\alpha\pi)}{\pi}\dfrac{(t-1)^{\alpha-1}}{(t-1)^{2\alpha}-2(t-1)^{\alpha}\cos(\alpha\pi)+1}\mathbb{1}_{(1,\infty)}(t)$

L: closed expression unknown

S: $\dfrac{1}{\pi}\dfrac{t^{\alpha}-c^{\alpha}}{t-c}$

T: $\dfrac{1}{\pi}\dfrac{\alpha t^{\alpha+1}-c(1+\alpha)t^{\alpha}+c^{1+\alpha}}{(t-c)^{2}}$

16.5   Inverse trigonometric functions

No	Function $f(\lambda)$	Comment
63	$\sqrt{\lambda}\arctan\left(\sqrt{\dfrac{a}{\lambda}}\right), \quad a > 0$	
64	$\sqrt{\lambda}\arctan\left(\sqrt{\dfrac{\lambda}{a}}\right), \quad a > 0$	

16.6   Hyperbolic functions

No	Function $f(\lambda)$	Comment
65	$\sqrt{\dfrac{\lambda}{2}\dfrac{\cosh^2(\sqrt{2\lambda})}{\sinh(2\sqrt{2\lambda})} - \dfrac{1}{4}}$	Appendix 1.5 of [68], Theorem 8.2 (v). See §16.12.2

Lévy (L), Stieltjes (S) and Thorin (T) representation measures

L: $\dfrac{1}{2} t^{-3/2} \gamma \left(\dfrac{3}{2}, at \right) = \dfrac{\sqrt{\pi}\, \mathrm{Erf}\left(\sqrt{at} \right)}{4 t^{3/2}} - \dfrac{e^{-at}}{2t}$

S: $\dfrac{1}{2\sqrt{t}}\, \mathbb{1}_{(0,a)}(t)$

T: $f \notin \mathcal{TBF}$

L: $\dfrac{2e^{-at}\sqrt{at} + \sqrt{\pi}\, \mathrm{Erfc}\left(\sqrt{at} \right)}{4 t^{3/2}}$

S: $\dfrac{1}{2\sqrt{t}}\, \mathbb{1}_{(a,\infty)}(t)$

T: $\dfrac{1}{2}\sqrt{a}\, \delta_a(dt) + \dfrac{1}{4\sqrt{t}}\, \mathbb{1}_{(a,\infty)}(t)\, dt$

Lévy (L), Stieltjes (S) and Thorin (T) representation measures

L: $\displaystyle\sum_{n=1}^{\infty} \dfrac{n^2 \pi^2}{16} \cos^2 \left(\dfrac{n\pi}{2} \right) e^{-n^2 \pi^2 t/8}$

S: $\displaystyle\sum_{n=1}^{\infty} \dfrac{1}{2} \cos^2 \left(\dfrac{n\pi}{2} \right) \delta_{n^2 \pi^2/8}(dt)$

T: $f \notin \mathcal{TBF}$

No	Function $f(\lambda)$	Comment
66	$\sqrt{\dfrac{\lambda}{2}}\,\dfrac{\sinh^2(\sqrt{2\lambda})}{\sinh(2\sqrt{2\lambda})}$	Appendix 1.6 of [68], Theorem 8.2 (v). See §16.12.2
67	$\dfrac{\sqrt{\lambda}}{2}\coth\left(\dfrac{1}{2\sqrt{\lambda}}\right)-\lambda$	Example 2.2 of [65], Theorem 8.2 (v)
68	$\sqrt{\lambda}\,\dfrac{a+\tanh(b\sqrt{\lambda})}{1+a\tanh(b\sqrt{\lambda})},\quad a,b>0$	Example 2, Section 2.3 of [231] and Theorem 8.2 (v)
69	$\dfrac{\pi\sqrt{\lambda}\,(\delta\sinh\left(2a\sqrt{\lambda}\right)-1)}{b^2\sinh^2\left(a\sqrt{\lambda}\right)-c^2\cosh^2\left(a\sqrt{\lambda}\right)},$ $a,b,c>0,\ \delta:=\dfrac{1}{2}\left(\dfrac{b}{c}-\dfrac{c}{b}\right)$	14.2(63) in [107], Theorem 8.2 (v)

Lévy (L), Stieltjes (S) and Thorin (T) representation measures

L: $\displaystyle\sum_{n=1}^{\infty} \frac{n^2\pi^2}{16} \sin^2\left(\frac{n\pi}{2}\right) e^{-n^2\pi^2 t/8}$

S: $\displaystyle\sum_{n=1}^{\infty} \frac{1}{2} \sin^2\left(\frac{n\pi}{2}\right) \delta_{n^2\pi^2/8}(dt)$

T: $f \notin \mathcal{TBF}$

L: $\displaystyle\sum_{n=1}^{\infty} \frac{1}{8\pi^4 n^4} e^{-t/(4\pi^2 n^2)}$

S: $\displaystyle\sum_{n=1}^{\infty} \frac{1}{2\pi^2 n^2} \delta_{1/4\pi^2 n^2}(dt)$

T: $f \notin \mathcal{TBF}$

L: closed expression unknown

S: $\dfrac{2a}{\pi\sqrt{t}\,\left(1 + a^2 + (1 - a^2)\cos\left(2b\sqrt{t}\right)\right)}$

T: $f \notin \mathcal{TBF}$

L: closed expression unknown

S: $\dfrac{1}{\sqrt{t}\,\left(b^2 \sin^2\left(a\sqrt{t}\right) + c^2 \cos^2\left(a\sqrt{t}\right)\right)}$

T: $f \notin \mathcal{TBF}$

No	Function $f(\lambda)$	Comment
70	$\log\left(\sinh(\sqrt{2\lambda})\right) - \log(\sqrt{2\lambda})$	1.431(2) of [134], 5.5(1) of [107], Corollary 6.3
71	$\log\left(\cosh(\sqrt{2\lambda})\right)$	1.431(4) of [134], 5.5(1) of [107], Corollary 6.3
72	$\log(\sqrt{2\lambda}) - \log\left(\tanh(\sqrt{2\lambda})\right)$	Entry 70, Theorem 8.2 (v)
73	$\sqrt{\lambda}\,\log\left(1 + b\tanh(a\sqrt{\lambda})\right),$ $a, b > 0$	14.2(71) of [107], Theorem 8.2 (v)

Lévy (L), Stieltjes (S) and Thorin (T) representation measures

---

L: $\displaystyle\sum_{n=1}^{\infty}\frac{1}{t}e^{-\pi^2 n^2 t/2}$

S: $\displaystyle\sum_{n=1}^{\infty}\frac{1}{t}\mathbb{1}_{(\pi^2 n^2/2,\,\infty)}(t)$

T: $\displaystyle\sum_{n=1}^{\infty}\delta_{\pi^2 n^2/2}(dt)$

L: $\displaystyle\sum_{n=1}^{\infty}\frac{1}{t}e^{-\pi^2(n-\frac{1}{2})^2 t/2}$

S: $\displaystyle\sum_{n=1}^{\infty}\frac{1}{t}\mathbb{1}_{(\pi^2(n-1/2)^2/2,\,\infty)}(t)$

T: $\displaystyle\sum_{n=1}^{\infty}\delta_{\pi^2(n-1/2)^2/2}(dt)$

L: $\displaystyle\sum_{n=1}^{\infty}\frac{1}{t}e^{-\pi^2(n-\frac{1}{2})^2 t/2}-\sum_{n=1}^{\infty}\frac{1}{t}e^{-\pi^2 n^2 t/2}$

S: $\displaystyle\sum_{n=1}^{\infty}\frac{1}{t}\mathbb{1}_{(\pi^2(n-1/2)^2/2,\,\pi^2 n^2/2)}(t)\,dt$

T: $f\notin\mathcal{TBF}$

L: closed expression unknown

S: $\displaystyle\frac{1}{2\pi\sqrt{t}}\log\left(1+b^2\tan^2(a\sqrt{t})\right)$

T: $f\notin\mathcal{TBF}$

No	Function $f(\lambda)$	Comment
74	$\sqrt{\lambda}\,\log\left(1 + b\coth(a\sqrt{\lambda})\right),$ $a,b > 0$	14.2(72b) of [107], Theorem 8.2 (v)

No	Function $f(\lambda)$	Comment
75	$\sqrt{\lambda}\,\log\left(b\sinh(a\sqrt{\lambda})\right.$ $\left. + c\cosh(a\sqrt{\lambda})\right) - a\lambda,$ $a > 0,\ b,c \geq 1$	14.2(68) in [107], Theorem 8.2 (v)

16.7   Inverse hyperbolic functions

No	Function $f(\lambda)$	Comment
76	$\sinh^{-1}(\lambda)$ $= \log(\sqrt{1 + \lambda} + \sqrt{\lambda})$	

Lévy (L), Stieltjes (S) and Thorin (T) representation measures

---

L: closed expression unknown

S: $\dfrac{1}{2\pi\sqrt{t}}\,\log\left(1+b^2\cot^2(a\sqrt{t})\right)$

T: $f\notin\mathcal{TBF}$

L: closed expression unknown

S: $\dfrac{1}{2\pi\sqrt{t}}\,\log\left(b^2\sin^2(a\sqrt{t})+c^2\cos^2(a\sqrt{t})\right)$

T: $f\notin\mathcal{TBF},\quad b\neq c$

Lévy (L), Stieltjes (S) and Thorin (T) representation measures

---

L: closed expression unknown

S: $\begin{cases}\dfrac{\arcsin(\sqrt{t})}{\pi t}, & 0<t<1\\[2ex]\dfrac{1}{2t}, & t\geq 1\end{cases}$

T: $\dfrac{1}{2\pi\sqrt{t(1-t)}}\,\mathbb{1}_{(0,1)}(t)$

No	Function $f(\lambda)$	Comment
77	$\cosh^{-1}(\lambda + 1)$ $= \log\left(\lambda + 1 + \sqrt{\lambda(\lambda + 2)}\right)$	Theorem 1.1 of [262], Corollary 6.3
78	$\dfrac{1}{2}\left(\cosh^{-1}(\lambda + 1)\right)^2$	Theorem 1.3 of [262], 5.15(8), (9) of [107]
79	$\sqrt{\dfrac{\lambda}{\lambda + 2}}\;\cosh^{-1}(\lambda + 1)$	Theorem 8.2 (v)
80	$\log\left(\sqrt{(\lambda + 1)^2 - 1}\right)$ $\quad - \log\left(\cosh^{-1}(\lambda + 1)\right)$	See §16.12.3 below: combine entries 70 and 77, Proposition 6.2 of [262]
81	$\log(\lambda + 1) - \log\left(\sqrt{(\lambda + 1)^2 - 1}\right)$ $\quad + \log\left(\cosh^{-1}(\lambda + 1)\right)$	See §16.12.3 below: combine entries 72 and 77, Proposition 6.2 of [262]

Lévy (L), Stieltjes (S) and Thorin (T) representation measures

L: $\dfrac{I_0(t)}{t}\, e^{-t}$

S: $\begin{cases} \dfrac{2\pi}{t}\, \sin^{-1}\left(\sqrt{\dfrac{t}{2}}\right), & 0 < t < 2 \\[2mm] \dfrac{1}{t}, & t \geq 2 \end{cases}$

T: $\begin{cases} \pi^{-1}(2t - t^2)^{-1/2}, & 0 < t < 2 \\ 0, & t \geq 2 \end{cases}$

L: $\dfrac{K_0(t)}{t} e^{-t}$

S: $\dfrac{\cosh^{-1}(t-1)}{t}\, \mathbb{1}_{(2,\infty)}(t)$

T: $(t^2 - 2t)^{-1/2}\, \mathbb{1}_{(2,\infty)}(t)$

L: closed expression unknown

S: $(t^2 - 2t)^{-1/2}\, \mathbb{1}_{(2,\infty)}(t)$

T: $f \notin \mathcal{TBF}$

L: $\dfrac{e^{-t}}{2t}\left(\cosh(t) - \displaystyle\int_0^\infty I_\nu(t)\, d\nu\right)$

S: closed expression unknown

T: unknown if in $\mathcal{TBF}$

L: $\dfrac{e^{-t}}{2t}\left(2 - \cosh(t) + \displaystyle\int_0^\infty I_\nu(t)\, d\nu\right)$

S: closed expression unknown

T: unknown if in $\mathcal{TBF}$

16.8   Gamma and related special functions

No	Function $f(\lambda)$	Comment
82	$\dfrac{\Gamma((\lambda + a)/(2a))}{\Gamma(\lambda/(2a))}, \quad a > 0$	Proposition 22 of [312]
83	$\dfrac{\Gamma(\alpha\lambda + 1)}{\Gamma(\alpha\lambda + 1 - \alpha)}, \quad 0 < \alpha < 1$	[51] (pp. 102–103)
84	$\dfrac{\lambda\,\Gamma(\alpha\lambda + 1 - \alpha)}{\Gamma(\alpha\lambda + 1)}, \quad 0 < \alpha < 1$	[51] (pp. 102–103)
85	$\lambda^{1-\nu}e^{\lambda}\,\Gamma(\nu, \lambda), \quad \nu \in \mathbb{R}$	(5.37) of [65]
86	$\lambda^{1-\nu}e^{a\lambda}\,\Gamma(\nu, a\lambda),$ $a > 0,\ 0 < \nu < 1$	14.2(17) in [107], 4.5(3) in [107], Theorem 8.2 (v)

Lévy (L), Stieltjes (S) and Thorin (T) representation measures

---

L: $\dfrac{a^{3/2}e^{2at}}{2\sqrt{\pi}\,(e^{2at}-1)^{3/2}}$

S: closed expression unknown

T: unknown if in $\mathcal{TBF}$

L: $\dfrac{e^{-t/\alpha}}{\Gamma(1-\alpha)(1-e^{-t/\alpha})^{1+\alpha}}$

S: closed expression unknown

T: unknown if in $\mathcal{TBF}$

L: $\dfrac{(1-\alpha)^2\,e^{t/\alpha}}{\alpha\,\Gamma(\alpha+1)\,(e^{t/\alpha}-1)^{2-\alpha}}$

S: closed expression unknown

T: unknown if in $\mathcal{TBF}$

L: $\dfrac{\Gamma(2-\nu)}{\Gamma(1-\nu)}\,(1+t)^{\nu-2}$

S: $\dfrac{e^{-t}t^{-\nu}}{\Gamma(1-\nu)}$

T: $f\notin\mathcal{TBF}$

L: $\dfrac{\Gamma(2-\nu)}{\Gamma(1-\nu)}\,(a+t)^{\nu-2}$

S: $\dfrac{t^{-\nu}e^{-at}}{\Gamma(1-\nu)}$

T: $f\notin\mathcal{TBF}$

No	Function $f(\lambda)$	Comment
87	$\lambda^\nu e^{a/\lambda}\,\Gamma(\nu, a/\lambda)$, $a > 0,\ 0 < \nu < 1$	14.2(19) and 4.5(29) in [107]
88	$\log\Gamma(\lambda + a + b) - \log\Gamma(\lambda + a)$ $\quad + \log\Gamma(a) - \log\Gamma(a + b)$, $a, b > 0$	9.2.3 of [67]
89	$\log\Gamma(1 + \lambda) - \lambda\log\lambda$ $\quad + \lambda^2\log\left(1 + \dfrac{1}{\lambda}\right)$	Theorem 1.3 of [34], Theorem 8.2 (v)
90	$\left(1 + \dfrac{1}{\lambda}\right)^\lambda \left(\Gamma(1 + \lambda)\right)^{1/\lambda}$	Theorem 1.3 of [34], Theorem 8.2 (v)
91	$\sqrt{\lambda}\left(\log\Gamma(\sqrt{\lambda}) - \left(\sqrt{\lambda} - \dfrac{1}{2}\right)\log\sqrt{\lambda}\right.$ $\quad \left. + \sqrt{\lambda} - \log\sqrt{2\pi}\right)$	(5.5) of [65], Theorem 8.2 (v)

Lévy (L), Stieltjes (S) and Thorin (T) representation measures

L: $\dfrac{2}{\Gamma(1-v)}(a/t)^{(v+1)/2}K_{v+1}(2\sqrt{at})$

S: $\dfrac{t^{v-1}e^{-a/t}}{\Gamma(1-v)}$

T: $\dfrac{(vt+a)\,t^{v-2}e^{-a/t}}{\Gamma(1-v)}$

L: $\dfrac{e^{-at}(1-e^{-bt})}{t(1-e^{-t})}$

S: $\displaystyle\sum_{n=1}^{\infty}\frac{1}{t}\mathbb{1}_{(n-1+a,\,n-1+a+b)}(t)$

T: $f\in\mathfrak{TBF}$ if, and only if, $b\in\mathbb{N}$

L: closed expression unknown

S: $\begin{cases}1-t, & 0<t<1 \\ 1-n/t, & n\le t<n+1,\ n=1,2,\dots\end{cases}$

T: $f\notin\mathfrak{TBF}$

L: closed expression unknown

S: $\dfrac{1}{\pi}\dfrac{t^{t-1}}{|1-t|^{t}|\Gamma(1-t)|^{1/t}}\sin\big(\pi\varphi(t)\big),\quad \varphi(t)=\begin{cases}1-t, & 0<t<1 \\ 1-n/t, & n\le t<n+1, \\ & n=1,2,\dots\end{cases}$

T: $f\notin\mathfrak{TBF}$

L: closed expression unknown

S: $\dfrac{1}{2\pi\sqrt{t}}\log\dfrac{1}{1-e^{-2\pi\sqrt{t}}}$

T: $f\notin\mathfrak{TBF}$

No	Function $f(\lambda)$	Comment
92	$\lambda \log(\sqrt{\lambda}) - \dfrac{\sqrt{\lambda}}{2} - \lambda\Psi_0(\sqrt{\lambda})$	(5.25a–c) of [65], Theorem 8.2 (v)
93	$\lambda^{3/2}\,\Psi_1(\sqrt{\lambda}) - \lambda - \dfrac{\sqrt{\lambda}}{2}$	(5.26a–c) of [65], Theorem 8.2 (v)
94	$-\lambda^2\Psi_2(\sqrt{\lambda}) - \lambda - \sqrt{\lambda}$	(5.33a–c) of [65], Theorem 8.2 (v)
95	$-\sqrt{\lambda} + \lambda\Psi_0\left(\dfrac{1}{2} + \dfrac{1}{2}\sqrt{\lambda}\right)$ $-\lambda\Psi_0\left(\dfrac{1}{2}\sqrt{\lambda}\right)$	14.2(73) in [107], Theorem 8.2 (v)
96	$\sqrt{\lambda}\left(\Psi_0\left(\dfrac{3}{4} + \dfrac{1}{2}\sqrt{\lambda}\right)\right.$ $\left. - \lambda\Psi_0\left(\dfrac{1}{4} + \dfrac{1}{2}\sqrt{\lambda}\right)\right)$	14.2(74) in [107]

Lévy (L), Stieltjes (S) and Thorin (T) representation measures

L: closed expression unknown

S: $\dfrac{1}{e^{2\pi\sqrt{t}} - 1}$

T: $f \notin \mathcal{TBF}$

L: closed expression unknown

S: $\dfrac{2\pi\sqrt{t}\,e^{2\pi\sqrt{t}}}{(e^{2\pi\sqrt{t}} - 1)^2}$

T: $f \notin \mathcal{TBF}$

L: closed expression unknown

S: $\dfrac{4\pi^2 t\,(e^{4\pi\sqrt{t}} + e^{2\pi\sqrt{t}})}{(e^{2\pi\sqrt{t}} - 1)^3}$

T: $f \notin \mathcal{TBF}$

L: closed expression unknown

S: $\operatorname{csch}(\pi\sqrt{t})$

T: $f \notin \mathcal{TBF}$

L: closed expression unknown

S: $\dfrac{\operatorname{sech}\left(\pi\sqrt{t}\right)}{\sqrt{t}}$

T: $f \notin \mathcal{TBF}$

No	Function $f(\lambda)$	Comment
97	$\dfrac{\lambda^2 \log(a\lambda)}{\log \Gamma(1+\lambda)} - \lambda - \dfrac{\lambda \log a}{(1-\gamma)(\lambda-1)}$, $a \geq 1, \gamma$: Euler's constant	Theorem 1.1 of [42] $(a \geq 1),(6)$–(7) of [40], Theorem 8.2 (v) $(a = 1)$
98	$\lambda + \dfrac{\lambda \log \Gamma(1+1/a)}{\lambda - 1/a} - \dfrac{\log \Gamma(\lambda+1)}{\log(a\lambda)}$, $a \geq 1$	Theorem 1.1 of [42] $(a \geq 1)$, Theorem 1.2 of [41], Theorem 8.2 (v) $(a = 1)$
99	$\lambda\, e^\lambda\, \mathrm{E}_n(\lambda), \quad n \in \mathbb{N}$	(5.70) of [65], Theorem 8.2 (v)
100	$-\lambda\, e^{a\lambda}\, \mathrm{Ei}(-a\lambda), \quad a > 0$	14.2(11) in [107], 4.5(2) in [107], Theorem 8.2 (v)
101	$\lambda\, e^{a\lambda}\big(\mathrm{Ei}(-ab-a\lambda) - \mathrm{Ei}(-a\lambda)\big)$, $a, b > 0$	14.2(12) in [107], Remark 8.2 (ii), Theorem 8.2 (v)

Lévy (L), Stieltjes (S) and Thorin (T) representation measures

L: closed expression unknown

S: $\displaystyle\sum_{n=1}^{\infty} t\, \frac{\log|\Gamma(1-t)| + (n-1)\log(at)}{(\log|\Gamma(1-t)|)^2 + (n-1)^2\pi^2}\, \mathbb{1}_{(n-1,n)}(t)$

T: $f \notin \mathcal{TBF}$

L: closed expression unknown

S: $\displaystyle\sum_{n=1}^{\infty} \frac{\log|\Gamma(1-t)| + (n-1)\log(at)}{t((\log(at))^2 + \pi^2)}\, \mathbb{1}_{(n-1,n)}(t)$

T: $f \notin \mathcal{TBF}$

L: $\dfrac{n}{(1+t)^{n+1}}$

S: $\dfrac{1}{(n-1)!}\, e^{-t}\, t^{n-1}$

T: $f \notin \mathcal{TBF}$

L: $\dfrac{1}{(a+t)^2}$

S: e^{-at}

T: $f \notin \mathcal{TBF}$

L: $\dfrac{1 - (b(t+a)+1)\, e^{-(t+a)b}}{(t+a)^2}$

S: $e^{-at}\, \mathbb{1}_{(0,b)}(t)$

T: $f \notin \mathcal{TBF}$

No	Function $f(\lambda)$	Comment
102	$-\lambda\, e^{a\lambda}\, \mathrm{Ei}(-ab - a\lambda)$	14.2(13) in [107], Remark 8.3 (ii), Theorem 8.2 (v)
103	$(-1)^{n+1}\lambda^{n+1}e^{a\lambda}\,\mathrm{Ei}(-a\lambda)$ $+ \displaystyle\sum_{k=1}^{n}(-1)^{n-k}(k-1)!\,a^{-k}\lambda^{n-k+1},$ $a > 0,\ n \in \mathbb{N}$	14.2(14) in [107], 4.5(3) in [107], Theorem 8.2 (v)
104	$\log(a\lambda) - e^{a\lambda}\,\mathrm{Ei}(-a\lambda),\quad a > 0$	14.2(18) in [107], 4.2(1) and 4.5(1) [107]
105	$2\lambda\cos(a\sqrt{\lambda})\,\mathrm{ci}(a\sqrt{\lambda})$ $-\, 2\lambda\sin(a\sqrt{\lambda})\,\mathrm{si}(a\sqrt{\lambda}),$ $a > 0$	14.2(20) in [107], 4.5(35) in [107], Theorem 8.2 (v)
106	$-2\sqrt{\lambda}\big(\sin(a\sqrt{\lambda})\,\mathrm{ci}(a\sqrt{\lambda})$ $+\cos(a\sqrt{\lambda})\,\mathrm{si}(a\sqrt{\lambda})\big),$ $a > 0$	14.2(21) in [107], 4.5(32) in [107], Theorem 8.2 (v)

Lévy (L), Stieltjes (S) and Thorin (T) representation measures

L: $\dfrac{(b(t+a)+1)\,e^{-(t+a)b}}{(t+a)^2}$

S: $e^{-at}\,\mathbb{1}_{(b,\infty)}(t)$

T: $f \notin \mathcal{TBF}$

L: $(n+1)!\,(a+t)^{-n-2}$

S: $t^n e^{-at}$

T: $f \notin \mathcal{TBF}$

L: $\dfrac{a}{t(t+a)}$

S: $t^{-1}(1-e^{-at})$

T: ae^{-at}

L: $3t^{-2}e^{a^2/(8t)}D_{-4}\left(\dfrac{a}{\sqrt{2t}}\right)$

S: $e^{-a\sqrt{t}}$

T: $f \notin \mathcal{TBF}$

L: $-\dfrac{a}{t^2}+\sqrt{\dfrac{\pi}{t^5}}\left(\dfrac{a^2}{4}+\dfrac{1}{2t}\right)e^{a^2/(4t)}\,\mathrm{Erfc}\left(\dfrac{a}{2\sqrt{t}}\right)$

S: $t^{-1/2}e^{-a\sqrt{t}}$

T: $f \notin \mathcal{TBF}$

No	Function $f(\lambda)$	Comment
107	$\lambda\left(\Gamma\left(1+\dfrac{1}{\lambda}\right)\right)^{\lambda}$	Theorem 3.2 of [6]
108	$\sqrt{\lambda}\,e^{a\lambda}\,\mathrm{Erfc}(\sqrt{a\lambda}),\quad a>0$	14.2(15) in [107], 4.5(3) in [107], Theorem 8.2 (v)
109	$\sqrt{\dfrac{\pi}{a}}\lambda - \pi\,\lambda^{3/2}\,e^{a\lambda}\,\mathrm{Erfc}(\sqrt{a\lambda}),$ $a>0$	14.2(16) in [107], 4.5(3) in [107], Theorem 8.2 (v)

16.9   Bessel functions

No	Function $f(\lambda)$	Comment
110	$-(\nu-1)\log 2 - \log\Gamma(\nu)$ $+\sqrt{\lambda}+\log K_{\nu}(\sqrt{\lambda}),\quad \nu>1/2$	Formula below (2.3) of [175]
111	$\sqrt{\lambda}\left(1-\dfrac{K_{\nu-1}(\sqrt{\lambda})}{K_{\nu}(\sqrt{\lambda})}\right),\quad \nu>1/2$	Entry 110 and Remark 8.3

Lévy (L), Stieltjes (S) and Thorin (T) representation measures

L, S: closed expression unknown   —   **T:**   unknown if in $\mathcal{TBF}$

L: $\dfrac{1}{2(t+a)^{3/2}}$

S: $\dfrac{e^{-at}}{\sqrt{\pi t}}$

T: $f \notin \mathcal{TBF}$

L: $\Gamma\left(\dfrac{5}{2}\right)(t+a)^{-5/2}$

S: $\sqrt{t}\, e^{-at}$

T: $f \notin \mathcal{TBF}$

Lévy (L), Stieltjes (S) and Thorin (T) representation measures

L, S: closed expression unknown

T: $\dfrac{1}{2\pi\sqrt{t}}\left(1 - \dfrac{2}{(\pi\sqrt{t})\left(J_\nu^2(\sqrt{t}) + Y_\nu^2(\sqrt{t})\right)}\right)$

L: closed expression unknown

S: $\dfrac{1}{\pi\sqrt{t}}\left(1 - \dfrac{2}{(\pi\sqrt{t})\left(J_\nu^2(\sqrt{t}) + Y_\nu^2(\sqrt{t})\right)}\right)$

T: $f \notin \mathcal{TBF}$

No	Function $f(\lambda)$	Comment
112	$(\nu - 1)\log 2 + \log \Gamma(\nu)$ $- \sqrt{\lambda} - \log K_\nu(\sqrt{\lambda}),$ $0 < \nu < 1/2$	Proof of Theorem 1 in [175]
113	$\sqrt{\lambda}\left(\dfrac{K_{\nu-1}(\sqrt{\lambda})}{K_\nu(\sqrt{\lambda})} - 1\right),$ $0 < \nu < 1/2$	Entry 112 and Remark 8.3 (iii)
114	$(\mu - \nu)(\log \sqrt{\lambda} - \log 2)$ $+ \log K_\mu(\sqrt{\lambda}) - \log K_\nu(\sqrt{\lambda}),$ $0 < \nu < \mu$	(4.2) of [176], Theorem 8.2 (ii)
115	$\sqrt{\lambda}\left(\dfrac{K_{\nu-1}(\sqrt{\lambda})}{K_\nu(\sqrt{\lambda})} - \dfrac{K_{\mu-1}(\sqrt{\lambda})}{K_\mu(\sqrt{\lambda})}\right),$ $0 < \nu < \mu$	Entry 114 and Remark 8.3 (iii)
116	$\sqrt{\lambda}\,\dfrac{K_{\nu-1}(\sqrt{\lambda})}{K_\nu(\sqrt{\lambda})}, \quad \nu \geq 0$	(1.3) of [173], Nicholson's integral formula (4.13.7) of [11]
117	$\sqrt{\lambda}\,\dfrac{I_{\nu+1}(\sqrt{\lambda})}{I_\nu(\sqrt{\lambda})}, \quad \nu > 0$	(2.2) of [175], Theorem 8.2 (v), Remark 8.3 (ii)

Lévy (L), Stieltjes (S) and Thorin (T) representation measures

L, S: closed expression unknown

T: $\dfrac{1}{\pi\sqrt{t}}\left(1-\dfrac{2}{\left(\pi\sqrt{t}\right)\left(J_\nu^2(\sqrt{t})+Y_\nu^2(\sqrt{t})\right)}\right)$

L: closed expression unknown

S: $\dfrac{1}{\pi\sqrt{t}}\left(\dfrac{2}{\left(\pi\sqrt{t}\right)\left(J_\nu^2(\sqrt{t})+Y_\nu^2(\sqrt{t})\right)}-1\right)$

T: $f\notin\mathcal{TBF}$

L, S: closed expression unknown

T: $\dfrac{1}{\pi^2 t}\left(\dfrac{1}{J_\nu^2(\sqrt{t})+Y_\nu^2(\sqrt{t})}-\dfrac{1}{J_\mu^2(\sqrt{t})+Y_\mu^2(\sqrt{t})}\right)$

L: closed expression unknown

S: $\dfrac{2}{\pi^2 t}\left(\dfrac{1}{J_\nu^2(\sqrt{t})+Y_\nu^2(\sqrt{t})}-\dfrac{1}{J_\mu^2(\sqrt{t})+Y_\mu^2(\sqrt{t})}\right)$

T: $f\notin\mathcal{TBF}$ in general

L, T: closed expression unknown

S: $\dfrac{2}{\pi^2 t}\dfrac{1}{J_\nu^2(\sqrt{t})+Y_\nu^2(\sqrt{t})}$

L: $2\displaystyle\sum_{n=1}^{\infty} j_{\nu,n}^2 e^{-j_{\nu,n}^2 t}$

S: $2\displaystyle\sum_{n=1}^{\infty} \delta_{j_{\nu,n}^2}(dt)$

T: $f\notin\mathcal{TBF}$

No	Function $f(\lambda)$	Comment
118	$\lambda^{1+(\nu-\mu)/2}\dfrac{I_\mu(\sqrt{\lambda})}{I_\nu(\sqrt{\lambda})},$ $-1<\nu<\mu\le 1+\nu$	Theorem 4.7 of [176], Theorem 8.2 (v), Remark 8.3 (ii)
119	$\sqrt{\lambda}\,\dfrac{I_\nu(\sqrt{\lambda})}{I_{\nu+1}(\sqrt{\lambda})+I_\nu(\sqrt{\lambda})},\;\; \nu>0$	Theorem 9 of [266], Theorem 8.2 (v)
120	$\sqrt{\lambda}\,\dfrac{I_{\nu+1}(\sqrt{\lambda})}{I_{\nu+1}(\sqrt{\lambda})+I_\nu(\sqrt{\lambda})},\;\; \nu>0$	Theorem 9 of [266], Theorem 8.2 (v)
121	$\lambda^{1-\mu/2}\dfrac{K_\nu(\sqrt{\lambda})}{K_{\nu+\mu}(\sqrt{\lambda})},$ $0<\mu<1,\; \nu\le-\mu$	(2.1) of [174], (67) on p. 97 of [106]
122	$\sqrt{\lambda}\,\dfrac{a_1 I_\nu(\sqrt{\lambda})+a_2 I_{\nu-1}(\sqrt{\lambda})}{a_3 I_\nu(\sqrt{\lambda})+a_4 I_{\nu-1}(\sqrt{\lambda})},$ $\nu>0,\; a_1,a_2,a_3,a_4>0$	Theorem 3.1 of [177]

Lévy (L), Stieltjes (S) and Thorin (T) representation measures

L: $-2 \sum_{n=1}^{\infty} j_{v,n}^{v+3-\mu} \dfrac{J_\mu(j_{v,n})}{J_v'(j_{v,n})} e^{-j_{v,n}^2 t}$

S: $-2 \sum_{n=1}^{\infty} j_{v,n}^{v+1-\mu} \dfrac{J_\mu(j_{v,n})}{J_v'(j_{v,n})} \delta_{j_{v,n}^2}(dt)$

T: $f \notin \mathcal{TBF}$

L: closed expression unknown

S: $\dfrac{1}{\pi\sqrt{t}} \dfrac{J_v^2(\sqrt{t})}{J_{v+1}^2(\sqrt{t}) + J_v^2(\sqrt{t})}$

T: $f \notin \mathcal{TBF}$

L: closed expression unknown

S: $\dfrac{1}{\pi\sqrt{t}} \dfrac{J_{v+1}^2(\sqrt{t})}{J_{v+1}^2(\sqrt{t}) + J_v^2(\sqrt{t})}$

T: $f \notin \mathcal{TBF}$

L: closed expression unknown

S: $\dfrac{1}{\pi t^{\mu/2}} \dfrac{J_{v+\mu}(\sqrt{t})Y_v(\sqrt{t}) - J_v(\sqrt{t})Y_{v+\mu}(\sqrt{t})}{J_{v+\mu}^2(\sqrt{t}) + Y_{v+\mu}^2(\sqrt{t})}$

T: unknown if in $\mathcal{TBF}$

L: closed expression unknown

S: $\dfrac{1}{\pi\sqrt{t}} \dfrac{a_1 a_3 J_v^2(\sqrt{t}) + a_2 a_4 J_{v-1}^2(\sqrt{t})}{a_3^2 J_v^2(\sqrt{t}) + a_4^2 J_{v-1}^2(\sqrt{t})}$

T: unknown if in $\mathcal{TBF}$

No	Function $f(\lambda)$	Comment
123	$\sqrt{\lambda}\,\dfrac{a_1 K_\nu(\sqrt{\lambda}) + a_2 K_{\nu-1}(\sqrt{\lambda})}{a_3 K_\nu(\sqrt{\lambda}) + a_4 K_{\nu-1}(\sqrt{\lambda})},$ $\nu \geq 0,\ a_1, a_2, a_3, a_4 > 0$	Theorem 3.3 of [177]
124	$2\lambda\, I_\nu\!\left(a\sqrt{\lambda}\right) K_\nu\!\left(a\sqrt{\lambda}\right),$ $a > 0,\ \nu \in \mathbb{R}$	14.3(22) in [107], Theorem 8.2 (v)
125	$\lambda^{1+\nu} e^{a\lambda}\, K_\nu(a\lambda),$ $a > 0,\ -1/2 < \nu < 1/2$	14.3(36) in [107]
126	$\sqrt{\lambda}\, e^{a\lambda}\, K_\nu(a\lambda),$ $a > 0,\ -1/2 < \nu < 1/2$	14.3(39) in [107]

Lévy (L), Stieltjes (S) and Thorin (T) representation measures

L: closed expression unknown

S:
$$\frac{a_1a_3\left[J_{\nu-1}^2(\sqrt{t})+Y_{\nu-1}^2(\sqrt{t})\right]+a_2a_4\left[J_\nu^2(\sqrt{t})+Y_\nu^2(\sqrt{t})\right]+\dfrac{2(a_2a_4+a_1a_3)}{\pi\sqrt{t}}}{\pi\left[a_4J_{\nu-1}^2(\sqrt{t})-a_3Y_\nu^2(\sqrt{t})\right]^2+\left[a_3J_\nu^2(\sqrt{t})+a_4Y_{\nu-1}^2(\sqrt{t})\right]^2}$$

T: unknown if in $\mathcal{TBF}$

L:
$$\frac{2a^{2\nu}\,\Gamma(3/2+\nu)}{\sqrt{\pi}\,\Gamma(1+\nu)}\,t^{-2(1+\nu)}\,{}_2F_1\left(\frac{1}{2}+\nu,\ \frac{3}{2}+\nu,\ 1+2\nu,\ -\frac{4a^2}{t^2}\right)$$

S: $J_\nu^2(at)$

T: $f\notin\mathcal{TBF}$

L:
$$\frac{2^{1+\nu}a^\nu\,\Gamma(3/2+\nu)}{\sqrt{\pi}\,\sec(\nu\pi)}\,t^{-3/2-\nu}(t+a)(t+2a)^{-3/2-\nu}$$

S: $\dfrac{1}{\sec(\nu\pi)}\,t^\nu e^{-at}I_\nu(at)$

T: unknown if in $\mathcal{TBF}$

L:
$$\frac{\cos(\nu\pi)}{2^{1+\nu}a^\nu(t+a)^{3/2+\nu}}\ \times$$
$$\times\left\{4^\nu(t+a)^{2\nu}\,\Gamma\left(\frac{3}{2}+\nu\right)\Gamma(\nu)\,{}_2F_1\left(\frac{3-2\nu}{4},\ \frac{5-2\nu}{4},\ 1-\nu,\ \frac{a^2}{(t+a)^2}\right)\right.$$
$$\left.+\,a^{2\nu}\,\Gamma\left(\frac{3}{2}+\nu\right)\Gamma(-\nu)\,{}_2F_1\left(\frac{3+2\nu}{4},\ \frac{5+2\nu}{4},\ 1+\nu,\ \frac{a^2}{(t+a)^2}\right)\right\}$$

S: $\dfrac{\cos(\nu\pi)}{\pi\sqrt{t}}\,e^{-at}K_\nu(at)$

T: $f\notin\mathcal{TBF}$

16.10 Miscellaneous functions

No	Function $f(\lambda)$	Comment
127	$\dfrac{\lambda^{1-\nu}}{(a-\lambda)^\mu} I_{1-\frac{\lambda}{a}}(\mu,\nu),$ $a>0,\ -\mu<\nu<1$	14.2(10) in [107], Lemma 2.1 in [79]
128	$_2F_1\left(1,\ 1-\nu,\ 2-\nu,\ -\dfrac{a}{\lambda}\right),$ $a>0,\ 0<\nu<1$	
129	$\lambda\,_2F_1\left(1,\ 1+\nu,\ 2+\nu,\ -\dfrac{\lambda}{a}\right),\quad a>0,\ 0<\nu<1$	
130	$\lambda^\nu\,_2F_1\left(\mu-1,\ \nu,\ \mu,\ 1-\dfrac{\lambda}{a}\right),\quad a>0,\ 0<\nu<\mu$	14.2(9) in [107], Lemma 2.1 in [79]
131	$-(2a)^\nu\,\Gamma(1-\nu)\dfrac{\Gamma(\nu+\frac{\lambda}{2a})}{\Gamma(1+\frac{\lambda}{2a})}\,_2F_1\left(1,\ \nu+\dfrac{\lambda}{2a},\ 1+\dfrac{\lambda}{2a},\ 1\right),$ $a>0,\ 0<\nu<1$	[98]

Lévy (L), Stieltjes (S) and Thorin (T) representation measures

L: $\dfrac{\sin(\nu\pi)}{\pi}\, a^{1-\nu-\mu}\, \Gamma(2-\nu)\, (at)^{-\frac{3-\nu-\mu}{2}}\, e^{\frac{at}{2}}\, W_{\frac{\nu-1-\mu}{2},\,\frac{2-\nu-\mu}{2}}(at)$

S: $\dfrac{\sin(\nu\pi)}{\pi}\, t^{-\nu}(a+t)^{-\mu}$

T: unknown if in $\mathfrak{TBF}$

L: closed expression unknown

S: $\dfrac{1-\nu}{a^{1-\nu}}\, t^{-\nu}\, \mathbb{1}_{(0,a)}(t)$

T: $f \notin \mathfrak{TBF}$

L: closed expression unknown

S: $\nu a^{\nu}\, t^{-\nu}\, \mathbb{1}_{(a,\infty)}(t)$

T: $a^{1-\nu}\, \delta_a(dt) + (1-\nu)\, t^{-\nu}\, \mathbb{1}_{(a,\infty)}(t)\, dt$

L: $\dfrac{\nu\, \Gamma(\mu)}{\Gamma(\mu-\nu)}\, a^{\frac{\nu+\mu-1}{2}}\, t^{-\frac{\nu+1-\mu}{2}}\, e^{\frac{at}{2}}\, W_{\frac{1-\mu-\nu}{2},\,\frac{2+\nu-\mu}{2}}(at)$

S: $\dfrac{\Gamma(\mu)\, a^{\mu-1}}{\Gamma(\nu)\, \Gamma(\mu-\nu)}\, t^{\nu-1}\, (a+t)^{1-\mu}$

T: unknown if in $\mathfrak{TBF}$

L: $\left(\dfrac{a}{\sinh(at)}\right)^{1+\nu}\, e^{(1-\nu)\,at}$

S: closed expression unknown

T: unknown if in $\mathfrak{TBF}$

No	Function $f(\lambda)$	Comment
132	$$(2a)^{\nu}\,\Gamma(1-\nu)\left(\frac{\Gamma(\mu)}{\Gamma(2-\mu)}\,{}_2F_1\left(1,\frac{\mu}{a},\frac{2-\mu}{a},1\right)\right.$$ $$\left.-\frac{\Gamma\!\left(\mu+\frac{\lambda}{2a}\right)}{\Gamma\!\left(2-\mu+\frac{\lambda}{2a}\right)}\,{}_2F_1\left(1,\mu+\frac{\lambda}{2a},2-\mu+\frac{\lambda}{2a},1\right)\right)$$ $$a>0,\ 0<\nu<1,\ \mu=\frac{1+\nu}{2}$$	[98]
133	$$\lambda\,\Phi_{\alpha,\beta}(\lambda),\quad \alpha>-\frac{1}{2},\quad -\frac{1}{2}<\beta\le\frac{1}{2},$$ $$\text{or:}\qquad \beta>-\frac{1}{2},\quad -\frac{1}{2}<\alpha\le\frac{1}{2}$$	Lemma 3.1 of [189]
134	$$\lambda^{(\mu+1)/2}\,e^{a\lambda/2}\,W_{-\frac{\mu+1}{2},-\frac{\mu}{2}}(a\lambda)$$ $$=a^{\mu+1}\,\lambda^{(\mu+1)/2}\,e^{a\lambda}\,\Gamma(-\mu,a\lambda),$$ $$a>0,\ \mu>0$$	[67, 268], Theorem 8.2 (v)
135	$$\lambda^{\mu+1/2}\,e^{a\lambda/2}\,W_{-\kappa,\mu}(a\lambda),$$ $$a>0,\ -1/2<\mu<\kappa+1/2$$	14.3(50) of [107], (9.221) of [134]
136	$$\lambda^{\kappa}\,e^{a\lambda/2}\,W_{-\kappa,\mu}(a\lambda),$$ $$a>0,\ |\mu|-1/2<\kappa<|\mu|+1/2$$	14.3(53) of [107], (9.222) of [134]

Lévy (L), Stieltjes (S) and Thorin (T) representation measures

L: $\left(\dfrac{a}{\sinh(at)}\right)^{1+v}$

S: closed expression unknown

T: unknown if in $\mathcal{TBF}$

L: closed expression unknown

S: $\dfrac{t^{\alpha+\beta}\,e^{-t}\,\Phi_{-\alpha,-\beta}(t)}{\Gamma\!\left(\alpha+\frac{1}{2}\right)\Gamma\!\left(\beta+\frac{1}{2}\right)}$

T: unknown if in $\mathcal{TBF}$

L: closed expression unknown

S: $\dfrac{a^{(\mu+1)/2}}{\mu\,\Gamma(\mu)}\,t^{\mu}e^{-at}$

T: $f\notin\mathcal{TBF}$

L: closed expression unknown

S: $\dfrac{t^{\mu-1/2}\,e^{-at/2}\,M_{\kappa,\mu}(at)}{\Gamma(2\mu+1)\,\Gamma\!\left(\kappa-\mu+\frac{1}{2}\right)}$

T: unknown if in $\mathcal{TBF}$

L: closed expression unknown

S: $\dfrac{t^{\kappa-1}\,e^{-at/2}\,W_{\kappa,\mu}(at)}{\Gamma\!\left(\kappa+\mu+\frac{1}{2}\right)\Gamma\!\left(\kappa-\mu+\frac{1}{2}\right)}$

T: unknown if in $\mathcal{TBF}$

No	Function $f(\lambda)$	Comment
137	$\lambda \dfrac{\Psi(a+1,b+1,\lambda)}{\Psi(a,b,\lambda)}, \quad a>0,\ b<1$	(1.4) of [176]
138	$1-\lambda^{-1/2}\dfrac{D_{-\nu-1}(\lambda^{-1/2})}{D_{-\nu}(\lambda^{-1/2})}, \quad \nu>0$	(1.3) of [176]
139	$\sqrt{\dfrac{\lambda}{2}}\,\Gamma\left(\dfrac{\lambda}{\kappa}+\dfrac{1}{2}\right)\dfrac{\left(D_{-\lambda/\kappa-1/2}\left(-a\sqrt{2\kappa}\right)\right)^2}{2\sqrt{\pi\kappa}},$ $\kappa>0,\ a\in\mathbb{R}$	p. 254 of [245]
140	$\lambda\dfrac{\Gamma(\frac{\lambda}{2\mu})\,{}_1F_1(\frac{\lambda}{2\mu},\beta,a)\,\Psi(\frac{\lambda}{2\mu},\beta,a)}{\Gamma(\beta)\,a^{1-\beta}},$ $a=2\kappa/\sigma^2,\ \beta=\nu+1$ $\mu>0,\ \nu\geq0,\ \sigma>0$	p. 257 of [245]
141	$\lambda e^{-a}\dfrac{\delta^2\,\Gamma(\frac{\lambda}{\gamma}+1)M_{\frac{\nu-1}{2}-\frac{\lambda}{\gamma}}(a)\,W_{\frac{\nu-1}{2}-\frac{\lambda}{\gamma}}(a)}{2\mu\,\Gamma(\nu+1)},$ $\nu=\dfrac{1}{2|\beta|},\ \gamma=2\mu|\beta|,\ a=\dfrac{\mu}{\delta^2|\beta|}$ $\beta<0,\ \delta>0,\ \mu\leq0$	pp. 260–261 of [245]

Lévy (L), Stieltjes (S) and Thorin (T) representation measures

L: closed expression unknown

S: $\dfrac{t^{-b}\, e^{-t}\, |\Psi(a, b, te^{i\pi})|^{-2}}{\Gamma(a+1)\,\Gamma(a-b+1)}$

T: unknown if in $\mathcal{TBF}$

L: closed expression unknown

S: $\dfrac{1}{\sqrt{2\pi}\,\Gamma(\nu+1)}\,\dfrac{|D_{-\nu}(it^{-1/2})|^{-2}}{t^{3/2}}$

T: unknown if in $\mathcal{TBF}$

L: closed expression unknown

S: $\displaystyle\sum_{n=0}^{\infty}\frac{1}{2^n\,n!}\left(\frac{\kappa}{\pi}\right)^{1/2}e^{-\kappa a^2}\left(H_n\left(a\sqrt{\kappa}\right)\right)^2 \delta_{\kappa(n+1/2)}(dt)$

T: unknown if in $\mathcal{TBF}$

L: closed expression unknown

S: $\displaystyle\sum_{n=0}^{\infty}\frac{n!\,\kappa}{\Gamma(\beta+n)}a^{\beta-1}L_n^{(\beta-1)}(a)\,\delta_{\kappa n}(dt)$

T: unknown if in $\mathcal{TBF}$

L: closed expression unknown

S: $\mathbb{1}_{\{\nu<0\}}\,\delta_0(dt)+\mathbb{1}_{\{\nu<-2\}}\displaystyle\sum_{n=1}^{[|\nu|/2]}\frac{2\,\Gamma(|\nu|)(|\nu|-2n)\,n!}{2^\nu\,\Gamma(1+|\nu|-n)}\left(L_n^{(|\nu|-2n)}\left(\tfrac{1}{2}\right)\right)^2\delta_{2n(|\nu|-n)}(dt)$

$+\dfrac{\sqrt{e}}{\pi^2}\dfrac{\Gamma(|\nu|)}{2^\nu}\left[W_{\frac{1-\nu}{2},\,\frac{i\sqrt{2t-\nu^2}}{2}}\left(\tfrac{1}{2}\right)\right]^2\left|\Gamma\left(\dfrac{\nu+i\sqrt{2t-\nu^2}}{2}\right)\right|^2\times$

$\times\sinh\left(\pi\sqrt{2t-\nu^2}\right)\mathbb{1}_{\{t>\nu^2/2\}}dt$

T: unknown if in $\mathcal{TBF}$

No	Function $f(\lambda)$	Comment
142	$$v(\lambda)\frac{\cosh^2\left(v(\lambda)\right) - \frac{\mu^2}{v^2(\lambda)}\sinh^2\left(v(\lambda)\right)}{2e^{2\mu}\sinh\left(2v(\lambda)\right)},$$ $$v(\lambda) = \sqrt{\mu^2 - 2\lambda}, \ \mu \in \mathbb{R}$$	p. 445 of [244]
143	$W(\lambda)$	Theorem 3.1 of [285]

Lévy (L), Stieltjes (S) and Thorin (T) representation measures

L: closed expression unknown

S: $\delta_0(dt) + \sum\limits_{n=1}^{\infty} \dfrac{e^{-2\mu}}{2\left[1+\left(4\mu^2/(\pi^2 n^2)\right)\right]} \left(\cos\left(\tfrac{\pi n}{2}\right) + \tfrac{2\mu}{\pi n}\, \sin\left(\tfrac{\pi n}{2}\right)\right)^2 \delta_{\frac{\mu^2}{2}+\frac{\pi^2 n^2}{8}}(dt)$

T: unknown if in $\mathcal{TBF}$

L: closed expression unknown

S: closed expression unknown

T: $\dfrac{1}{\pi} \left(\dfrac{d}{dt}\, \mathrm{Im}\, W(-t + i\,0)\right) \mathbb{1}_{(e^{-1},\infty)}(t)$

16.11 $\mathcal{CBF}$s given by exponential representations

In this section we explain a method which allows us to construct many complete Bernstein functions in exponential representation. Recall the following theorem from Chapter 6.2:

Theorem 6.17. *For* $f \in \mathcal{CBF}$*, there exist a real number* γ *and a function* η *on* $(0, \infty)$ *taking values in* $[0, 1]$ *such that*

$$f(\lambda) = \exp\left(\gamma + \int_0^\infty \left(\frac{t}{1+t^2} - \frac{1}{\lambda+t}\right) \eta(t)\, dt\right), \quad \lambda > 0. \tag{6.21}$$

Conversely, any function of the form (6.21) *is in* $\mathcal{CBF}$*. Thus there is a one-to-one correspondence between* $\mathcal{CBF}$ *and the set*

$$\Gamma = \left\{(\gamma, \eta) \; : \; \gamma \in \mathbb{R} \text{ and } \eta : (0, \infty) \xrightarrow{measurable} [0, 1]\right\}. \tag{6.22}$$

In order to find new complete Bernstein functions it is enough to identify 'simple' functions $\eta : (0, \infty) \to [0, 1]$ such that the integral

$$\int_0^\infty \left(\frac{t}{1+t^2} - \frac{1}{\lambda+t}\right) \eta(t)\, dt$$

can be explicitly calculated. Then we can combine the Tables 16.2–16.10 of complete Bernstein functions with Theorems 6.2 and 6.17.

Recall that every complete Bernstein function f admits a Stieltjes representation

$$f(\lambda) = a + b\lambda + \int_{(0,\infty)} \frac{\lambda}{\lambda+t}\, \sigma(dt), \quad \lambda > 0, \tag{6.7}$$

$(a, b \geq 0$ and σ satisfies $\int_{(0,\infty)} (1+t)^{-1}\, \sigma(dt) < \infty)$ as well as a Nevanlinna–Pick representation

$$f(\lambda) = \alpha + \beta\lambda + \int_{(0,\infty)} \left(\frac{t}{1+t^2} - \frac{1}{\lambda+t}\right)(1+t^2)\, \rho(dt) \tag{6.17}$$

(ρ is a finite measure on $(0, \infty)$ with $\int_0^1 t^{-1}\rho(dt) < \infty$ and $\alpha, \beta \geq 0$ such that $\alpha \geq \int_{(0,\infty)} t^{-1}\rho(dt)$). The measure σ in (6.7) is called the Stieltjes measure of f and the measure ρ in (6.17) is called the Nevanlinna–Pick measure of f. The quantities appearing in the formulae (6.7) and (6.17) are related as follows:

$$a = \alpha - \int_0^\infty \frac{1}{t}\rho(dt), \quad b = \beta, \quad \text{and} \quad \sigma(dt) = \frac{1+t^2}{t}\rho(dt).$$

Now we can inspect the Tables 16.2–16.10 of complete Bernstein functions to see which entry f has a representation measure $(1 + t^2)\, \rho(dt) = t\sigma(dt)$ which is of the form $\eta(t)\, dt$ with some $\eta : (0, \infty) \to [0, 1]$. If so, $g(\lambda) := \exp(f(\lambda))$ is another complete Bernstein function.

No	Function $g(\lambda)$	Parameters	Density $\eta(t)$	Comments
1	$\left[1 + \dfrac{a}{\lambda}\right]^\lambda$	$0 < a \leq 1$	$t\,\mathbb{1}_{(0,a)}(t)$	Entry 27
2	$\dfrac{b(\lambda + a)}{a(\lambda + b)}$	$0 < a < b$	$\mathbb{1}_{(a,b)}(t)$	Entry 28
3	$e^{1/a}\left[1 + \dfrac{\lambda}{a}\right]^{-\frac{1}{\lambda}}$	$a \geq 1$	$\dfrac{1}{t}\,\mathbb{1}_{(0,a)}(t)$	Entry 29
4	$\dfrac{(\lambda+a)^{\lambda+a}}{\lambda^\lambda\, a^a}$	$a \in (0,1]$	$t \wedge a$	Entry 31
5	$\dfrac{(\lambda+b)^{\lambda+b}}{(\lambda+a)^{\lambda+a}}\,\dfrac{a^a}{b^b}$	$b \in (a, a+1]$ $a > 0$	$(t-a)^+ \wedge (b-a)$	Entry 32
6	$\left[\dfrac{a}{\lambda}\right]^{\frac{\lambda}{a-\lambda}}$	$a > 0$	$\dfrac{t}{a+t}$	Entry 33
7	$\left[\dfrac{a}{\lambda}\exp\left[\dfrac{\pi\lambda}{2a}\right]\right]^{\frac{\lambda}{a^2+\lambda^2}}$	$a \geq \dfrac{1}{2}$	$\dfrac{t}{t^2 + a^2}$	Entry 34
8	$\left[\left[\dfrac{\lambda}{a}\right]^\lambda \exp\left[\dfrac{\pi a}{2}\right]\right]^{\frac{\lambda}{a^2+\lambda^2}}$	$a > 0$	$\dfrac{t^2}{t^2 + a^2}$	Entry 35

No	Function $g(\lambda)$	Parameters	Density $\eta(t)$	Comments
9	$\left[1+\dfrac{1}{\lambda}\right]^{\lambda(1+\lambda)} e^{-\lambda}$		$t(1-t)\mathbb{1}_{(0,1)}(t)$	Entry 45
10	$e^{\lambda}\left[\dfrac{\lambda}{\lambda+1}\right]^{\lambda^2}$		$t^2\mathbb{1}_{(0,1)}(t)$	Entry 46
11	$\dfrac{\log\left[1+\frac{1}{a}\right]}{\log\left[1+\frac{1}{a+\lambda}\right]}$	$a>0$	$\left[\dfrac{1}{2}-\dfrac{\arctan\left[\frac{1}{\pi}\log\frac{a-t+1}{t-a}\right]}{\pi}\right]\mathbb{1}_{[a,a+1]}(t)$ $+\,\mathbb{1}_{(a+1,\infty)}(t)$	Entry 51
12	$\dfrac{\lambda}{\log\frac{1+(1+a)\lambda}{1+a\lambda}}$	$a>0$	$\left[\dfrac{1}{2}+\dfrac{\arctan\left[\frac{1}{\pi}\log\frac{(1+a)t-1}{1-at}\right]}{\pi}\right]\mathbb{1}_{[\frac{1}{a+1},\frac{1}{a}]}(t)$ $+\,\mathbb{1}_{(\frac{1}{a},\infty)}(t)$	Entry 52
13	$1+2\lambda+2\sqrt{\lambda(1+\lambda)}$		$\dfrac{2}{\pi}\arcsin(\sqrt{t})\,\mathbb{1}_{(0,1)}(t)+\mathbb{1}_{[1,\infty)}$	Entry 57
14	$\exp\left[\sqrt{\lambda}\arctan\left[\sqrt{\dfrac{a}{\lambda}}\right]\right]$	$a\le 4$	$\dfrac{\sqrt{t}}{2}\mathbb{1}_{(0,a)}(t)$	Entry 63
15	$\exp(\lambda e^{\lambda}\mathrm{E}_n(\lambda))$	$n=1,\dots,6$	$\dfrac{1}{(n-1)!}e^{-t}t^n$	Entry 99

No	Function $g(\lambda)$	Parameters	Density $\eta(t)$	Comments
16	$\exp(\lambda e^{a\lambda}\mathrm{Ei}(-ab-a\lambda))$ $\times \exp(-\lambda e^{a\lambda}\mathrm{Ei}(-a\lambda))$	$a \geq \dfrac{1}{e},$ $b > 0$	$t\,e^{-at}\,\mathbb{1}_{(0,b)}(t)$	Entry 101
17	$\exp(-\lambda e^{a\lambda}\mathrm{Ei}(-ab-a\lambda))$	$a \geq \dfrac{1}{e},$ $b > 0$	$t\,e^{-at}\,\mathbb{1}_{(b,\infty)}(t)$	Entry 102
18	$e^{(-1)^{n+1}\lambda^{n+1}e^{a\lambda}\mathrm{Ei}(-a\lambda)}$ $\times e^{\sum_{k=1}^{n}(-1)^{n-k}(k-1)!a^{-k}\lambda^{n-k+1}}$	$a \geq \dfrac{n+1}{e}$	$\dfrac{1}{(n-1)!}\,e^{-t}\,t^{n}$	Entry 103
19	$\dfrac{a\lambda}{\exp(e^{a\lambda}\mathrm{Ei}(-a\lambda))}$	$a > 0$	$1 - e^{-at}$	Entry 104
20	$\exp\!\left[2\lambda\cos(a\sqrt{\lambda})\,\mathrm{ci}(a\sqrt{\lambda})\right]$ $\times\exp\!\left[-2\lambda\sin(a\sqrt{\lambda})\,\mathrm{si}(a\sqrt{\lambda})\right]$	$a > \dfrac{2}{e}$	$t\,e^{-a\sqrt{t}}$	Entry 105
21	$\exp\!\left[-2\sqrt{\lambda}\sin(a\sqrt{\lambda})\,\mathrm{ci}(a\sqrt{\lambda})\right]$ $\times\exp\!\left[-2\sqrt{\lambda}\cos(a\sqrt{\lambda})\,\mathrm{si}(a\sqrt{\lambda})\right]$	$a > \dfrac{1}{e}$	$\sqrt{t}\,e^{-a\sqrt{t}}$	Entry 106
22	$\exp\!\left[\sqrt{\lambda}\,e^{a\lambda}\,\mathrm{Erfc}(\sqrt{a\lambda})\right]$	$a > \dfrac{1}{2\pi e}$	$(t/\pi)^{1/2}\,e^{-at}$	Entry 108

No	Function $g(\lambda)$	Parameters	Density $\eta(t)$	Comments
23	$\exp\left[\sqrt{\dfrac{\pi}{a}}\,\lambda - \pi\lambda^{3/2}e^{a\lambda}\operatorname{Erfc}(\sqrt{a\lambda})\right]$	$a > \dfrac{3}{2e}$	$t^{3/2}e^{-at}$	Entry 109
24	$e^{\arctan(1/a)}\left[\dfrac{a}{a+\lambda}\right]^{1/\lambda}$	$a > 1$	$\dfrac{1}{t}\,\mathbb{1}_{[a,\infty)}(t)$	
25	$4e^{-1/\lambda}(1+\lambda)^{1/\lambda^2}$		$\dfrac{1}{t}\,\mathbb{1}_{[1,\infty)}(t)$	
26	$e^{-\pi/4}\lambda^{-\frac{\lambda}{\lambda-1}}$		$\dfrac{t}{1+t}$	
27	$e^{1/2}e^{-\frac{\lambda\pi}{2+2\lambda^2}}\lambda^{-\frac{1}{1+\lambda^2}}$		$\dfrac{1}{1+t^2}$	
28	$\dfrac{a(1-a)(\lambda-1)^2}{(\lambda^a-1)(\lambda^{1-a}-1)}$	$a\in(0,1)$	$\dfrac{1}{\pi}\left[\arctan\dfrac{(t^a+t^{1-a})\sin a\pi}{1-t-(t^a-t^{1-a})\cos a\pi}\right]\mathbb{1}_{(0,1)}(t)$ $+\dfrac{1}{\pi}\left[\pi-\arctan\dfrac{(t^{-a}+t^{-1+a})\sin a\pi}{1-t^{-1}-(t^{-a}-t^{-1+a})\cos a\pi}\right]\mathbb{1}_{(1,\infty)}(t)$	[15]
29	$\left[\dfrac{\lambda}{1+\lambda}\right]^{1+\lambda}(1+\lambda)^{1+1/\lambda}$		$(1-t)\mathbb{1}_{[0,1]}(t) + (1-t^{-1})\mathbb{1}_{(1,\infty)}(t)$	[141]
30	$\lambda^a\left[\dfrac{1+\lambda}{2}\right]^{1-2a}$	$a\in[0,1]$	$a\mathbb{1}_{[0,1]}(t) + (1-a)\mathbb{1}_{(1,\infty)}(t)$	[142]

16.12   Additional comments

16.12.1   Entries 7–10, 12, 29–31, 40, 49–52, 57, 58, 60, 61

The entries 7–10, 12, 29–31, 40, 49–52, 57, 58, 60 and 61 are taken from [50] and [187]. Since these papers are written from a probabilistic point of view, we give a brief sketch how the representation measures can be derived from the information given there.

Let $\mathbb{G}$ be a strictly positive random variable with distribution function F and assume that

$$\int_0^\infty (1 \wedge t)\, \mathbb{E}[e^{-t\mathbb{G}}]\, \frac{dt}{t} < \infty.$$

Define

$$g(\lambda) := \exp\left\{ -\int_0^\infty (1 - e^{-\lambda t})\, \mathbb{E}[e^{-t\mathbb{G}}]\, \frac{dt}{t} \right\}.$$

An application of the Frullani integral identity, cf. [50, (1.64), p. 323] yields

$$g(\lambda) = \exp\left\{ -\mathbb{E}\left[\log\left(1 + \frac{\lambda}{\mathbb{G}}\right) \right] \right\} = \exp\left\{ -\int_{(0,\infty)} \log\left(1 + \frac{\lambda}{t}\right) F(dt) \right\}.$$

Thus,

$$f(\lambda) := -\log g(\lambda) = \int_0^\infty (1 - e^{-\lambda t})\, \mathbb{E}[e^{-t\mathbb{G}}]\, \frac{dt}{t} = \int_{(0,\infty)} \log\left(1 + \frac{\lambda}{t}\right) F(dt)$$

is contained in $\mathcal{TBF}$ with the Thorin measure $\tau(dt) = F(dt)$. By Remark 8.3 (iii), $\lambda f'(\lambda)$ is in $\mathcal{CBF}$ with Stieltjes measure equal to the Thorin measure of f, namely $F(dt)$. By [50, Thm. 1.7, Point 3] we have

$$f'(\lambda) = \mathbb{E}\left(\frac{1}{\lambda + \mathbb{G}} \right),$$

hence,

$$\lambda f'(\lambda) = \mathbb{E}\left(\frac{\lambda}{\lambda + \mathbb{G}} \right) = \int_{(0,\infty)} \frac{\lambda}{\lambda + t}\, F(dt).$$

We illustrate the method for the family of random variables $\mathbb{G}_\alpha$, $0 \leq \alpha \leq 1$. Their densities are given in formulae (3.3), (3.4), (3.5) and (3.6) of [187]:

$$f_{\mathbb{G}_\alpha}(t) = \frac{\alpha \sin(\pi\alpha)}{(1-\alpha)\pi} \frac{t^{\alpha-1}(1-t)^{\alpha-1}}{(1-t)^{2\alpha} - 2(1-t)^\alpha t^\alpha \cos(\pi\alpha) + t^{2\alpha}}\, 1_{[0,1]}(t),$$

$$0 < \alpha < 1,$$

$$\mathbb{G}_1 \text{ is uniformly distributed on } [0,1],$$

$$\mathbb{G}_0 = \frac{1}{1 + \exp(\pi C)}, \quad C \text{ is a standard Cauchy random variable.}$$

The basic formula is [187, (3.2), p. 41] saying that for $0 \le \alpha \le 1$

$$\mathbb{E}(e^{-\lambda \Gamma_1(\mathbb{G}_\alpha)}) = \exp\left\{-\int_0^\infty (1 - e^{-\lambda t})\, \mathbb{E}[e^{-t\mathbb{G}_\alpha}]\, \frac{dt}{t}\right\}.$$

In [187] explicit formulae for the Laplace exponent of $\Gamma_1(\mathbb{G}_\alpha)$ are provided, i.e. explicit formulae for the function g defined above with $\mathbb{G}$ standing for any member of the family $\mathbb{G}_\alpha$. The distribution $F(dt)$ can now be computed from the density $f_{\mathbb{G}_\alpha}(t)$.

16.12.2   Entries 20, 65 and 66

The entries 20, 65 and 66 are from [68]. We give an explanation for 65, which comes from [68, Appendix 1.5, pp. 121–122]. Take $a = 0$, $b = 2$, $x = y = 1$, replace α in [68] by λ. Then the formula for the λ-Green function reads

$$\frac{\cosh^2(\sqrt{2\lambda})}{\sqrt{2\lambda}\,\sinh(\sqrt{2\lambda})} = G_\lambda(1,1) = \int_0^\infty e^{-\lambda t}\, p(t,1,1)\, dt,$$

where

$$p(t,1,1) = \frac{1}{2}\left(\frac{1}{2} + \sum_{n=1}^\infty e^{-n^2\pi^2 t/8} \cos^2\left(\frac{n\pi}{2}\right)\right) = \int_{[0,\infty)} e^{-ts}\, \tilde\sigma(ds),$$

and

$$\tilde\sigma = \frac{1}{4}\delta_0 + \sum_{n=1}^\infty \frac{1}{2}\cos^2\left(\frac{n\pi}{2}\right)\delta_{n^2\pi^2/8}.$$

Thus,

$$G_\lambda(1,1) = \int_0^\infty e^{-\lambda t}\, dt \int_{[0,\infty)} e^{-ts}\, \tilde\sigma(ds) = \int_{[0,\infty)} \frac{1}{\lambda + s}\, \tilde\sigma(ds).$$

Hence,

$$\frac{\cosh^2(\sqrt{2\lambda})}{\sqrt{2\lambda}\,\sinh(\sqrt{2\lambda})} - \frac{1}{4} = \int_{[0,\infty)} \frac{1}{\lambda + s}\, \sigma(ds)$$

with

$$\sigma = \sum_{n=1}^\infty \frac{1}{2}\cos^2\left(\frac{n\pi}{2}\right)\delta_{n^2\pi^2/8}.$$

16.12.3   Entries 80 and 81

Entry 80 (and, similarly, 81) can be obtained in the following way. Entry 70 shows that

$$f_1(\lambda) = \log\left(\sinh(\sqrt{2\lambda})\right) - \log(\sqrt{2\lambda})$$

is in $\mathcal{CBF}$. From entry 78 we know that

$$f_2(\lambda) = \frac{1}{2}\left(\cosh^{-1}(1+\lambda)\right)^2$$

is also in $\mathcal{CBF}$. Therefore, $f_1 \circ f_2 \in \mathcal{CBF}$. A straightforward computation gives that $f_1 \circ f_2 = f$. Proposition 6.2 of [262] gives the Lévy density $m(t)$.

Appendix

In this appendix we collect a few concepts and results which are not used in a uniform way or which are particular to one branch of mathematics.

A.1 Vague and weak convergence of measures

Throughout this section E will be a locally compact separable metric space equipped with its Borel σ-algebra $\mathscr{B}(E)$. We write $\mathscr{M}^+(E)$ for the set of all locally finite, inner regular Borel measures on E, i.e. the *Radon measures* on E. Most of the time we will use Greek letters $\mu, \nu, \ldots$ to denote elements in $\mathscr{M}^+(E)$; $\mathscr{M}_b^+(E)$ are all finite measures, i.e. all $\mu \in \mathscr{M}^+(E)$ such that $\mu(E) < \infty$. Full proofs for the topics in this section can be found in Bauer [26, Sections 30, 31] or Schwartz [329, vol. 3, Chapter V.13].

Definition A.1. A sequence $(\mu_n)_{n \in \mathbb{N}}$ in $\mathscr{M}^+(E)$ converges *vaguely* to $\mu \in \mathscr{M}^+(E)$, if

$$\lim_{n \to \infty} \int_E u(x) \, \mu_n(dx) = \int_E u(x) \, \mu(dx) \quad \text{for all } u \in C_c(E). \tag{A.1}$$

A sequence of finite measures $(\nu_n)_{n \in \mathbb{N}}$ in $\mathscr{M}_b^+(E)$ converges *weakly* to $\nu \in \mathscr{M}_b^+(E)$, if

$$\lim_{n \to \infty} \int_E u(x) \, \nu_n(dx) = \int_E u(x) \, \nu(dx) \quad \text{for all } u \in C_b(E). \tag{A.2}$$

We denote the closure of $C_c(E)$ with respect to the uniform norm $\|\cdot\|_\infty$ by $C_\infty(E)$; these are the continuous functions vanishing at infinity. Note that $(C_\infty(E), \|\cdot\|_\infty)$ is a Banach space. It is not hard to see that we can replace (A.1) by

$$\lim_{n \to \infty} \int_E u(x) \, \mu_n(dx) = \int_E u(x) \, \mu(dx) \quad \text{for all } u \in C_\infty(E). \tag{A.3}$$

Remark A.2. If E is a compact topological space, $\mathscr{M}^+(E) = \mathscr{M}_b^+(E)$, and vague convergence and weak convergence coincide. Weak convergence is, in general, *not* the weak convergence used in functional analysis. Topologically, the space of signed finite Radon measures $\mathscr{M}_b(E) := \{\mu - \nu : \mu, \nu \in \mathscr{M}_b^+(E)\}$ can be identified with the topological dual of $C_\infty(E)$. Thus, *vague* convergence is actually the topological weak* convergence $\sigma(\mathscr{M}_b(E), C_\infty(E))$.

Remark A.3. Let $(\mu_t)_{t\geq 0}$ be a family of measures in $\mathcal{M}^+(E)$ such that the function $t \mapsto \int_E u(x)\,\mu_t(dx)$ is measurable for every $u \in C_c(E)$, and let κ be a measure on $[0,\infty)$ such that

$$\Lambda(u) := \int_{[0,\infty)} \int_E u(x)\,\mu_t(dx)\,\kappa(dt) < \infty \qquad\qquad (A.4)$$

for all $u \in C_c(E)$. Then Λ defines a positive linear functional on $C_c(E)$ which we may identify with a measure $\lambda \in \mathcal{M}^+(E)$. Formally,

$$\lambda = \int_{[0,\infty)} \mu_t\,\kappa(dt)$$

and we call this a *vague integral*, provided that (A.4) holds for all $u \in C_c(E)$. Using standard approximation techniques it is easy to show that

$$\lambda(B) = \int_{[0,\infty)} \mu_t(B)\,\kappa(dt), \qquad B \in \mathcal{B}(E).$$

We have the following relation between vague convergence and weak convergence of measures.

Theorem A.4. *A sequence of measures* $(\mu_n)_{n\in\mathbb{N}}$, $\mu_n \in \mathcal{M}_b^+(E)$, *converges weakly to* μ *if, and only if, it converges vaguely to* μ *and if the total masses converge to the total mass of the limit*
$$\lim_{n\to\infty} \mu_n(E) = \mu(E).$$

A vaguely convergent sequence of measures $(\mu_n)_{n\in\mathbb{N}}$, $\mu_n \in \mathcal{M}^+(E)$, is necessarily *vaguely bounded*, i.e. for each $u \in C_c(E)$ or for each compact set $K \subset E$ we have

$$\sup_{n\in\mathbb{N}} \int u\,d\mu_n < \infty \quad\text{or}\quad \sup_{n\in\mathbb{N}} \mu_n(K) < \infty,$$

but some kind of converse is also true.

Theorem A.5. *A subset* $\mathcal{F} \subset \mathcal{M}^+(E)$ *is vaguely relatively compact if, and only if,* $\mathcal{F}$ *is vaguely bounded.*

This means, in particular, that every vaguely bounded sequence of Radon measures has at least one vaguely convergent subsequence. The Banach–Alaoglu theorem for the dual pair $(C_\infty(E), \mathcal{M}^+(E))$ can thus be written as

Corollary A.6 (Banach–Alaoglu). *Every ball* $\{\mu \in \mathcal{M}_b^+(E) \ : \ \mu(E) \leq r\}$ *of radius* $r > 0$ *is vaguely, or weak*, compact.*

Weak convergence can also be characterized via the *portmanteau theorem*.

Theorem A.7. *A sequence of measures* $(\mu_n)_{n\in\mathbb{N}}$, $\mu_n \in \mathcal{M}_b^+(E)$, *converges weakly to a measure* $\mu \in \mathcal{M}_b^+(E)$ *if one, hence all, of the following equivalent conditions are satisfied:*

(i) $\limsup_{n\to\infty} \mu_n(F) \leq \mu(F)$ *for all closed sets* $F \subset E$;

(ii) $\liminf_{n\to\infty} \mu_n(G) \geq \mu(G)$ *for all open sets* $G \subset E$;

(iii) $\lim_{n\to\infty} \mu_n(B) = \mu(B)$ *for all Borel sets* $B \subset E$ *with* $\mu(\partial B) = 0$.

If μ_n, μ *are finite measures on the real line* $\mathbb{R}$, *we can add a further equivalence in terms of their distribution functions:*

(iv) $\lim_{n\to\infty} \mu_n(-\infty, x] = \mu(-\infty, x]$ *for all* $x \in \mathbb{R}$ *such that* $\mu\{x\} = 0$.

Recall that $F(x) := \mu(-\infty, x]$ is a distribution function and that we have a one-to-one relation between finite measures on the real line and all bounded, non-decreasing and right-continuous functions $F : \mathbb{R} \to \mathbb{R}$. In the context of vague convergence and weak convergence the classical Helly's selection theorem becomes:

Corollary A.8 (Helly). *If* $(F_n)_{n\in\mathbb{N}}$ *is a sequence of distribution functions which is uniformly bounded, i.e.* $\sup_{n\in\mathbb{N}, x\in\mathbb{R}} |F_n(x)| < \infty$, *then there exists a distribution function* $F(x)$ *and a subsequence* $(F_{n_k})_{k\in\mathbb{N}}$ *such that*

$$\lim_{k\to\infty} F_{n_k}(x) = F(x) \quad \text{at all continuity points } x \text{ of } F.$$

The corresponding sequence of finite measures, $(\mu_{n_k})_{k\in\mathbb{N}}$ *converges vaguely to the finite measure* μ *induced by* F.

It converges weakly if, and only if, $\lim_{R\to\infty} \sup_k [F_{n_k}(R) - F_{n_k}(-R)] = 0$, *or equivalently, if* $\lim_{k\to\infty} \mu_{n_k}(\mathbb{R}) = \mu(\mathbb{R})$.

Lemma A.9. *Let* $(\nu_n)_{n\in\mathbb{N}}$ *be a sequence of finite measures on* $[0, \infty)$. *Then the vague limit* $\nu = \lim_{n\to\infty} \nu_n$ *exists if, and only if,* $\lim_{n\to\infty} \mathscr{L}(\nu_n; x) = g(x)$ *exists for all* $x > 0$. *If this is the case,* $g(x) = \mathscr{L}(\nu; x)$.

The measures converge weakly if, and only if, $\lim_{n\to\infty} \mathscr{L}(\nu_n; x) = g(x)$ *for all* $x \geq 0$. *If this is the case,* $g(x) = \mathscr{L}(\nu; x)$.

Proof. The assertion on vague convergence is a consequence of the closedness of the set $\mathcal{CM}$, see the proof of Theorem 1.6. If, in addition,

$$\mathscr{L}(\nu_n, 0) = \nu_n[0, \infty) \xrightarrow{n\to\infty} \nu[0, \infty) = \mathscr{L}(\nu; 0)$$

holds, we can use Theorem A.4 to get weak convergence.   □

A.2   Hunt processes and Dirichlet forms

In this section we collect some basic definitions and results pertinent to Hunt processes and the theory of Dirichlet forms. Let E be a locally compact separable metric space and let E_∂ denote its one-point compactification. If E is already compact, then ∂ is added as an isolated point. The Borel σ-algebras on E and E_∂ will be denoted by $\mathscr{B} = \mathscr{B}(E)$ and $\mathscr{B}_\partial := \mathscr{B}(E_\partial)$, respectively.

Let $(\Omega, \mathscr{F})$ be a measurable space. A *stochastic process* with time-parameter set $[0, \infty)$ and state space E is a family $X = (X_t)_{t \geq 0}$ of measurable maps $X_t : \Omega \to E$. A *filtration* on $(\Omega, \mathscr{F})$ is a non-decreasing family $(\mathscr{G}_t)_{t \geq 0}$ of sub-σ-algebras of $\mathscr{F}$. The filtration $(\mathscr{G}_t)_{t \geq 0}$ is said to be right-continuous if $\mathscr{G}_t = \mathscr{G}_{t+} := \bigcap_{s > t} \mathscr{G}_s$ for every $t \geq 0$. The stochastic process $X = (X_t)_{t \geq 0}$ is *adapted* to the filtration $(\mathscr{G}_t)_{t \geq 0}$ if for every $t \geq 0$, X_t is $\mathscr{G}_t$ measurable. A *stopping time* with respect to the filtration $(\mathscr{G}_t)_{t \geq 0}$ is a mapping $T : \Omega \to [0, \infty]$ such that $\{T \leq t\} \in \mathscr{G}_t$ for each $t \geq 0$. The σ-algebra $\mathscr{G}_T$ is defined as $\mathscr{G}_T = \{F \in \mathscr{F} : F \cap \{T \leq t\} \in \mathscr{G}_t \quad \text{for all } t \geq 0\}$. Let $\mathscr{F}_\infty^0 := \sigma(X_s : s \geq 0)$ and for $t \geq 0$, $\mathscr{F}_t^0 := \sigma(X_s : 0 \leq s \leq t)$.

Definition A.10. Let $X = (X_t)_{t \geq 0}$ be a stochastic process on $(\Omega, \mathscr{F})$ with state space $(E_\partial, \mathscr{B}_\partial)$, and let $(\mathbb{P}_x)_{x \in E_\partial}$ be a family of probability measures on $(\Omega, \mathscr{F})$. The family $X = \big((X_t)_{t \geq 0}, (\mathbb{P}_x)_{x \in E_\partial}\big)$ is called a *Markov process* on $(E, \mathscr{B})$ if the following conditions hold true

(M1) $x \mapsto \mathbb{P}_x(X_t \in B)$ is $\mathscr{B}(E)$ measurable for every $B \in \mathscr{B}(E)$.

(M2) There exists a filtration $(\mathscr{G}_t)_{t \geq 0}$ on $(\Omega, \mathscr{F})$ such that X is adapted to $(\mathscr{G}_t)_{t \geq 0}$ and
$$\mathbb{P}_x(X_{t+s} \in B \,|\, \mathscr{G}_t) = \mathbb{P}_{X_t}(X_s \in B), \quad \mathbb{P}_x\text{-a.s.} \tag{A.5}$$
for every $x \in E, t, s \geq 0$ and $B \in \mathscr{B}(E)$.

(M3) $\mathbb{P}_\partial(X_t = \partial) = 1$ for all $t \geq 0$.

The Markov process X is called *normal* if, in addition,

(M4) $\mathbb{P}_x(X_0 = x) = 1$ for all $x \in E$.

The condition (M2) is called the *Markov property* of X with respect to the filtration $(\mathscr{G}_t)_{t \geq 0}$.

Let $\mathscr{M}_1(E_\partial)$ denote the family of all probability measures on E_∂. For $\mu \in \mathscr{M}_1(E_\partial)$ and $\Lambda \in \mathscr{F}$ we define $\mathbb{P}_\mu(\Lambda) := \int_{E_\partial} \mathbb{P}_x(\Lambda) \, \mu(dx)$. For $\mu \in \mathscr{M}_1(E_\partial)$ we denote by $\mathscr{F}_\infty^\mu$ the completion of $\mathscr{F}_\infty^0$ with respect to $\mathbb{P}_\mu$ and $\mathscr{F}_t^\mu$ is the $\mathbb{P}_\mu$-completion of $\mathscr{F}_t^0$ within $\mathscr{F}_\infty^\mu$; finally, set $\mathscr{F}_t := \bigcap_{\mu \in \mathscr{M}_1(E_\partial)} \mathscr{F}_t^\mu$ where $t \in [0, \infty]$. Then $(\mathscr{F}_t)_{t \geq 0}$ is called the *minimal completed admissible filtration*. If X has the Markov property with respect to $(\mathscr{F}_{t+}^0)_{t \geq 0}$, then the filtration $(\mathscr{F}_t)_{t \geq 0}$ is right-continuous, see e.g. [123, Lemma A.2.2].

Let X be a Markov process with respect to a filtration $(\mathscr{G}_t)_{t\geq 0}$. Then X is called a *strong Markov process* with respect to $(\mathscr{G}_t)_{t\geq 0}$ if the filtration $(\mathscr{G}_t)_{t\geq 0}$ is right-continuous and

$$\mathbb{P}_\mu(X_{T+s} \in B \mid \mathscr{G}_T) = \mathbb{P}_{X_T}(X_s \in B), \quad \mathbb{P}_\mu\text{-a.s.,} \tag{A.6}$$

for every $(\mathscr{G}_t)_{t\geq 0}$-stopping time T such that $\mathbb{P}_\mu(T < \infty) = 1$, and for all probability measures $\mu \in \mathscr{M}_1(E_\partial)$, $B \in \mathscr{B}_\partial$, $s \geq 0$. A Markov process X is said to be *quasi left-continuous* if for any sequence $(T_n)_{n\in\mathbb{N}}$ of $(\mathscr{G}_t)_{t\geq 0}$-stopping times increasing to a stopping time T it holds that

$$\mathbb{P}_\mu\left(\lim_{n\to\infty} X_{T_n} = X_T, \, T < \infty \right) = \mathbb{P}_\mu(T < \infty) \quad \text{for all } \mu \in \mathscr{M}_1(E_\partial). \tag{A.7}$$

Definition A.11. Let X be a normal strong Markov process on $(E, \mathscr{B}(E))$ with respect to the filtration $(\mathscr{G}_t)_{t\geq 0}$ satisfying the following condition

(M5) (i) $X_t(\omega) = \partial$ for every $t \geq \zeta(\omega)$, where $\zeta(\omega) = \inf\{t \geq 0 : X_t = \partial\}$ is the *lifetime* of X;

(ii) For each $t \geq 0$, there exists a map $\theta_t : \Omega \to \Omega$ such that $X_t \circ \theta_s = X_{t+s}$, $s \geq 0$;

(iii) For each $\omega \in \Omega$, the sample path $t \mapsto X_t(\omega)$ is right-continuous on $[0, \infty)$ and has left limits on $(0, \infty)$.

If X is also quasi left-continuous, then X is called a *Hunt process*.

The process X is a Hunt process if, and only if, X is a strong Markov process and quasi left-continuous with respect to the minimal completed admissible filtration $(\mathscr{F}_t)_{t\geq 0}$, see e.g. [123, Theorem A.2.1].

Let X be a Markov process with respect to the filtration $(\mathscr{G}_t)_{t\geq 0}$. For $x \in E, t \geq 0$ and $B \in \mathscr{B}(E)$ define the *transition function of the Markov process* X by

$$P_t(x, B) := \mathbb{P}_x(X_t \in B). \tag{A.8}$$

Then P_t is a transition kernel on $(E, \mathscr{B}(E))$, i.e. $x \mapsto P_t(x, B)$ is $\mathscr{B}(E)$ measurable for all $B \in \mathscr{B}(E)$, and $B \mapsto P_t(x, B)$ is a sub-probability measure on $\mathscr{B}(E)$ for all $x \in E$. For a measurable function $\phi : E \to \mathbb{R}$ we set $P_t\phi(x) := \int_E \phi(y)P_t(x, dy)$ whenever the integral makes sense. The Markov property of X implies that $(P_t)_{t\geq 0}$ is a *Markov transition function*, i.e. for every $\mathscr{B}(E)$ measurable non-negative function $\phi : E \to E$,

$$P_t P_s\phi = P_{t+s}\phi, \quad t, s \geq 0. \tag{A.9}$$

Note that $P_t\phi(x) = \mathbb{E}_x(\phi(X_t)\mathbb{1}_{\{t<\zeta\}})$, where $\mathbb{E}_x$ denotes the expectation with respect to the probability measure $\mathbb{P}_x$. If we agree that every function ϕ on E is extended to E_∂ by letting $\phi(\partial) = 0$, then we can write $P_t\phi(x) = \mathbb{E}_x(\phi(X_t))$. Note that the right-continuity of $t \mapsto X_t(\omega)$ for all ω implies that $(t, \omega) \mapsto X_t(\omega)$ is also measurable, hence $(t, \omega) \mapsto \mathbb{1}_B(X_t(\omega))$ is measurable for every $B \in \mathcal{B}$. By Fubini's theorem it follows that $t \mapsto P_t(x, B) = \mathbb{E}_x \mathbb{1}_B(X_t)$ is also measurable. The Laplace transform of the transition function

$$G_\lambda(x, B) = \int_0^\infty e^{-\lambda t} P_t(x, B)\, dt, \quad x \in E,\ B \in \mathcal{B}(E),\ \lambda > 0, \qquad (A.10)$$

is called the *resolvent kernel of the Markov process* X.

The transition function $P_t(x, B)$ is said to be a *Feller transition function* if every P_t maps $C_\infty(E)$ into $C_\infty(E)$ and if $\lim_{t \to 0} P_t \phi(x) = \phi(x)$ for all $x \in E$ and for all $\phi \in C_\infty(E)$. If $(P_t)_{t \geq 0}$ is a Feller transition function on E, then there exists a Hunt process on E having $(P_t)_{t \geq 0}$ as its transition function, compare [123, Theorem A.2.2.] or [60, Theorem I.9.4.]).

From now on we assume that m is a positive Radon measure on $(E, \mathcal{B}(E))$ with full support, i.e. $\operatorname{supp}(m) = E$. The transition function $(P_t)_{t \geq 0}$ is called m-*symmetric* if for all non-negative measurable functions ϕ and ψ and for all $t \geq 0$,

$$\int_E \phi(x) P_t \psi(x)\, m(dx) = \int_E \psi(x) P_t \phi(x)\, m(dx). \qquad (A.11)$$

A Hunt process X with an m-symmetric transition function P_t is called an m-*symmetric Hunt process*. For m-symmetric transition functions we see, using Cauchy's inequality,

$$\int_E (P_t \phi(x))^2\, m(dx) \leq \int_E \phi(x)^2\, m(dx), \quad \text{for all } \phi \in \mathcal{B}_b(E) \cap L^2(E, m).$$

This implies that P_t can for all $t \geq 0$ be uniquely extended to a contraction operator on $L^2(E, m)$ which we denote by T_t. Thus, $T_t : L^2(E, m) \to L^2(E, m)$. It is easy to see that $(T_t)_{t \geq 0}$ is a semigroup on $L^2(E, m)$, i.e. $T_{t+s} = T_t T_s$ for all $t, s \geq 0$. In the same way we can extend the resolvent kernels G_β, $\beta > 0$, to bounded operators on $L^2(E, m)$ which we denote by R_β. The operators $(R_\beta)_{\beta > 0}$ satisfy the *resolvent equation*

$$R_\alpha \phi - R_\beta \phi = (\beta - \alpha) R_\alpha R_\beta \phi, \quad \alpha, \beta > 0, \quad \phi \in L^2(E, m). \qquad (A.12)$$

Remark A.12. The families $(T_t)_{t \geq 0}$ and $(R_\beta)_{\beta > 0}$ are a priori defined on the underlying Banach space, while $(P_t)_{t \geq 0}$ and $(G_\beta)_{\beta > 0}$ are given by the process and have a pointwise meaning. It is possible to show, cf. [123, Theorem 4.2.3], that we have $T_t u = P_t u$ and $R_\beta u = G_\beta u$ m-almost everywhere on E.

Assume that X is an m-symmetric Hunt process with transition function $(P_t)_{t \geq 0}$. Then it holds that

$$\lim_{t \to 0} P_t \phi(x) = \lim_{t \to 0} \mathbb{E}_x \phi(X_t) = \phi(x) \quad \text{for all } x \in E \text{ and all } \phi \in C_c(E).$$

This implies, see e.g. [123, Lemma 1.4.3], that the semigroup $(T_t)_{t \geq 0}$ is *strongly continuous*: $L^2\text{-}\lim_{t \to 0} T_t \phi = \phi$ for every $\phi \in L^2(E, m)$. Similarly, the resolvent $(R_\lambda)_{\lambda > 0}$ is strongly continuous in the sense that $L^2\text{-}\lim_{\lambda \to \infty} \lambda R_\lambda \phi = \phi$ for every $\phi \in L^2(E, m)$.

The *(infinitesimal) generator* A of a strongly continuous semigroup $(T_t)_{t \geq 0}$ on $L^2(E, m)$ is defined by

$$\begin{cases} A\phi := \lim_{t \to 0} \frac{T_t \phi - \phi}{t} \\ \mathfrak{D}(A) := \{\phi \in L^2(E, m) \ : \ A\phi \text{ exists as a strong limit}\}. \end{cases} \tag{A.13}$$

Let $(G_\lambda)_{\lambda > 0}$ be a strongly continuous resolvent on $L^2(E, m)$. It follows from the resolvent equation (A.12) and the strong continuity that G_λ is invertible. The *generator* A of $(G_\lambda)_{\lambda > 0}$ on $L^2(E, m)$ is defined by

$$\begin{cases} A\phi := \lambda \phi - G_\lambda^{-1} \phi \\ \mathfrak{D}(A) := G_\lambda(L^2(E, m)). \end{cases} \tag{A.14}$$

Note that the resolvent equation shows that A and $\mathfrak{D}(A)$ are independent of $\lambda > 0$.

Lemma A.13. *The generator of a strongly continuous resolvent is a negative semi-definite self-adjoint operator. Moreover, the generator of a strongly continuous semi-group coincides with the generator of its resolvent.*

Since the operator $-A$ is positive semi-definite and self-adjoint we can use the spectral calculus to define its non-negative square root $(-A)^{1/2}$ with the domain $\mathfrak{D}((-A)^{1/2})$.

Definition A.14. Let X be an m-symmetric Hunt process on E, $(T_t)_{t \geq 0}$ the corresponding strongly continuous semigroup on $L^2(E, m)$, and A the generator of the semigroup. The *Dirichlet form* of X is the symmetric bilinear form $\mathcal{E}$ on $L^2(E, m)$ defined by

$$\begin{cases} \mathcal{E}(\phi, \psi) = \langle (-A)^{1/2} \phi, (-A)^{1/2} \psi \rangle_{L^2} \\ \mathfrak{D}(\mathcal{E}) = \mathfrak{D}((-A)^{1/2}). \end{cases} \tag{A.15}$$

Here $\langle \cdot, \cdot \rangle_{L^2}$ denotes the inner product in $L^2(E, m)$.

The form $\mathcal{E}$ is a closed symmetric form on $L^2(E, m)$. For $\beta > 0$ let

$$\mathcal{E}_\beta(\phi, \psi) := \mathcal{E}(\phi, \psi) + \beta \langle \phi, \psi \rangle_{L^2}, \quad \phi, \psi \in \mathfrak{D}(\mathcal{E}).$$

Then $(\mathfrak{D}(\mathcal{E}), \mathcal{E}_\beta)$ is a Hilbert space. Since the semigroup $(T_t)_{t \geq 0}$ is Markovian, the form $\mathcal{E}$ is also Markovian, cf. [123, Theorem 1.4.1], which means that all normal contractions operate on $\mathcal{E}$, see [123, p. 5]. The Dirichlet form $\mathcal{E}$ of X is called *regular* if there exists a subset $\mathscr{C}$ of $\mathfrak{D}(\mathcal{E}) \cap C_c(E)$ such that $\mathscr{C}$ is dense in $\mathfrak{D}(\mathcal{E})$ with respect to $\mathcal{E}_1$ and dense in $C_c(E)$ with respect to the uniform norm.

Clearly, $R_\lambda(L^2(E,m)) = G_\lambda(L^2(E,m)) = \mathfrak{D}(A) \subset \mathfrak{D}(\mathcal{E})$. Therefore the connection between the resolvent $(R_\lambda)_{\lambda>0}$ and the Dirichlet form $\mathcal{E}$ can be expressed by

$$\mathcal{E}_\lambda(R_\lambda\phi, \psi) = \langle\phi, \psi\rangle_{L^2}, \quad \phi \in L^2(E,m), \; \psi \in \mathfrak{D}(\mathcal{E}). \tag{A.16}$$

For an open set $B \subset E$ let

$$\mathrm{Cap}(B) := \inf\{\mathcal{E}_1(\phi,\phi) : \phi \geq 1 \; m\text{-a.e. on } B\}, \tag{A.17}$$

and for an arbitrary $A \subset E$,

$$\mathrm{Cap}(A) := \inf\{\mathrm{Cap}(B) : B \text{ open}, \; B \supset A\}. \tag{A.18}$$

Then $\mathrm{Cap}(A)$ is called the 1-*capacity* or simply the *capacity* of the set A. Let $A \subset E$. A statement depending on points in A is said to hold *quasi everywhere* (*q.e.* for short) if there exists a set $N \subset A$ of zero capacity such that the statement holds for all points in $A \setminus N$.

For any open subset D of E, let $\tau_D = \inf\{t > 0 : X_t \notin D\}$ be the first exit time of X from D. The process X^D obtained by killing the process X upon exiting from D is defined by

$$X_t^D := \begin{cases} X_t, & t < \tau_D, \\ \partial, & t \geq \tau_D. \end{cases}$$

Then X^D is again an m-symmetric Hunt process on D. The Dirichlet form associated with X^D is $(\mathcal{E}^D, \mathfrak{D}(\mathcal{E}^D))$ where $\mathcal{E}^D = \mathcal{E}$ and

$$\mathfrak{D}(\mathcal{E}^D) = \{u \in \mathcal{F} : u = 0 \; \mathcal{E}\text{-q.e. on } E \setminus D\}, \tag{A.19}$$

cf. [123, pp. 153–154].

Let $(P_t^D)_{t\geq 0}$ and $(G_\lambda^D)_{\lambda>0}$ denote the transition semigroup and resolvent of X^D. For $\lambda > 0$ let

$$H_\lambda^D \phi(x) := \mathbb{E}_x\big(e^{-\lambda\tau_D}\phi(X_{\tau_D})\big).$$

Then for every $\phi \in L^2(E,m)$ such that $\phi = 0$ q.e. on $E \setminus D$, it holds that

$$G_\lambda\phi(x) = G_\lambda^D\phi(x) + H_\lambda^D G_\lambda\phi(x) \quad \text{for } x \in D, \tag{A.20}$$

and $\mathcal{E}_\lambda(G_\lambda^D\phi, H_\lambda^D G_\lambda\phi) = 0$. This is the $\mathcal{E}_\lambda$-orthogonal decomposition of $G_\lambda\phi$ into $G_\lambda^D\phi \in \mathfrak{D}(\mathcal{E}^D)$ and its complement, cf. [123, Sections 4.3 and 4.4].

Definition A.15. A measure μ on $(E, \mathscr{B}(E))$ is called *smooth* if it charges no set of zero capacity and if there exists an increasing sequence $(F_n)_{n\in\mathbb{N}}$ of closed sets such that $\mu(F_n) < \infty$ for all $n \in \mathbb{N}$ and

$$\lim_{n\to\infty} \mathrm{Cap}(K \setminus F_n) = 0$$

for every compact set K.

Every Radon measure on $(E, \mathscr{B}(E))$ that charges no set of zero capacity is smooth, see [123, p. 81].

For a Borel set $B \subset E$, let $T_B = \inf\{t > 0 : X_t \in B\}$ be the *hitting time* of X to B. A set $N \subset E$ is called *exceptional* for X if there exists a Borel set $\tilde{N} \supset N$ such that $\mathbb{P}_m(T_{\tilde{N}} < \infty) = 0$.

Definition A.16. Let $C = (C_t)_{t \geq 0}$ be a stochastic process with values in $[0, \infty]$. Then C is called a *positive continuous additive functional* (PCAF) of X if

(a) C is adapted to the minimal completed admissible filtration $(\mathscr{F}_t)_{t \geq 0}$

and if there exists a set $\Lambda \in \mathscr{F}_\infty$ and an exceptional set $N \subset E$ such that

(b) $\mathbb{P}_x(\Lambda) = 1$ for all $x \in E \setminus N$,

(c) $\theta_t \Lambda \subset \Lambda$ for all $t > 0$,

(d) $t \mapsto C_t(\omega)$ is continuous on $[0, \zeta(\omega))$ for all $\omega \in \Lambda$,

(e) for all $\omega \in \Lambda$ we have

$$\begin{cases} C_0(\omega) = 0, & t = 0, \\ C_t(\omega) < \infty, & t < \zeta(\omega), \\ C_t(\omega) = C_{\zeta(\omega)}(\omega), & t \geq \zeta(\omega), \end{cases}$$

(f) $C_{t+s}(\omega) = C_t(\omega) + C_s(\theta_t \omega), \quad s, t \geq 0.$

Two positive continuous additive functionals $C^{(1)}$ and $C^{(2)}$ are *equivalent* if for each $t > 0$ and for quasi every $x \in E$ we have $\mathbb{P}_x(C_t^{(1)} = C_t^{(2)}) = 1$.

Let $C = (C_t)_{t \geq 0}$ be a positive continuous additive functional of X. For a Borel function $\phi : E \to [0, \infty)$ and $\lambda > 0$ define

$$U_C^\lambda \phi(x) := \mathbb{E}_x \left(\int_{[0,\infty)} e^{-\lambda t} \phi(X_t) \, dC_t \right).$$

The PCAF C and the smooth measure μ are said to be in *Revuz correspondence* if for every $\lambda > 0$ and all non-negative Borel functions ϕ and ψ it holds that

$$\langle \phi, U_C^\lambda \psi \rangle_{L^2} = \int (G_\lambda \phi) \psi \, d\mu. \tag{A.21}$$

In this case, the measure μ is called the *Revuz measure* of the PCAF C. The condition (A.21) is one of several equivalent conditions describing the Revuz correspondence; for the others we refer to [123, Theorem 5.1.3].

Theorem A.17. *The family of all equivalence classes of positive continuous additive functionals of X and the family of all smooth measures are in one-to-one correspondence under the Revuz correspondence.*

Bibliography

[1] Abramowitz, M. and Stegun, I. A., *Handbook of Mathematical Functions* (5th edn.), Dover, New York 1968. *300*

[2] Akhiezer, N. I., *The Classical Moment Problem*, Oliver and Boyd, Edinburgh 1965. *90, 91, 193*

[3] Akhiezer, N. I. and Glazman, I. M., *Theory of Linear Operators in Hilbert Space I, II*, Frederick Ungar, New York 1961–63. (Unabridged reedition in one volume: Dover, New York 1993.) *90, 197*

[4] Akhiezer, N. I. and Kreĭn, M. G., On a generalization of the lemmas of Schwarz and Löwner, *Zap. Mat. Otd. Fiz.-Mat. Fak. i Har'kov. Mat. Obšč.* **23** (1952), 95–101 (in Russian). *91*

[5] Alzer, H. and Berg, C., Some classes of completely monotonic functions, *Ann. Acad. Sci. Fenn., Math.* **27** (2002), 445–460. *312, 314*

[6] Alzer, H. and Berg, C., Some classes of completely monotonic functions II, *Ramanujan J.* **11** (2006), 225–248. *350*

[7] Ameur, Y., Kaijser, S. and Silvestrov, S., Interpolation classes and matrix monotone functions, *J. Operator Theory* **57** (2007), 409–427. *199*

[8] Anderson, W. N. and Trapp, G. E., A class of monotone operator functions related to electrical network theory, *Linear Algebra Appl.* **15** (1976), 53–67. *199*

[9] Ando, T., Concavity of certain maps on positive definite matrices and applications to Hadamard products, *Linear Algebra Appl.* **26** (1979), 203–241. *107*

[10] Ando, T., Means of positive linear operators, *Math. Ann.* **249** (1980), 205–224. *199*

[11] Andrews, G. E., Askey, R. and Roy, R., *Special Functions*, Encycl. Math. Appl. vol. **71**, Cambridge University Press, Cambridge 1999. *352*

[12] Aronszajn, N. and Donoghue, W. F., On exponential representations of analytic functions in the upper half plane with positive imaginary part, *J. Anal. Math.*, **5** (1956), 321–385. *91*

[13] Aoyama, T., Lindner, A. and Maejima, M., A new family of mappings of infinitely divisible distributions related to the Goldie-Steutel-Bondesson class, *Electron. J. Probab.* **15** (2010), 1119–1142. *157*

[14] Aoyama, T., Maejima, M. and Rosiński, J., A subclass of type G selfdecomposable distributions on $\mathbb{R}^d$, *J. Theor. Probab.* **21** (2008), 14–34. *157, 158*

[15] Audenaert, K., Cai, L. and Hansen, F., Inequalities for quantum skew information, *Lett. Math. Phys.* **85** (2008), 135–146. *370*

[16] Balakrishnan, A. V., An operational calculus for infinitesimal generators of semi-groups, *Trans. Am. Math. Soc.* **91** (1959) 330–353. *253, 254*

[17] Balakrishnan, A. V., Fractional powers of closed operators and the semigroups generated by them, *Pacific J. Math.* **10** (1960) 419–437. *253, 254*

[18] Bañuelos, R., Intrinsic ultracontractivity and eigenfunction estimates for Schrödinger operators, *J. Funct. Anal.* **100** (1991), 181–206. *267*

[19] Bañuelos, R. and Kulczycki, T., The Cauchy process and the Steklov problem, *J. Funct. Anal.* **211** (2004), 355–423. *255*

[20] Barndorff-Nielsen, O. E., Maejima, M. and Sato, K., Some classes of multivariate infinitely divisible distributions admitting stochastic integral representation, *Bernoulli* **12** (2006), 1–33. *135, 157, 158*

[21] Barndorff-Nielsen, O. E. and Thorbjørnsen, S., Lévy laws in free probability, *Proc. Natl. Acad. Sci. USA* **99** (2002), 16568–16575. *135*

[22] Barndorff-Nielsen, O. E. and Thorbjørnsen, S., Lévy processes in free probability, *Proc. Natl. Acad. Sci. USA* **99** (2002), 16576–16580. *135, 157*

[23] Bass, R. F., *Probabilistic Techniques in Analysis*, Springer, Prob. Appl., New York 1995. *264*

[24] Bass, R. F., *Diffusions and Elliptic Operators*, Springer, Prob. Appl., New York 1998. *264*

[25] Bass, R. F. and Burdzy, K., Lifetimes of conditioned diffusions, *Probab. Theory Relat. Fields* **91** (1992), 405–444. *267*

[26] Bauer, H., *Measure and Integration Theory*, de Gruyter, Studies in Math. **26**, Berlin 2001. *374*

[27] Beckenbach, E. F. and Bellman, R., *Inequalities*, Ergebnisse Math. Grenzgeb. (NF) vol. **40**, Springer, Berlin 1961. *255*

[28] Becker, R., *Convex Cones in Analysis*, Hermann, travaux en cours vol. **67**, Paris 2006. *15*

[29] Bendat, J. and Sherman, S., Monotone and convex operator functions, *Trans. Am. Math. Soc.* **79** (1955), 58–71. *198, 199*

[30] Bendikov, A. and Maheux, P., Nash-type inequalities for fractional powers of non-negative self-adjoint operators, *Trans. Am. Math. Soc.* **359** (2007), 3085–3097. *255*

[31] Berenstein, C. A. and Gay, R., *Complex Variables. An Introduction*, Springer, Grad. Texts Math. vol. **125**, New York 1997. *6*

[32] Berg, C., The Stieltjes cone is logarithmically convex, in: Laine, I., Lehto, O. and Sorvali, T. (eds.), *Complex Analysis, Joensuu 1978*, Springer, Lect. Notes Math. vol. **747**, Berlin 1979, 46–54. *19, 90, 107*

[33] Berg, C., Quelques remarques sur le cone de Stieltjes, in: Hirsch, F. and Mokobodzki, G. (eds.), *Séminaire de Théorie du Potentiel, Paris, No. 5*, Springer, Lect. Notes Math. vol. **814**, Berlin 1980, 70–79. *19, 91, 107*

[34] Berg, C., Integral representation of some functions related to the Gamma function, *Mediterranean J. Math.* **1** (2004), 433–439. *67, 342*

[35] Berg, C., Stieltjes–Pick–Bernstein–Schoenberg and their connection to complete monotonicity, in: Mateu, J. and Porcu, E., *Positive Definite Functions. From Schoenberg to Space-Time Challenges*, Dept. of Mathematics, University Jaume I, Castellon, Spain, 2008. *19, 21, 90, 91, 107, 299*

[36] Berg, C., Boyadzhiev, K. and deLaubenfels, R., Generation of generators of holomorphic semigroups, *J. Austral. Math. Soc. (A)* **55** (1993), 246–269. *216, 253*

[37] Berg, C., Christensen, J. P. R. and Ressel, P., Positive definite functions on abelian semigroups, *Math. Ann.* **223** (1976), 253–272. *47*

[38] Berg, C., Christensen, J. P. R. and Ressel, P., *Harmonic Analysis on Semigroups*, Springer, Grad. Texts Math. vol. **100**, New York 1984. *13, 34, 35, 46, 47*

[39] Berg, C. and Forst, G., *Potential Theory on Locally Compact Abelian Groups*, Springer, Ergeb. Math. Grenzgeb. vol. **87**, Berlin 1975. *13, 19, 21, 34, 35, 45, 46, 47, 66, 67, 165, 176, 178, 232*

[40] Berg, C. and Pedersen, H. L., A completely monotone function related to the Gamma function, *J. Comput. Appl. Math.* **133** (2001), 219–230. *346*

[41] Berg, C. and Pedersen, H. L., Pick functions related to the Gamma function, *Rocky Mountain J. Math.* **32** (2002), 507–525. *346*

[42] Berg, C. and Pedersen, H. L., A one-parameter family of Pick functions defined by the Gamma function and related to the volume of the unit ball in n-space, *Proc. Am. Math. Soc.* **139** (2011), 2121–2132. *346*

[43] Bernstein, S., Sur la définition et les propriétés des fonctions analytiques d'une variable réelle. *Math. Ann.* **75** (1914), 449–468. *13*

[44] Bernstein, S., *Leçons sur les propriétés extrémales et la meilleure approximation des fonctions analytiques d'une variable réelle*, Gauthier-Villars, Collection de Monographies sur la Théorie des Fonctions (Collection Borel), Paris 1926. *13*

[45] Bernstein, S., Sur les fonctions absolument monotones, *Acta Math.* **52** (1929), 1–66. *2, 13*

[46] Bertoin, J., Application de la théorie spectrale des cordes vibrantes aux fonctionnelles additives principales d'un brownien réfléchi, *Ann. Inst. Henri Poincaré, Probab. Stat.* **25** (1989), 307–323. *297*

[47] Bertoin, J., *Lévy Processes*, Cambridge University Press, Camb. Tracts Math. vol. **121**, Cambridge 1996. *21, 33, 34, 51, 66, 67, 166, 269*

[48] Bertoin, J., Regenerative embeddings of Markov sets, *Probab. Theory Relat. Fields* **108** (1997), 559–571. *178*

[49] Bertoin, J., Subordinators: examples and applications, in: *Lectures on Probability and Statistics, Ecole d'Eté de Probabilités de Saint-Flour XXVII-1997*, Springer, Lect. Notes Math. vol. **1717**, Berlin 1999, 1–91. *21, 33, 66, 253*

[50] Bertoin, J., Fujita, T., Roynette B. and Yor, M., On a particular class of self-decomposable random variables: the durations of Bessel excursions straddling independent exponential times, *Probab. Math. Stat.* **26** (2006), 315–366. *306, 308, 316, 320, 324, 326, 371*

[51] Bertoin, J. and Yor, M., On subordinators, self-similar Markov processes and some factorizations of the exponential random variable, *Electron. Commun. Probab.* **6** (2001), 95–106. *340*

[52] Beurling, A., Sur quelques formes positives avec une application à la théorie ergodique, *Acta Math.* **78** (1946), 319–334. Also in [76, vol. 2], 71–86. *46*

[53] Beurling, A., On the spectral synthesis of bounded functions, *Acta Math.* **81** (1948), 1–14. Also in [76, vol. 2], 93–106. *47*

[54] Beurling, A. and Deny, J., Dirichlet spaces, *Proc. Natl. Acad. Sci.* **45** (1958), 208–215. Also in [76, vol. 2], 201–208. *47*

[55] Bhatia, R., *Matrix Analysis*, Springer, Grad. Texts Math. vol. **169**, New York 1996. *198*

[56] Bhatia, R., *Positive Definite Matrices*, Princeton University Press, Princeton Ser. Appl. Math., Princeton (NJ) 2007. *198*

[57] Bihari, I., A generalization of a lemma of Bellman and its application to uniqueness problems of differential equations, *Acta Math. Hungarica* **7** (1956), 81–94. *255*

[58] Bingham, N. H., Goldie, C. M. and Teugels, J., *Regular Variation*, Cambridge University Press, Encycl. Math. Appl. vol. **27**, Cambridge 1987. *34*

[59] Blumenthal, R. M., Some relationships involving subordination, *Proc. Am. Math. Soc.* **10** (1959), 502–510. *253*

[60] Blumenthal, R. M. and Getoor, R. K., *Markov Processes and Potential Theory*, Academic Press, Pure Appl. Math. vol. **29**, New York 1968. *262, 265, 267, 379*

[61] Bochner, S., *Vorlesungen über Fouriersche Integrale*, Akademische Verlagsgesellschaft, Mathematik und ihre Anwendungen in Monographien und Lehrbüchern vol. **12**, Leipzig 1932. (Unaltered reprint: Chelsea, New York 1948.) *47*

[62] Bochner, S., Diffusion equation and stochastic processes, *Proc. Natl. Acad. Sci. U.S.A.* **35** (1949) 368–370. Also in: Gunning, R. C. (ed.), *Collected Papers of Salomon Bochner (4 vols.)*, Am. Math. Soc., Providence (RI) 1992, vol. 2, 465–467. *33, 68, 253*

[63] Bochner, S., *Harmonic Analysis and the Theory of Probability*, University of California Press, California Monogr. Math. Sci., Berkeley 1955. *21, 27, 33, 34, 68, 135, 212, 253, 254*

[64] Bogdan, K., Byczkowski, T., Kulczycki, T., Ryznar, R., Song R. and Vondraček, Z., *Potential Analysis of Stable Processes and Its Extensions*, Springer, Lect. Notes Math. vol. **1980**, Berlin, 2009. *267*

[65] Bonan-Hamada, C. and Jones, W. B., Stieltjes continued fractions and special functions: a survey. *Commun. Anal. Theory Contin. Fract.* **12** (2004), 5–68. *332, 340, 342, 344, 346*

[66] Bondesson, L., Classes of infinitely divisible distributions and densities, *Z. Wahrscheinlichkeitstheor. verw. Geb.* **57** (1981), 39–71. Correction and Addendum, *Z. Wahrscheinlichkeitstheor. verw. Geb.* **59** (1982), 277. *116, 129*

[67] Bondesson, L., *Generalized Gamma Convolutions and Related Classes of Distributions and Densities*, Springer, Lect. Notes Stat. vol. **76**, New York 1992. *116, 127, 129, 130, 320, 328, 342, 360*

[68] Borodin, A. N. and Salminen, P., *Handbook of Brownian Motion – Facts and Formulae (2nd ed.)*, Birkhäuser, Probab. Appl., Basel 2000. *312, 330, 332, 372*

[69] Bouleau, N., Quelques résultats probabilistes sur la subordination au sens de Bochner, in: *Seminar on Potential Theory, Paris, No. 7*, Springer, Lect. Notes Math. vol. **1061**, Berlin 1984, 54–81. *257*

[70] Bourbaki, N., *Elements of Mathematics: Topological Vector Spaces Chapters 1–5*, Springer, Berlin 1987. *250*

[71] Brauer, A. and Rohrbach, H., *Issai Schur Gesammelte Abhandlungen (3 vols.)*, Springer, Berlin 1973. *402*

[72] Brown, M., Further monotonicity properties for specialized renewal processes, *Ann. Probab.* **9** (1981), 891–895. *171*

[73] Butzer, P. L. and Berens, H., *Semi-Groups of Operators and Approximation*, Springer, Grundlehren Math. Wiss. vol. **145**, Berlin 1967. *254*

[74] Carlen, E., Kusuoka, S. and Stroock, D. W., Upper bounds for symmetric Markov transition functions, *Ann. Inst. Henri Poincaré, Probab. Stat.* Supplement au no. **23.2** (1987), 245–287. *255*

[75] Carathéodory, C., Über die Winkelderivierten von beschränkten analytischen Funktionen, *Sitzungsber. Preuß. Akad. Wiss. Berlin. Phys.-math. Kl.* **32** (1929), 39–54. Also in: *C. Carathéodory: Gesammelte mathematische Schriften (5 Bde.)*, C. H. Beck, München 1954–57, vol. III, 185–204. *90*

[76] Carleson, L. et al. (eds.), *Collected Works of Arne Beurling (2 vols.)*, Birkhäuser, Contemporary Mathematicians, Boston 1989. *386*

[77] Cauer, W., The Poisson integral for functions with positive real part, *Bull. Am. Math. Soc.* **38** (1932), 713–717. *90*

[78] Chatterji, S. D., Commentary on the papers [150] and [151], in: [152], 163–171 and 222–235. *14*

[79] Chaudhry, M. A., Laplace transforms of certain functions with applications, *Int. J. Math. Sci.* **23** (2000), 99–102. *358*

[80] Chen, M.-F., *Eigenvalues, Inequalities, and Ergodic Theory*, Springer, Probability and Its Applications, London 2005. *255*

[81] Chen, Z.-Q. and Song, R., Two-sided eigenvalue estimates for subordinate processes in domains, *J. Funct. Anal.* **226** (2005), 90–113. *255*

[82] Chen, Z.-Q. and Song, R., Continuity of eigenvalues of subordinate processes in domains, *Math. Z.* **252** (2006), 71–89. *256*

[83] Chen, Z.-Q. and Song, R., Spectral properties of subordinate processes in domains, in: *Stochastic Analysis and Partial Differential Equations, Contemp. Math.* vol. **429**, Am. Math. Soc., Providence (RI) 2007, 77–84. *248, 255*

[84] Choquet, G., *Lectures on Analysis vol. II*, W. A. Benjamin, Math. Lect. Notes Ser., New York, 1969. *15*

[85] Choquet, G., Deux examples classiques de représentation intégrale, *Enseign. Math.* **15** (1969), 63–75. *2, 15*

[86] Chung, K. L., *Lectures on Boundary Theory for Markov Chains*, Princeton University Press, Ann. Math. Stud. vol. **65**, Princeton 1970. *178*

[87] Chung, K. L., *Lectures from Markov Processes to Brownian Motion*, Springer, Grundlehren Math. Wiss. vol. **249**, New York 1982. *262, 266, 267*

[88] Davies, E. B., *One-Parameter Semigroups*, Academic Press, London Math. Soc. Monogr. vol. **15**, London 1980. *197, 198, 200, 204, 253*

[89] Davies, E. B., *Heat Kernels and Spectral Theory*. Cambridge University Press, Camb. Tracts Math. vol. **92**, Cambridge 1989. *247, 261*

[90] Davies, E. B. and Simon, B., Ultracontractivity and the heat kernel for Schrödinger operators and Dirichlet Laplacians, *J. Funct. Anal.* **59** (1984), 335–395. *261, 267*

[91] DeBlassie, R. D., Higher order PDEs and symmetric stable processes, *Probab. Theory Relat. Fields* **129** (2004), 495–536. Correction: *Probab. Theory Relat. Fields*, **33** (2005), 141–143. *243, 255*

[92] de Branges, L., *Hilbert Spaces of Entire Functions*, Prentice-Hall, Ser. Mod. Anal., Englewood Cliffs (NJ) 1968. *91*

[93] Dellacherie, C. and Meyer, P.-A., *Probabilités et potentiel: chapitres IX à XI. Théorie discrète du potentiel*, Hermann, Actual. sci. ind. **1410**, Publ. Inst. Math. Univ. Strasbourg t. **XVIII**, Paris 1983. *46, 47, 67*

[94] Dellacherie, C. and Meyer, P.-A., *Probabilités et potentiel: chapitres XII à XVI. Théorie des processus de Markov*, Hermann, Actual. sci. ind. **1417**, Publ. Inst. Math. Univ. Strasbourg t. **XIX**, Paris 1987. *67*

[95] Doetsch, G., Sätze vom Tauberschen Charakter im Gebiet der Laplace- und Stieltjes-Transformation, *Sitzungsber. Preuß. Akad. Wiss. Berlin. Phys.-math. Kl.* (1930), 144–157. *19*

[96] Doetsch, G., *Handbuch der Laplace-Transformation (3 Bde.)*, Birkhäuser, Math. Reihe Bde. **14, 15, 19**, Basel 1950–56. *13*

[97] Donati-Martin, C. and Yor, M., Some explicit Krein representations of certain subordinators, including the Gamma process, *Publ. Res. Inst. Math. Sci.* **42** (2006), 879–895. *284*

[98] Donati-Martin, C. and Yor, M., Further examples of explicit Krein representations of certain subordinators, *Publ. Res. Inst. Math. Sci.* **43** (2007), 315–328. *284, 358, 360*

[99] Donoghue, W. F., *Distributions and Fourier Transforms*, Academic Press, Pure Appl. Math. vol. **32**, New York 1969. *40*

[100] Donoghue, W. F., *Monotone Matrix Funcions and Analytic Continuation*, Springer, Grundlehren Math. Wiss. vol. **207**, New York 1974. *90, 198, 199*

[101] Doob, J. L. and Koopman, B., On analytic functions with positive imaginary parts, *Bull. Am. Math. Soc.* **40** (1934), 601–606. *179, 197*

[102] Dubourdieu, M. J., Sur un théorème de M. S. Bernstein relatif à la transformation de Laplace–Stieltjes, *Compos. Math.* **7** (1940), 96–111. *2, 14*

[103] Dubuc, S., Le théorème de Bernstein sur les fonctiones complètement monotones, *Canad. Math. Bull.* **12** (1969), 517–520. *15*

[104] Duren, P. L., *Theory of H^p Spaces*, Academic Press, Pure Appl. Math. vol. **38**, New York 1970. (Unabridged and corrected republication: Dover, Mineola (NY), 2000.) *91*

[105] Dym, H. and McKean, H. P., *Gaussian Processes, Function Theory, and the Inverse Spectral Problem*, Academic Press, Probab. Math. Stat. vol. **31**, New York 1976. *272, 275, 276, 284, 292, 297*

[106] Erdélyi, A. et al., *Higher Transcendental Functions 1, 2, 3 (The Bateman Manuscript Project)*, McGraw-Hill, New York, 1953. *300, 354*

[107] Erdélyi, A. et al., *Tables of Integral Transforms 1, 2 (The Bateman Manuscript Project)*, McGraw-Hill, New York 1954. *13, 300, 308, 312, 318, 320, 328, 332, 334, 336, 338, 340, 342, 344, 346, 348, 350, 356, 358, 360*

[108] Ethier, S. E. and Kurtz, T. G., *Markov Processes: Characterization and Convergence*, Wiley, Ser. Probab. Stat., Hoboken (NJ) 1985. *253*

[109] Faa di Bruno, C. F., Note sur une nouvelle formule du calcul differentiel, *Quart. J. Math.* **1** (1855), 359–360. *27*

[110] Fan, K., Les fonctions définies-positives et les Fonctions complètement monotones. Leurs applications au calcul des probabilités et à la théorie des espaces distanciés, *Mém. Sci. Math.* **CXIV** (1950), 1–48. *14*

[111] Faraut, J., Puissances fractionnaires d'un noyeau de Hunt, *Sém. Brelot–Choquet–Deny*, 10e année (1965/66), Exposé no. 7, pp. 1–13. *33, 34, 253*

[112] Faraut, J., Semi-groupes de mesures complexes et calcul symbolique sur les générateurs infinitésimaux de semigroupes d'opérateurs, *Ann. Inst. Fourier* **32.1** (1970), 89–210. *178, 253*

[113] Farkas, W., Jacob, N. and Schilling, R. L., Function spaces related to continuous negative definite functions: ψ-Bessel potential spaces, *Diss. Math.* **CCCXCIII** (2001), 1–62. *253*

[114] Federbush, P., Partially alternate derivation of a result of Nelson, *J. Math. Phys.* **10** (1969), 50–52. *255*

[115] Fejér, L., Trigonometrische Reihen und Potenzreihen mit mehrfach monotoner Koeffizientenfolge, *Trans. Am. Math. Soc.* **39** (1936), 18–59. *14*

[116] Feller, W., Completely monotone functions and sequences, *Duke Math. J.* **5** (1939), 661–674. *2, 14*

[117] Feller, W., On a generalization of Marcel Riesz' potentials and the semi-groups gen-
       erated by them, *Meddelanden Lunds Univ. Mat. Sem. Supplement M. Riesz* (1952),
       73–81. *253*

[118] Feller, W., *An Introduction to Probability Theory and Its Applications, vol. II (2nd
       ed.)*, Wiley, Wiley Ser. Prob. Math. Stat., New York 1971. *14, 33, 66*

[119] Fourati, S. and Jedidi, W., Private communication, August 2009, September 2010 and
       November 2011. *34, 108*

[120] Fourati, S. and Jedidi, W., Some remarks on the class of Bernstein functions and some
       sub-classes, *Preprint* December 2011. *34, 108*

[121] Fritzsche, B. and Kirstein, B., *Ausgewählte Arbeiten zu den Ursprüngen der Schur-
       Analysis. Gewidmet dem großen Mathematiker Issai Schur (1875–1941)*, Teubner,
       Teubner-Archiv zur Mathematik vol. **16**, Stuttgart 1991. *392, 399, 400, 402*

[122] Fukushima, M., On an L^p-estimate of resolvents of Markov processes, *Publ. RIMS
       Kyoto Univ.* **13** (1977), 191–202. *255*

[123] Fukushima, M., Oshima, Y. and Takeda, M., *Dirichlet Forms and Symmetric Markov
       Processes*, de Gruyter, Studies in Math. **19**, Berlin 1994. *243, 269, 377, 378, 379, 380,
       381, 382*

[124] Furuta, T., Logarithmic order and dual logarithmic order, in: Kérchy, L. et al. (eds.),
       Recent Advances in Operator Theory and Related Topics (Szeged, 1999), Oper. Theory
       Adv. Appl. vol. **127**, Birkhäuser, Basel 2001, 279–290. *326*

[125] Furuta, T., An operator monotone function $\frac{t\log t - t + 1}{\log^2 t}$ and strictly chaotic order, *Linear
       Algebra Appl.* **341** (2002), 101–109. *322, 326*

[126] Furuta, T., Concrete examples of operator monotone functions obtained by an elemen-
       tary method without appealing to Löwner integral representation, *Linear Algebra Appl.*
       429 (2008), 972–980. *199, 308, 322, 324, 326*

[127] Gibilisco, P., Hansen, F. and Isola, T., On a correspondence between regular and non-
       regular operator monotone functions, *Linear Algebra Appl.* **430** (2009), 2225–2232.
       108

[128] Glover. J., Rao, M., Šikić, H. and Song, R., Γ-potentials, in: Gowrisankaran, K. et al.
       (eds.), *Classical and Modern Potential Theory and Applications. Proceedings of the
       NATO Advanced Research Workshop on Classical and Modern Potential Theory and
       Applications, Chateau de Bonas, France, July 1993*, Kluwer, Dordrecht 1994, 217–
       232. *267*

[129] Glover, J., Pop-Stojanovic, Z., Rao, M., Šikić, H., Song, R. and Vondraček, Z., Har-
       monic functions of subordinate killed Brownian motions, *J. Funct. Anal.* **215** (2004),
       399–426. *267*

[130] Gnedenko, B. W. and Kolmogorov, A. N., *Limit Distributions for Sums of Independent
       Random Variables*, Addison-Wesley, Reading (MA) 1954. *54, 55, 67*

[131] Gneiting, T., On the Bernstein–Hausdorff–Widder conditions for completely mono-
       tone functions, *Expo. Math.* **16** (1998), 181–183. *15*

[132] Gomilko, A., Haase, M. and Tomilov, Y., Bernstein functions and rates in mean ergodic theorems for operator semigroups, to appear in *J. Anal. Math. 107*

[133] Gong, F.-Z. and F.-Y. Wang, Functional inequalities for uniformly integrable semigroups and application to essential spectrums, *Forum Math.* **14** (2002), 293–313. *255*

[134] Gradshteyn, I. S. and Ryzhik, I. M., *Table of Integrals, Series and Products (4th corrected and Enlarged Ed.)*, Academic Press, San Diego 1980. *27, 213, 214, 300, 334, 360*

[135] Gravereaux, J. B., Probabilité de Lévy sur $\mathbb{R}^d$ et équations différentielle stochastiques linéaires, *Séminaire de Probabilités 1982*, Université de Rennes 1, Publications de Séminaires de Mathématiques, 1–42 *157*

[136] Gripenberg, G., Londen, S.-O. and Staffans, O., *Volterra Integral Equations and Functional Equations*, Cambridge University Press, Encycl. Math. Appl. vol. **34**, Cambridge 1990. *14*

[137] Groh, J., Über eine Klasse eindimensionaler Markovprozesse, *Math. Nach.* **65** (1975), 125–136. *297*

[138] Gross, L., Logarithmic Sobolev inequalities, *Am. J. Math.* **97** (1975), 1061–1083. *255*

[139] Hahn, H., Über die Integrale des Herrn Hellinger und die Orthogonalinvarianten der quadratischen Formen von unendlich vielen Veränderlichen, *Monatsh. Math. Phys.* **23** (1912), 161–224. Also in: Schmetterer, L. and Sigmund, K., *Hans Hahn: Gesammelte Abhandlungen – Collected Works (3 vols.)*, Springer, Wien 1995–1997, vol. 1, 109–172. *197*

[140] Hamburger, H., Über eine Erweiterung des Stieltjesschen Momentenproblems I, II, III, *Math. Ann.* **81** (1920), 235–319 **82** (1921), 120–164 and 168–187. *89*

[141] Hansen, F., Selfadjoint means and operator monotone functions, *Math. Ann.* **256** (1981), 29–35. *370*

[142] Hansen, F., Characterizations of symmetric monotone metrics on the state space of quantum systems, *Quantum Inf. Comput.* **6** (2006), 597–605. *370*

[143] Hansen, F., Some operator monotone functions, *Linear Algebra Appl.* **430** (2009), 795–799. *308*

[144] Hansen, F. and Pedersen, G. K., Jensen's inequality for operators and Löwner's theorem, *Math. Ann.* **258** (1982), 229–241. *198, 199*

[145] Hardy, G. H., *Divergent Series*, Oxford University Press, Oxford 1949. *13*

[146] Hartman, P., On differential equations and the function $J_\mu^2 + Y_\mu^2$, *Am. J. Math.* **83** (1961), 154–188. *14*

[147] Harzallah, Kh., Fonctions opérant sur les fonctions définies-négatives réelles, *C. R. Acad. Sci. Paris* **260** (1965), 6790–6793. *34, 254*

[148] Harzallah, Kh., Fonctions opérant sur les fonctions définies-négatives à valeurs complexes, *C. R. Acad. Sci. Paris* **262** (1966), 824–826. *34, 254*

[149] Harzallah, Kh., Fonctions opérant sur les fonctions définies-négatives, *Ann. Inst. Fourier* **17.1** (1967), 443–468. *33, 34, 254*

[150]   Hausdorff, F., Summationsmethoden und Momentfolgen. I, II, *Math. Z.* **9** (1921), I: 74–109, II: 280–299. Also in [152], 107–162. *13, 14, 387*

[151]   Hausdorff, F., Momentenprobleme für ein endliches Intervall, *Math. Z.* **14** (1923), 220–248. Also in [152], 193–221. *14, 387*

[152]   Hausdorff, F., *Gesammelte Werke. Bd. IV (Analysis, Algebra und Zahlentheorie)*, Springer, Berlin 2001. *14, 387, 392*

[153]   Hawkes, J., On the potential theory of subordinators, *Z. Wahrscheinlichkeitstheor. verw. Geb.* **33** (1975), 113–132. *178*

[154]   Heinz, E., Beiträge zur Störungstheorie der Spektralzerlegung, *Math. Ann.* **123** (1951), 415–438. *198*

[155]   Hellinger, E., Neue Begründung der Theorie quadratischer Formen von unendlichvielen Veränderlichen, *J. reine angew. Math.* **136** (1909), 210–271. *197*

[156]   Herglotz, G., Über Potenzreihen mit positivem, reellen Teil im Einheitskreis, *Ber. Sächs. Ges. Wiss. Leipzig, Math.-Phys. Kl.* **63** (1911), 501–511. Also in: Schwerdtfeger, H. (ed.), *Gustav Herglotz: Gesammelte Schriften*, Vandenhoeck & Rupprecht, Göttingen 1979, 247–257, and in [121], 9–19. *90*

[157]   Herz, C. S., Fonctions opérant sur les fonctions définies-positives, *Ann. Inst. Fourier* **13** (1963), 161–180. *254*

[158]   Hiai, F. and Kosaki, H., Means for matrices and comparison of their norms, *Indiana Univ. Math. J.* **48** (1999), 899–936. *199*

[159]   Hilbert, D., Grundzüge einer allgemeinen Theorie der linearen Integralgleichungen, *Nachr. Göttinger Akad. Wiss. Math. Phys. Klasse* (1904), 49–91. (Reprinted in: Hilbert, D., *Grundzüge einer allgemeinen Theorie der linearen Integralgleichungen*, Teubner, Leipzig 1912, and Chelsea, New York 1953.) *46*

[160]   Hille, E. and Phillips, R. S., *Functional Analysis and Semi-Groups (2nd ed)*, Am. Math. Soc., Colloquium Publ. vol. **XXXI**, Providence (RI), 1957. *253*

[161]   Hirsch, F., Sur une généralisation d'un théorème de M. Ito, *C. R. Acad. Sci. Paris, Sér. A* **271** (1970), 1236–1238. *253*

[162]   Hirsch, F., Intégrales de résolvantes et calcul symbolique, *Ann. Inst. Fourier* **22.4** (1972), 239–264. *19, 90, 107, 178, 211, 253*

[163]   Hirsch, F., Transformation de Stieltjes et fonctions opérant sur les potentiels abstraits, in: Faraut, J. (ed.), *Théorie du Potentiel et Analyse Harmonique*, Springer, Lect. Notes Math. vol. **404**, Berlin 1974, 149–163. *90, 107, 178, 211, 253, 254*

[164]   Hirsch, F., Familles d'opérateurs potentiels, *Ann. Inst. Fourier* **25.3–4** (1975), 263–288. *165, 172, 178, 253*

[165]   Hirsch, F., Extension des propriétés des puissances fractionnaires, in: *Séminaire de Théorie du Potentiel de Paris, No. 2*, Springer, Lect. Notes Math. vol. **563**, Berlin 1976, 100–120. *90, 253*

[166]   Hirsch, F., Private communication, March 2007. *90*

[167] Hirschman, I. I. and Widder, D. V., *The Convolution Transform*, Princeton University Press, Math. Ser. vol. **20**, Princeton (NJ) 1955. *20, 116*

[168] Hoffmann, K., *Banach Spaces of Analytic Functions*, Prentice-Hall, Ser. Mod. Anal., Englewood Cliffs (NJ) 1962. *91*

[169] Horn, R., Infinitely divisible matrices, kernels, and functions, *Z. Wahrscheinlichkeitstheor. verw. Geb.* **8** (1967), 219–230. *67*

[170] Horn, R. and Johnson, C., *Matrix Analysis*, Cambridge University Press, Cambridge 1990. *198*

[171] Horn, R. and Johnson, C., *Topics in Matrix Analysis*, Cambridge University Press, Cambridge 1991. *198*

[172] Ikeda, N., Kotani, S. and Watanabe, S., On the asymptotic behaviors of transition densities in generalized diffusion processes, *Seminar on Probability* **41** (1975), 103–120 (in Japanese). *297*

[173] Ismail, M. E. H., Bessel functions and the infinite divisibility of the Student *t*-distribution, *Ann. Probab.* **5** (1977), 582–585. *352*

[174] Ismail, M. E. H., Integral representations and complete monotonicity of various quotients of Bessel functions, *Canad. J. Math.* **29** (1977), 1198–1207. *354*

[175] Ismail, M. E. H., Complete monotonicity of modified Bessel functions, *Proc. Am. Math. Soc.* **108** (1990), 353–361. *350, 352*

[176] Ismail, M. E. H. and Kelker, D. H., Special functions, Stieltjes transforms and infinite divisibility, *SIAM J. Math. Anal.* **10** (1979), 884–901. *352, 354, 362*

[177] Ismail, M. E. H. and May, C. P., Special functions, infinite divisibility and transcendental equations, *Math. Proc. Camb. Phil. Soc.* **85** (1979), 453–464. *354, 356*

[178] Itô, K. and McKean, H. P. (Jr.), *Diffusion Processes and Their Sample Paths*, Springer, Grundlehren Math. Wiss. vol. **125**, Berlin 1974 (2nd printing). *129, 270, 278, 283, 297*

[179] Itô, M., Sur les sommes de noyaux de Dirichlet, *C. R. Acad. Sci. Paris, Sér. A* **271** (1970), 937–940. *253*

[180] Itô, M., Sur une famille sous-ordonée au noyau de convolution de Hunt donné, *Nagoya Math. J.* **51** (1973), 45–56. *67, 165*

[181] Itô, M, Sur la famille sous-ordonée au noyau de convolution de Hunt II, *Nagoya Math. J.* **53** (1974), 115–126. *67, 68, 165*

[182] Izumino, S. and Nakamura, N., Operator monotone functions induced from Löwner-Heinz inequality and strictly chaotic order, *Math. Inequal. Appl.* **7** (2004), 103–112. *199*

[183] Jacob, N., *Pseudo Differential Operators and Markov Processes (3 vols)*, Imperial College Press, London 2001–2006. *34, 90, 187, 253, 254, 255*

[184] Jacob, N. and Schilling, R. L., Lévy-type processes and pseudo-differential operators, in: Barndorff-Nielsen, O. E. et al. (eds.), *Lévy Processes: Theory and Applications*, Birkhäuser, Boston 2001, 139–168. *187*

[185] Jacob, N. and Schilling, R. L., Function spaces as Dirichlet spaces (about a paper by Maz'ya and Nagel), *Z. Anal. Anwendungen* **24** (2005), 3–28. *254, 299*

[186] Jacobsthal, E., Mittelwertbildung und Reihentransformation, *Math. Z.* **6** (1920), 100–117. *14*

[187] James, L. F., Roynette, B. and Yor, M., Generalized Gamma convolutions, Dirichlet means, Thorin measures, with explicit examples, *Probability Surveys* **5** (2008), 346–415. *306, 316, 324, 326, 328, 371, 372*

[188] Jayakumar, K. and Pillai, R. N., Characterization of Mittag-Leffler distribution, *J. Appl. Statist. Sci.* **4** (1996), 77–82. *328*

[189] Jones, W. B. and Shen, G., Asymptotics of Stieltjes continued fraction coefficients and applications to Whittaker functions, in: Lange, L. J., Berndt, B. C. and Gesztesy, F. (eds.), *Continued Fractions: From Analytic Number Theory to Constructive Approximation (Columbia, MO, 1998)*, Am. Math. Soc., Contemp. Math. **236**, Providence (RI) 1999, 167–178. *360*

[190] Julia, G., Extension nouvelle d'un lemme de Schwarz, *Acta math.* **42** (1920), 349–355. *90*

[191] Jurek, Z. J., Relations between *s*-selfdecomposable and selfdecomposable measures, *Ann. Probab.* **13** (1995), 592–608. *157, 158*

[192] Jurek, Z. J., The random integral representation hypothesis revisited: new classes of *s*-selfdecomposable laws in: Chuong, N. M., Nirenberg L. and Tutschke, W. (eds.), *Abstract and applied analysis, Proceedings of the International Conference held in Hanoi, August 17–21, 2002*, World Sci. Publ., River Edge, NJ, 2004, 479–498. *158*

[193] Jurek, Z. J. and Vervaat, W., An integral representation for seldecomposable Banach space valued random variables, *Z. Wahrsch. Verw. Gebiete* **62** (1983), 247–262. *157, 158*

[194] Kac, I. S. and Kreĭn, M. G., Criteria for discreteness of the spectrum of a singular string, *Izv. Vyss. Ucebn. Zaved. Mat.* **2** (1958), 136–153 (in Russian). *292*

[195] Kac, I. S. and Kreĭn, M. G., *R*-functions – analytic functions mapping the upper helfplane into itself, *Am. Math. Soc. Translat., II. Ser.* **103** (1974), 1–18. *90*

[196] Kac, I. S. and Kreĭn, M. G., On the spectral functions of the string, *Am. Math. Soc. Translat., II. Ser.* **103** (1974), 19–102. *272, 275, 276, 297*

[197] Kahane, J.-P., Sur les fonctions de type positif and négatif, in: *Seminar on harmonic analysis, Publications du Séminaire de Mathématiques d'Orsay*, Paris 1979. *254*

[198] Kasahara, Y., Spectral theory of generalized second order differential operators and its applications to Markov processes, *Japan J. Math., New Series* **1** (1975), 67–84. *297*

[199] Kasahara, Y., Kotani, S. and Watanabe, H., On the Green function of one-dimensional diffusion processes, *Publ. Res. Inst. Math. Sci.* **16** (1980), 175–188. *297*

[200] Kato, T., *Perturbation Theory for Linear Operators (2nd ed.)*, Springer, Grundlehren Math. Wiss. vol. **132**, Berlin 1980. *179, 197*

[201] Keilson J., Log-concavity and log-convexity in passage time densities of diffusion and birth-death processes, *J. Appl. Probab.* **8** (1971), 391–398. *130*

[202] Kendall, D. G., Extreme-point methods in stochastic analysis, *Z. Wahrscheinlichkeits-theor. verw. Geb.* **1** (1963), 295–300. *15*

[203] Kent, J. T., Eigenvalue expansions for diffusion hitting times, Research Report No. 3, Department of Statistics, University of Leeds, 1979. (Extended version of [204]) *130, 297*

[204] Kent, J. T., Eigenvalue expansions for diffusion hitting times, *Z. Wahrscheinlichkeits-theor. verw. Gebiete* **52** (1980), 309–319. *130, 297, 395*

[205] Kent, J. T., Convolution mixtures of infinitely divisible distributions, *Math. Proc. Camb. Phil. Soc.* **90** (1981), 141–153. *129*

[206] Kent, J. T., The spectral decomposition of a diffusion hitting time, *Ann. Probab.* **10** (1982), 207–219. *297*

[207] Khintchine, A. Ya., *Predel'nye teoremy dlja summ nesavisimykh slutschainykh velitschin (Limit Theorems for Sums of Independent Random Variables – in Russian),* GONTI, Moscow 1938. *67*

[208] Kim, P., Song, R. and Vondraček, Z., Two-sided Green function estimates for killed subordinate Brownian motions, *Proc. London Math. Soc.* **104** (2012), 927–958. *91*

[209] Knight, F. B., Characterization of the Lévy measures of inverse local times of gap diffusion, in: *Seminar on Stochastic Processes 1981*, Birkhäuser, Progr. Probab. Stat. vol. **1**, Boston 1981, 53–78. *270, 297*

[210] Knight, F. B., On the path of an inert object impinged on one side by a Brownian particle, *Probab. Theory Relat. Fields* **121** (2001), 577–598. *320*

[211] Knopp, K., Mehrfach monotone Zahlenfolgen, *Math. Z.* **22** (1925), 75–85. *14*

[212] Komatu, Y., Über analytische Funktionen, die in einer Halbebene positiven reellen Teil besitzen, *Kodai Math. Sem. Rep.* **22** (1970), 219–230. *90*

[213] Koosis, P., *The Logarithmic Integral. Vol. I*, Cambridge University Press, Stud. Adv. Math. vol. **12**, Cambridge 1988. *34*

[214] Korányi, A., On a theorem of Löwner and its connections with resolvents of selfadjoint transformations, *Acta Sci. Math. Szeged* **17** (1956), 63–70. *108, 199, 403*

[215] Korányi, A. and Sz.-Nagy, B., Operatortheoretische Behandlung und Verallgemeiner-ung eines Problemkreises in der komplexen Funktionentheorie, *Acta Math.* **100** (1958), 171–202. *108, 199*

[216] Kotani, S., On a generalized Sturm–Liouville operator with a singular boundary, *J. Math. Kyoto Univ.* **15** (1975), 423–454. *297*

[217] Kotani, S. and Watanabe, S., Krein's spectral theory of strings and generalized diffusion processes, in: Fukushima, M. (ed.), *Functional Analysis in Markov Processes*, Springer, Lect. Notes Math. vol. **923**, Berlin 1982, 235–259. *270, 292, 297*

[218] Köthe, G., *Topological Vector Spaces I*, Springer, Grundlehren Math. Wiss. vol. **159**, Berlin 1969. *233*

[219] Kreĭn, M. G., Concerning the resolvents of an Hermitian operator with the deficiency index (m, m), *Dokl. Acad. Nauk SSSR* **52** (1946), 651–654. *90*

[220] Kreĭn, M. G., Solution of the inverse Sturm–Liouville problem, *Dokl. Akad. Nauk SSSR* **76** (1951), 21–24. *90, 297*

[221] Kreĭn, M. G., Determination of the density of a nonhomogeneous cord by its frequency spectrum, *Dokl. Akad. Nauk SSSR* **76** (1951), 345–348. *90, 297*

[222] Kreĭn, M. G., On inverse problems for a non-homogeneous cord, *Dokl. Akad. Nauk SSSR* **82** (1952), 669–672. *297*

[223] Kreĭn, M. G., On a generalization of an investigation of Stieltjes, *Dokl. Akad. Nauk SSSR* **87** (1952), 881–884. *297*

[224] Kreĭn, M. G., On some cases of effective determination of the density of an inhomogeneous cord from its spectral function, *Dokl. Akad. Nauk SSSR* **93** (1953), 617–620. *284, 297*

[225] Kreĭn, M. G. and Nudelman, A. A., *The Markov Moment Problem and Extremal Problems*, Am. Math. Soc., Translat. Math. Monogr. vol. **50**, Providence (RI) 1977. *91*

[226] Krylov, V. I., On functions regular in a half-plane, *Math. Sb.* **6** (1939), 95–138. Reprinted in: *Am. Math. Soc., Translat. II. Ser.* **32** (1963), 37–81. *85, 91*

[227] Kubo, F. and Ando, T., Means of positive linear operators, *Math. Ann.* **246** (1980), 205–224. *107*

[228] Küchler, U., Über parabolische Funktionen eindimensionaler Quasidiffusionsprozesse, in: *Information Theory, Statistical Decision Functions, Random Processes*. Trans. 8th Prague Conf. vol. B, Prague 1978 (1978), 23–28. *297*

[229] Küchler, U., Some asymptotic properties of the transition densities of one-dimensional quasidiffusions, *Publ. Res. Inst. Math. Sci.* **16** (1980), 245–268. *297*

[230] Küchler, U., On parabolic functions of one-dimensional quasidiffusions, *Publ. Res. Inst. Math. Sci.* **16** (1980), 269–287. *297*

[231] Küchler, U., On sojourn times, excursions and spectral measures connected with quasi-diffusions, *J. Math. Kyoto Univ.* **26** (1986), 403–421. *332*

[232] Küchler, U. and Lauritzen, S. L., Exponential families, extreme point models and minimal space-time invariant functoins for stochastic processes with stationary and independent increments, *Scand. J. Statist.* **16** (1989), 237–261. *129*

[233] Kwaśnicki, M., Spectral analysis of subordinate Brownian motions on the half-line, *Studia Math.* **206** (2011), 211–271. *91, 107, 254*

[234] Kwaśnicki, M., Małecki, J. and Ryznar, M., Suprema of Lévy processes, to appear in *Ann. Probab. 91*

[235] Kyprianou, A. E. and Rivero, V., Special, conjugate and complete scale functions for spectrally negative Lévy processes, *Electron. J. Probab.* **13** (2008), 1672–1701. *178*

[236] Kyprianou, A. E., Rivero, V. and Song. R., Convexity and smoothness of scale functions and de Finetti's control problem, *J. Theor. Probab.* **23** (2010), 547–564. *178*

[237] Landau, E. and Valiron, G., A deduction from Schwarz's Lemma, *J. London Math. Soc.* **4** (1929), 162–163. *90*

[238] Langer, H., Partzsch, L. and Schütze, D, Über verallgemeinerte gewöhnliche Dif-
       ferentialoperatoren mit nichtlokalen Randbedingungen und die von ihnen erzeugten
       Markov-Prozesse, *Publ. Res. Inst. Math. Sci.* **7** (1972), 659–702. *297*

[239] Lax, P. D., *Functional Analysis*, Wiley, New York 2002. *8, 15, 19, 47, 183, 197*

[240] Lengyel, B. A., On the spectral theorem for selfadjoint operators, *Acta Litt. Sci. Szeged*
       9 (1939), 174–186. *179, 183, 197*

[241] Levin, B. Ja., *Distribution of zeros of entire functions*, Am. Math. Soc., Translat. Math.
       Monogr. vol. **5**, Providence (RI) 1964. (German translation: Lewin, B. J., *Nullstel-
       lenverteilung ganzer Funktionen*, Akademie-Verlag, Mathematische Lehrbücher und
       Monographien, II. Abt., vol. **14**, Berlin 1962.) *34*

[242] Lévy, P., Détermination générale des lois limits, *C. R. Acad. Sci. Paris* **203** (1936),
       698–700. Also in: Dugué, D. (ed.), *Œuvres de Paul Lévy (6 vols.)*, Gauthier-Villars,
       Paris 1973–1980, vol. III, 265–266. *67*

[243] Lévy, P., *Théorie de l'Addition des Variables Aléatoires*, Gauthier-Villars, Monogra-
       phies des Probabilités Fasc. **I**, Paris 1937. *67*

[244] Linetsky, V., On the transition densities for reflected diffusions, *Adv. Appl. Probab.* **37**
       (2005), 435–460. *364*

[245] Linetsky, V., Spectral methods in derivative pricing, in: Birge, J. R. and Linetsky,
       V., *Handbooks in Operations Research and Management Science vol. 15: Financial
       Engineering*, Elsevier/North-Holland, Amsterdam 2007, 223–300. *362*

[246] Loève, M., *Probability Theory 1 (4th ed.)*, Springer, Graduate Texts Math. vol. **45**,
       New York 1977. *67*

[247] Lorch, L., Muldoon M. E. and Szego, P., Higher monotonicity properties of certain
       Sturm–Liouville functions. III, *Can. J. Math.* **22** (1970), 1238–1265. *14*

[248] Löwner, K., Über monotone Matrixfunktionen, *Math. Z.* **38** (1934), 177–216. Also
       in: Bers, L. (ed.), *Charles Loewner: Collected Papers*, Birkhäuser, Contemporary
       Mathematicians, Boston 1988, 65–104. *14, 108, 198*

[249] Loewner, K., Some classes of functions defined by difference or differential inequali-
       ties, *Bull. Am. Math. Soc.* **56** (1950), 308–319. *14*

[250] Lukacs, E., A characterization of stable processes, *J. Appl. Probab.* **6** (1969), 409–418.
       157

[251] Lukacs, E., *Characteristic Functions (2nd ed.)*, Hafner, New York 1970. *67*

[252] Lukeš, J., Malý, J., Netuka, I. and Spurný, J., *Integral Representation Theory. Applica-
       tions to Convexity, Banach Spaces and Potential Theory*, de Gruyter, Studies in Math.
       vol. **35**, Berlin 2010. *47*

[253] Maejima, M. and Nakahara, G., A note on new classes of infinitely divisible distribu-
       tions on $\mathbb{R}^d$, *Electron. Commun. Probab.* **14** (2009), 358–371. *157, 158*

[254] Maejima, M. and Rosiński, J., The class of type G distributions on $\mathbb{R}^d$ and related
       subclasses of infinitely divisible distributions, *Demonstratio Math.* **34** (2001), 251–
       266. *157*

[255]   Maejima, M. and Rosiński, J., Type G distributions on $\mathbb{R}^d$, *J. Theoret. Probab.* **15** (2002), 323–341. *157, 158*

[256]   Maejima, M. and Sato, K., The limits of nested subclasses of several classes of infinitely divisible distributions are identical with the closure of the class of stable distributions, *Probab. Theory Relat. Fields* **145** (2009), 119–142. *158*

[257]   Maejima, M. and Ueda, Y., Compositions of mappings of infinitely divisible distributions with applications to finding the limits of some nested subclasses, *Electron. Commun. Probab.* **15** (2010), 227–239. *158*

[258]   Mahajan, A. and Ross, D. K., A note on completely and absolutely monotone functions, *Can. Math. Bull.* **25** (1982), 143–148. *14*

[259]   Markushevich, A. I., *Theory of Functions of a Complex Variable (2nd ed., three volumes in one)*, Chelsea, New York 1977. *85, 287*

[260]   Martínez-Carracedo, C. and Sanz-Alix, M., *The Theory of Fractional Powers of Operators*, Elsevier, North Holland Math. Studies vol. **187**, Amsterdam 2001. *253*

[261]   Mathias, M., Über positive Fourier-Integrale, *Math. Z.* **16** (1923), 101–125. *47*

[262]   Matsumoto, H., Nguyen, L. and Yor, M., Subordinators related to the exponential functionals of Brownian bridges and explicit formulae for the semigroups of hyperbolic Brownian motions, in: Buckdahn, R. et al. (eds.), *Stochastic Processes and Related Topics, Proceedings of the 12th Winter School, Siegmundsburg, Germany, February 27–March 4, 2000*, Taylor & Francis, London 2002, 213–235. *306, 338, 373*

[263]   Mattner, L., Bernstein's theorem, inversion formula of Post and Widder, and the uniqueness theorem for Laplace transforms, *Expo. Math.* **11** (1993), 137–140. *2*

[264]   Mercer, J., Functions of positive and negative type, and their connection with the theory of integral equations, *Phil. Trans. Royal Soc. London* **209A** (1909), 415–446. *46*

[265]   Meyer, P.-A., *Probabilités et potentiel*, Hermann, Actual. Sci. Ind. **1318**, Publ. Inst. Math. Strasbourg t. **XIV**, Paris 1966. *15*

[266]   Miller, K. S. and Samko, S. G., Completely monotone functions, *Integr. Transf. Spec. Funct.* **12** (2001), 389–402. *354*

[267]   Molchanov, S. A. and Ostrovski, E., Symmetric stable processes as traces of degenerate diffusion processes, *Theory Probab. Appl.* **14** (1969), 128–131. *284*

[268]   Nadarajah, S. and Kotz, S., On the Laplace transform of the Pareto distribution, *Queueing Syst.* **54** (2006), 243–244. *360*

[269]   Nakamura, Y., Classes of operator monotone functions and Stieltjes functions, in: *The Gohberg Anniversary Collection, vol. II: Topics in Analysis and Operator Theory*, Operator Theory: Advances and Applications, vol. **41**, Dym, H. et al. (eds.), Birkhäuser, Basel 1989, 395–404. *90, 107*

[270]   Nelson, E., A functional calculus using singular Laplace integrals, *Trans. Am. Math. Soc.* **88** (1958), 400–413. *68, 253*

[271]   Neumann, J. von, Allgemeine Eigenwerttheorie Hermitescher Funktionaloperatoren, *Math. Ann.* **102** (1929), 49–131. Also in: Taub, A. H. (ed.): *Collected Works of John von Neumann (6 vols.)*, Pergamon, Oxford, vol. 2, 3–85. *197*

[272] Nevanlinna, R., Über beschränkte Funktionen, die in gegebenen Punkten vorgeschrie-
       bene Werte annehmen (Finnish), *Ann. Acad. Sci. Fenn. (A)* **13** (1920), 1–71. *89*

[273] Nevanlinna, R., Asymptotische Entwicklungen beschränkter Funktionen und das
       Stieltjes'sche Momentenproblem, *Ann. Acad. Sci. Fenn. (A)* **18** (1922), 1–53. *89, 90*

[274] Nevanlinna, R., Über beschränkte analytische Funktionen, *Ann. Acad. Sci. Fenn. (A)*
       32 (Lindelöf-Festschrift) (1929), 1–75. Also in [121], 22–49, 99–171. *89, 90*

[275] Nevanlinna, F. and Nieminen, T., Das Poisson–Stieltjessche Integral und seine Anwen-
       dung in der Spektraltheorie des Hilbert'schen Raumes, *Ann. Acad. Sci. Fenn. (A)* **207**
       (1955), 1–38. *179, 197*

[276] Nguyen, V. T., Multiply self-decomposable probability measures on Banach spaces,
       Studia Math. **66** (1979), 161–175. *67*

[277] Nguyen, V. T., Universal multiply self-decomposable probability measures on Banach
       spaces, *Probab. Math. Statist.* **3** (1982), 71–84. *67*

[278] Nguyen, V. T., Fractional calculus in probability, *Probab. Math. Statist.* **5** (1984), 173–
       189. *67*

[279] Nguyen, V. T., An alternative approach to multiply self-decomposable probability
       measures on Banach spaces, *Probab. Theory Relat. Fields* **72** (1986), 35–54. *67*

[280] Nollau, V., Über Potenzen von linearen Operatoren in Banachschen Räumen, *Acta Sci.
       Math.* **28** (1967), 107–121. *253*

[281] Nollau, V., Über den Logarithmus abgeschlossener Operatoren in Banachschen Räu-
       men, *Acta Sci. Math.* **30** (1969), 161–174. *253*

[282] Nollau, V., Über gebrochene Potenzen infinitesimaler Generatoren Markovscher Über-
       gangswahrscheinlichkeiten I, II, *Math. Nachr.* **65** (1975), 235–246 and **72** (1976), 99–
       107. *253*

[283] Oberhettinger, F. and Badii, L., *Tables of Laplace Transforms*, Springer, New York
       1973. *312*

[284] Ôkura, H., Recurrence and transience criteria for subordinated symmetric Markov pro-
       cesses, *Forum Math.* **14** (2002), 121–146. *34, 197, 245*

[285] Pakes, A. G., Lambert's *W*, infinite divisibility and Poisson mixtures, *J. Math. Anal.
       Appl.* **378** (2011), 480–492. *364*

[286] Pazy, A., *Semigroups of Linear Operators and Applications to Partial Differential
       Equations*, Springer, Appl. Math. Sci. vol. **44**, New York 1983. *200, 210, 228, 253*

[287] Perron, O., *Die Lehre von den Kettenbrüchen*, B. G. Teubner, Leipzig, 1913. (2nd ed.
       1929, 3rd ed. 1954) *19*

[288] Pestana, D. D. and Mendonça, S., Higher-order monotone functions and probability
       theory, in: Hadjisavras, N. et al. (eds.), *Generalized Convexity and Generalized Mono-
       tonicity*, Springer, Lecture Notes in Econ. Math. Sys. vol. **502** part 2, Berlin 2001,
       317–331. *47*

[289] Petrov, V. V., *Sums of Independent Random Variables*, Springer, Ergeb. Math. Grenz-
       geb. 2. Ser. vol. **82**, Berlin 1975. *67*

[290] Petrov, V. V., *Limit Theorems of Probability Theory*, Clarendon Press, Oxford Stud. Probab. vol. **4**, Oxford 1995. *67*

[291] Phelps, R. R., *Lectures on Choquet's Theorem*, Van Nostrand, Math. Studies vol. **7**, Princeton (NJ) 1966. *8, 15*

[292] Phillips, R. S., On the generation of semigroups of linear operators, *Pacific J. Math.* **2** (1952), 343–369. *204, 210, 253, 254*

[293] Pick, G., Über die Beschränkungen analytischer Funktionen, welche durch vorgegebene Funktionswerte bewirkt werden, *Math. Ann.* **77** (1916), 7–23. Also in [121], 77–93. *89, 198*

[294] Pick, G., Über die Beschränkungen analytischer Funktionen durch vorgegebene Funktionswerte, *Math. Ann.* **78** (1917), 270–275. *89*

[295] Pollard, H., Note on the inversion of the Laplace integral, *Duke Math. J.* **6** (1940), 420–424. *14*

[296] Prudnikov, A. P., Brychkov, Yu. A. and Marichev, O. I., *Integrals and Series Volumes 4 & 5*, Gordon and Breach, New York 1992. *13*

[297] Prüss, J., *Evolutionary Integral Equations and Applications*, Birkhäuser, Monogr. Math. vol. **87**, Basel 1993. *14, 34, 89*

[298] Pustyl'nik, E. I., On functions of a positive operator, *Math. USSR Sbornik* **47** (1984), 27–42. (Russian original edn.: *Mat. Sb.* **119**(161) (1982), 32–47) *216, 253, 254*

[299] Qi, F. and Chen, C.-P., A complete monotonicity property of the gamma function, *J. Math. Anal. Appl.* **296** (2004), 603–607. *67*

[300] Qi, F. and Guo, B.-N., Complete monotonicities of functions involving the gamma and digamma functions, *RGMIA Research Reports Collection* **7**.1 (2004), paper no. 8. (Available from http://rgmia.org/papers/v7n1/minus-one.pdf) *67*

[301] Reed, M. and Simon, B., *Methods of Modern Mathematical Physics IV: Analysis of Operators*, Academic Press, San Diego (CA) 1978. *251*

[302] Ressel, P., Laplace-Transformation nichtnegativer und vektorwertiger Maße, *Manuscripta Math.* **13** (1974), 143–152. *47*

[303] Reuter, G. E. H., Über eine Volterrasche Integralgleichung mit totalmonotonem Kern, *Arch. Math.* **VII** (1956), 59–66. *89, 107*

[304] Revuz, D. and Yor, M., *Continuous Martingales and Brownian Motion (3rd ed.)*, Springer, Gundlehren math. Wiss. vol. **293**, Berlin 1999 (corr. 3rd printing 2005). *283*

[305] Rocha-Arteaga, A. and Sato, K., *Topics in infinitely divisible distributions and Lévy processes*, Aportaciones Matemaáticas Investigacion **17**, Sociedad Matemática Mexicana, México, 2003. *157*

[306] Rogers, L. C. G., Wiener-Hopf factorization of diffusions and Lévy processes, *Proc. London Math. Soc.* **47** (1983), 177–191. *254*

[307] Rogers, L. C. G. and Williams, D., *Diffusions, Markov Processes and Martingales. Vol. 2: Itô Calculus*, Cambridge University Press, Cambridge Math. Libr., Cambridge 2000. *269*

[308] Rosenblum, M. and Rovnyak, J., *Hardy Classes and Operator Theory*, Oxford University Press, New York 1985. *91, 108, 199*

[309] Rossberg, S.-J., Jesiak, B. and Siegel, G.: *Analytic Methods of Probability Theory*, Akadamie-Verlag, Math. Lehrbücher Monogr. II: Math. Monogr. vol. **67**, Berlin 1985. *67*

[310] Royall, N. N., Laplace transforms of multiply monotone functions, *Duke Math. J.* **8** (1941), 546–558. *14*

[311] Royer, G., Une initiation aux inégalités de Sobolev logarithmiques, Société Mathématique de France, Cours Spécialisés No. **5**, Paris 1999. *255*

[312] Salminen, P., Vallois, P. and Yor, M., On the excursion theory for linear diffusions, *Jap. J. Math.* **2** (2007), 97–127. *340*

[313] Saloff-Coste, L., *Aspects of Sobolev-Type Inequalities*, Cambridge University Press, London Math. Soc. Lecture Notes **289**, London 2002. *255*

[314] Sato, K., *Lévy Processes and Infinitely Divisible Distributions*, Cambridge University Press, Stud. Adv. Math. vol. **68**, Cambridge 1999. *67, 166, 253*

[315] Sato, K., Class *L* of multivariate distributions and its subclasses, *J. Multivariate Anal.* **10** (1980), 207–232. *67*

[316] Sato, K., Two families of improper stochastic integrals with respect to Lévy processes, *ALEA Lat. Am. J. Probab. Math. Stat.* **1** (2006), 47–87. *157, 158*

[317] Sato, K., Fractional integrals and extensions of selfdecomposability, in: Barndorff-Nielsen, O. et al. (eds.), *Lévy Matters I*, Springer, Lecture Notes in Math., vol. **2001**, Berlin 2010, 1–92. *13, 15, 152, 157, 158*

[318] Sato, K., Description of limits of ranges of iterations of stochastic integral mappings of infinitely divisible distributions, *ALEA Lat. Am. J. Probab. Math. Stat.* **8** (2011), 1–17. *158*

[319] Sato, K. and Yamazato, M., Stationary processes of Ornstein-Uhlenbeck type, in: Itô, K. and Prokhorovj. V. (eds.), *Probability theory and mathematical statistics, Fourth USSR-Japan Symposium, Tbilisi, 1982*, Springer, Lecture Notes in Math., vol. **1021**, Berlin 1983, 541–551. *157*

[320] Schilling, R. L., *Zum Pfadverhalten von Markovschen Prozessen, die mit Lévy-Prozessen vergleichbar sind*, PhD-thesis, Universität Erlangen 1994. *33, 34, 90, 107, 198, 216, 253, 254*

[321] Schilling, R. L., On the domain of the generator of a subordinate semigroup, in: Král, J. et al. (eds.), *Potential Theory – ICPT 94. Proceedings Intnl. Conf. Potential Theory, Kouty (CR) 1994*, de Gruyter, Berlin 1996, 449–462. *90, 107, 216, 253*

[322] Schilling, R. L., Subordination in the sense of Bochner and a related functional calculus, *J. Austral. Math. Soc. (A)* **64** (1998), 368–396. *90, 107, 220, 253, 254*

[323] Schilling, R. L., *Measures, Integrals and Martingales*, Cambridge University Press, Cambridge 2005. *214*

[324] Schilling, R. L. and Wang, J., Functional inequalities and subordination: Stability of Nash and Poincaré inequalities, to appear in *Math. Zeitschrift* DOI: 10.1007/s00209-011-0964-x. *34, 255*

[325] Schoenberg, I. J., Metric spaces and completely monotone functions, *Ann. Math.* **29** (1938), 811–841. Also in: de Boor, C. (ed.), *I. J. Schoenberg: Selected Papers (2 vols.)*, Birkhäuser, Contemporary Mathematicians, Boston 1988, vol. **1**, 115–145. *15, 33, 46, 67, 212, 254*

[326] Schoenberg, I. J., On absolutely convex functions (abstract), *Bull. Am. Math. Soc.* **47** (1941), 389. *12*

[327] Schur, I., Über Potenzreihen, die im Inneren des Einheitskreises beschränkt sind. I, II, *J. reine angew. Math.* **147** (1917), 205–232 and **148** (1918), 122–145. Also in [71, vol. 2], 137–164, 165–188, and in [121], 22–49, 50–73. *89*

[328] Schur, I., Über lineare Transformationen in der Theorie der unendlichen Reihen, *J. reine angew. Math.* **151** (1921), 79–111. Also in [71, vol. 2], 289–321. *13*

[329] Schwartz, L., *Analyse I–IV (nouvelle éd. corr.)*, Hermann, Collection enseignement des sciences t. **42–45**, Paris 1993–97. *374*

[330] Simon, B., *Convexity: An Analytic Viewpoint*, Cambridge University Press, Camb. Tracts Math. vol. **187**, Cambridge 2011. *47, 199*

[331] Song, R. and Vondraček, Z., Harnack inequality for some classes of Markov processes, *Math. Z.* **246** (2004), 177–202. *262*

[332] Song, R. and Vondraček, Z., Potential theory of special subordinators and subordinate killed stable processes, *J. Theor. Probab.* **19** (2006), 817–847. *178, 267*

[333] Song, R. and Vondraček, Z., Some remarks on special subordinators, *Rocky Mt. J. Math* **40** (2010), 321–337. *178*

[334] Sparr, G., A new proof of Löwner's theorem on monotone matrix functions, *Math. Scand.* **47** (1980), 266–274. *199*

[335] Srivastava, H. M. and Tuan, V. K., A new convolution theorem for the Stieltjes transform and its application to a class of singular integral equations, *Arch. Math.* **64** (1995), 144–149. *20*

[336] Stampacchia, G., Le problème de Dirichlet pour les équations elliptiques du second ordre à coefficients discontinus, *Ann. Inst. Fourier* **15** (1965), 189–258. *255*

[337] Stein, E. M. and Weiss, G., *Introduction to Fourier Analysis on Euclidean Spaces (2nd corrected printing)*, Princton University Press, Princeton Math. Ser. vol. **32**, Princeton (NJ) 1975. *213*

[338] Steutel, F. W., *Preservation of Infinite Divisibility Under Mixing and Related Topics*, Mathematisch Centrum, Math. Center Tracts vol. **33**, Amsterdam 1970. http://oai.cwi.nl/oai/asset/13052/13052A.pdf *129, 178*

[339] Steutel, F. W. and van Harn, K., *Infinite Divisibility of Probability Distributions on the Real Line*, Marcel Dekker, Pure Appl. Math. vol. **259**, New York 2004. *67, 116, 129, 178*

[340] Stieltjes, T. J., Recherches sur les fractions continues, *C. R. Acad. Sci.* **118** (1894), 1401–1403. Also in [342, vol. 2], 402–405. *19*

[341] Stieltjes, T. J., Recherches sur les fractions continues, *Ann. Fac. Sci. Toulouse* **8** (1894), 1–122 and **9** (1895), 1–47. Also in [342, vol. 2], 406–570 with english translation 609–745. *19*

[342] Stieltjes, T. J., *Œvres Complètes – Collected Papers (2 vols.)*, Springer, Berlin, 1993. *19, 403*

[343] Stone, M. H., *Linear Transformations in Hilbert Space and Their Applications to Analysis*, Am. Math. Soc., Colloquium Publ. vol. **15**, New York 1932. *197*

[344] Sz.-Nagy, B., Remarks on the preceding paper of A. Korányi (i.e., [214]), *Acta Sci. Math. Szeged* **17** (1956), 71–75. *199*

[345] Tanabe, H., *Equations of Evolution*, Pitman, Monogr. Math. vol. **6**, London 1979. *253*

[346] Teke, S. P. and Deshmukh, S. R., Inverse renewal thinning of Cox and renewal processes, *Stat. Probab. Letters* **78** (2008), 2705–2708. *107*

[347] Thorin, O., On the infinite divisibility of the Pareto distribution, *Scand. Actuarial J.* (1977), 31–40. *116, 129*

[348] Thorin, O., On the infinite divisibility of the lognormal distribution, *Scand. Actuarial J.* (1977), 121–148. *116, 129*

[349] Tomisaki, M., Comparison theorems in generalized diffusion process, *Mem. Fac. Sci., Kyushu Univ., Ser. A* **XXX** (1976), 247–256. *297*

[350] Uchiyama, M., Operator monotone functions which are defined implicitly and operator inequalities, *J. Funct. Anal.* **175** (2000), 330–347. *199*

[351] Uchiyama, M., Inverse functions of polynomials and orthogonal polynomials as operator monotone functions, *Trans. Am. Math. Soc.* **355** (2003), 4111–4123. *199*

[352] Uchiyama, M., Operator monotone functions and operator inequalities, *Sugaku Expo.* **18** (2005), 39–52. *199*

[353] Uchiyama, M., A new majorization between functions, polynomials and operator inequalities, *J. Funct. Anal.* **231** (2006), 221–244. *107, 199*

[354] Uchiyama, M., A new majorization induced by matrix order, in: Ando, T. et al. (eds.), *Recent Advances in Operator Theory and Applications. Proceedings Intnl. Workshop on Operator Theory and Applications (IWOTA), Seoul, Korea 2006*, Birhäuser, Oper. Theory Adv. Appl. vol. **187**, Basel 2009, 211–216. *107*

[355] Uchiyama, M., Majorization and some operator monotone functions, *Linear Algebra Appl.* **432** (2010), 1867–1872. *107*

[356] Uchiyama, M. and Hasumi, M., On some operator monotone functions, *Integral Equations Oper. Theory* **42** (2002), 243–251. *199*

[357] Urbanik, K., Slowly varying sequences of random variables, *Bull. Acad. Polon. Sci. Sér. Sci. Math. Astronom. Phys.* **20** (1972), 679–682. *67*

[358] Urbanik, K., Limit laws for sequences of normed sums satisfying some stability conditions, in: Krishnaiah, P. R. (ed.), *Multivariate analysis, III (Proc. Third Internat. Sympos., Wright State Univ., Dayton, Ohio, 1972)*, Academic Press, New York 1973, 225–237. *67*

[359] van Haeringen, H., Completely monotonic and related functions, *J. Math. Anal. Appl* **204** (1996), 389–408. *14*

[360] van-Harn, K. and Steutel, F. W., Generalized renewal sequences and infinitely divisible lattice distributions, *Stoch. Proc. Appl.* **5** (1977), 47–55. *178*

[361] van Herk, C. G. G., A class of completely monotonic functions, *Compos. Math.* **9** (1951), 1–79. *15, 19, 107*

[362] Varopoulos, N. Th., Saloff-Coste, L. and Coulhon, T., *Analysis and Geometry on Groups*, Cambridge University Press, Cambridge Tracts in Math. vol. **100**, Cambridge 1992. *233, 255*

[363] Walter, W., *Differential and Integral Inequalities*, Ergeb. Math. Grenzgeb. vol. **55**, Springer, Berlin 1970. *255*

[364] Wang, F.-Y., *Functional Inequalities, Markov Semigroups and Spectral Theory*, Science Press, Beijing 2005. *233, 241, 242, 255*

[365] Watanabe, S., On time-inversion of one-dimensional diffusion processes, *Z. Wahrscheinlichkeitstheor. verw. Geb.* **31** (1975), 115–124. *297*

[366] Wendland, H., *Scattered Data Approximation*, Cambridge University Press, Cambridge Monographs on Applied and Computational Mathematics **17**, Cambridge, 2005. *47*

[367] Westphal, U., Ein Kalkül für gebrochene Potenzen infinitesimaler Erzeuger von Halbgruppen und Gruppen von Operatoren. Teil I: Halbgruppenerzeuger, *Compositio Math.* **22** (1970), 144–149. *253, 254*

[368] Widder, D. V., Necessary and sufficient conditions for the representation of a function as a Laplace integral, *Trans. Am. Math. Soc.* **33** (1931), 851–892. *2, 14*

[369] Widder, D. V., The iterated Stieltjes transform, *Proc. Natl. Acad. Sci. U.S.A.* **23** (1937), 242–244. *19*

[370] Widder, D. V., *The Laplace Transform*, Princeton University Press, Princeton Math. Series vol. **6**, Princeton (NJ), 1946. *13, 14, 19, 175*

[371] Widder, D. V., *An Introduction to Transform Theory*, Academic Press, Pure Appl. Math. vol. **42**, New York, 1971. *13, 14*

[372] Williamson, R. E., Multiply monotone functions and their Laplace transforms, *Duke Math. J.* **23** (1956), 189–207. *11, 14, 15*

[373] Wolfe, S. J., On a continous analogue of the stochastic difference equation $X_n = \rho X_{n-1} + B_n$, *Stoch. Proc. Appl.* **12** (1982), 301–312. *156, 157, 158*

[374] Wolff, J., Sur une généralisation d'un théorème de Schwarz, *C. Rend. Acad. Sci.* **182** (1926), 918–920 and **183** (1926), 500–502. *90*

[375] Woll, J. W., Homogeneous stochastic processes, *Pacific J. Math.* **9** (1959), 293–325. *33, 253*

[376] Yamazato, M., Characterization of the class of upward first passage time distribution of birth and death processes and related results, *J. Math. Soc. Japan* **40** (1988), 477–499. *130*

[377] Yamazato, M., Hitting time distributions of single points for 1-dimensional generalized diffusion processes, *Nagoya Math. J.* **119** (1990), 143–172. *130, 297*

[378] Yamazato, M., Characterization of the class of hitting distributions of 1-dimensional generalized diffusion processes, in: Shiryaev, A. N. et al. (eds.), *Probability Theory and Mathematical Statistics. Proceedings of the 6th USSR-Japan symposium, Kiev, USSR, 5–10 August, 1991*, World Scientific, Singapore 1992, 422–428. *298*

[379] Yosida, K., *Functional Analysis (6th ed.)*, Springer, Grundlehren Math. Wiss. vol. **123**, Berlin 1980. *90, 178, 191, 197, 200, 209, 253*

[380] Zhan, X., *Matrix Inequalities*, Springer, Lect. Notes Math. vol. **1790**, Berlin 2002. *199*

Index

Overview of the classes of distributions, Laplace transforms and exponents

measure	Laplace transform	Laplace exponent	references
ID	log-$\mathcal{CM}$	extended $\mathcal{BF}$	5.6, 5.10, 5.11
ID, $\pi[0,\infty) \leq 1$	log-$\mathcal{CM}$, $f(0+) \leq 1$	$\mathcal{BF}$	5.8
SD	—	$\mathcal{BF}$, $\mu(dt) = m(t)\,dt$, $t\,m(t)$ non-increasing	5.17
SD_∞	—	$\mathcal{PBF}$	5.21, 5.22
BO	—	$\mathcal{CBF}$	9.1, 9.2
ME	$\mathcal{S}$, $f(0+) \leq 1$	$\mathcal{CBF}$, $\sigma(dt) = \frac{\eta(t)}{t}\,dt$, $0 \leq \eta(t) \leq 1$	9.4, 9.5
GGC	log-$\mathcal{S}$, $f(0+) \leq 1$	$\mathcal{TBF}$	8.7, 9.10, 9.11
CE	$e^{-c\lambda}\prod_n \left(1 + \frac{\lambda}{b_n}\right)^{-1}$	$\mathcal{TBF}$, $\tau = \sum_n \delta_{b_n}$	(9.8), 9.16
Exp	$\left(1 + \frac{\lambda}{b}\right)^{-1}$	$\mathcal{TBF}$, $\tau = \delta_b$	9.16

Integral representations for various classes of Bernstein functions

$\mathcal{BF}$
$$f(\lambda) = a + b\lambda + \int_{(0,\infty)} (1 - e^{-\lambda t})\,\mu(dt) \tag{3.2}$$

$\mathcal{CBF}$
$$f(\lambda) = a + b\lambda + \int_{(0,\infty)} (1 - e^{-\lambda t})\,m(t)\,dt, \qquad m \in \mathcal{CM} \tag{6.1}$$

$$= a + b\lambda + \int_{(0,\infty)} \frac{\lambda}{\lambda + t}\,\sigma(dt) \tag{6.7}$$

$$= \alpha + \beta\lambda + \int_{(0,\infty)} \frac{\lambda t - 1}{\lambda + t}\,\rho(dt) \tag{6.13}$$

$$= \exp\left(\gamma + \int_0^\infty \left(\frac{t}{1+t^2} - \frac{1}{\lambda + t}\right)\eta(t)\,dt\right), \qquad \eta(t) \in [0,1] \tag{6.21}$$

$\mathcal{TBF}$
$$f(\lambda) = a + b\lambda + \int_{(0,\infty)} (1 - e^{-\lambda t})\,m(t)\,dt, \qquad t \cdot m(t) \in \mathcal{CM} \tag{8.1}$$

$$= a + b\lambda + \int_{(0,\infty)} \log\left(1 + \frac{\lambda}{t}\right)\tau(dt) \tag{8.2}$$

$$= a + b\lambda + \int_0^\infty \frac{\lambda}{\lambda + t}\,\frac{w(t)}{t}\,dt, \qquad w \text{ non-decreasing} \tag{8.3}$$

$(a, b, \alpha, \beta \geq 0, \gamma \in \mathbb{R}$ — all measures and densities are such that the integrals converge$)$